Close to 1.6 billion people, including 60 million Indigenous people, depend on mixed-age, mixed-species forests for food, shelter, fuel, and livelihoods. Nine out of 10 land-based species of animals and plants rely not only upon their canopies, deadfalls, and roots, but also upon the myriad companion vines, ferns, understory shrubs, and fungal forests. To see forests as obstructions to development, as national parks, or as something to be timbered or removed is blind. To understand them as living providers of everything we need, while regulating water cycles, preventing soil erosion, and helping to keep our climate stable, is the wisdom of *Coppice Agroforestry*.

—Albert Bates, ESG advisor, NOAH ReGen, developer,
Cool Lab concept, author, *The Biochar Solution*

After a long career as a forester, tree farmer, and forestry educator, I sometimes mistakenly think that there isn't that much more to learn. Mark's book has shaken me wide awake again and the creative ideas are flowing. This book— full of photos and illustrations that are as captivating as the text—has started a new journey on how I think about trees, woods, and my beloved profession of forestry.

—Brett Chedzoy, regional extension forester,
Cornell Cooperative Extension

Anyone passionate about trees and forests, and the functional role they can have in our lives and vocations is sure to be delighted with this tome of information, spanning the ancient uses of coppice and pollard to the practical ways modern forestry and biology inform decision making. Krawczyk's writing is a pleasure to consume and the breadth of information is a testament to his lifetime of devotion to the topic. The abundance of pictures, diagrams, and charts helps any reader chart a course to successful coppicing and envision a future where resprout agriculture persists in landscapes everywhere.

—Steve Gabriel, farmer, author, *Silvopasture*,
and co-author *Farming the Woods*

After reading this, I will never look at a woody plant in the same way again. Rich in history, culture, botany, and practice, this remarkable work shows exceptional scholarship, dedication, and experience. Krawczyk, who has bridged the Atlantic world, is a worthy successor to Rackham, and just the interpreter of coppice North Americans need to create their own thriving backwoods industries. Ten years in the making, it could not be more timely. Buy it. Read it. Live it.

—Peter Bane, author, *The Permaculture Handbook*,
executive director, Permaculture Institute of North America

Coppice Agroforestry provides an essential reference guide to the farmer, rancher, woodlot owner, arboriculturist, and gardener interested in re-establishing the once widespread practice of coppice woodland management for the making of pathways, mats, furniture, hurdles, traps, snares, baskets and boats, and generating biomass energy production. These systems that so beautifully maintain the renewal capacity of the earth while garnering many wood products, provide the basis for a new vision of human-nature interactions that will go a long way toward healing the earth while honoring and celebrating our rich ancestral heritage in both the New and Old Worlds.

—M. Kat Anderson, author, *Tending the Wild*

Coppicing is an essential component of climate-friendly agriculture and land management. *Coppice Agroforestry* is a comprehensive guide, based on science and practical experience, to a truly perennial system for producing diverse wood products. Krawczyk's book will be inspiring readers to coppice for decades to come.

—Eric Toensmeier, author, *The Carbon Farming Solution*,
co-author, *Edible Forest Gardens*

COPPICE
AGROFORESTRY

TENDING TREES FOR PRODUCT, PROFIT, & WOODLAND ECOLOGY | **MARK KRAWCZYK**

new society
PUBLISHERS

Cover design by Diane McIntosh.

Cover Images: Main (top) and bottom right (fence) © iStock; bottom left—William Bode; bottom centre—Ammy Martinez.

Illustrations by William Bode. Sidebar background: Adobestock 58151515

Printed in Canada. First printing June 2022. Second printing November 2023.

This book is intended to be educational and informative. It is not intended to serve as a guide. The author and publisher disclaim all responsibility for any liability, loss or risk that may be associated with the application of any of the contents of this book.

Inquiries regarding requests to reprint all or part of *Coppice Agroforestry* should be addressed to New Society Publishers at the address below. To order directly from the publishers, order online at www.newsociety.com

Any other inquiries can be directed by mail to:
New Society Publishers
P.O. Box 189, Gabriola Island, BC V0R 1X0, Canada
(250) 247-9737

LIBRARY AND ARCHIVES CANADA CATALOGUING IN PUBLICATION

Title: Coppice agroforestry : tending trees for product, profit, & woodland ecology / Mark Krawczyk.

Names: Krawczyk, Mark, author.

Description: Includes bibliographical references and index.

Identifiers: Canadiana (print) 20220200203 | Canadiana (ebook) 20220200254 | ISBN 9780865719705 (softcover) | ISBN 9781550927641 (PDF) | ISBN 9781771423601 (EPUB)

Subjects: LCSH: Coppice forests. | LCSH: Coppice forests—Management. | LCSH: Agroforestry. | LCSH: Coppice forest ecology.

Classification: LCC SD387.C68 K73 2022 | DDC 634.9/9—dc23

Funded by the Government of Canada   Financé par le gouvernement du Canada

New Society Publishers' mission is to publish books that contribute in fundamental ways to building an ecologically sustainable and just society, and to do so with the least possible impact on the environment, in a manner that models this vision.

To our collective ancestors, human and nonhuman—
May we learn to cherish, hone, and build on the skills and wisdom
you've gifted us so future generations may enjoy safe, healthy, and fulfilling lives
rooted in reverence, stewardship, and connection to creation.

And to the trees—
Among our greatest teachers, nurturers, and companions
on this magical globe. May we cultivate relationships of reciprocity,
gratitude, and stewardship by listening, adapting, and tending in a
mutually beneficial way.

And Amina—
Our newborn connection to human succession.
May our generation help yours manifest a more
caring, attentive, engaged, and bountiful tomorrow.

Contents

Acknowledgments

As the days, weeks, months, and years pass with seemingly greater and greater speed, the limits of one human life grow increasingly clear. This project has marked more than a decade of my brief window on this planet, and it's been touched by the lives, insights, and experiences of many people.

I feel immeasurably blessed having had the privilege to stand on the shoulders of many giants during my lifetime. I owe a deep debt of gratitude to many of the pioneers in the realm of ecological design who forever transformed the way I have come to see the world. Bill Mollison and David Holmgren, Masanobu Fukuoka, John Todd, Vandana Shiva, J. Russell Smith, P.A. Yeomans, Darren Doherty, Allan Savory, Elaine Ingham, and so many more, who's vision, life work, and inspiration helped show me that a different world is possible, and necessary, and we all have a place in making it a reality.

In honing my own vision and learning how to be a more conscious, skillful, and engaged human, I had the amazing privilege of working alongside several mentors in the realm of agroforestry, natural building, traditional woodworking, and permaculture design. In my early 20s, I spent close to four years traveling the US and Europe, interning, work trading, and apprenticing with these incredible people, learning techniques for holistic, healthy living but perhaps more importantly, gaining inspiration from the many ways these trailblazing practitioners had come to view their place in the world around them.

In chronological order: Jerome Osentowski, Dan and Cynthia Hemenway, Ianto Evans and Linda Smiley, Kiko Denzer and Hannah Field, Rick Hazard, Ben Law, Peter Bane and Drew and Louise Langsner. Thank you from the bottom of my heart. Your willingness to open your homes and lives to young learners has had an impact far greater than you will ever likely realize, and in many ways, I owe parts of the person I've become to the inspiration you shared with me.

I also need to specifically call out Ben Law. The nine months I spent at Prickly Nut Wood was one of the most impactful experiences of my life. I will forever carry with me the memories of my time there; the joy of hard work (and sometimes harder play); the beauty, integrity, and honor of your lifestyle; and the habitat and community you've cultivated around yourself.

Thank you sincerely. The impact you've made on the world resounds far and wide.

Dave Jacke. This project emerged from a germ of an idea we shared more than 10 years ago. We've been part of each other's lives through several major turning points; travelled, dreamed, and pondered together; excitedly discussed hyper-specific details from bud types, to pollarding best practice, the prehistoric origins of coppice, and the balance and flow of carbohydrates in woody plants throughout the year; and built a strong and lasting friendship.

You amassed a remarkable amount of data, and put in countless hours crafting it into something we both know our world will benefit from. As complex as it may have been, ultimately, I'm thankful for our process. Your tenacity, eagerness, precision, and passion helped expand the scope of this book in numerous ways. And you spearheaded the conception of several fundamental concepts that have guided the overarching structure of the book, including:

- the flow of the historical narrative in chapter 1 (along with a detailed dive into the manorial system and the commons and their relevance to coppicing),
- a comprehensive "bud typology" framework organizing and describing the scope and origins of woody resprouts, and
- the systems framework that has ultimately come to guide the organization of the second half of the book.

Dave has also created and/or spearheaded almost the entirety of the species database that has helped inform our understanding of what species coppice and why. The curated species selection tables in appendix 2 represent the tip of the iceberg of what probably amounts to thousands of hours of research, collation, and cross-referencing. (And a very special thanks also to Daniel Plane and Aaron Guman who made significant contributions to the massive research and organizational work that helped build these datasets.)

I'm grateful for the work we've created together, humbled by your contributions, and thankful for our friendship.

To the close to 300 Kickstarter backers who believed in this project from the beginning and gave this project the boost it needed. Your contributions helped me gain a much deeper understanding of the potential for resprout silviculture in different climates and contexts across the US and Europe and supported the thousands of hours of research that has gone into this book's extensive literature review, species lists, and manuscript. I hope this book proves to be the resource you've been waiting for. I sincerely appreciate your patience, eagerness, and generosity.

And what's life without community? I feel blessed to have cultivated personal and professional relationships with many people who've shaped my perspective, helped make me who I am, and allowed me to enjoy the deep value of personal connection. So, I'm probably missing a few here but, in alphabetical order: Jared Aldrich, Danielle Allen, Harry Atkinson and Lisa Marchetti, Jason and Jill Baldwin, John Bauer, Mike Blazewicz, Edmund Brown and

Normandy Alden, Kay Cafasso, Ben Dana, Lisa DePiano, Skip Dewhirst, Ben and Erica Falk, Andrew Faust, Lisa Fernandes, Claude Genest, Meghan Giroux, Dan and Amanda Goosen, Susie Grey, Owen Hablutzel, Chris Jackson, Andy and Rose Kihn, Matt Kubacki, Geoff Lawton, David Ludt, Keith Morris, Bill Murphy, Cornelius Murphy, Jono Neiger, Sasha Rabin, Deva Racusin and Mary Niles, Elijah, Naomi and Micah, Jay Renshaw, Tim and Kat Rieth, Erik Schellenberg, Grant Schultz, Paul Schwartzkopf and Alissa White, Marcus Smart, Connor Stedman, Jesse Watson, Forrest White, and Phil White.

A big part of what made this book an especially rich experience was in connecting with many practitioners from different parts of the world. Learning from and seeing how different people decide to solve their own unique design challenges helped move this project from the theoretical to the practical. I only have space to name names, but thank you to all of you for your vision, energy, willingness to share, and hard work. If only we could propagate more of you vegetatively…

Brock Dolman, Robert Kourik, M. Kat Anderson, Frank Lake, Tom Ward, Tomi Hazel, Margaret "Pegg" Mathewson, Ish Shalom, Rick Valley, Mark Shepard, Leon Carrier, Phil Rutter, Helen Read, Coates Wetlands and Willows, Martin Crawford, Michael Ashby, Chantal Chevrier and family, Peter Szabo, Stjepan Dekanic, Georgi Hinkov, Tzvetan Zlatanov, Ivaylo Velichkov, Pande Trojkov, Theocharis Zagkas, Martinka & Ilija Chilimanova, Kathrin Bateman, Kim Almeida, Steve Gabriel, Sean Dembrosky, Brett Chedzoy, Dale Hendricks, Shana Hanson, Ed Gilman, Mark Angelini, Loren Luyendyk, Jim Jones, Tom Girolamo, Emmet Van Driesche. A special thanks to Kevin T. Smith, PhD, Supervisory Plant Physiologist with the USDA Forest Service Northern Research Station for his review of chapter 2 and helpful suggestions and ground truthing corrections.

This book would not be what it has become without the remarkable talents of my close friend and colleague William Bode. He is responsible for creating all of this book's illustrations. I thoroughly enjoyed our back and forth for the last four months of this project. Your ability to turn ideas and concepts into beautiful clear graphics left me proud and deeply impressed. I am so grateful for our friendship and the parallel paths we've followed over the past decade and beyond. I am so astonished by your artistic skill, your patient listening, the open, understanding way you respond to comments and accept feedback, and the beautiful simplicity of the lives you and Corinne lead. I hope this is just another step in many more collaborations and that readers fully appreciate the elegant clarity of the works of art you've contributed.

Unless otherwise noted, the photos in this book came from photos I've taken on my travels and of my own projects over the past two decades.

To Rob West and the folks at New Society Publishers for believing in the project and helping make it a reality. I knew from our first interaction that you would be a perfect fit for this project, Rob. I appreciate your vision, your feedback and guidance, and your willingness to adapt and give space to this resource so that it might achieve its full potential and help ignite and inform new and evolving relationships with woody plants in North America and beyond.

I would never be where I am today if it weren't for the loving and supportive family that have shaped who I am and given me the space, trust, and resources to explore the life I've felt called to lead. I know it may not always make sense, and it's also not always easy, but I hope you realize that even in the difficult times, it brings me great joy and fulfillment. My grandparents, Mike and Ruth Field and Leonard and Agnes Krawczyk, who helped connect me to my ancestry and nurtured and supported our family. Uncle Jerry, Dave, Margaret, and cousins, despite the distance between us, the connections we share as part of a biological community are unbreakable and deeply special. My sister Carol who has been a caring companion and confidant through it all. I'm so impressed with the person you are, the important work you do, and the beautiful family you've grown. And thank you to Erik for the companionship and skills you share and 'lil

Weston, your (and our) collective connection to the next generation.

Ammy Martinez, you are the love of my life. You've been with me through the most challenging and rewarding part so far. I'm so deeply inspired by your positive outlook, your strength, diligence and hard work, your kindness and fun-loving spirit, and your fierce and loving loyalty. I'm so proud of what we've done together, so deeply appreciative of having been able to spend this past decade together, and so excited to see where the next few coppice rotations bring us. Thank you for supporting and believing in me.

And last, place. "The land." And specifically the piece of the Earth we now call home. You give my life relevance and teach me every day. Thank you for nourishing us and connecting us to a place in a way I've never known before.

We are of course only inheriting it short term. Deep gratitude to the Western Abenaki who stewarded, tended, and learned from this place for countless generations. We hope to channel our love, care, and vision into our tending of this little patch of the globe and continue to leave it better for generations to follow.

Preface

If one process has dominated this book's narrative arc and the journey it's taken to get here, it's **succession**. (This is a good time to mention that definitions for terms in bold can be found in the glossary.) This book began as an outgrowth of a weekend workshop Dave Jacke and I taught in Massachusetts back in 2009. As we built the curriculum for our program, it became clear that major gaps in the existing literature left a niche waiting to be colonized, and the structure of our workshop provided a comprehensive foundation for the manual we wished we had.

This project has undergone many waves of growth, deconstruction, and reorganization. It certainly was not a linear process, and at many times, these phases lead to confusion, misdirection, and overwhelm. And at the same time, there have been numerous instances of profound understanding and emergent clarity. Connections to people, places, and practices that I may have otherwise never had the chance to enjoy. A newfound context, understanding, and appreciation for these management tools that probably would not have emerged any other way. And so just like the ridges and valleys of life, I find myself grateful to recollect this journey for the process as much as the end product.

Woodlands are complex systems with life cycles nearly impossible for the average human mind to comprehend. But there are patterns that ring true in all of our lived experiences. We know more than anything that change is the one true constant, and with that in mind, I'm eager to plant this literary seed to see how it grows, evolves, and finds relevance in this global ecosystem.

One remarkable characteristic of resprout **silviculture** systems is their human scale. They transform vegetation that may otherwise grow 150 feet tall into a shrub-type form that we might harvest repeatedly at 25 feet. And we can carry out most of our management with basic tools that our ancestors have used for millennia. **Resprout silviculture** systems help bring us into relationship with forest ecosystems on a timescale that we can really relate to. We can begin to measure our own lives by the number of coppice rotations we get to participate in.

My journey on this path is now entering its third rotation. As a young permaculture devotee back in 2001 on a cross-country educational wanderlust, chasing people, places, and skills that I knew I would need to live the life I dreamed of, I was fortunate enough to pick up a copy of Ben Law's *The Woodland Way*. Captivated by his narrative, I placed myself in a lifestyle intimately connected to the forests around me. A seasonal life of deep reverence and reciprocity. I wrote an old-fashioned letter to Ben asking if he took apprentices, and just my luck, I spent the fall, winter and spring of 2002-2003 in his more than 160-year-old sweet chestnut coppice in southeast England, living the dream that I'd so recently become aware of.

I returned to the States forever changed. A vision of resprout silviculture planted deeply in my imagination but with little outlet to share it and no land base to experiment. I became enamored with "green"/traditional woodworking and for close to eight years lived in a compact semi-subterranean basement apartment in Burlington, VT, building my teaching and design capabilities, sharing my vision of coppicing with students and clients, and actively searching for a place to put my ideas into practice. This early succession in my professional life entered a new phase as Dave Jacke and I decided to embark on this project together. And the past 10 plus years have marked a new cycle in my own life and vision.

During that time, I found a piece of the Earth to call home and met the love of my life. We built our homestead from the ground up, planted thousands of trees, and have cultivated a life together along with a deep relationship with place. And many of the ideas in this book we've been putting into practice.

And now, this book's release marks the next stage of succession—another rotation in the coppice cycle of life. And I'm so grateful that you're a part of it. Here, more than two decades of dreams and lived experience made manifest will hopefully help inspire you to pursue a newfound relationship with the wondrous resilience of woody plants. And they'll teach you as you shape them. And we'll all get to learn from one another. Because, in the end, it's just one more wave of succession washing over civilization.

I've done my best to craft a narrative that's comprehensive, accurate, and concise. A resource that explains not just what but why and how. And that said, I'm still left with many questions. And we're all learners on this journey. Please do your due diligence as you put these concepts into practice. Coppice happens—whether we want it to or not. Woody plants have been doing it long before humans walked the planet. But as you develop your designs, take your time, realizing that we're all collectively exploring the modern state of this ancient art.

I hope you find the tools you need here to guide and manifest your vision. I firmly believe it is our birthright to ethically steward the places we call home. May your hands be guided by gratitude, knowledge, reverence, and the intuition we inherit from the countless generations of ancestors who've all sought solutions to these same fundamental problems.

With gratitude, humility, and excitement for the future,

—Mark Krawczyk

And a brief note: Ben Law begins his book with a note that I feel is also relevant here. In many places in the text, I've used the term "woodsman" to describe practitioners engaged with woodland management. "Man" in this case is not so much a reference to gender but more an acknowledgment of its Latin etymological root, *manus* or "hand." As in "the hand of the woods." We are nature working. Embrace your role as an extension of your ecosystem and do it well.

Foreword

I find coppice agroforestry in its myriad forms fascinating—practically, historically, culturally, horticulturally, and ecologically. This has been true ever since Philip Stewart's article "Coppicing with Standards" in the Spring 1981 edition of *CoEvolution Quarterly* magazine first introduced me to a whole realm of cultural ecology that I never even knew existed. Coppicing blends so many intriguing elements: forgotten practices, hidden history, unacknowledged importance for human evolution, high regenerative potential, and botanical magic! Stewart's piece lit my fire—a fire stoked initially by my childhood experiences with resprouting woody plants, my passion for ecological design, and my visceral eco-culture imaginings. Unfortunately, over the following decades I found little more than tidbits to feed that fire in the pre-internet literature, and no practitioners on this continent. My coppicing fire burned down to coals. I fed them as I could to keep them alive. Then a wind in the form of Mark Krawczyk woke the embers and added a large load of fuel on top.

Mark is a woodsman. You can tell by his tools and his hands, as well as his creations. Traits integral to that include his mostly patient and unassuming demeanor, his wry and dry wit, his confident humility, his observant questioning, and his way of walking and working the land. Mark is no flash in a pan. He steadily accumulates skills and knowledge the way trees build wood: layer-by-layer, ring-by-ring, and year-by-year. Like a tree, he seems to build stronger wood in reaction to setbacks and mistakes. You may be able to knock Mark down, but he is likely to pop back up again, perhaps more vigorously than before. The resonance between who Mark is, or has become, and the field of coppice agroforestry is anything but a mistake. It reflects Mark's intent. That resonance, and Mark's resilience, were key forces that drove this book forward to completion.

Mark and I met in 2009 through the permaculture network here in New England, USA. His traditional woodworking experience and months coppicing with Ben Law in Britain nearly a decade earlier immediately sparked my interest. Our shared pursuits in ecological design gave us common ground and a common language. Soon thereafter, Mark and I decided to run a coppice workshop in Western Massachusetts so we could work and learn together while sharing what we knew with others. During that collaboration we asked each other many questions the other could answer, and many more we could not. We wished there was a solid resource that gave us the information we sought, but there wasn't. Hence this book.

We began writing *Coppice Agroforestry* together. With a huge boost from a successful Kickstarter campaign in late 2010 (thank you!), we traveled, met practitioners and trees, researched, read, observed, played with woody plants, gathered data, and wrote. Mark was to be primary author, and me the primary literature research and database guy. We made progress. We hit bumps and roadblocks and overcame them. Our lives took their turns and twists, the project had its ebbs and flows. Eventually, for a host of reasons, it became clear that I needed to step away from the project. Mark carried on to completion. I feel glad he did, also for many reasons—key among them: the big pickle in which we as a species find ourselves.

We live at a time when the consequences of the industrial revolution that began in the 18th century are culminating in transformational cultural and ecological crises. Underpinned and supercharged by the discovery, development, and mass deployment of fossil fuels, that revolution disintegrates our social systems, corrupts our politics, and quite literally kills our planet. How might we move forward?

Taking in the sweep of human history can offer great meaning in such times. What is the big picture and how do we fit into it? What led to the current predicament? What pieces of culture have we lost that might again

hold adaptive value? While I will not attempt to fully answer to these questions here, they provide critical context for this book. The long and short of the situation as I see it is: we are sleep-running through an inflection point where humanity's continued reliance on *ancient hydrocarbons* will destroy us and everything we hold dear. We must rapidly create economies, cultures, lifestyles, and landscapes based largely upon *recent carbohydrates* if many of us are to survive, much less thrive.

"Recent carbohydrates"—an odd turn of phrase, but an essential frame. Plants, of course, produce carbohydrates when they photosynthesize, using water, sunlight, and carbon from the air. Carbohydrates form the fundamental building blocks and primary currency of all life on Earth. Indeed, carbohydrates were the building blocks of the ancient hydrocarbons, too, before geologic time, heat, and pressure had their way with them. While many carbohydrate sources will be useful in a post-fossil, "neo-carbon" world, this book focuses on woody plants, and for good reason.

Wood formed the backbone of human materials and energy cultures around the planet for millennia. We don't hear much about this because the evidence has mostly decomposed. Still, it is true. Without wood we would never have come this far as a species. The reciprocal relationship between humans and woody plants began long before recorded history. The associated arts evolved to the point that, as this book makes clear, we could never have had an industrial revolution without them. And, that very revolution caused the demise of the culture and horticulture that supported the revolution in the first place. Coppice agroforestry withered in the industrial era. Even so, the ways our ancestors managed woody plants and converted wood into valuable tools and products hold relevance for us now.

As I develop my own homestead, I keep finding uses for coppice materials and practices, all of which Mark discusses here. From mulch, to kindling, to stakes, bean poles, mushroom logs, and edible tree-leaf production, woody resprouts come in handy. And this is only the beginning. Attentive, imaginative, and forward-thinking people will see much value for a neo-carbon world. I find it both ironic and highly appropriate that the coppice culture that helped birth industry may, in turn, resprout to midwife post-industrial cultures. I also find it appropriate that an in-depth treatment and attempt at re-imagining coppice culture from the ground up would arise here in the so-called New World.

Most heirs of the colonizers of North America have little or no awareness of the coppicing traditions of our forbears, or those of the First Peoples here. That is a blessing and a curse. The curse is easy to understand: we have so much to learn in so many interrelated realms of endeavor. Woody plant systems grow and respond to our interactions fairly slowly. We know relatively little about the coppicing behaviors of North American species. Coppice skills and knowledge take significant practice to hone. Ignorance of Indigenous land management practices is very far from bliss. The

blessings? Well, let's just say that tradition and experience often limit one's perceptions. Few resources on the topic exist, first of all, but most coppicing literature doesn't go deep or do more than retell the ways things were once done. Meanwhile, our times demand massive innovations.

Our context is vastly different from the pre-industrial world. I believe we can design and create well beyond the coppicing legacies our ancestors gave us. We have the opportunity to reframe and recreate the field for our current cultural contexts. Hence the phrase "resprout silviculture": let's broaden the terms of discussion! Let's open our minds to all the possible ways that we might manage woody resprouts for a host of products and values! Let's create new uses of resprouting woodies for a neo-carbon world, beyond what Mark and I and maybe anyone else has yet imagined!

Coppice Agroforestry goes deeper and further into this subject than any reference I have seen so far. The book's integration of research, theory, and practice offer trailheads that lead into vast forests of discovery and lifetimes of patient, observant, quiet, productive, and regenerative adventure. I look forward to joining with Mark and you in moving the field of resprout silviculture, the conversations, and the culture, forward.

Dave Jacke
Montague, Massachusetts, Nipmuc &
Pocumtuc Territory
February, 2022

Introduction: What Is Coppice?

There is not a more noble and worthy husbandry than this.

— John Evelyn[1]

The lives of humans and woody plants have been inextricably linked for as long as our species has populated the planet. Even in modern times, when we are so dramatically dissociated from the resources and landscapes that support us, it's remarkably enlightening to reflect on all the ways trees influence our lives and our experience of the world around us.

Trees provide...
- Shelter
- Building materials
- Shade
- Fuel: for heating, cooking, electricity
- Industrial production
- Transportation
- Food: for humans, livestock, mammals, birds, insects, fungi, and microbes
- Craft materials
- Erosion control
- Climate stabilization
- Water cycling
- Wildlife habitat
- Medicine
- Enclosures
- Soil building processes and biomass
- Air filtration
- Beauty and inspiration

It's hard to imagine life on Earth without the myriad benefits proffered by trees. In fact, life as we know it would not be possible without them.

Keeping all this in mind, take a moment to reflect on your relationship (and more broadly, our cultural relations) with the forests that support us. For the vast majority of us in the one-third or "developed" world, that relationship is virtually unconscious. Most of us have no idea where the wood that shelters us, keeps us warm, cleans and oxygenates our air, and stabilizes our soil actually comes from, and even more importantly, how it's produced and managed. Historically, only a privileged few were affluent enough to detach themselves from their relationship with woodlands for anything

1

FIGURE I.1: Imagine a world without these majestic creatures.

pay testament to the widespread suitability of this land management technique. As we'll explore in later chapters, when well-managed and maintained, coppice woodlands and their biological community are exceedingly healthy, robust, and resilient, host a broad diversity of species, age classes, and forms, and yield an array of forest products for human use.

This book attempts to reconnect our culture with the woodlands that we are part of. While our focus rests primarily on coppice woodland management and other related means of symbiotic silviculture, a much deeper theme underlies the nuts and bolts of system design, establishment, and maintenance. Our real goal is to help initiate and inspire the revival of a woodland lifestyle—a life lived in direct and complementary relationship with the forests that nurture us. This is no small task, but I believe that to bring about the change we wish to see in the world, we first need to envision what it looks like. So please accept this invitation to reconnect, restore, and reinhabit our true home in a way that benefits all beings.

WHAT IS IT?

The term "coppice," both a noun and verb, comes from the Old French word *copeiz*, which today translates to *couper*, meaning "to cut."[2] Additional sources connect the word to the Greek *kolaphos*, "blow," via the Latin verb *colpare*, "to cut with a blow."[3] While etymology can be enlightening, it doesn't offer much in the way of insight into the practice itself.

Coppice management is an ancient silvicultural technology where broad-leaved woody

other than hunting and recreation. Industrial culture has enabled this disconnect to become the norm, but today our need to reconnect with our woodlands grows ever clearer.

Coppice forestry is an ancient silvicultural practice that provides one of the best living examples of a symbiotic, cooperative relationship between humans and forest ecosystems. While the modern ecoforestry movement also places humans as active and beneficial participants in the landscape, the longevity of coppice woodlands, stools, and pollards the world over

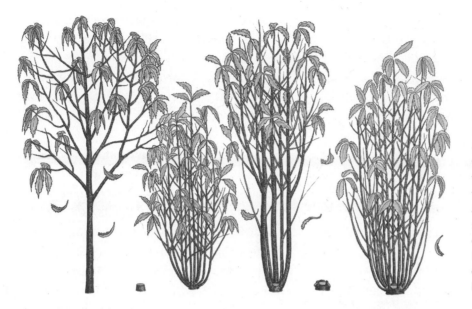

FIGURE I.2: Two harvest rotations in the life of a coppice stool. Starting at the left, we see a young tree, coppiced for the first time, allowed to resprout and grow to a harvestable size, when it's cut back and coppiced once again. The life cycle of a coppiced tree or shrub may theoretically be indefinite.

plants are cut on cycles of 1 to over 40 years during dormancy and allowed to regenerate from the stump. These stump sprouts develop into a new crop of poles harvested during the next felling cycle.

This vegetative regeneration or "stump sprouting" is a common ecological process often found along roadsides and utility lines, where regular clearing by road crews and utility companies establishes "coincidental coppice." Often, for many people, this incredible expression of biological vigor is a nuisance. We were trying to clear the land after all. But only in a culture flooded with cheap energy can we afford to view an abundant self-renewing resource as waste. Yet when we recognize the immense potential of the humble stump sprout, we can develop production systems that are largely self-maintaining, which is the essence of coppice agroforestry.

Dissecting Our Definition

The ability to form a permanent structural stem with perennating organs (buds) well above the ground differentiates most **woody plants** from their herbaceous kin. This adaptation affords woody plants the opportunity to shade out herbaceous competition, avoid many herbivores (especially the big ones), and occupy a vertical niche that extends potentially hundreds of feet skyward. Broadly speaking, the term "woody plants" refers to plants with rigid stems high in compounds called lignin and cellulose. However, plants may be called "woody" even if they don't make actual wood. In this book the term "woody" will refer to tissues made of, or plants that make, actual wood in the strict sense: the secondary xylem lying beneath the bark of a tree or shrub, the hard fibrous material forming the main substance of the trunk or branches of shrubs and trees, composed *primarily* of lignin and cellulose.

Generally speaking, most broad-leaved woody plants (**angiosperms**)—trees and shrubs that have wide leaves as opposed to needles and bear their seeds in fruits instead of cones—will coppice, meaning they respond to cutting during dormancy with vigorous resprouting come spring. Some of the best-known examples include maple, ash, oak, linden, hazel, willow, and poplar. The vast majority of broad-leaved woody plants are deciduous, meaning they lose their leaves during the dormant season. Conversely, "evergreens" retain their foliage throughout the year. These trees are typically conifers, which bear cones containing seeds and have needle-like leaves. Most conifers do not coppice in the true sense of the word, although parallel forms of management that we discuss later do allow us to leverage resprouting in conifers.

Under coppice management, woody plants are cut on cycles of 1 to 40 or more years. Traditional felling cycles—the frequency of harvest—revolved around the desired size of materials for a particular product and the species being managed. Poles harvested for fence posts 3 to 6 inches (7.5 to 15 cm) in diameter require a longer cycle than shoots used for 1-to-2-inch (2.5 to 5 cm) diameter thatching spars or pea sticks. We generally consider coppice stands harvested on 1-to-5-year cycles "short-rotation" and those exceeding 5 years "long-rotation."

The *timing* of felling operations is critical to the health and vigor of shoot regrowth. While many species will still resprout after cutting during spring and summer, historically, coppicing was carried out during the winter months once trees had gone dormant. Winter felling is desirable for several reasons. During dormancy, trees experience considerably less stress, and insect, fungal, and bacterial populations are quite low, reducing the likelihood of disease and infection. Also, first-year shoots from a freshly felled **stool** (the stump of a coppiced tree that's managed for resprouts) are tender and pithy and benefit from a long growing season to harden off before the arrival of autumnal frosts that can damage or kill young sprouts.

Table I.1: Top genera/species for coppicing and pollarding

Latin Name	Common Name
Acer spp.	Maple
Castanea spp.	Chestnut
Cornus spp.	Dogwood
Corylus spp.	Hazel
Eucalyptus spp.	Eucalyptus
Fraxinus spp.	Ash
Morus spp.	Mulberry
Platanus spp.	Sycamore
Populus spp.	Poplar/Aspen/Cottonwood
Prosopis spp.	Mesquite
Quercus spp.	Oak
Robinia pseudoacadia	Black Locust
Salix spp.	Willow
Tilia spp.	Linden/Basswood
Ulmus spp.	Elm

Traditional winter felling and extraction work, especially on frozen ground, dramatically reduces the overall impact on understory vegetation and soil structure while improving extraction routes. The heavy work of tree felling and extraction is well-suited to the cold temperatures of winter, while craft and value-added work was typically performed in spring and summer, creating a varied and seasonally balanced livelihood.

Many coppice sprouts emerge from **preventitious** (dormant) **buds** embedded beneath the bark, capable of prolific new growth following disturbance. Often these buds develop in the spring at the base of a recently felled tree. We call this remnant stump a coppice stool. This contrasts with the suckering tendency of some species, where shoots or suckers emerge from a tree's root system. Cherry, aspen, beech, and black locust (*Prunus* spp., *Populus* spp., *Fagus grandifolia*, and *Robinia pseudoacacia*) are examples of species prone to suckering following disturbance. While the new shoots emerge from a different portion of the tree, we'll discuss both stump-sprouting and suckering species throughout this book. And then of course, there's pollarding—a training system for woody plants that manages sprouts high up on the stem in the crown where they're out of reach of livestock and wildlife—hedgelaying, and shredding. All of these techniques add diversity to our continuum of resprout silvicultural practices.

So, with a basic understanding of what coppicing is, let's now look at why it emerged historically as a widespread silvicultural practice that has benefitted human cultures and how it can best be adapted to meet our needs today.

WHY COPPICE?

To understand the evolution of coppice management, we must begin with some familiarity of the historical context. We'll explore this fascinating history in greater depth in the next chapter, but a moment's worth of historical insight offers many clues about the circumstances that shape the way people manage forests.

Imagine living in the late Stone Age with the same basic needs as modern humans of food, shelter, warmth. Now imagine harvesting and processing the raw materials to provide for these needs using stone tools without any means of transport beyond human power. It's in this context that coppice forest management systems evolved to match the tools and energy available.

Because coppice forestry produces poles of a regular dimension that can be easily processed and used in craft, small-diameter **polewood** was far better suited to the resources available at the time. In British silvicultural lingo, the terms "wood" and "timber" refer to two different forms of forest products. "Wood" describes small-diameter polewood, while "timber" refers to full-size trees, grown for lumber. For Neolithic humans right through to the later peasant class of the Middle Ages, wood was actually a much more valuable and useful material than timber as it could be worked up with simple, readily made tools. In contrast, timber required specialized metal axes for hand-hewing timbers or long, complex metal saws for energy-intensive pit-sawing in which a pair of sawyers worked together, one above and one below the suspended log in an excavated pit to rip it lengthwise into boards.

Countless ancient coppice and pollard specimens exist throughout Europe, having lived for

Pit-sawing

FIGURE I.3: For much of human history, lumber was an exceedingly high-value and labor-intensive product. Here two sawyers are working in tandem, "pit-sawing" a log. Understanding this helps make clear why polewood has been so valuable to human cultures throughout history.

centuries beyond their typical natural life span as a direct result of this management. Research proves that coppicing dramatically increases the life span of woody plants, often by a factor of three or more. This is probably best evidenced by the Greek word *kouri* used to describe a pollarded tree. Essentially describing a process of "keeping young," the super-intensive pruning that is coppicing and pollarding removes most of the mineral- and nutrient-demanding biomass of the plant's aerial parts, stimulating healthy, vigorous new growth. As human management of these ancient trees wanes, many otherwise healthy trees have begun to lose their productive vigor, which is a testament to the critical role humans play in maintaining overall system health.

Coppice and related forms of woodland management formed the backbone of numerous societies around the globe for centuries, that is until humans discovered new and abundant fuel sources: coal and oil. Yet, despite this history, the practice is virtually unknown to North Americans of European descent. Why is this the case?

WHY IS IT NONEXISTENT HERE?

We know that sprout-based management was a part of the lives and land management practices of many Indigenous North Americans, but the majority of those traditions were extinguished in the wake of colonization, displacement, and in some cases, genocide.

At the same time, it's difficult to say with certainty why coppicing traditions disappeared from the European silvicultural tool kit as they

The Pollard Cycle

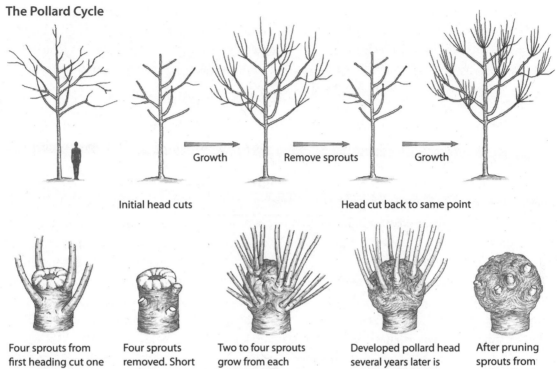

Growth → Remove sprouts → Growth

Initial head cuts   Head cut back to same point

Four sprouts from first heading cut one year after pruning.

Four sprouts removed. Short stubs left to retain buds.

Two to four sprouts grow from each removed sprout. Pollard head is developed.

Developed pollard head several years later is considerably wider in diameter than stem.

After pruning sprouts from pollard head.

FIGURE I.4: Like coppicing, pollarding draws on woody plants' ability to sprout following intentional disturbance. In the case of pollarding, the harvest occurs well aboveground, usually out of reach of wildlife and livestock. These trees can be managed for polewood or tree hay. Here we see a young tree that's pruned to shape to optimize plant architecture, followed by the final pollard cuts that initiate the knobs from which new sprouts will originate in future harvest cycles. Adapted from Gilman, 2002, p. 150.

landed on this continent. Given the abundance of mature forest blanketing the landscape, it stands to reason that a resource-conscious system for woody biomass production was probably not on the top of the priority list. Faced with winters far more challenging than much of Western Europe, and little in the way of resources beyond the tools and possessions they brought or made themselves, land clearing, food production, and shelter construction presumably demanded any and all waking hours.

Pioneers struggling to carve out a niche in a landscape dominated by majestic forests likely found the woody resource of eastern North America to be as much of an obstacle as an asset. Most colonists had probably never seen trees of North American size, having lived in the shadow of centuries of resource overextraction and depletion in their native land. These forests must have appeared an inexhaustible resource, with land clearing for agriculture a much more pressing need, and fuel and construction material shortage a rapidly fading memory. There was likely little need to intentionally develop coppice systems as coppice regrowth would have often been a thorn in the side of agricultural development with stump sprouts vigorously reclaiming cleared pastures and farm fields. Somewhere between the struggle for survival and the explosion of industrial culture, European settlers lost the legacy of coppice agroforestry.

Today, with the exception of scattered experiments in short-rotation biomass production, willow harvest for basket materials, and potentially unknown systems managed by homesteaders, foresters, and farmers, coppicing is less than a cultural memory. It has been long

FIGURE I.5: Neatly stacked cordwood piles in a springtime meadow near Limoges, France.

forgotten. But as the circumstances that shape our lives and the long-term viability of our civilization rapidly shift, the value of complementary forest management strategies grows ever clearer.

WHY DO IT TODAY?

Reasons for developing coppice woodlands in North America, and the world over, for that matter, are at least as numerous as the staggering number of products that can be produced with coppice wood. Considering the pivotal role coppice played in the stabilization and expansion of many cultures in the face of extensive resource depletion, it would also seem an appropriate response to many of the same problems in our modern world.

The following are some of the benefits of coppice that we'll explore in depth.

Home and Community Scale Energy Security

On a farm/homestead/community scale, intensively managed coppice woodlands could enable individuals to produce small-diameter cordwood to become self-sufficient in home heating. This increases community resilience, reduces or even eliminates the need to import fuel, engages the populace in a direct relationship with the management of their environment, and helps engage them in the understanding and implications of their energy consumption.

Local Livelihoods and a Culture of Craft

Probably one of the most valuable yields coppice woodlands could provide is a renewed local economy based on home-scale production of useful and necessary products. As our current globalized economic system self-destructs before our eyes, we must begin to think about strategies to create a meaningful, right livelihood that enables people to express their skills and creativity and contribute to the renewal of our communities.

It seems clear that we need to build a new modern economy around the production of goods and management of land-use systems that serve the people who use them. In the very same way that coppice sustained a skilled and independent class of craftspeople and land managers historically on the European continent, could we see a similar evolution of a productive community of citizens providing products for neighbors and community members today? I believe so. In fact, I believe it is and will be essential!

While this is no small task to design and implement, human history proves time and again that we cannot separate ecology from economy.

Preserving Native Forests

By concentrating intensive woodland management within coppice **cants** and field edges, we could help reduce the pressure placed on our "natural" forests to provide for our resource needs. By no means do I propose that coppice systems replace modern forestry systems, nor to slight the inspiring, responsible, and cutting-edge forest management strategies that are becoming more and more common. Coppice could provide a valuable complement to conventional forestry management and, in some cases, help to concentrate production in

intensively managed systems and protect and preserve our ever-shrinking natural forests.

Leaving a Legacy

Upon establishment, coppice woodlands provide useful materials that will shelter, warm, nurture, and employ us for generations. When I traveled to England to apprentice with Ben Law in his copse of sweet chestnut, I was participating in the harvest of coppice stools that were planted over 160 years ago and that may continue to yield for centuries.

Probably the biggest challenge we face is the considerable investment in time and energy needed to transform fields and forests into silvopasture and coppice cants. But imagine being part of a future generation that inherits a thriving cant of black locust poles that can be used to heat your home and provide a steady supply of one of the world's most rot-resistant woods for building projects. If all we do is lament the resources we lack, we'll never set the stage for the generation that gives thanks each season when they commence the cutting of a cant that their great-great-grandparent nurtured and established.

Why Not?

We know that our planet's life-support systems are in dramatic decline, along with the hydrocarbon economy built on oil and coal. It's up to us to help build a new "carbohydrate economy" equipped with a toolbox stocked with regenerative and productive land-use strategies thousands of years in the making. I believe that we can work to restore our landscapes, build resilience in our communities, and take responsibility for our needs through the intensive design, development, and management of coppice woodlands. And we know that we can have a heck of a lot of fun doing it!

In navigating these waters, we face considerable obstacles, namely, a culture that has forgotten how to provide for its most basic needs and a worldview that expects and demands instant results. While the map is incomplete, we're well acquainted with our destination. I hope that this book will serve as a compass, helping point us in the right direction and inspiring others to dip their oars into the sea alongside us.

PURPOSE OF THIS BOOK

To my knowledge, there is no existing practical handbook on coppice system design and establishment. Of the resources that do at least begin to fill this void, we find a general revisitation of traditional systems that don't necessarily integrate modern ecological thinking and design.

This book strives to inform and inspire a new generation of woodland managers and a renewed experience of "participatory ecology." It draws together much of the existing literature on coppice systems, rural woodland-based economies, ecological design, and projected yields into a single resource that makes a compelling case for the establishment of dynamic,

On the Ethics of Coppice Agroforestry

For most of us, a primary objective in developing resprout silviculture systems lies in meeting our needs though conscious woodland management—a goal that can take many forms. Coppicing is a tool. And it's not always the best tool in all scenarios. The objective of this book isn't so much to promote coppicing, but rather to provide readers with information and tools to make good decisions when considering how to best meet their needs while increasing ecosystem health, diversity, and function.

Diverse, healthy, productive, intact forests are often poor candidates for coppice conversion. If our goal is to cultivate productive ecosystems while restoring healthy landscape functions, *disturbed landscapes* are our best places to intervene. If you're looking for tools to manage mature high forest stands, first explore the range of silvicultural practices at your disposal—like uneven-aged selection management, patch cuts, or seed tree treatments (more on this in chapter 4). On the other hand, old fields, backyards, low-grade forests, vacant lots, treeless pastures, farm fields and riparian buffers, and early-successional field edges are the types of landscapes well-suited to coppice woodland and silvopasture conversion. Remember—it's all a matter of context, and universal solutions rarely exist. Silvicultural practices are powerful tools for landscape transformation, and with them comes great responsibility. Apply them with care and forethought.

multigenerational, self-regenerating silvicultural systems.

I hope you will use this book to assess the state of forest land for coppice conversion, design your own multifunctional copses and silvopasture systems, develop new ideas for engaging, productive livelihoods, and pave the way for a reinterpretation of this ancient forestry system. I hope this book will serve as a critical tool inspiring the emergence of resprout silviculture systems throughout North America and that you'll share your experiences and research so that we can co-create a culture of educated, skilled practitioners. Let's begin by exploring the reasons why people have used coppicing over the past 8,000 years.

Part I

History, Biology, Ecology, Systems, and Economy

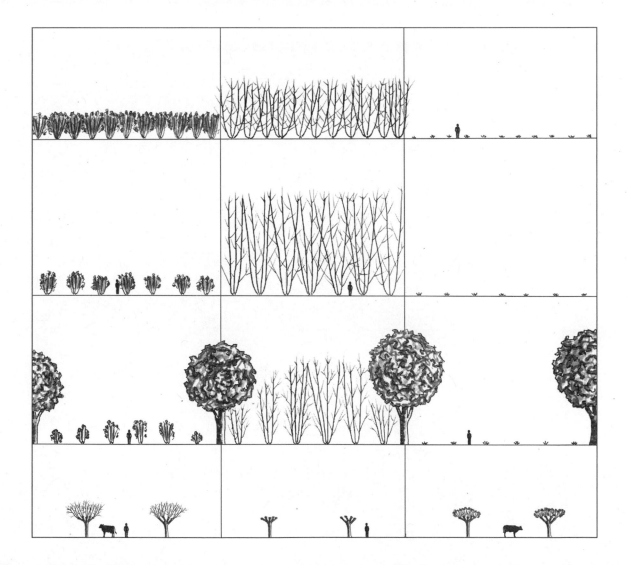

Chapter 1:
A Cultural History of Coppice Agroforestry

For there is hope of a tree, if it be cut down, that it will sprout again,
and that the tender branch thereof will not cease.

— The Book of Job, ca. 600 BCE

Once the economic reason supporting it disappears,
no rural pattern survives in a healthy condition for long.

— Roger Miles, *Forestry in the English Landscape*[1]

Many mysteries lie deep in human history and prehistory. Ever wondered about the day-to-day realities of our ancestors from the modern era through prehistory? From clearing land for agriculture; to procuring fuel for warmth, cooking, and industrial production; to creating shelter, tools, and crafts; and weathering the effects of widespread shortage, the story of civilization has often run parallel to the story of their woodlands. While we'll be forever left to ponder much of this story, historical records have left us scattered clues suggesting that the humble stump sprout has played a central role in the development, sustenance, and expansion of human cultures the world over since the Stone Age.

Take, for example, the fact that woody resprouting was once so common that it played a key role in the mythology of the Israelites. In fact, a derivative term, the Hebrew word

netzer, meaning sprout or shoot, was used to name Nazareth, the childhood home of Jesus Christ, and to signify the Messiah himself. "The prophets, in speaking about the destruction and re-emergence of Israel, used the metaphor of Israel being like a tree that had been cut down, but which would sprout again. Israel would be led by a messianic figure called 'the branch.'"[2] Yet, though coppice was used to symbolize a figure central to much of Western civilization, our culture has largely forgotten about this ancient practice. Remembering this knowledge, and reviving and reinventing these practices for our times, could well play a key role in humanity's future.

Humans have relied on forest resources for thousands, if not millions, of years. Cultures both ancient and modern require adequate wood supplies to meet a range of critical needs, to develop and expand. As civilizations grew, the means by which they met these needs changed as they used their woodlands—or used them up. On every inhabited continent, if human cultures didn't learn to sustainably use forest resources, their civilization didn't last, or they had to buy, beg, borrow, or steal wood from their neighbors. Archaeological work around the world provides evidence of forest management systems built on woody resprouts.

This chapter illustrates how humans have harnessed woody plants' sprouting ability to meet their fundamental needs for millennia and how this relationship has in turn shaped their cultures. We see this relationship expressed in the crafts, buildings, lifestyles, land use and ownership patterns, livelihoods, and economics of societies stretching from prehistory to the modern era.

Yet, coppicing and coppice craft have mostly disappeared in the wake of industrialization. Clearly, Roger Miles is right: as the need and economic demand for woody sprouts declined, the practices vanished, and the rural landscape changed, leaving their legacy shrouded in mystery. Understanding the historical relationships between human civilizations and resprouting trees and shrubs can help us envision how modern coppice management and the polewood economy might find a valued place in our culture, both today and in the future.

This chapter focuses on the coppice history of Europe, especially Britain, and North America. The burgeoning field of **historical ecology** informs this understanding of European landscape and land-use history. The

availability of this information, along with the ecological similarities, make this information especially relevant to us here in North America. Most English-language resources describe British resprout silvicultural traditions. While this limits the geographical and cultural extent of our survey, these stories and examples illustrate at least some of the patterns, events, and dynamics that have driven woodland management during the past several thousand years all over the world.

We begin by exploring prehistoric coppice husbandry in Europe and North America, then follow its evolution through Roman and medieval times, primarily in Europe, and conclude with a discussion of coppice in the modern era. What role did resprouts play in human history and cultural evolution? What were the ecological and social milieus within which coppice played a central part, and how did these milieus and resprout management influence each other? Besides the general historical interest, these past realities can teach us about ourselves and our current context and help inform useful design directions.

Remember, however, that even when historical records are available, the information is scant. "It is often quite difficult to discover the nature of many of the traditional practices of woodland management, such as coppicing, pollarding, woodland grazing, making temporary arable fields, and the use of fire. This is partly because their very prevalence and normality meant that authors felt they did not need to comment on them."[3] By presenting even a modest sense of the realities on the ground through history, we can begin to envision wider possibilities.

COPPICE: AN ESSENTIAL PREHISTORIC RESOURCE

We begin with evidence of the use of coppice materials in several parts of very early Europe. Onto this, we enjoy a glimpse of North American land management and coppice use to further flesh out the prehistoric use of resprout silviculture.

The Landscapes of Prehistoric Europe and Early Evidence of Coppice

About 13,000 years ago, at the end of the last Ice Age, glaciers covered most of northern Europe. Tundra or shrub tundra primarily occupied the unglaciated belt south of the ice sheet and north of the Alps, while open, semi-arid woodlands covered the area south of the Alps, containing varying mixtures of oak (*Quercus* spp.) and pine (*Pinus* spp.), along with junipers (*Juniperus* spp.), goosefoots (family *Chenopodiaceae*), and rhododendrons and their allies (order *Ericales*). "Only in the east was forest vegetation extensive; and here various forest types with varying proportions of *Picea, Pinus, Betula,* and *Alnus* were present" (spruce, pine, birch, and alder).[4] A rapidly shifting climate disrupted these vegetation patterns around 8000 BCE (10,000 years ago) as the glaciers melted, marking the end of the Paleolithic Era and the beginning of the Mesolithic.

By 6000 BCE, forests covered the vast majority of mainland Europe,[5] and the British Isles were well into a period of relatively stable **wildwood** (primeval or old-growth forest) dominance.[6] At least, this is what **palynological** evidence (spores, pollen, and certain algae trapped in sediments) has led many to believe. The presence of high percentages of arboreal pollen (from trees) in the pollen record, as compared to grasses and herbs, implies that the Mesolithic landscape was mature **high forest**— what we tend to think of as a "climax" closed canopy forest.[7]

Any peoples living there were probably mostly nomadic, and subsisted by hunting, gathering, trapping, and fishing, like typical Mesolithic cultures. The Mesolithic Era ended, and the Neolithic began, when people started living in long-term settlements, making polished rather than knapped stone tools, and practicing agriculture using cereal grains and domesticated animals. In Europe, that usually involved clearing forests. This is essentially the history as described by Iversen's Landnam Theory (1941). "Landnam" translates to "taking of the land"[8] and suggests that roughly 5,000 years ago humans began to clear the forests of Northwest and Central Europe to make way for agriculture. Their livestock prevented forest regeneration, creating a more open, park-like landscape.[9]

Our survey of coppice history in Europe begins right around this time period, the Late Mesolithic. And since we may only rely on theories to understand what the landscape looked like pre-agriculture, let's also take a brief look at the fascinating work of Dutch ecologist and ornithologist Frans Vera as an alternative to Landnam. In his book *Grazing Ecology and Forest History*, Vera proposes:

> The natural vegetation (of the lowlands of Western and Central Europe—and eastern North America for that matter) consists of a mosaic of large and small grasslands, scrub, solitary trees and groups of trees, in which the indigenous fauna of large herbivores is essential for the regeneration of the characteristic trees and shrubs of Europe. The **wood pasture** can be seen as the closest modern analogy for this landscape.[10]

Framing the foundation of his theory, Vera cites several inconsistencies in the pollen record, namely, the relative abundance of pedunculate oak (*Quercus robur*), sessile oak (*Q. petraea*), and hazel (*Corylus avellana*) in Central and Western Europe over the course of 9,000 years. These three species are unable to survive and reproduce in closed canopy forests. In other words, they do not tolerate shade. So their consistent presence over time would seem to imply that these landscapes must have contained much larger gaps to enable their regeneration.

In short, Vera suggests that populations of large herbivores including aurochs (*Bos primigenius*), tarpan or European wild horse (*Equus przewalski gmelini*), European bison (*Bison bonasus*), red deer (*Cervus elaphus*), roe deer (*Capreolus capreolus*), beaver (*Castor fiber*), and wild boar (*Sus scrofa*) all acted as the primary disturbance agents in these Late- and Postglacial ecosystems. Their impact created a landscape mosaic of "mantle and fringe" vegetation, comprised of grassland, scrub, trees, and groves. These open landscapes enabled light-demanding species like oak and hazel to thrive, sheltered from browsing herbivores by the protective cover of thorny mantle vegetation [blackthorn (*Prunus spinosa*), common hawthorn and English hawthorn (*Crataegus monogyna* and *C. laevigata*), guelder rose (*Viburnum opulus*), common privet (*Ligustrum vulgare*), dogwood (*Cornus sanguinea*), wild apple (*Malus sylvestris*), wild pear (*Pyrus pyraster*), wild cherry (*Prunus avium*), rowan (*Sorbus aucuparia*), and many species of roses (*Rosaceae*)].[11]

So, according to Vera, coppice with standards-type ecosystems were the direct

FIGURE 1.1: A modern example of the emerging scrub and mantle vegetation on an overgrazed pasture in Vermont, USA. The thorny shrubby regrowth protects the emergence of longer-lived mast-producing tree species.

FIGURE 1.2

Timeline of the History and Evolution of Coppicing Explored in This Book

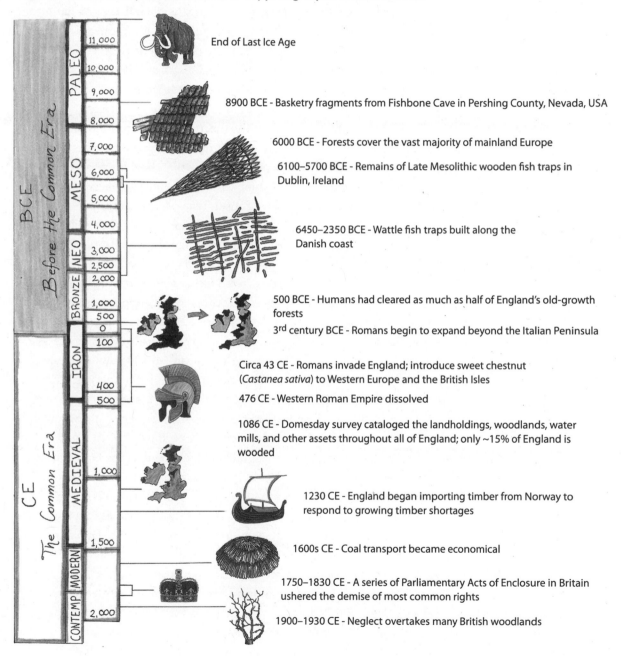

End of Last Ice Age

8900 BCE - Basketry fragments from Fishbone Cave in Pershing County, Nevada, USA

6000 BCE - Forests cover the vast majority of mainland Europe

6100–5700 BCE - Remains of Late Mesolithic wooden fish traps in Dublin, Ireland

6450–2350 BCE - Wattle fish traps built along the Danish coast

500 BCE - Humans had cleared as much as half of England's old-growth forests

3rd century BCE - Romans begin to expand beyond the Italian Peninsula

Circa 43 CE - Romans invade England; introduce sweet chestnut (*Castanea sativa*) to Western Europe and the British Isles

476 CE - Western Roman Empire dissolved

1086 CE - Domesday survey cataloged the landholdings, woodlands, water mills, and other assets throughout all of England; only ~15% of England is wooded

1230 CE - England began importing timber from Norway to respond to growing timber shortages

1600s CE - Coal transport became economical

1750–1830 CE - A series of Parliamentary Acts of Enclosure in Britain ushered the demise of most common rights

1900–1930 CE - Neglect overtakes many British woodlands

result of early humans harvesting the wooded scrub for firewood and allowing seedlings and young trees to grow on to maturity within the protection of the mantle vegetation.[12] This contrasts with the prevailing theory that, as humans created gaps within the dense primeval forest, a shrub layer gradually developed that they later came to manage as coppice.[13]

However history proceeded, these anthropogenic disturbances occurred at different times in different places, but it appears that the use of woody resprouts and their husbandry likely predated the invention of agriculture in the region.

Earliest Coppice Evidence: Hazel Fish Traps in Ireland and Denmark

Some of the first indications of coppice utilization in Europe date to just after post-glacial climates and forest cover appears to have stabilized. In 2004, archaeologists monitoring a development project along the docks of Dublin, Ireland, unearthed the remains of Late Mesolithic wooden fish traps. This passive fishing system consisted of a network of weirs built at low tide to guide fish swimming along the shore at high tide into traps for collection once the tide receded. The archeologists uncovered five well-preserved fish traps, stakes and **wattle** sections—flexible stems horizontally woven between wooden uprights—along an ancient shoreline. These represent some of the earliest-known relics of European fishing culture, dating to as early as 6100 to 5700 BCE (7,700 to 8,100 years ago), at least 1,200 years before agriculture arrived there.[14, 15]

So, what's the relation to coppicing? Well, these fish traps were built using round small-diameter rods, mostly of hazel, though European alder (*Alnus glutinosa*), European ash (*Fraxinus excelsior*), and common dogwood (*Cornus sanguinea*) may also have been used. Most of the stakes were made from similarly sized 8-to-9-year-old coppice growth, cut using small stone axes. The sheer volume of rods and the consistency in their size and growth rate suggests with some confidence that these materials originated as stump sprouts. To fully appreciate this, you must consider the relative difficulty of sourcing a significant volume of straight, flexible weaving material from untended woodland. While one will certainly find quality weavers here and there, you'd be hard-pressed to find thousands upon thousands of consistently sized shoots in one location were it not for some type of sprout-inducing disturbance. Also, the dates of the artifacts spanned 200 years, suggesting sustained coppice usage near this particular fish trap site, and implies sustained husbandry or even planned management of the woodland over a significant period.[16]

Similarly, along the long the Danish coast, researchers have so far uncovered around 1,500 Mesolithic and Neolithic sites both above and below sea level. Key among these are multitudes of fish traps similar to the Dublin finds with an age range of at least 6450 to 2350 BCE, along with larger, more elaborate fishing structures and massive volumes of fish bones.[17]

For example, the Danish Nekselø island site contains "an unusual concentration" of Meso- and Neolithic wooden fishing structures, including the longest Stone Age weir found in Europe. This Neolithic fish trap spanned at least 820 feet (250 m), stood in water 13 to 16 feet (4 to 5 m) deep, and required 6,000 to 7,000 good-quality straight hazel rods up to 13 feet (4 m) long, plus hundreds of longer poles, up to 20 feet (6 m) in length. The production of such huge quantities of material would have been a serious undertaking, and the vastness of the installation implies a considerable organizational capacity and technical skill, with the wattle construction so tight that an adult index finger couldn't fit between the wattle's gaps. This density could only be reached through the use of long, perfectly straight stakes, unlikely to be obtained in any quantity without utilizing coppiced materials.[18]

These tight-knit wattle fences were probably used to catch eel during their autumn migrations, a type of eel trapping that continued in Denmark until the end of the 19th century! These ancient weirs have been found along much of Denmark's coast[19] and, as in Ireland, were probably also used elsewhere.

The high dependence on fish and fish traps strongly implies that these Mesolithic and Neolithic cultures learned early on how to procure coppice, mostly of hazel, to produce the materials they needed to build and maintain the traps as a core part of their sustenance.

Ancient Coppice Husbandry in Britain's Somerset Levels

In the British Isles, evidence points to an intensification of human settlement at around 4000 BCE. These early Neolithic peoples cleared old-growth forests with stone axes to prepare land for cultivation. They probably cultivated among the stumps, later turning the land to grazing.[20] Deforestation continued there through the Neolithic (4000 to 2500 BCE) and Bronze Ages (2500 to 800 BCE), peaking in the first few hundred years of the Iron Age (800 BCE until the Romans invaded circa 43 CE). Historical ecologist Oliver Rackham estimates that humans had cleared as much as half of England's old-growth forests by 500 BCE.[21] Given that many of Britain's tree species resprout after cutting, Neolithic peoples probably had plenty of exposure to coppice material. Similar clearance patterns presumably occurred throughout Europe, though perhaps not as extensively as in Britain. How did coppicing fit into their cultures and ecosystems?

In 1964, archaeological examinations of a wooden trackway buried in lowland peat deposits of southwestern England revealed traces of a highly developed Neolithic settlement. Located in a hydrologically isolated, ecologically rich bottomland landscape known as the Somerset Levels, an ancient human agrarian culture subsisted on the diverse resources available there. Inhabiting and navigating this landscape of elevated hill and island settlements demanded a network of access routes throughout the seasonally inundated lowlands whose foundations remained preserved in peat until modern times. Decades of research into the quality and character of the materials they used to build these ancient trackways reveal a sophisticated culture of active coppice woodcraft and husbandry.[22]

Throughout the Somerset Levels lie the preserved remnants of at least 11 different tracks from the Neolithic and Bronze Ages, all exhibiting the use of coppiced wood in their construction.

Probably the best-known is a wooden access way, discovered by Raymond Sweet, known as the Sweet Track. This linked the Polden Hills with the rock island of Westhay located in the middle of a reed swamp. Radiocarbon dating

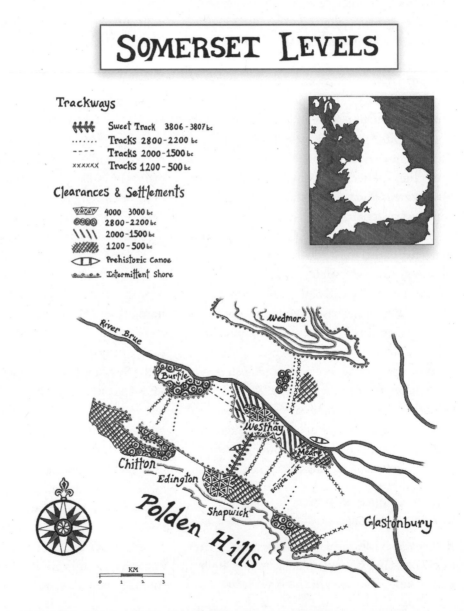

SOMERSET LEVELS

Trackways

┼┼┼┼	Sweet Track	3806 - 3807 bc
....	Tracks	2800 - 2200 bc
- - - -	Tracks	2000 - 1500 bc
xxxxxx	Tracks	1200 - 500 bc

Clearances & Settlements

▽▲▽▲▽	4000 3000 bc
◎◎◎◎	2800 - 2200 bc
\\\\	2000 - 1500 bc
▨▨▨	1200 - 500 bc
⟨▢▢⟩	Prehistoric Canoe
●-●-●-	Intermittent Shore

Wedmore

River Brue

Burtle

Westhay

Meare

Chitton

Edington

Eclipse Track

Shapwick

Glastonbury

Polden Hills

KM
0 1 2 3

FIGURE 1.3: The Somerset Levels is a uniquely diverse landscape that hosts some of the earliest archaeological evidence of humans' use of coppiced materials. Here we see the distribution of some of the key archaeological finds in the area, namely a succession of wooden tracks connecting hilltop settlements separated by marsh and seasonally inundated floodplain.

and dendrochronological studies have fixed the date of construction at 3806 or 3807 BCE.[23] Containing almost 6,560 feet (2 km) of split oak planks, longitudinal rails, and 10,000 sharpened pegs, the scale of the Sweet Track suggests the presence of an extensive, interconnected, and cooperative community. The mostly oak, but also ash (*Fraxinus* spp.) and lime (*Tilia* spp.), planks up to 10 feet (3 m) long, 16 inches (40 cm) wide, and 2 inches (5 cm) thick came from trees up to 400 years old and 3.3 feet (1 m) in diameter. These were felled and split radially using only stone axes and wooden mallets and wedges—quite an art. The rails consisted of long straight poles of mostly hazel and alder, up to 20 feet (6.1 m) long, that may have come from secondary growth on tree stumps.[24] The straight sharpened pegs were made of ash, oak, and lime/linden that didn't fork or branch, suggesting they were derived from "deliberately or fortuitously coppiced" material.[25] The construction implies that the builders used coppice as well as **standard** (single-stemmed trees allowed to grow to maturity) trees along the Polden Hills, and that they had been doing so for at least 120 years prior to building the Sweet Track.[26] Numerous other trackways of different lengths, ages, and types of construction have also been discovered in the Levels, including the Post Track, which runs parallel to the Sweet Track, and is 30 years older. However, a track built about 1,000 years after the Sweet Track provides some of the best evidence of intentional coppice silviculture in the Levels.

The Middle Bronze Age Eclipse Track between the Polden Hills and Meare Island was

Figure 1.4: The Sweet Track was an elevated walkway fashioned using oak planks and coppice-derived stakes that connected hilltop settlements in this seasonally inundated lowland habitat.

a 1.24-mile (2 km) path made of roughly 1,000 woven hurdles that required at least 45,000 hazel rods to build. A project of this scale would have required an estimated 7 to 11 acres (3 to 4.5 ha) of mature coppice or the annual production of 59 to 88 acres (24 to 36 ha) coppiced on an 8-year rotation.[27]

The Walton Heath Track dates to between 2700 and 2300 BCE. It took 5,000 to 6,000 rods 8 feet (2.5 m) long and 0.6 inches (15 mm) in diameter to make about 40 immensely strong woven wooden hurdles to cover the soft peat of a 197-foot (60 m) portion of this route. Made in the traditional Somerset manner, these hurdles were fabricated by weaving long straight shoots of hazel between heavier poles to make panels 6.6 to 9.9 feet (2 to 3 m) long and over 3.3 feet (1 m) wide.[28] The rods bore many signs of having been coppiced: heels at the base of the rod where the shoot was cut or pulled from the stump or stool; narrow growth with no side branches; straight shoots that grew directly upwards to compete for light in a dense stand; rings showing rapid growth aided by an already-developed root system; and the fact that 85% of the rods were from a single species, hazel.[29]

J.M. Coles believes that the Walton Heath hurdle construction demonstrates these people's development of a hazel coppice with oak standards-type system. They did not follow the more modern technique of regular coppice rotations, instead felling the properly sized materials when anywhere from 3 to 8 years old. They would selectively harvest hazel rods from stools when 0.75 to 1 inch in diameter (18 to 26 mm), leaving smaller shoots to grow and mature. Coles also points out that this appears to be the origin of the woven hurdle, which today, 5,000 years later, continues to be a common craft industry in the region![30, 31]

Remember that wood's ability to withstand the ravages of time makes archaeological discoveries of ancient woodwork very rare. Like the undersea conditions in Denmark and Ireland, the Levels' peatlands provided ideal conditions to preserve wooden infrastructure for 5,000 years. Coles points out that archaeologists often fail to acknowledge the role wood plays in prehistoric cultures due to its relative absence in the archaeological record—yet, every human tool of the past for every endeavor from house building to mining to weaving and fiber spinning required wood at some stage.[32] The same was true with Paleolithic and Mesolithic tools, though often only stone or bone parts survive.

Coles calling the Neolithic Somerset woodlands "coppice-with-standards" raises interesting questions. Some authors claim the coppice-with-standards system was not in use as a *system* until the 16th century.[33] Draw-felling stools—harvesting individual rods from stools at need, and leaving widely spaced standard trees up to 300 to 400 years old, differs significantly from a system where entire coppice stands are cut at once in a regular cycle with rotationally managed standard trees. In the end, we don't know exactly what these people were doing, but clearly they were onto something. How advanced and systematic was their husbandry? Did they have long-term intentions, or were they short-term opportunists?

We will probably never fully understand Neolithic peoples' intimacy with their woodlands or the sophistication of their management. But based on their selection and use of different species for different functions in the Sweet Track, and the quality of their craftsmanship, we can at least infer that our prehistoric ancestors had developed considerable knowledge and skill in the art of woodcraft.

Coppicing Elsewhere in Europe

A number of studies support coppice or pollard husbandry during the Neolithic in other parts of modern Britain along with the Netherlands, the Alps, and Scandinavia.[34] Some of these also indicate that people favored the thinning, selection, and harvest of individual rods from stools, rather than clear-felling an entire stand. Pollarding for leaf fodder is also at least implied, if not clearly supported, by both charcoal remains and what are believed to be leaf-cutting knives dating from during or before the Iron Age in Sweden.[35]

In the Montafon region of western Austria, palynological evidence also points to coppice use among Neolithic peoples, if not intentional cultivation or husbandry, but the overall land-use pattern seems different than in Britain. The Montafon valley bottoms host deciduous tree species known to coppice (beech *Fagus sylvatica*, oak, hornbeam *Carpinus betulus*, alder, and ash *Fraxinus excelsior*). The initial valley clearance was followed by 50 years of regeneration of those species plus light-demanding, **early successional species**, including birch (*Betula* spp.), hazel, and aspen (*Populus* spp.). One hundred years of sustained woodcutting followed. These people used sprouts of the previously cut trees, as well as conifer wood, to process ores from nearby mines into bronze.[36]

This research suggests, though, that human clearance patterns in western Austria were patchier than Britain, and that long periods of abandonment and forest regeneration occurred between occupation events. The settlements' irregular nature probably related to patchiness in the mountainous landscape, with clearings occurring where soils and microclimates were best, or lay near ore deposits. The occupation and abandonment cycles appear to relate to climatic changes, with abandonment and forest regeneration occurring when people left because the region got too cold or dry, and people returning when the mountain valleys warmed or moistened—or forest regeneration would again support habitation.[37] Similar patterns appear in Slovenia.[38] Indeed, there is evidence that ebbs and flows of population and land-use intensity occurred all over Europe in the Mesolithic, the Neolithic, and at least into the Early Middle Ages for various reasons.[39] These human tides certainly would have affected the character of woodlands and the coppice resources they provided.

It seems clear, then, that small-diameter woody resprouts provided essential resources for human survival, either patchily and periodically or intensively and consistently.

While further digging and a multilingual literature review could reveal more evidence, it appears certain that coppicing provided key resources that strongly shaped the cultures of prehistory. The evidence demonstrates the use of woody resprouts in parts of prehistoric Europe. It also supports the idea that active and intentional silviculture occurred much earlier than we may have thought, becoming increasingly sophisticated and larger in scale over time, even in prehistory. Indeed, given the evidence that coppice husbandry predated the invention of agriculture in at least some areas, one could argue that coppicing may have taught people enough to help them invent agriculture. It may be possible that coppicing was, at least in some regions, the first form of agriculture. Further, complex constructions such as the Sweet Track and fish weirs would have required the development of cooperative social systems and the ability to cultivate large areas of woodland in defined and complex ways over multiple generations. This suggests the ability to develop and teach the next generation craftsmanship and woodland horticulture.

Resprouts in Precontact North America

Much of the cultural history of Indigenous people in North America has been lost since contact through the ravages of disease, forced migration, assimilation, and genocide.[40] Due to the relatively late expansion of Europeans into what is now the western United States, some Indigenous ways of living and knowing have continued to the present. M. Kat Anderson has studied the land management practices of Indigenous peoples in what is now California in some detail. While this section draws strongly from her work, the evidence suggests similar cultural patterns occurred elsewhere on the continent, though much knowledge has been lost or is guarded carefully. So, how did Indigenous North Americans manage and use woody resprouts? What role did resprouts play in their societies, which were deeply embedded in specific ecosystems?

Horticultural Societies in California

Social scientists commonly label Indigenous peoples as either "hunter-gatherer" or "agriculturalist." The hunter-gatherer model envisions a mobile lifestyle utilizing harvested resources but otherwise implies passivity in interacting with and managing ecosystems. And the agriculturalist model assumes a focused effort to domesticate and manage plants and animals—often seen as the only way human cultures modify their environment.[41] Both models oversimplify and represent two extremes of human ecological intervention rather than a more comprehensive spectrum of human-landscape engagement.[42]

For thousands of years, Indigenous North Americans have engaged in intimate relationships with their ecosystems, meeting their needs while increasing the land's ecological health, diversity, and abundance.[43] Many Indigenous land management practices influence resources at the organism, population, plant community, and landscape levels,[44] sometimes all at once with a single interaction. Rather than "management" in the Western sense that implies control, Indigenous systems tend or care for plants and animals by establishing deep relationships of reciprocity.[45] Using practices like burning, coppicing, and pruning, and often simply by harvesting in the proper way, Indigenous practices create evolving ecosystems with heightened biological diversity that supported the flora and fauna that sustained them by efficiently producing high-quality food and building, craft, and hunting materials.

Indigenous lifeways in California and the world over revolve around annual cycles, demonstrating a deep knowledge of the rhythmic patterns and needs of other species and the habitats in which these species live. This often includes seasonal migrations and interactions relating to phenology—periodic events in species' life cycles—and biological patterns rather than haphazard attempts to locate game and wild foods.

The term **traditional ecological knowledge** describes the assemblage of information and experience that has enabled Indigenous people to culture the ecosystem. In this context, it is perhaps most accurate to describe Indigenous Californians not as hunter-gatherers or as agriculturalists but as "horticultural" people who have nurtured, fostered, tended, shifted, and shaped plant populations and communities to better meet their needs, mostly without wholesale tillage. Such horticultural societies have likely influenced ecosystems across the globe, as similar proto-agricultural practices may have been practiced almost universally in Africa, Europe, and Asia for 30,000 to 50,000 years.[46] Anderson notes that "much of what we consider wilderness today (in California) was in fact shaped by Indian burning, harvesting, tilling, pruning, sowing, and tending."[47]

Managed Disturbance

Indigenous people in California have employed horticultural practices including pruning, tilling, cutting, burning, and sowing that serve as small, intermediate, and landscape-scale disturbance events, thereby shaping the ecosystems they rely on. These anthropogenic disturbances helped give rise to the majestic landscapes of precontact California.

Over time these ecosystems came to rely on regular human disturbance for optimal growth, reproduction, and survival, while producing high-quality craft materials—straight, supple, vigorous young growth that emerged from tree and shrub stumps and roots in the wake of fire.

Coppicing with Fire

Fire was the most powerful and widespread tool employed by Indigenous people in precontact California and across North America. An estimated 5.6 to 13 million acres (2.25 to 5.26 million ha) burned each year in California as a result of lightning and human-set fires.[48] Fire has numerous beneficial effects on ecosystem health and productivity, key among these is that *fire stimulates woody plants to coppice.*

Coppicing by burning and pruning induces rapid elongation of young vegetative shoots useful for craft, and removes dead and diseased wood, increasing plant vitality. Burning *after* cutting also appears to stimulate shoot growth in some species, such as sourberry (*Rhus trilobata*) and bigleaf maple (*Acer macrophyllum*).

Some of the most prominent species coppiced using fire in California include chokecherry (*Prunus virginiana* var. *demissa*), redbud (*Cercis occidentalis*), sourberry, red willow (*Salix laevigata*), deer brush (*Ceanothus integerrimus*), California button willow (*Cephalanthus occidentalis* var. *californicus*), black oak (*Quercus kelloggii*), blue oak (*Q. douglasii*), interior live oak (*Q. wislizenii*), and flannelbush (*Fremontodendron californicum*).

Numerous archival photographs capture the preponderance of straight, supple young coppice shoots in Indigenous crafts, weapons, and baskets.[49] Evidence suggests that Indigenous Californians used these strategies for at least several thousand years before Europeans arrived.[50] In fact, it's possible that still-living redbud or sourberry shrubs have been cultivated for basket sprouts by generations of Indigenous women for hundreds or even thousands of years.[51]

Northwestern California's Yurok people burned at a frequency appropriate for each purpose: hazelnut (*Corylus cornuta* var. *californica*) every 2 years for basketry materials; under tanoaks (*Notholithocarpus densiflorus*) for brush control every 3 years; for elk fodder every 4th or 5th year; in the redwoods for brush and downed fuel control every 3 to 5 years.[52] Burning also stimulates low, young, tender vegetative regrowth that wildlife find more accessible, palatable, and nutritious. Increasing the quantity and quality of forage therefore induces larger, healthier game populations.[53]

Pruning

Many Indigenous tribes also regularly pruned trees and shrubs and harvested young shoots

FIGURE 1.5: A patch of woods in western Oregon post fire.

and sprouts from redbud, sourberry, buck brush (*Ceanothus cuneatus*), maple, dogwood (*Cornus* spp.), sumac (*Rhus* spp.), mock orange (*Philadelphus lewisii*), and California button willow after leaf fall. Chemehuevi, Washoe, and Pauite weavers harvested young willow shoots from fall until early spring.[54] The Yuki and Pomo cut redbud branches in winter or early spring to produce basketry material for harvest the following fall. The Pomo pruned gray willow (*Salix exigua*) for basketry materials and to encourage the plant to extend its range by developing lateral underground runners.[55]

Cultural Products

Indigenous people in California and elsewhere used young woody shoots to craft items for many cultural uses including basketry, ceremonial items, clothing, cordage, games, musical instruments, cages and traps, structures, tools and utensils, and weapons.[56]

Some tribes made snowshoes from several-years-old oak and serviceberry (*Amelanchier* spp.) coppice shoots.[57] Indigenous women often used dense, stout digging sticks of mountain mahogany (*Cercocarpus betuloides*), serviceberry, or buckbrush to pry open the soil surface and harvest edible bulbs, corms, rhizomes, taproots, and tubers.[58] The Mono used black oak sprouts for spoons and 3-to-4-year-old flannelbush shoots for tongs to cook acorn mush.

One-to-two-year-old shoots from shrubs like button willow, mock orange, gooseberry (*Ribes* spp.), willow, alder, snowberry (*Symphoricarpos albus*), spicebush (*Calycanthus occidentalis*), and hazelnut produce quality arrows. Arrow-makers burned or pruned the patch at least a year or 2 before harvesting the shoots. For larger weapons like fish harpoons and spears, sprouts were left to grow for several years.[59]

Coppiced rods from the native hazelnut served a multitude of functions including baskets and cradleboards for carrying babies, while some tribes twisted new growth to make rope and fish traps.[60] The Miwok and Yokuts bent 2-year-old sprouts into looped sticks used to stir acorn mush. Most weavers actually prefer red shoots arising from burnt hazel stools as opposed to those that have been cut. Many tribes crafted musical instruments like flutes and clapper sticks from 1-to-4-year-old elderberry shoots, preferring elderberry sprouts that emerged after burning because they often had elongated growth between leaf nodes.[61]

Baskets played a central role in the culture and worldviews of Indigenous Californians since ancient times. Basketry fragments from Fishbone Cave in Pershing County, Nevada, date to 8900 BCE.[62] Baskets require an immense volume of uniform high-quality material. According to Ruth Merrill, Indigenous Californians used 78 plant species from 36 different families in their basketry traditions.[63] Weavers sought long, straight, abundantly available, colorful (containing reddish anthocyanin pigments), flexible materials, free of bark blemishes, consistent in diameter, and free of lateral branches.[64] A Western Mono cradleboard required 500 to 675 straight sourberry sticks, while a larger cone-shaped burden basket needed as many as 1,200! To make a dozen or so baskets in one year, a weaver would need to harvest as many as 10,000 stems during the winter, and a village of 100 residents could require as many as 250,000 cultivated stems per season to supply a team of 25 weavers.[65]

Table 1.1: Ages of resprouts used in select indigenous Californian's cultural products

Species	Item	Age	Tribes
Cercis orbiculata (redbud)	Basket warp and weft	1 year	Western Mono, Foothill Yokuts, Southern and Central Miwok
Ceanothus cuneatus (buck brush)	Basket warp Basket rim stick	1 year 1–3 years	Western Mono, Foothill Yokuts, Southern and Central Miwok
Cornus spp. (Dogwood)	Basket warp	1 year	Foothill Yokuts, Southern and Central Miwok
Corylus cornuta var. *californica* (hazelnut)	Basket warp Basket weft	1 year 1 year	Southern and Central Miwok
Acer macrophyllum (bigleaf maple)	Basket weft	1 year	Southern and Central Miwok
Quercus spp.	Looped stirring stick	1–3 years	Western Mono, Foothill Yokuts, Southern Miwok
Salix spp.	Fish trap	1 year	Paiute, Foothill Yokuts, Southern Miwok
Salix spp.	Fish weir	1 year	Western Mono, Paiute, Tubatulabal, Foothill Yokuts
Fremontodendron californicum (flannel bush)	Cordage	1–3 years	Western Mono and Chukchansi Yokuts
Sambucus mexicana	Flute	1–3 years	Foothill Yokuts, Southern and Central Miwok
Fremontodendron californicum (flannel bush)	Cooking tongs	3–4 years	Western Mono
Quercus spp. (oak) and *Cercocarpus betuloides* (Mountain Mahogany)	Digging stick	2–4 years 4–6 years	Western Mono, Foothill Yokuts, Tubatulabal, Southern Miwok

Adapted from Anderson, 2006, page 219–220.

Most structures built by Indigenous Californians also required saplings, especially those with a hemispherical shape. Lightweight pole frames usually supported layers of bark, mats of tule (*Scirpus* spp.) or willow sprouts, brush thatch, or earth. Indigenous women built houses requiring a frame made from a dozen long willow poles, bundles of cattail leaves, willow withies for woven mats, and strings of sagebrush (*Artemisia* spp.) bark or strips of old cloth to fasten the structure together.[66]

This modest glimpse into the oft-forgotten legacy of North American Indigenous people reveals an interconnected, integrated relationship with the plants and landscapes of which they were a part. These Indigenous lifeways demonstrate how human lifestyles, land use, and culture can all respond to and shape ecological realities. We see how their needs for food, weapons, structures, ceremonial and craft materials, and the simple yet powerful horticultural tools at their disposal shaped their engagement with their world and influenced their kinship with nature. The traditional ecological wisdom built into Indigenous horticulture minimized labor and maximized leverage—each act had multiple benefits. Imagine what a horticultural society might look like in the modern world and what it would take for us to internalize a worldview that inspires us to *tend* rather than dominate nature.

Having examined coppicing in prehistoric Europe and North America, let's return to Europe to explore how Neolithic coppice systems evolved in intensity, scope, and sophistication as their cultures developed. As we enter historical times, the written record offers increasing detail and clarity about the development of coppice.

COPPICE IN EUROPE: FROM SUBSISTENCE RESOURCE TO MARKET COMMODITY

The word "resource" comes from the old French feminine past participle resourdre, *which meant "to rise again."*[67]

European cultures and landscapes changed rapidly and radically over the past 2,000 years. Small, tribal, subsistence agrarian communities in patchy, increasingly pastoral landscapes became complex market-driven industrial cultures with megacities. Initially, woodland management practices probably varied from place to place and innovation spread slowly, but two forces worked towards homogenization: the Roman Empire and modern industrialization. Until the mid- to late-1800s, coppice played key roles in sustaining European peoples.

The Roman Empire brought the first wave of European industrialization as it began to expand beyond the Italian Peninsula in the 3rd century BCE. It cross-fertilized cultures and spread species, tools, and woodland management techniques. This, in addition to growing population, spurred land-use innovation and intensification to meet people's food, energy, and material needs.

After the chaos following the fall of the Empire, the medieval manorial or feudal system took hold. Old traditions of common rights to subsistence use of the landscape were overlaid with a ruling baronial class that owned that land. Over centuries, the landlords enclosed the commons and instituted exclusive private property rights. Meanwhile, a quiet and generally unacknowledged medieval industrial revolution took place bringing developments in energy, agriculture, mining, labor, architecture, mathematics, reason, experimental science, the environment, and pollution.[68]

Privatization provided emerging market economies with a hungry, mobile, former-peasant labor force progressively divorced from the land. The increasing use of coal and new manufacturing and transportation technologies initiated a steady lengthening in woodland rotation cycles and a shift away from firewood and craft production.[69] Standard trees over coppiced underwood in intensive rotations provided the charcoal and other materials fueling a third wave of industrialization. However, coppice systems worked themselves out of a job: they enabled iron-based technologies to bring fossil fuels to market, destroying coppice systems' economic basis. So, woody resprouts evolved from a community subsistence resource to a market commodity before losing their value as either. All in just 20 centuries.

We find the earliest written coppice references in the Roman Empire. What can we learn from Roman, medieval, and early modern woodland culture? And how did this history lead to the eventual collapse of coppicing?

Roman Legacies: Industrial-scale Copses and Sweet Chestnut

In the first known Latin agricultural treatise, written around 200 BCE, Roman statesman,

soldier, and agriculturalist Cato the Elder wrote that estates should develop self-sufficiency in wood, meeting their needs with on-site resources. He describes coppice establishment and cultivation strategies and the installation of willow and poplar groves.[70] Cato believed that willow and coppice wood were among the top nine most profitable crops, and recommends that farmers "plant elms and poplars round the borders of the farm and along the roads to give you leaves for the sheep and cattle, and timber when you need it."[71]

In the first century CE, Lucius Junius Columella and Pliny the Elder also wrote about profitable, practical land management in Italy. They both advised landowners to grow their own oak or sweet chestnut (*Castanea sativa*) vineyard stakes, Columella suggesting coppice rotations of 7 and 5 years respectively, and Pliny 10- and 7-year cutting cycles. In his *Historia Naturalis* (ca 77–79 CE), Roman author, philosopher, and naturalist Pliny the Elder describes osier (willow) cultivation for withies. And while these three authors illustrate the value of coppice wood during Roman times, it appears that the woods were appreciated for their fodder and pasture perhaps even more so than for wood products.[72] As they advanced across Europe, the Romans brought these traditions and knowledge of the yields and labor required for coppice maintenance.[73]

Keep in mind, these elder statesmen had to encourage tree planting and coppicing for a reason: the Romans had severely deforested their homelands.[74] Before the Roman invasion in 43 CE, Britain was a country of mostly pastoral landscapes with an unknown amount of woodlands and **wood pastures**. These illiterate, Celtic-speaking Iron Age people lived in agrarian households around hillforts and small hamlets organized into tribal kingdoms.[75] The Romans introduced a literate, hierarchical, highly organized culture of industrial-scale development. Their cities featured bridges; large timber-framed buildings; glass, lead, and iron industries, and more, all of which required huge amounts of wood.

Henry Cleere estimates that just one Roman military ironworks in Britain's Weald region produced 550 tons of iron goods annually. Oliver Rackham estimates that sustaining this one ironworks alone would have required charcoal from 23,000 acres (9,308 ha) of coppice woodland! Considering that this was just one of many ironworks in the Weald, it appears the Roman impact on the landscape may have exceeded that of the region's iron industry at its 17th-century peak![76]

Roman Britain would probably have consumed much of its woodland. However, it would have been worse without the regenerative capacity of coppice, along with one of Rome's most important silvicultural legacies: the spread of sweet chestnut to almost every place they occupied. While evidence indicates that Roman copses were mixed-species affairs, not monoculture plantations, sweet chestnut ranks as one of the most important coppice species in Europe.[77] Nevertheless, few records remain of what actually happened in Britain during the Empire—or throughout the rest of Europe, for that matter—because of the mayhem during and after its long decline.

Medieval Woodlands: Composition, Patterns, and Protection

The post-Roman Early Middle Ages (500-1000 CE) was a period of much chaos and migration. Overall, the centuries after the Western Roman Empire dissolved in 476 CE saw Europe's population increase alongside a technological revolution that helped intensify agriculture and silviculture through the development of mills and factories for fulling cloth, tanning leather, crushing wheat, and milling flour among other things.[78] However, this overall growth occurred erratically, with episodes of expansion and regression in different areas at different times. For example, after the Romans left Britain ca 410 CE, the population dropped from about 5 million to 1.5 million in 250 years, even with the immigration and eventual dominance of Germanic Anglo-Saxons. Let's examine how these changes affected the character of the British landscape, the woodland management systems they employed, the products that drove this management, and the overlying social structures.

While some lowland forests of Northwestern and Central Europe and Britain had been wooded since the Mesolithic, they had probably all been affected by humans by 500 CE, if not earlier,[79, 80] and the majority were probably cultivated in some way. Through the post-Roman turmoil, many of these same woodlands continued to provide vital necessities for centuries using mostly Neolithic techniques.[81] This makes sense: a few years of neglect won't hurt a copse or wood pasture. It's easily returned to use by harvesting what has grown in the interim. Woodlands' perennial nature stabilizes both ecosystems and human society.

In 1066, William the Conqueror was crowned King of England. Five years later, he completed the Norman Conquest, bringing sweeping change to English society, including record keeping that gives us greater insight into the workings of woodland culture at the time. The Domesday survey of 1086 cataloged the landholdings, woodlands, water mills, and other assets throughout all of England. It provides baseline data on the English landscape. We don't know in any detail what it was like before.

Britain's Landscape and Woodland Composition at Domesday

According to Oliver Rackham, the Domesday Book reveals that in 1086 only about 15% of England was wooded. Only half the settlements with Domesday data had any woodland associated with them. Almost 75% of England's counties had under 20% woodland, some with essentially zero (though The Weald, which means "woodland" in Old English, had a guesstimated 70%). Domesday also makes it plain that no wildwood remained in the whole of England at the time. The vast majority of the English landscape in 1086 was pastoral. Millennia of pastoralism and centuries of Roman industrialization and war, among other things, had taken their toll, but then it stabilized for a while.

Despite this scarcity, those woodlands that existed appear to have retained their size and shape for centuries. Conversion to other uses was rare. Woodlands often persist not because they occupy sites well-suited to tree growth but because the sites work poorly for other uses—mainly crop production or pasture. Many woods thus survived on landscapes too steep, rocky, wet, or otherwise poor to support arable crops. Sites distant from habitations and mineral deposits were also less prone to be cut. In addition, contemporary social structures stabilized those land-use patterns for centuries because locals strongly resisted the loss

FIGURE 1.6: European settlement patterns and woodland distribution still remain etched in the modern landscape.

of their ancestral wooded commons, which they depended upon for survival. "Woods in England in 1086 were part of the cultural landscape: every wood belonged to some person or community, and was used. But the use seems not to have been so intensive as in later centuries."[82]

When populations were low and land was managed primarily through common rights and communal use, boundaries of ownership and defined management directions for particular areas may have been less clear. Imagine a landscape populated by a range of uncoordinated users with diverse needs, intentions, uses, and interactions. These ecosystems would experience greater "chatter," leading to less order, consistency, and clarity of successional direction and habitat structure. It seems clear that increasing social order would lead to systematization and continuity of management across regions.

It appears medieval woodsmen seem to have done little to record or modify the species composition of their woods. Woodlands usually contained a mix of species and age classes of trees—perhaps the result of previous draw-felling of individual rods as needed. However, timber (big logs from large trees) and wood (small-diameter material from young trees or coppice sprouts) were differentiated clearly, because different people had the rights to use each, even in the same woodland.[83] Though little was recorded about Medieval woodland husbandry, there were several general kinds of woodlands, each providing a suite of products for subsistence and market.

Medieval Woodland Systems and Products

What follows gives a sense of these medieval woodland systems and their respective products, particularly in Britain. We'll discuss each system in greater detail in chapter 4. There's some disagreement in the literature as to which systems predominated at different times. This reflects the nature of the historical beast: each woodland (and each region) has its own history. Generalizing is difficult.

Copses: Fuelwood, Polewood, Craft Materials

Having been used for millennia, copses—patches of managed coppice stools—were well established by 1086. European copses primarily yielded fuelwood, charcoal, and materials for fencing, basketry, and other crafts, though all parts of felled trees and the ecosystems as a whole served some need. Let's briefly examine two of these primary coppice products: fuelwood and charcoal.

Remember, wood was essentially the only widely available fuel during the Middle Ages. And it was necessary not only to heat uninsulated living spaces but also for cooking and industrial production. Small-diameter coppice growth yielded fuelwood well-suited to the needs of the people of the time, easy to harvest and transport, and requiring minimal processing.

Charcoal, on the other hand, drove industrial development from the Bronze Age to the early modern era. This energy-dense, hot-burning, value-added wood product enabled people to melt copper, tin, iron, lead, gold, silver, and more, while significantly reducing the volume and weight of the original wood so it could be transported more easily and economically.

And beyond these "mass-market" products, historical accounts from the Middle Ages record monetary exchanges for tree branches, bark, *loppium et chippium*, twigs, and even leaves.[84] Local regulations allowed livestock to graze within copses after shoots grew beyond animals' reach. People seasonally gathered acorns, hazelnuts, chestnuts, bracken ferns, and linden tree **bast** (inner bark used to make cordage). In parts of Germany, people

FIGURE 1.7: Making charcoal in Gloucestershire, UK. Photo courtesy of the Museum of English Rural Life, University of Reading, used by permission.

sometimes burned cants after felling, planting a rye crop between the stools.[85] We find few specifics on the character and development of these multifunctional systems, but they were adapted to local circumstances, with numerous variations on the theme.

It appears most medieval copses were of mixed species and age groups. As population and demand for materials grew over time, the selective felling traditions of the past began to shift to clearcutting even-aged cants. In 1086, rotations were short, averaging 4 to 8 years, depending on needs.[86] A host of factors drove these cycles, namely commoners' needs in any given season. At the same time, specialized craft-based livelihoods came to rely on coppice growth for a variety of different products.

Crafts, Stationary, and Itinerant

Coppice crafts in the Middle Ages and early modern period took two primary forms: stationary crafts whose specialized tools and workshops required raw materials be brought to them, and itinerant crafts that required a few easily transported tools, set up in temporary workshops in the woods. Some woodsmen were employees of small factories or estates, rotationally harvesting their woodlands and converting raw materials into products.[87] The bodgers (woodturners), wattle hurdle makers, thatching spar makers (thatching spars hold thatching material onto a roof), and charcoal burners are a few of the better-known mobile craftspeople.[88]

While itinerant craftspeople did not own woodland, the contemporary socioeconomic

Figure 1.8: Traditional woodland work was as much a way of life as it was a livelihood.

structure supported their roaming lives and livelihoods, ensuring coppice was harvested on a fairly consistent rotation. Many craftspeople attended annual auctions held by large estates and other woodland owners.[89] The craftsperson with the winning bid moved into the woodland, set up a temporary shelter, felled the stand, and converted it into product. Landowners kept careful records of these transactions, detailing financial arrangements and agreed-upon extraction routes.[90] Bodgers traditionally built lightweight A-frame shelters using coppice poles covered with bundles of thatch, topped off with a thick layer of wood shavings. Often built in a single day, these huts reportedly stayed warm and draft-free.[91]

FIGURE 1.9: Gate hurdle maker and hut in Surrey, UK, possibly near Elstead. Note the wood shavings covering the hut, the stack of completed hurdles behind him, and the large hewing hatchet in his left hand. Photo courtesy of the Museum of English Rural Life, University of Reading, used by permission.

Coppice With Standards: Timber, Polewood, Fuel, Craft Materials

There lies some debate as to when coppice-with-standards systems were first conceived. As we mentioned earlier, Coles suggests that the inhabitants of the Somerset Levels maintained these systems in prehistoric times. However, Rotherham reports that the first records of coppice-with-standards management in Yorkshire, England, date to 1421,[92] while Machar states that the system was developed in 17th-century France.[93] These disparities may reflect the challenge in clearly identifying when harvest from tended forest stands developed into intentional management with a targeted structure, regular rotation intervals, and a long-term vision. Nonetheless, it appears that medieval British woodlands were largely managed as coppice-with-standards systems.[94]

That coppice with standards would increasingly dominate the wooded landscape between the 15th and 18th centuries makes sense. A two-layered ecosystem provides greater potential yields than simple coppice. Coppice-with-standards systems produce timbers from the standard trees for building and naval needs, polewood from the coppice, and **mast** (acorns, seeds and fruit) to feed pigs and wild game.[95, 96] Occasionally these systems also included a third class of older coppice stools left to grow for two or three underwood rotations, providing even more product diversity. In addition, the herb layer may have provided medicinal and edible species, along with the potential to hunt and trap. Standard trees generally could not comprise more than 25% of the total stand area (roughly 20 trees between 120 and 150 years old per acre or 50 per hectare) without dramatically compromising underwood production.[97]

Wood Pastures: Fodder, Fuelwood, Craft Materials, and Livestock

Though their form has varied through history, wood pastures integrate pollarded tree yields [polewood (~10-to-15-year rotation) or fodder (~2 to 7 years)] in pasturelands. Some of the best-known wood pastures are the Spanish **dehesa** (or Portuguese **montado**)—savannah-type ecosystems with widely spaced trees (often oaks) supporting pigs with seasonal pasture and abundant mast.

These productive landscapes were similarly valued in Britain. So much so, that in their separate studies of Yorkshires' woodland histories based on Domesday Book records, Ian Rotherham and Melvyn Jones estimated 95% to 100% of Yorkshire woodland in 1086 was wood pasture.[98, 99] Below we offer a brief overview of two wood pastures, Epping Forest and Burnham Beeches, to illustrate the scale and management of these common resources.

FIGURE 1.10: Ash pollards in a public park in Madrid, Spain, once managed as a wood pasture.

Figure 1.11: A fresco in Elmelunde Church on Denmark's island of Møn, painted in early 16th century by an unknown artist. It depicts a peasant climbing into a shredded tree and harvesting leaf hay. Photo released under CC BY-SA 4.0; commons. wikimedia.org/wiki/File:Elmelunde_Fresco_2017_03.jpg

Figure 1.12: Ancient beech pollards at Burnham Beeches outside London, UK.

The 6,000-acre (2,428 ha) Epping Forest near Essex, England, was an industrial-scale wood pasture operation helping meet London's fuelwood needs. The forest once contained over 500,000 English oak, European beech, and hornbeam pollards.[100] Each of its 32 manors were governed by a lord who granted specific individuals commercial rights to the woods, some extending back 1,000 years. As part of the commons system, villagers could lop fuelwood from pollards between All Saint's Day (November 1) through Saint George's Day (April 23) until one of the Acts of Enclosure, the Epping Forest Act of 1878, abolished commoners' lopping rights in order to create public recreation space. Today the forest contains an estimated 140,000 pollards.[101]

And Burnham Beeches' 200-acre (80 ha) wood pasture, located in a London suburb, once held as many as 3,000 pollards managed for fuelwood production. As in other wooded commons, villagers grazed cattle, pigs, horses, and sheep, and collected both turf and firewood for fuel. They also coppiced oak and beech within the commons, at one point converting two portions into hazel coppice for hurdles, wattle, and thatching spars. Communal use of the Beeches declined during the 19th century, virtually disappearing by 1879 when the estate was broken up and sold. Today the site includes 540 living pollards, estimated at 330 to 420 years old, many of which have not been cut for nearly 200 years.[102]

Wood pasture management occurred throughout Central Europe until widespread 19th-century wood shortages led to reforms that virtually banned the practice completely.[103] Some farmers still occasionally employ pollarding and shredding in Norway, where managing elm, linden, and birch for leaf fodder was once common, even earning its own specialized vocabulary.[104]

Hedgerows: Secure Boundaries, Polewood, Fodder, Food, Herbs

Hedgerows—belts of trees and shrubs partitioning fields—provided living fences, sheltered fields, connected habitats, and produced firewood, craft material, and wild foods. They were ubiquitous in the medieval landscape and originated as either linear remnants of cleared woods, intentionally planted field boundaries, or by default under the cover of existing fencelines.[105] They contained timber trees and coppiced, pollarded, and shredded trees.

Hedging is an old practice. Oliver Rackham writes that hedges in Germany and the Netherlands date back to the Neolithic.[106] Julius Caesar is credited with the first written reference to managed hedgerows in his report on the Battle of Gaul in 57 BCE. He describes the hedges of the Nervii tribe used to protect cattle from other tribes along the Belgium-France border. By cutting and laying small trees bound with thorn and brambles, they created living, stock-proof barriers.[107] Roman writers Columella, Palladius Rutilius, and Siculus Flaccus all describe elaborate techniques to maintain hedgerows.[108]

While hedgerows served as multipurpose barriers enclosing livestock, mounting pressures on wood supplies and livestock forage made it

Figure 1.13: Man laying a hedge in Leominster, UK at a Farmers Weekly competition, January 1958. Photo courtesy of the Museum of English Rural Life, University of Reading, used by permission.

increasingly important to protect woodlands from thieves and errant livestock.

Wood's Value and Woodland Protection

In the Middle Ages, small-diameter wood was more useful and accessible than large-diameter timber, so in some ways it had more value.[109] In South Wales during the 4th century CE, a single hazel stool was worth the same amount as 3.75 sheep![110] Wood's value and essential role in medieval life made it all the more important to clearly delineate property boundaries and protect the resource. Lacking readily available maps, clear field borders distinguished property. These boundaries required an immense amount of labor—a testament to the value contained within.[111] Thomas (1998) writes that, in one case, woodland enclosure maintenance costs equaled 25% of the total wood sales revenue.[112]

Virtually all British woodlands more than 100 years old are surrounded by some type of boundary earthwork, and many of the permanent woodland boundaries that remain date from the Early Middle Ages or earlier. The woodland boundaries of England's Chalkney Wood, for example, date back to the Iron Age.[113] Boundaries usually took several forms: **woodbanks**, stone walls and rows, or **lynchets**.

Woodbanks consist of a ditch excavated along the outer perimeter of the woodland with spoils banked up along the inner boundary. They're found in England, the Netherlands, Germany, France, Romania, Belgium, Scotland, Wales, Hungary, Denmark, the Czech Republic, Austria, and Ireland. Often capped with a living or dead hedge, they made access nearly impossible for most animals and limited cart access to locked gates along roads, reducing theft. Some woods were bounded by stone walls or rows using stone from nearby fields. These served more as boundary markers than protective enclosures.

Frans Vera suggests that the concave shape of many early woodbanks as depicted in some of Oliver Rackham's historical maps may help substantiate his theory that coppice systems were created from the concentrically expanding scrub groves of prehistory. The curved pattern of the woodbank preserved the original vegetation's shape.[114]

Coppice, coppice with standards, wood pastures, hedgerows, and the structures that protected these resources collectively are the primary systems medieval Europeans used to meet their most fundamental needs for wood. In this evolving physical landscape patterning, we see perhaps the most important drivers of woodland character, use and management—the socioeconomic systems of the time.

FIGURE 1.14: Woodbanks lining a woodland ride, southeast England.

The Manorial System: Private Ownership and Wooded Commons

Over the next few pages, we'll be diving deep into the complex history of medieval Britain. While this may feel overwhelming, it offers crucial insights into the relationships between humans and their natural resources and may offer inspiration as we move towards the "design" of multigenerational woodlands. In the development of the manorial system, growing tensions between market pressures and subsistence demands for woodland resources become clear.

Despite the complex strata deposited on the British Isles by the Romans, Vikings, Anglo-Saxons, and Normans, "the Celtic substratum persisted" in British culture.[115] One key aspect, present since the Neolithic and the Iron Age, was shared or "common" use of woodlands for subsistence needs by the "common" people. Between the 5th and 11th centuries, the Anglo-Saxons built a solid governmental structure, brought into full flower by succeeding Normans who laid the manorial system over the previous society.[116] Here, we focus specifically on the wooded commons and common rights in general: how they affected woodland management, and how they interacted and conflicted with ownership by the baronial class. This social structure enabled these societies to survive under great population and resource pressures for centuries, and it ultimately helps us understand how coppice systems died out in the modern era.

When William the Conqueror invaded from Normandy and took over England in 1066, he co-opted the Anglo-Saxon power structure and established his own, seizing lands owned by the previous aristocracy and granting them to his own loyalists. He thereby bestowed wealth, privilege, taxation and governing powers, and management duties to a baronial class of landowners: lords or landlords. For them to retain their lands and make real their support, each landlord had to pay **fealty** and give service to the king. They then either employed people to work their land directly or let it to tenants required to pay rent in money, goods, or services. Or, in like fashion, they granted lands to loyalists who then owed them fealty for the privilege.

This means the landlords depended upon the peasants for basic goods and services, and the peasants were allowed to meet basic subsistence needs. These lower classes had common rights to the landlord's lands and resources, and had to pay him rent, taxes, and/or fealty. These common rights had probably carried down through generations of unrecorded history, and lived in memory and tradition, changing as needed.

All this social structure divides up resources, defining who has access to, responsibility for, and decision-making authority over them. This is one way to reduce competition and social conflict, or at least channel it. These common rights patterns and structures were rarely written down but nevertheless formed serious social contracts. They also varied greatly from place to place, as we might expect in a society with limited mobility and communications, diverse landscape characteristics, and millennia of social structure and landscape coevolution. Despite this, we do find a number of common patterns.

Recall the historical distinction between timber and wood. Large timber trees were laborious to fell and move but necessary for large structures. Landlords usually owned the timber and either leased the right to cut it or directly employed people to fell, process, and transport the logs. The king, when needed, might command that such timbers be called into service for national defense or to build ships, cathedrals, or castles, usually at a cost. The availability of these materials at times limited construction or caused social and economic distress.

The social constructs around wood varied. While the lord of the manor often owned a portion of the available wood, some access was allotted as a customary right to tenants of the estate (known as serfs or "villeins"). These rights to wood were described as **bote** (from Middle or Old English: help, relief, benefit, advantage, profit, remedy) and specifically included "hedgebote, firebote, housebote, cartbote, gatebote, stilebote," and more, each word describing the right to use wood for a certain purpose. **Estovers** is a related term, in this case generally meaning an allowance of wood made to a tenant or the freedom of a tenant to take necessary wood from land they occupy. Land "owned" by the landlord, given to the lord by the king, who won that right through warfare, which was affirmed by the Church and "the divine right of Kings"—all despite the fact that the "commoners" and their

ancestors had lived there for thousands of years before them. The wooded commons was one type of common land. Pasture and agricultural land had similar structures, as did hunting rights.

The Merovingian Franks began using the term "*forestis*" in the Early Middle Ages. There lies some debate as to the original meaning of the term, but many believe it derives from the Latin *foris* or *foras* and means "outside" or "outside the settlement." So, on the European continent, forestis (or forestes) described both a place—the wilderness including the vegetation, water, and wild animals present—and a legal concept based on the Roman law *Codex Iustinianus X*. It essentially stated that resources lacking a clear owner became the property of the king or lord. So the "forest" wasn't so much a dense woodland (in fact, most was open land) but rather a place where the king owned the entirety of the uncultivated ecosystem, and it was forbidden for anyone to pasture animals, cultivate crops, or harvest wood without permission. This permission was granted by officials known as *forestarii*, entrusted to oversee the use of the forestis and pass judgement on relevant matters. Commoners could earn the right to harvest firewood and building materials, graze livestock, create fields, and cultivate crops at a scale adequate to meet the needs of the household, but they had to prove it was necessary and then contribute a *medem* or share of the harvest.[117]

Although similar, the concept took a slightly different form in England. Upon conquering England, William the Conqueror instituted the Forest Law, thereby protecting the king's right to all wild animals, especially deer. So a forest described a place subject to Forest Law, and it differed from the law of the continent in that Forest Law also applied to places that did have a clear owner—the manors granted to the English nobility by the king.[118] Offenders who hunted the king's deer could be put to death, and collecting estovers without permission was also a serious crime, enforced by sheriffs.[119] Kings continually enlarged their forests to even include settled towns, pastures, and woods, seizing common rights in the process and angering landlords who also lost resources and income. Over several centuries, the forests were a source of many conflicts between barons and royals, not to mention between royals and common rights holders (commoners). Today, however, the word "forest" has lost this rich cultural history and meaning, and refers to land covered in trees.

Some of the earliest regulations within the forests date to the 6th and 7th centuries. What were then known as "fruitful" trees—initially oak, beech, wild apple, wild pear, wild cherry and serviceberry, and later whitebeam (*Sorbus* spp.), chestnut, walnut (*Juglans regia*), hazelnut, and alder buckthorn (*Rhamnus frangula*)—all required permission from the "forest court" before cutting, with severe punishments for non-compliance. At least in the German-speaking parts of Europe, non-fruit bearing trees, as well as shrubs and dead trees, could be used by commoners without court permission. "In France, a distinction was made between *bois vif* and *mort bois*. The mort bois could be collected to meet the people's needs. The bois vif were trees that bore fruit, such as oak, but also species that did not bear fruit, such as hornbeam, aspen, sycamore, and birch, which could not be freely used."[120]

It appears grazing was unregulated until the 13th century, when new restrictions aimed to protect young coppice shoots during the first 3 to 9 years of growth. This further necessitated the compartmentalization of coppice stands so new growth could be more easily protected from livestock. By the 15th century, fodder harvest was prohibited altogether in many parts of Western and Central Europe.[121]

These medieval norms surrounding land tenure and use were varied, complex, dynamic, and overlaid one upon another, so that several individuals might retain rights to different material resources on the same land. Different people often had rights to use a woodland's underwood and timber.[122] Also, villagers usually had access to pollard shoots to feed livestock, while the lord owned the pollard **bollings** (trunks) for timber.[123] One can imagine the possible disputes surrounding damage to the bolling following fodder harvests, or the landlord wanting to cut the bollings for timber when villagers needed the fodder. It might be possible that, in a single woodland, one person would hold ownership of the land, several the rights to woodbote divided among themselves, others with grazing rights or rights of **pannage**—that is, allowing pigs to eat acorns or beechnuts for fattening in the fall—not to mention rights of wildcrafting of herbs, hunting rabbits and pheasants, and so on. These rights varied depending on the specifics of each landlord, village, and common, but typically the lord of the manor owned the land and the timber, conferring grazing and woodcutting rights to the commoners to meet their basic needs.

This social structure had a huge impact on woodland management before, during, and after the Middle Ages. It limited the systematization of woodland management. With so many overlapping needs, uses, and users, it was difficult to control the harvest and management of common resources. But it's probable that this system maximized the yields and the diversity of yields from the overall system because more people were vested in different resources, managing and harvesting them more completely than just a few people ever could.

Many landlords disliked the common rights system. It limited their ability to manage "their" land and also limited profits. The lords spent centuries attempting to dislodge the commons from medieval and modern societies. This created many a social upheaval, protest, riot, massacre, civil war, and Robin Hood figure, whether it be Robin Hood himself dodging the sheriff in Sherwood Forest (the King's land? or a commons?) or the Zapatistas of Chiapas more recently.

Those conflicts were partly responsible for the creation and several reaffirmations of the Magna Carta, and its lesser-known companion the Carta de Foresta (Charter of the Forest) in the 13th century and later. The Magna Carta granted legal rights to free men and became foundational for democratic legal systems around the world. The Carta de Foresta granted economic subsistence rights and held a line against aristocratic seizure of common lands. These two charters arose simultaneously because it was understood that the legal rights of free people would inevitably erode unless they also had economic rights. Rights to wood through coppice management were at the time the very foundation of economic rights and survival, hence the Charter of the Forest. The Magna Carta survived mostly intact for many centuries, while the Carta de Foresta fell victim to the sustained attacks on both charters from "possessioners" and despots immediately following their enactment.

Within this medieval social context of strong, but vulnerable, common rights arose the pressures caused by population growth and limited wood supplies, especially in urban and semi-urban communities. At this point in history, wood still was the primary or only source of energy and building and craft material for most of Europe, if not the world. It was an essential resource for all people.

Wood Shortages: Necessity Is the Mother

Some parts of Europe were still heavily forested by the end of the 10th century. Agricultural expansion appears to have been the greatest driver of woodland clearance in Europe. Swelling populations required a growing food supply. To feed the masses, English agriculture shifted from mixed production towards intensive grain culture. Grain demand grew, and people came to value even minimal yields more than wood or pasture,[124] causing further agricultural land clearance. With just 15% woodland cover remaining in 1086, England's population more than doubled in the 12th and 13th centuries.[125] Despite the Great Plague of the mid-14th century that killed 30% to 60% of Europe's population,[126] reports suggest increasing wood shortages by the 15th century with consequent price increases.[127]

Poor transport infrastructure compounded these shortages and price increases. Growing communities' demand drained local wood resources, and access to further resources was limited.[128] At one point, wood prices in northern France were so high the poor used rented coffins to bury their dead! Dutch fuelwood supplies had been virtually exhausted by the 17th century. Using peat as a substitute, it's estimated that people there burned the equivalent of roughly 1,976,000 acres (800,000 ha) of coppice wood around 1650.[129]

England began importing timber from Norway in 1230, stimulating regional resource competition and wars, while exploration further afield led to another wave of globalization and colonization. The period of extreme and prolonged cold between 1550 and 1850 known as the Little Ice Age further strained the continent's dwindling wood supply. The pattern continued—increased woodland clearance and further reduction in wood availability.

Yet these stressors inspired cultural innovations and transformations. In some parts of Europe, reforestation efforts aimed to boost wood supply, spurring the development of the field of horticulture. Innovations in wood-burning stove design appeared.[130] British ecological historian E. E. Green writes, "It is reasonable to assume that at one time all trees which were within easy reach of human habitation were essential working trees, i.e., pollarded, shredded or coppiced."[131] Thus wood shortages helped maintain woodlands in the short term by making coppicing a viable business. At the same time, wood shortages increased the demand for coal, setting up the demand and supply for fossil energy sources that eventually undermined woodland enterprises.

These increases in wood value and demand spurred a shift in the mindset around coppice resources, from seeing wood as a subsistence resource to a market commodity. And these growing markets increased stress on the common rights system of old. It left landowners more interested in privatizing renewable coppice wood for making money—increasing the pressure to eliminate common rights.

Enclosure: Extinguishing Common Rights

The story of the struggle between peasants' ancient common subsistence rights and the culture of private property, aristocracy, and commodity markets for profit is long and convoluted. This history shows up throughout English culture of the past 800 years, running through the work of Shakespeare, Renaissance ballads, derisive slang, key words and concepts used in many fields, and so on. Suffice to say that, over time, private property won (not without accommodations to the commoners) and the **enclosure movement** succeeded. This changed everything for coppice husbandry.

From the 13th through the mid-19th centuries, enclosure struggles raged. Commoners fought long and hard, giving ground slowly, at least initially. But in the end, a host of forces undermined commoning—wood shortages and land clearance, the rise of market economies, increased fossil fuel use, and so on. According to Peter Linebaugh, even the American colonies and the Atlantic slave trade played a role: "The enclosure movement and the slave trade ushered industrial capitalism into the modern world."[132] Peeling peasants off the land was necessary to create a mobile labor force to power both the new market economies and the spread of the British, Dutch, French, Spanish, and Portuguese empires alongside the world's first corporations.

Ultimately, the passage of a series of Parliamentary Acts of Enclosure in Britain between 1750 and 1830 ushered the demise of most common rights,[133] though it took hundreds of separate acts to achieve the objective.[134] The Parliamentary Acts of Enclosure

Table 1.2: Enclosure acts and acreage enclosed

Ruler	Years	Number of Acts	Acres Enclosed
Queen Anne	1704-1714	2	1,439 (583 ha)
George I	1714-1727	16	17,960 (7,271 ha)
George II	1727-1760	220	318,778 (129,060 ha)
George III	1760-1797	1532	2,804,197 (1,135,302 ha)

A brief timeline showing the increasing drive to encourage the enclosure and privatization of what was once common land by British rulers during the 18th century. Adapted from MacLean, 2006, page 16.

required landowners to mark newly enclosed lands with boundary ditches planted with hedgerows along the inside bank. As a result, between 1750 and 1850, Britons planted roughly 200,000 miles of hawthorn hedges, equal to the total amount planted during the previous 500 years.[135]

To expedite establishment, some planted "quickset" hedges, a tactic described by Norden in 1607. To plant a quickset hedge, one wound mixed seeds of oak, hawthorn, and ash within a rough straw rope, burying it along the top of a bank.[136] Industrious wildcrafters also earned an income gathering wild seedlings for transplanting. But even that wasn't enough. These hedgerows would have required a billion or more plants. Hence, enclosure gave rise to 18th-century tree nurseries.

Land newly enclosed by hedges became private property, or mostly so. Commoners who had suffered severe economic privations over centuries but could nonetheless subsist on common resources became a landless working class with no means of subsistence except wage labor and the market economy. Many migrated into cities to find work.

They say history is written by the victors; the bad blood from centuries of commons-vs-private-property struggles left stigmas attached to commoning. Many of the thousands of witch burnings in 1600s England involved women persecuted for taking their rightful estovers of common.[137] Being called a "commoner" is still considered derisive. And the word "villain," now describing an evil character or criminal, has its origins in "villein" that described the vast majority of peasants in the manorial system: tenant farmers, paying rent, subsisting on common resources but legally bound to serve the lord and unable to leave the manor without the lord's permission. That villeins would stand up for their rights again and again, despite being trod upon for centuries, is admirable. Nonetheless, stigmas fell on commoning and villeins as also happened with coppice agroforestry to some degree. The struggle between privatization and common rights continues around the world even today and still affects woodland management in a host of ways.

Without common rights and commoners' subsistence uses, landlords were able to control woodland management with little, if any, outside intervention. This allowed them to focus on producing marketable commodities like charcoal and parts for craft industries, dramatically simplifying wooded commons' products and social structure. From the 13th century on, coppice rotations became more regulated and stands were managed following established cycles. The aim was to increase wood production[138] while also trending towards longer rotations for higher value products. Social simplification allowed—indeed probably required—ecosystem simplification. With fewer users and managers, management systems and woodland architecture probably had to be streamlined. But this also presumably reduced the chatter in the ecosystems, allowing woodsmen to better understand the ecosystems' dynamics and more easily direct succession.

COPPICE IN THE EMERGING MARKET ECONOMY

Enclosure marked a major turning in the history of resprout silviculture. The material goods needed by society and the ways they were created and distributed transformed in short order. Commoning's decline extinguished a complex social system that managed multilayered wooded ecosystems for diverse yields. Few individual landowners could use or manage the woodlands as fully as the commons economy. As fossil energy strengthened, energy prices fell relative to labor costs, rendering complex commons management more and more uneconomical. It was too labor-intensive. Markets for a few products began to drive woodland management. Resprout silviculture adapted to the times.

Charcoal for Industry Boosts Coppice Production

Industry was the largest single market for coppice wood. Until the 15th century, ironmasters processed iron in a bloomery, a mound-shaped construction of alternating layers of crushed ore and charcoal built on a stone hearth and covered with a layer of clay. Bellows delivered air inside the bloomery, producing softened wrought iron.[139]

In parts of West Yorkshire, copses were cut for charcoal on a 16-to-18-year rotation. By the 16th century, the iron industry's explosion had nearly exhausted local timber supplies. Between 1545 and 1547, the Worth Ironworks in Sussex consumed 9,000 cords (32,620 cubic meters) of wood. In an effort to conserve mature wood supplies, parliament developed regulations that prohibited charcoal production from "mature wood" (timber), thereby requiring the use of coppice.[140]

And counterintuitively, once this law passed, the steady demand for manufacturing fuel actually led industry to sustain woodland management. According to Hammersley, "Ironmasters… did not plow up woodlands or uproot them, neither did they nibble the young shoots; most of them wanted to protect their investment and maintain their profits, and that needed fuel for the future as well as the present."[141] In fact, woodland owners often partnered directly with ironworks.

The Lorn (Argyll) furnace in Wales, the longest-lived furnace in the UK, sustained production until 1876, maintaining 10,000 acres (4,047 ha) of oak coppice.[142] The woods in and around Sheffield, in North Yorkshire, England, one of the most richly wooded industrial cities in Northern Europe, were largely preserved to meet the fuelwood demands of nearby industrial landowners.[143] Meanwhile, on the 7th Earl of Shrewsbury's estate at Wentworth in Yorkshire, the rise in profits from charcoal production led to the decline of woodland grazing: "By 1723 the annual profits from the coppice were seen as sustainable and relatively risk free. Charcoal production was of vital importance in the development of the Industrial Revolution."[144] And it wasn't long before the next stage in industrial development again transformed the coppice's role in history.

Though coal had been developed as a heating and industrial fuel during the Middle Ages, it didn't become widely available until the 17th century.[145] As the Industrial Revolution progressed and coal became cheaper and more available, the relevance of coppice management in Europe declined precipitously. Coal eventually replaced charcoal in industry, destroying coppice wood's main market across the continent. The declining cost of energy relative to labor also undermined woodland management's economic feasibility.

FIGURE 1.15: The traditional practice used to make charcoal. A carefully and strategically constructed core of cordwood, covered by layers of soil. The charcoal burner needed to remain vigilant throughout the entire burn, often 24 to 48 hours, to ensure the shifting pile doesn't create gaps that cause the material to ignite. This process fueled industrial production from early history until the Industrial Revolution. Original caption: "Preparing a new charcoal pit for coaling. C.A. Masie, collier (left), and David Daniels putting wet leaves and then dirt over the lapwood so that the cordwood will burn in a deficiency of oxygen and produce charcoal." Source: commons.wikimedia.org.

Shifting Products, Markets, and Management

So, during this transition into the early modern period (1500 to 1750), enclosure, increased urbanization, and the rise of coal reduced the demand for small-diameter coppice wood. In 18th-century Germany, changes in wood-burning technology brought a similar shift in demand from small-diameter sticks to easily split cordwood. Production shifted and rotations doubled or even tripled, yielding larger stems on rotations as long as 30 to 50 years.[146] Stricter felling rotations stabilized material supplies and income streams. The traditional practice of draw-felling individual rods from stools had long passed, and by early modern times whole cants were usually clear-cut to create a patchy, even-aged, rotational management system.

For example, typical coppice-with-standards woods during the Early Middle Ages (around 1000 CE) were harvested on short irregular 4-to-8-year cycles, with standards felled after 25 to 70 years (up to 100).[147] In Hayley Wood in eastern England, coppice felling rotations shifted from 7 years during the 1300s, to 10 to 11 years by 1584, and 15 years by 1765. Despite a great deal of regional variation, this trend was typical throughout the UK and much of Europe. Pre-1500, woodlands were rarely cut on cycles longer than 9 years. By the 18th century, rotations under 9 years were rare, with 15-to-20-year rotations not uncommon.[148]

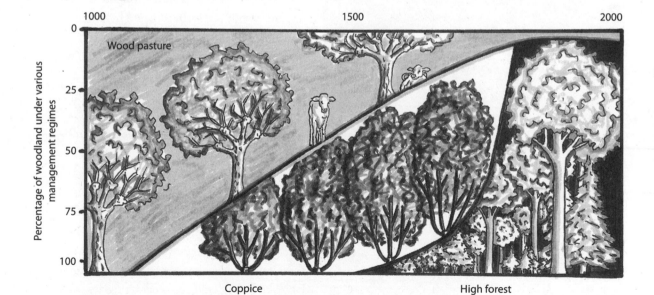

FIGURE 1.16: This timeline demonstrates the gradual shift in the concentration and distribution of wood pasture, coppice, and high forest in southwest Yorkshire, UK, between 1000 and 2000 CE. Adapted from Jones, 1998, p. 57.

In southwest Yorkshire, wood pasture comprised virtually 100% of the region's woodland between 1000 and 1100 CE. "Spring woods"—both copses and coppice with standards—then began to displace wood pasture. The coppice was mostly used for firewood and charcoal. By the middle 1400s, coppice woods comprised 50% of the wooded land, growing to 66% by 1800 with wood pasture declining to a mere 12% of southwest Yorkshire's wooded acreage. And meanwhile, about 25% of the county's woods had become high forest, which began to increase exponentially. During the ensuing century, **high forest** came to dominate the

Figure 1.17: A craftsman carefully stacks finished wooden tent pegs to dry in the woods, ca. 1941. Note the outdoor workshop in the background. Photo courtesy of the Museum of English Rural Life, University of Reading, used by permission.

region's wooded acreage, coppice was all but gone, and wood pasture covered the sparse acreage that remained. This trend from coppice and wood pasture towards high forest appears to have occurred across most of Europe.

The early 19th-century drive towards industrial production rapidly transformed many crafts and livelihoods. The standardization of crafts led many craftspeople to transition from subsistence-based livelihoods to full-time work in cottage industries, factories, and large estates. They became part of a specialized workforce, breaking the process into individual parts. The bodgers produced turned chair parts in the nearby woods, receiving piecework compensation from the factory. Workshops housed bottomers, benchmen, and framers who shaped the seats, bows, and spindles and assembled the finished product.[149] This created inexpensive products available to a larger market, but it undermined the livelihoods of the self-employed craftsperson.

Woodsman and author Raymond Tabor suggests that disorganization within craft networks created unstable markets that ultimately eroded the traditional trades. Lacking price controls and incentives to keep producers from flooding markets, the economics of coppice crafts varied wildly, making it difficult for craftspeople to earn a living.[150] The need to earn cash to pay taxes and purchase food, goods, and services caused a division of labor, reducing crafts to separate trades, each focused on one part of the flow of materials: production, harvest, processing, assembly, and marketing.[151] These compounding factors

Děvín Wood, Czech Republic

In 2011, I spent close to three months in Europe, visiting coppice woodlands and meeting with foresters, craftspeople, and historical ecologists. Péter Szabó, PhD, of the Institute of Botany, Academy of Sciences of the Czech Republic, in Brno, kindly offered me a guided day visit to Děvín Wood, a woodland with a rich history and usage patterns that largely mirror those we've just discussed. Spanning 640 acres (260 ha) on a small hill in southeastern Czech Republic, Děvín Wood belonged to an estate that included about 15 settlements governed from the nearby town of Mikulov from the 13th to 20th centuries. The community largely subsisted on internal resources, skills, and trade. The overlords possessed legal authority over village tenants, who relied on their decisions and actions. The overlords guided management of the estate woodlands until after World War II, when the Communist regime nationalized private landholdings.[152]

Czech woodland records originate in the 15th century, tracking wood sales, coppice and timber harvests, and underwood and timber trees. Děvín Wood enjoyed stable management as a coppice-with-standards system until the years following World War II, though like many medieval woodlands, the rotation length and standard tree density adapted to changing material demands and social orders.

Until the 16th century, Děvín Wood's primary product was firewood for village tenants, with an average harvest cycle of 7 years during the late 14th and early 15th centuries. Around 1500, the overlords elected to retain ownership of all standard trees,

leasing neighboring villages rights to the underwood for a set amount called *holzgeld* (wood-money). With time, conflicts between the overlord and tenants arose as the overlord increased the number of standard trees, reducing underwood yields. By 1692, the coppice cycle had increased to 12 years, with just 0.2 standards per acre (0.5 per ha). By 1808, standard tree density had increased to 13 per acre (33 per ha) with a 30-year coppice rotation, save a 42-acre (17 ha) parcel leased to tenants where the cycle remained 13 years. The coppice rotation remained 30 years until 1948 when it grew to 40 years under Communist control. The story of Děvín Wood helps illustrate these common themes in wood product demand and evolving management practices.

Figure 1.18: A view of the hilltop estate at Děvín Wood in the southeastern Czech Republic.

virtually extinguished many traditional crafts. Subsistence-based common rights had all but disappeared from woodland culture, displaced by market-driven mass production.

THE NEAR EXTINCTION OF COPPICING

The decline of European coppice systems occurred at different rates in different places, but in general, routine coppice management largely disappeared from the woodlands of Western Europe in just a century or less.

Between 1900 and 1930, neglect overtook many British woodlands, with felling ceasing in some places as early as 1870 because of fossil energy and rural electrification. Sweet chestnut poles were still in demand by the hop industry that required long, straight, rot-resistant vine supports grown on 8-to-10-year rotations, but by the 1950s, most coppice woodlands (along with their craft and fuel markets) had entered a period of dramatic contraction.[153, 154] These woodlands were converted to other uses.

Table 1.3: Total coppice woodland in the UK from 1905-1980

Year	Acres	Hectares
1905	575,755	233,000
1947	350,900	142,000
1965	74,130	30,000
1980	98,800	40,000

The distribution of coppice woodlands declined dramatically in the UK during the early to mid-20th century, only recently beginning to rebound. Adapted from Peterken, 1992, p. 9.

Similar factors brought similar results elsewhere in Europe. During the mid-20th century in Spain, lengthening rotations, state forest administrations supplanting local custom and lordly control, shifting rural socioeconomics, growing urban migration, and widespread fossil fuel availability led to near abandonment of woodland management across large portions of the countryside.[155]

Nearly two centuries of detailed forest management plans clearly document changes in northeast Switzerland's woodland history. In 1825, coppice or coppice-with-standards covered upwards of two-thirds of the area's forests. By the end of the 19th century, simple coppicing had almost completely disappeared, while coppice-with-standards systems survived until the mid-20th century. In subsequent years, many stands were converted into high forest.[156]

Agricultural intensification, increases in pasture productivity and stocking rates, land abandonment, urbanization, road construction, and a shift towards confined livestock management all led to the disappearance of wood pastures in the 19th century.[157] In some places, woodlands had been used to maintain fertility for field crops. Managed woods in Germany's Neidlingen Valley may have provided an estimated 75% of the nitrogen and 90% of the phosphorus needed to support arable agriculture until 1500. These nutrients came in the form of collected manure, leaf litter, wood ash, charcoal, and fodder residues.[158] The pressures placed on rural landscapes to support growing populations increased forest nutrient export, undermining the litter and nutrient cycles that sustain woodland fertility, reducing the soil's acid-neutralizing capability, and accelerating soil acidification.[159] Livestock confinement also sparked an increase in woodland litter collection for fodder, especially in dry years, further robbing these landscapes of their fertility.[160] Foresters warned against the dangers of this nutrient redistribution, but in the end, they had little impact. Ultimately, the longstanding tradition of wood pasture management became a relic in much of Western Europe.

Ironically, British hedgerows suffered similar pressures. Although planted en masse centuries earlier because of commons enclosure,

as technology advanced, hedgerows impeded privatization. Within 15 years of World War II's end, the tractor almost completely replaced the horse as the primary farm management tool. This required larger contiguous fields. The British Ministry of Agriculture encouraged hedgerow removal, offering grant money to support land clearance.[161] About 140,000 miles (224,000 km) of hedgerow were lost between 1945 and 1970—roughly a quarter of all British hedgerows. Dramatic declines in the farm labor pool (falling from 2.5 million workers to 500,000 between 1940 and 1994) eroded the available workforce, rendering hedgerow

Modern-Era Coppicing in Southeastern Europe

Southeastern Europe stands at the crossroads between several ancient civilizations and three continents. Humans began utilizing the region's forest resources several thousand years ago, although forest exploitation didn't peak until the 18th and 19th centuries. A number of factors contributed to forest management intensification: private land consolidation; emerging commodity and monetary-based relationships; the growth of agriculture, craft, and industry; and population growth and expansion. Together, these changes brought a fairly rapid transformation of high forest stands into coppice woodlands or pasture,[162] the opposite of the trend in Western Europe at the time.

With increasing forest utilization, Croatia, Bulgaria, and later Serbia developed forestry laws (in 1769, 1870, and 1891, respectively) that established

guidelines intended to develop sustainable forest management practices. These regulations yielded some positive results, but they could not overcome the effects of widespread poverty and regional wars during the late-19th and early-20th centuries. These stressors created a cycle of forest degradation that began with overharvesting, later transitioning into irregular coppice woodlands, brushland, and ultimately, barren landscapes. Nonetheless, in Bulgaria, coppice managed on 15-to-30-year rotations was common until the 1950s. In Macedonia, oak and beech coppice management continues on cycles of up to 50 years, while black locust (Robinia pseudoacacia) stands are harvested on up to 30-year cycles.[163] Had it not been for the forests' resprout abilities, it seems these regions would have been completely deforested.

maintenance a demanding financial burden for farmers. The seasonal hedgelaying craft once undertaken by full-time farm workers declined precipitously.[164]

The Rise of Modern Forestry

The same conditions that undermined coppice agroforestry may have set the stage for the invention of modern forestry. Some suggest that modern forestry could not have arisen until enclosure had removed agriculture from the woods by displacing subsistence peasantry, conferring woodland ownership and management control to either state or private entities. The removal of hunting rights for the aristocracy was also vital as "princely numbers of deer imperiled the natural regrowth,"[165] as was the ability to easily and cheaply transport and process large logs. Together, these factors, along with the rise of the scientific method, ushered the creation of the science of silviculture.

Academic study of forests began in Europe (specifically Germany, Austria, and later France) during the 1700s, and science-based forest management became common during the latter part of the century.[166] Some of the first examples of modern forestry in Germany drew directly from coppice traditions by dividing up a forest into equivalent compartments and rotationally clearcutting one per year.[167] In 19th-century Germany, forestry evolved from an observation-based practice steeped in tradition to a scientific discipline. This paradigm shift spurred a transition towards timber-based wood production and the disintegration of grazing and woodland management. Forestry grew to focus on monitoring trees' annual and cumulative growth increments as indicators of site productivity. By regularly measuring yields, foresters could theoretically develop practical economic valuations and plan for long-term sustained management.[168]

As modern forestry and scientific horticulture developed further, planned management led to the emergence of plantations, the epitome of industrial forestry. The practice spread so rapidly that, by the 1930s, much of Germany's forest had been converted. The rise of plantations reflected a management shift towards a system aiming to increase and intensify production that much more closely resembles agriculture. Like agriculture, plantation systems reproduce the highly productive stages of early succession,[169] require soil disturbance to plant seedlings, and tend towards monocultures.

This tumultuous era in European woodland history found a markedly different expression on the North American continent as colonists emigrated in search of new lives and abundant resources.

Coppice in Colonial and Industrial North America?

Few if any written records document the development and maintenance of coppice stands to sustain fuel and wood supplies in New England or virtually anywhere in North America during the colonial era. However, anecdotal evidence does indicate coppice supplied early America with wood products.[170] Colonists in Williamsburg, Virginia, harvested garden stakes, trellises and other garden architecture, livestock fodder, and firewood from coppiced and pollarded ash (*Fraxinus americana*), sycamore (*Platanus occidentalis*), and vitex (*Vitex agnus-castus*).[171] The tendency of many native trees to sprout after cutting led to coppicing for fuelwood and charcoal harvests, despite "haphazard management."[172] However, coppice seems to have mostly disappeared during the centuries following the colonization of the eastern United States. Whatever practices the colonists may have once used lie shrouded in succession, with advancing forests consuming nearly all evidence. It appears the woodland history of colonial North America is mostly a story of clearance, destruction, and profiteering, with little emphasis on sprout management.

Early historical narratives describe pre-colonial North America as a majestic, abundant landscape that probably reflected conscious Indigenous management, primarily using fire, similar to that in California. Many travelers described abundant crops of wild berries and game, beautiful forests of easy passage, and an open "parklike" landscape during early voyages to this continent.[173] Yet early European explorers largely sought to export goods to Europe for profit, calling them "merchantable commodities." In the UK, wood for heating and building had become increasingly scarce, so the extensive timber of North America proved invaluable in feeding the hungry British market.[174]

Colonists cleared massive swaths of forest to create homesteads, fields, and pastures; to harvest fuel and building materials; and to generate income. Over the course of 2 to 4 summers, American families cleared up to 3 acres of woods with an ax each season to carve out their homesteads. They used the material to build cabins, erect fencing, and keep warm through the winter.[175] As they cleared, early settlers often battled the incessant stump sprouts. Colonists typically felled trees in the late summer months to discourage resprouting, then burned the remaining wood and leaves during the driest parts of the following May. This tended to kill the stumps, preventing sprouting so crops could grow unhindered. Settlers in the wooded regions of 18th-century America often grew subsistence crops, deriving income from selling wood[176] or potash from burnt woody debris. In 1717, reports claimed New England farm laborers could "burn and process four acres of forest to produce eight tons of potash worth £40 to £60 per ton"[177] that was used to manufacture gunpowder and soap.

Colonists' ravenous demand for fuel left even larger openings than homestead clearing. During the 17th century, "a typical New England household probably consumed as much as 30 or 40 cords (109 to 145 m^3) of firewood per year… a stack of wood four feet wide, four feet high and three hundred feet long; obtaining such a woodpile meant cutting more than an acre of forest each year."[178] Between 1630 and 1800, New England devoured over 260 million cords (942 million m^3) of firewood, and in 1880, annual American fuelwood consumption peaked at 4.1 million cords (14.9 million m^3).[179] Much of that consumption resulted from wildly inefficient combustion and building construction. Homes built from stone and wood leaked considerable air and had no insulation. Many people burned wood in open fireplaces requiring four to five times as much wood as the cast-iron stoves used by Pennsylvania Germans. New England burned up to 18 times more wood for fuel as it cut for lumber in 1800.[180]

As in Europe, demand for industrial fuel created another drain on New England forest resources. Processing a single ton of finished iron required 250 bushels of charcoal, requiring 6 cords of wood (~22 m³). As iron furnaces sprang up in Rhode Island, western Massachusetts, and Connecticut, each producing 500 tons of iron or more per year, it's easy to see how 16,000,000 acres (6,475,000 ha) of New England forest had been cut by 1840, with many parts of the region 80% cleared.[181]

Despite the relative lack of clear documentation of the use of coppicing in early American colonial development, several sources suggest it was more commonplace than we might otherwise recognize. In *Traditional Woodland Crafts*, Raymond Tabor claims that coppice was once a widespread strategy for fuelwood production on this continent.[182] William Cronon writes that both oak and birch woods were usually coppiced, ready for harvest once again after as little as 14 years.[183] Some reports state that parts of Massachusetts, Connecticut, northern New Jersey, and New York's southern Hudson Valley were repeatedly cut for fuelwood and charcoal for the burgeoning industries of the northeastern seaboard.[184] Early land records in northwestern Connecticut indicated that chestnut comprised between 4% and 15% of the forest, but after well over a century of repeated cutting, the species comprised about 60% of forest in the same area by 1908.[185] "Coppice cutting was a major reason that chestnuts, prolific sprouters, increased their share of New England forests following European settlement."[186] So, once again as with much of our study of the history of coppicing, we find more scattered traces of evidence that occasionally contradict one another. We may never know the full story.

The large-scale abandonment of New England hill farms and westward migration in the mid-1800s enabled the return of the forest; however, the resultant cover was but a shred of

FIGURE 1.19: Many landscapes often unintentionally maintain a coppice-like architecture due to past land-use patterns including grazing, clear-cutting, brush hogging, controlled burns, and other forms of disturbance. All this to say that we have plenty of models to work from if we simply take the time to observe how succession directs woodland regrowth in the bioregions we call home.

the majestic Indigenous-managed growth blanketing the region just 150 years prior.[187]

One hundred years later, 40% of New England forestland was less than 20 years old, but these second- or third-growth stands no longer expressed the same growth potential as their forebears. The practice of **high-grading** forest stands (taking the best and leaving the rest), ongoing since 1600, had robbed returning forests of the genetics of the best-quality trees. Forest clearance on hillsides and unchecked grazing caused erosion of millennia worth of topsoil accumulation. And the general abandonment of forestland left overcrowded stands of young trees requiring significant thinning to achieve their true potential.[188]

Much of North America shares a similar story. We inherit this legacy today.

REVIVAL? COPPICING FOR THE 22ND CENTURY

To "revive" a practice it must have once been commonplace. In the US, a "coppice revival" really amounts to testing, developing, and adopting the practice, in addition to further uncovering and applying Indigenous land management strategies. In Europe, on the other hand, the culture of coppicing and coppice crafts lives on, and coppice woodlands still cover an estimated 57 million acres (23 million ha),[189] about 2.3% of Europe's land mass. There, the decline in coppicing and wood pasture management has left a legacy of neglected, *overstood coppice* (coppice growth well past its prime and in decline) awaiting demand for wood products and a restorative vision.

As markets reemerge for these products, so does the potential to reinvigorate these traditions. Now nearly four decades in the making, the growing rediscovery of British woodland traditions has begun to usher in a new era of coppice culture.

European Revival

Some trace the origins of British coppice renewal to the release and reprinting of several books in the 1970s, including the 1974 reprint of Herbert Edlin's classic *Woodland Crafts in Britain*, Percy Blandford's *Country Craft Tools*, and John Jones's *Crafts from the Countryside* the following year. Later, Ivan Sparkes' *Woodland Craftsmen* and Jack Hill's *The Complete Practical Book of Country Crafts* further helped build this movement.[190]

Around the same time, the once-endangered hedgelaying craft was embraced by conservation organizations for the benefits they offer wildlife. Hedgerow removal has largely slowed in recent years as new hedges are planted for their ecological value. Today grant programs incentivize and encourage hedgerow maintenance and installation.

Britain's forested landscape is deeply fragmented. Woodlands over 124 acres (50 ha) are uncommon, and those over 248 acres (100 ha) are rare.[191] British conservation organizations have also come to recognize the ecological benefits coppice management offers invertebrates, birds, small mammals, and ground flora. The development of the Greenwood Trust in Telford, England, in 1984 helped crystallize a partnership between conservationists and craftspeople working to restore and reestablish coppice to create habitat while also making use of the wood products.[192] Additionally, numerous British groups emerged in the 1980s and 90s that focused on the preservation and perpetuation of traditional crafts and coppice management.[193]

In parts of the UK, long-valued products like hop poles, baskets, and charcoal still drive coppice management, while in other areas, markets have begun to reemerge during the past few decades. The chestnut coppice industry in Kent and East Sussex, still plods along, relatively intact, and the chestnut paling industry continues to employ folks whose families have been involved with coppicing for generations.[194]

In West Sussex and Hampshire, the hazel coppice industry has resprouted to varying degrees. These regions' once-extensive stands suffered during the second half of the 20th century with the decline of the hurdle industry. Today, coppice in Hampshire continues to grow.[195] Depending on the quality, the value of uncut British hazel coppice (in 2001) fluctuated between £50 and £350/acre ($80-550/acre or £124-865/hectare).[196]

While other parts of Europe still possess marked remnants of the coppiced trees that once dominated the landscape, they're little more than a shred of the landscape that once was.[197]

Fodder-producing pollard trees still grow in many parts of Scandinavia, though many have suffered neglect in recent years. Finland's Åland Islands alone contain over 40,000 pollards.[198] Today the majority of remnant Finnish and Swedish pollard meadows lie within nature reserves, managed as living museums rather than working landscapes.[199]

In modern Spain, coppice stands occupy 20% of the total forest area and 40% of the nation's hardwood forests. Three oak species, *Quercus ilex, Q. faginea,* and *Q. pyrenaica,* constitute 82% of all coppice woodlands, although many of these stands are much lower-yielding, overstocked, fire-prone, older, and shadier than if they were properly managed. Due to the lack of access to markets, silvopasture researchers have begun to suggest slowly reducing stand density, converting coppice woods into high forest.[200]

Coppice woodlands still constitute considerable portions of forested landscapes in parts of Southeastern Europe (48% of total forests in Bulgaria, 59% in Macedonia, 65% in Serbia).[201] Twenty-two percent of Croatia's forests are coppice stands,[202] 18% of which is Turkey oak (*Quercus cerris*), black locust, and sweet chestnut.[203] In Albania, 74% of the oak forests are coppice woods, with European beech and pubescent oak (*Quercus pubescens*) comprising about half of these woodlands.[204]

During the first decade of the 21st century, a number of foresters and research scientists formed CForSEE, Coppice Forests in Southeastern Europe—Multifunctional Management of Coppice Forests. The network connects research and management institutions, enabling them to share resources, experiences, challenges, and successes, while learning how to economically manage coppice woodlands in the 21st century. Recognizing growing interest in coppicing, largely due to growing demand for biomass energy, the network aims to help landowners invest in improved woodland management, developing coppice-with-standards type systems that profitably yield fuelwood and timber.

However, a number of challenges cause Southeastern European researchers to question the long-term viability of coppice management in the region. Many of today's woodlands developed from remnants of intensive forest cutting during the 18th and 19th centuries, rather than as intentionally coppiced stands. Hence, they often exhibit low productivity and poor stem quality, vitality, health, and age structure.[205] Many of these low-quality stands occupy high-quality sites that could produce more valuable timber yields. And finally, many privately owned woodlots are particularly small, averaging as little as 2.5 acres (1 ha), making sustainable management difficult.[206]

Today, many foresters in Southeastern Europe work to transform healthy beech- and oak-dominated coppice stands on good sites with relatively high productivity into high forests of seed origin. Many of these efforts have been unsuccessful due to poor natural regeneration. Foresters have also tried to reconstruct low-quality coppice stands on poor or otherwise degraded sites by clear-cutting the existing vegetation and substituting higher-quality species. Often these efforts fared poorly as coppice stump sprouts frequently out-competed planted seedlings. Therefore, many woodlands planned for transformation to high forest will instead remain as coppice.[207]

Table 1.4: Total area and percentage of forested land in coppice in select European countries (2009).

Country	Total Forested Area (ha)	Coppice Woodlands (ha)	Percentage of Total
Albania	942,000	405,000	43
Austria	3,992,000	70,000	2
Bulgaria	3,700,000	1,750,000	47
Croatia	2,403,000	512,000	21
France	14,470,000	6,822,000	47
Macedonia	948,000	557,000	59
Greece	2,512,000	1,640,000	65
Hungary	1,702,000	501,000	29
Italy	6,013,000	3,397,000	56
Serbia (not including Kosovo)	2,252,000	1,456,000	65

Adapted from Stajić et al., 2009.

As we look to integrate resprout silviculture into our forest management practices here in North America, we would do well to learn from these patterns in Europe. There are scattered examples of resprout silviculture techniques employed by people across this continent at scales large and small—backyard stands of basket willow; rustic furniture makers; self-reliant homesteaders and farmers growing their own fuel, fencing, and fodder; horticulturalists producing ornamental **woody cuts** for the floral industry; on up to foresters and restoration ecologists using fire to manage fuel loads and regenerate forest stands and farmers and researchers producing biomass at an industrial scale. Perhaps we're on the verge of yet another silvicultural renaissance where sustainability, regenerative silviculture, and backyard self-sufficiency and entrepreneurialism converge to revitalize our neglected forest resources.

RECAPITULATION: CRAFTING REGENERATIVE RESPROUT SILVICULTURE

So, we see small steps towards recreating an active and engaged coppice culture. This brief exploration of the history of woodland management in Europe and the United States helps illustrate woodlands' vital role in civilizations' stability throughout the past several thousand years. Humans rely on wood, and the ways we procure it speak volumes about a culture's values, needs, tools, technology, goals, economy, and worldview.

Our ancestors' cultures relied on polewood yields and silvopastoral landscapes managed for their resprouts. Their crafts, architecture, and lifestyles all evolved to utilize the products these systems provided. Modern transitions towards large-scale, industrial, globalized production and economic markets have displaced our need for rapidly renewable polewood resources. But these materials can once again become relevant to our lives.

In much of North America, we are blessed with abundant high forest. At the same time, we're surrounded by damaged landscapes in need of human care and attention, along with growing populations who need the basics of life in an economy increasingly incapable of meeting those needs. Coppice management invites us to reinvent our relationship with wood, woodlands, and one another. We'll likely never know with certainty how our ancestors transformed ancient woodlands into coppice woods and wood pastures. We must recreate these strategies ourselves. We do not know exactly how they tended these systems, and even if we did, we live on a different continent than our European forbears, with different species and different conditions. Nor do we know how our Indigenous brothers and sisters created the mosaics of coppiced successional woodlands that they tended long before European colonization. But everything we need to do has been done before, in some fashion.

Our path may well parallel the evolution of coppice agroforestry the first time around: learning which species sprout, how well they sprout, what their wood is good for. We have models of many of the systems used throughout coppicing history, though we lack many details. We'll need to adapt these models to our own culture, economy, and ecosystems. While today we presumably have more information on the biology and ecology of resprouting in woody plants than our ancestors, it still will take time to perfect the art. Those of us alive today can help lay the groundwork, so that our children may continue to build on this legacy.

And perhaps even more important than the most promising species, systems, rotation length, etc., for coppice systems to find true relevance in our lives today and in the future, we must discover how it complements a land-based culture emerging from the shadows of globalized industry and consumerism. In so doing, we have an opportunity to share and cultivate our visions, skills, and passions. If we consider coppice management as part of a system, we can begin to imagine how we might best apply it to the landscapes and communities we inhabit. It's an exciting time. And we have so much to learn. Let's begin by exploring the anatomy and biology of woody plants so that we understand how to best ally ourselves with the stools and resprouts that form the foundation of coppice systems.

FIGURE 1.20: Wattle hurdle maker in Surrey, UK. Photo courtesy of the Museum of English Rural Life, University of Reading, used by permission.

Chapter 2

The Anatomy and Physiology of Woody Plants

It is the plant that is always right.

—N. Woodhead, quoted by Adrian Bell[1]

Knowing how a plant reproduces is the first step to encouraging its growth in wild settings or in domestic culture.[2]

—Richard Cech

To be good stewards of our coppice stools and pollarded trees, we must start with a firm grasp of the anatomy (structure) and physiology (function) of woody plants. This understanding improves our ability to design, implement, and manage these systems and properly harvest and add value. The causes of sprouting, the various forms it takes, and the reasons they evolved teach us to manage plants effectively so they produce useful materials with minimal stress. Constructive information includes:

- the forms woody plants take;
- the organs, tissues, and cells that comprise them;
- how they grow;
- their life cycles and energy allocation;
- their adaptations and responses to wounding and disturbance;
- the nature and physiology of different types of sprouting; and

- the impacts of sprouting on individual plants, the uses of their wood and how to best manage them.

All of these factors relate to plants' ecology and evolutionary history, which we'll explore in the coming pages. Through all of this, remember that the plants themselves are our teachers. We are the students. *The plant is always right.* These concepts and ideas provide something of a map, but the plants are the territory. They will guide, train, and test us as we engage with them.

WOODY PLANT ORGANS AND TISSUES

Trees, as we have come to know them, first began to evolve over 200 million years ago.[3] The ability to form a long-lasting structural stem with "perennating" organs (that survive from year to year) well above the ground differentiates woody plants from their herbaceous relatives. While this adaptation enables them to colonize the space above lower-lying plants, it has many consequences in terms of their form, cellular structure, and energy dynamics. Let's explore woody plant anatomy, beginning with a quick review of their primary organs, growth forms, and the tissues they're composed of. Along the way, we'll discuss how this understanding can inform our coppice design and management.

Primary Organs: Crown, Trunk, and Roots

Plants drive terrestrial ecosystems by making solar energy available to themselves and other life forms. Through the miracle of photosynthesis, they combine carbon dioxide (CO_2) with water to capture, store, and transport sunlight within the chemical bonds of different types of sugars. They burn some of these sugars for metabolic energy and use them to create their own tissues. They even exchange them through their roots with other organisms for various favors. Plant sugars and tissues form the basis for virtually all organic life. Plants typically acquire the sunlight, CO_2, and water they need for photosynthesis from two sources: the sky and the earth. Connecting these two resource pools is the main function of all plant anatomy.

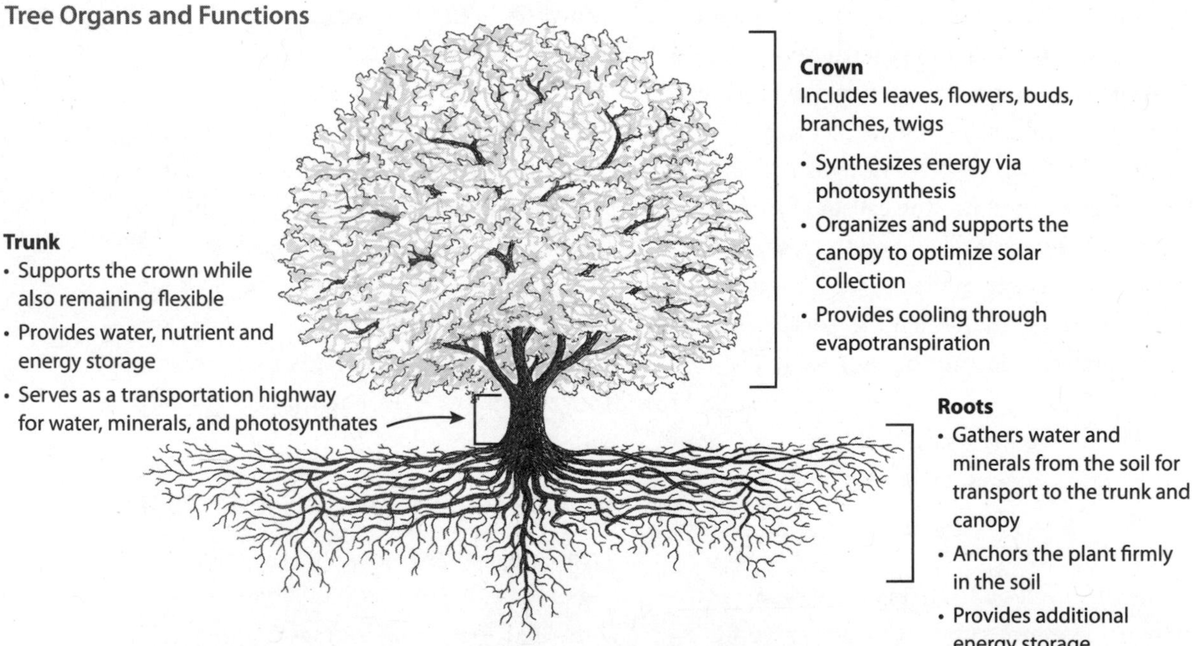

Tree Organs and Functions

Crown
Includes leaves, flowers, buds, branches, twigs

- Synthesizes energy via photosynthesis
- Organizes and supports the canopy to optimize solar collection
- Provides cooling through evapotranspiration

Trunk
- Supports the crown while also remaining flexible
- Provides water, nutrient and energy storage
- Serves as a transportation highway for water, minerals, and photosynthates

Roots
- Gathers water and minerals from the soil for transport to the trunk and canopy
- Anchors the plant firmly in the soil
- Provides additional energy storage

FIGURE 2.1: Woody plants are comprised of three primary organs: the crown, the trunk, and the roots. Each plays an essential role in their structural integrity and life processes. The illustration is a composite of a white mulberry (*Morus alba*) branching pattern and root system.

Woody plants' three primary parts or "organs"—the **crown** (or canopy), the **trunk** (or stem, bole, or main axis), and the root system—each perform specific essential functions that support plant growth. We'll start with a brief look at their main functions and follow with details on their anatomy and physiology.

The Crown: Sky Gatherer

A woody plant's crown is a network of interlinked shoots. The crown typically includes the upper portions of the trunk or trunks, along with **branches, twigs,** and their attendant leaves, flowers, buds, and other structures.

The crown is the armature that organizes leaves in order to maximize absorbed solar energy and CO_2. Plants can only convert about 2% of ambient sunlight into stored chemical energy, so their overall form needs to minimize energy use for limb construction and maintenance. Each leaf must also connect to its source of water. A branching pattern accesses, collects, and distributes dispersed resources very efficiently. Each species' specific canopy responses to a host of factors including competition, herbivory, and damage by wind, ice, drought, and other forces of nature stimulate incredibly diverse structural patterns.

The Trunk: Essential Connector and Structural Supporter

Photosynthesis requires water, and plant growth requires mineral elements from the soil. Meanwhile, photosynthates (the products of photosynthesis) sustain all the plant's cells, along with a range of phytochemicals that plants require throughout to function. So, plants must distribute water, nutrients, photosynthates, and phytochemicals throughout their body.

Woody plants' rigid stems make these connections possible and allow their crowns to rise above lower-lying light competitors and terrestrial herbivores. They transport sugars and other biomolecules from where they are made to where they are used or stored through the inner bark, or **phloem**, via the **sap** (water within the plant, containing dissolved nutrients and plant chemicals). At the same time, leaves transpire water to cool their surfaces, flowing upwards from the roots through the youngest layers of wood (the **xylem**). This water carries with it essential mineral nutrients absorbed from the soil.

At the same time, the trunk must bear the weight of the crown and the water it contains, while remaining limber enough to flex with the winds and other stressors without breaking. Older layers of wood inside the trunk no longer transport water, but they may continue to provide structural support. Most of the cells within the trunk are dead, but living cells also provide reserve water, mineral, and photosynthate storage for the following year and serve as insurance against catastrophe.

The Root System: Earth Gatherer, Foundation, and Anchor

Plant roots serve many functions. Primary among them is gathering water and nutrients from the soil and delivering them to the trunk for transport to the crown. Most plants also exude photosynthates and other chemicals to feed fungi and other soil organisms in their root zone that make water and/or inorganic soil minerals available to roots in exchange. Though roots and their microbial associates supply these essential elements to the tree, they are "captive" organs, meaning they cannot make food themselves. They require a continuous exchange with the canopy.[4] And similarly, the canopy and trunk cannot exist without the roots.

Like the canopy, root systems tend to exhibit a branching pattern, gathering dispersed resources in the soil and transporting them efficiently to the trunk. However, soil and water resources are not distributed as evenly as sunlight and air, so roots often vary in growth pattern depending on the site and the plant's genetic predisposition.

Roots anchor plants physically and energetically. Structurally, root systems must bear the weight of the crown and the trunk, stabilizing the plant as it moves in the wind, transmitting these forces to the ground while keeping the plant upright. Energetically, roots store energy as carbohydrates both for future growth and as a reserve against disaster.

Trees transition from roots to trunk in a zone at the base of the plant called the **root collar**, usually just below ground level.[5] You can identify the root collar based on changes in texture, color, and pattern from stem bark to root bark. For many species, the root collar is central to post-disturbance coppice-type sprouting responses. We'll discuss this in greater detail later.

Together, the crown, trunk, and roots comprise the whole organism. Yet each species—and each individual plant—expresses these three primary organs in different fundamental patterns. We call these patterns plant "form."

Woody Plant Forms: Overall Aboveground Architecture

A plant's form helps categorize their architecture to aid in species identification, characterize their adaptive strategies and behaviors, and inform their optimal use and management. Despite the diversity of woody plant forms, there's little agreement about how to define or categorize them—or even what "form" actually means (some authors use "habit" to describe what we call here form). This diversity of opinion and physical expression make consistency difficult.

Nonetheless, in this book, **form** describes a plant's physical size and overall aboveground architecture, while **habit** refers to the above- and belowground patterns of growth within that form. Consider form a snapshot of an undisturbed plant's architecture while habit provides more detail, indicating a plant's normal growth processes and behaviors. Let's first discuss form and examine habits after discussing woody plants' basic tissues.

Woody plants have adapted to ground-level herbaceous competition and herbivore damage by evolving perennial tissues that lift their buds well above the ground. Although woody plants include trees, shrubs, and woody vines, vines lie outside the scope of this book. In the botanical world, the rules are often riddled with exceptions. While there are no 100% accurate definitions of "trees" and "shrubs," most botanists consider trees to have one main elongated woody stem and a well-defined crown well above the ground. They usually mature at a height above 33 feet (10 m), but they may range from 10 feet (3 m) to upwards of 300 feet (91 m).

Shrubs typically produce multiple stems that originate belowground or at or very near ground level. Most shrubs are shorter than most trees, but here's where things get fuzzy. Single-stem woody plants topping out under 10 feet (3 m) don't really function as trees in the ecosystem, and multistemmed woody plants taller than 33 feet (10 m) don't really function as shrubs.

As you can see, these definitions combine both plant height and stem pattern. While there's still some disagreement about these definitions in the literature, here we'll use plant height and stem pattern as our two primary distinguishing criteria. Why does this matter?

A plant's form includes its characteristic patterns of sprouting, elongation, budding, and branching—its innate aboveground strategy of situating perennating buds and photosynthetic organs. These patterns can tell us what to expect from a plant's behavior. For example, a plant that normally branches close to the ground is unlikely to grow tall and unbranched, at least not without some effort, whereas a single-stemmed tree forming multiple stump sprouts after cutting is no longer a single-stemmed "tree." Each shoot will likely tend to individually express the plant's predetermined genetic pattern.

Of course, plants don't care what we call them. These categories can help us understand them, but only as fuzzy patterns, not rules. So, although a woody plant's form gives us some information about its size and growth pattern, understanding its habit offers even more insights into their patterns of growth and how that can best inform management. We discuss habit in greater detail later in this chapter. First, we'll break down the parts and processes that contribute to woody plant growth.

The crown, trunk, and roots all grow from the same kinds of tissues—meristems—together forming the whole organism. Understanding the basics of meristem growth and function helps us anticipate new growth, carefully prune and harvest, and appreciate the nature of new wood development.

Meristems: Tissues Capable of Division

"Meristem" derives from the Greek *meristos*, which means "divisible" or "able to divide into parts." These clusters of **undifferentiated plant cells** divide to form new cells that then differentiate to create different types of plant tissues. Meristems embody the very germ of plant life. Woody plants grow via two main types of meristems: **apical meristems** and **lateral meristems**.

Apical Meristems

Apical meristems typically lie at the *apex* of a shoot or root. They're the growing tips where woody plants explore and begin to occupy space. As an aboveground apical meristem grows, the cells it leaves behind differentiate to form stems, leaves, flowers, thorns, buds, and so on. Belowground apical meristems produce root tissues, root hairs, etc. So the growth of a single apical meristem creates the most fundamental organizational units of plant architecture, a **shoot** or root. Therefore, the crown develops from a network of individual shoots, each formed by a single apical meristem. The same is true for the roots: an apical meristem forms a single root, and the root system results from multiple linked roots. We call the types of new tissues created by apical meristem extension **primary growth**.

A single shoot consists of one apical meristem, the stem it has formed, and all of its other botanical structures—leaves, buds, flowers, fruits, spines, tendrils, etc. The places along the shoot where leaves attach are called **nodes**. We call the lengths of stem between nodes **internodes**. As a shoot grows, the apical meristem's undifferentiated cells divide and differentiate into miniature sets of leaves, flowers, tendrils, thorns, etc. Each apical meristem typically leaves these nodes behind one by one as it grows, each fully developing its leaves and other structures. Meanwhile, over the course of hours or days, the internodes also lengthen, extending the shoots into new space. At the core of each shoot, we find the **pith**—the trace that remains following apical meristem division. Encircling the pith is primary xylem (wood), lateral meristem (cambium), primary phloem (more to come on these), and lastly, epidermis, or first-year outer bark.

A **bud** is an undeveloped embryonic shoot that includes differentiated parts such as roots, shoots, leaves, or flowers and is usually covered with modified leaves in the form of scales. We can characterize bud types in several ways—based on their location, the timing of their formation, and their function. If we solely consider bud location, woody plants form two main types during undisturbed growth—**terminal buds** and **lateral buds** (which also may eventually include **epicormic** buds that maintain a direct connection to the pith but lie along the stem and are usually triggered by some form of disturbance).

Functionally speaking, we can organize buds into three main categories: "just regular old buds" that form at the apices of shoots and

FIGURE 2.2: Axillary (lateral) and terminal buds on a first-year sweet chestnut (*Castanea sativa*) coppice sprout.

in leaf axils, **disturbance anticipation buds** (or preventitious buds, see glossary entry), and "de novo" buds (adventitious). During undisturbed growth, the regular old bud types include terminal buds, axillary buds, and additional root-shoot buds. Here we define a "terminal bud" as a dormant apical meristem at a shoot's tip that will extend the parent shoot's growth along the same axis once growth resumes. Terminal buds typically contain **preformed** structures—that is, all the physical structures that will be expressed during the next phase of elongation. They can overwinter and resume growth the following year, or in some species, they can form as a pause between two or more growth flushes in the same growing season.

Axillary buds (often called lateral buds) form at the leaf **axils**—the nodes where leaves attach to the stem—along a shoot's axis as it extends. They're capable of forming a new shoot and helping the plant occupy even more space. Axillary buds tend to follow unique growth patterns for different species. Some may begin growing immediately after forming (called **sylleptic** growth, most common in tropical species), while some remain dormant until the following season (**proleptic** growth). They can also become a **latent** or **dormant bud** that remains viable, yet grows just enough to remain embedded in the bark. Most of these buds appear to only remain viable for a few years.

Additional root-shoot buds give rise to the growth and development of root suckers. These buds form on and within healthy, uninjured roots and generate new vegetative shoots that are clonal, often in response to a wound or other disturbance. Root-shoot buds' growth usually keep pace with root cambial expansion and retain a bud trace that extends to the center of the root. Among woody plants, root

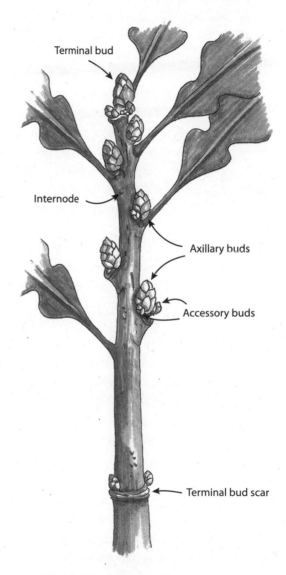

FIGURE 2.3: An oak twig showing the location and relative scale of axillary, accessory, and terminal buds.

suckering is relatively rare, although a number of commonly occurring species possess this capability. Some of the better-known species include aspens/poplars, sweetgum, American beech, several *Prunus* species including the wild plum (*Prunus americana*) and choke cherry (*Prunus virginiana*), black locust, sumac species, *Alnus incana*, and several *Viburnum* species.

Initially all these new tissues are green and tender, but as the season progresses, they lignify and harden to fend off herbivores and harsh weather. After their initial flush of elongation, shoots can only grow in length through continued apical meristem growth. As the growing season comes to an end, growth processes stop, and the plant prepares to go quiescent, if not fully dormant.

Later in this chapter, we'll discuss the two main types of buds that contribute to resprouting: "disturbance anticipation" or preventitious and "de novo" or adventitious buds.

Lateral Meristems: Cambium

The **vascular cambium** and **cork cambium** are woody plants' two types of lateral meristems. (We'll discuss the cork cambium in the next section.) Sandwiched between the bark and wood (or pith in a young shoot), the vascular cambium is a thin cylindrical layer of living undifferentiated cells. Lateral meristem cells divide inwards, forming new wood (xylem), and outwards, forming inner bark (phloem). This **secondary thickening** increases the stem's diameter, forming wood. This means that once formed, wood cells remain stationary, as the stem increases in thickness.

Because lateral meristems are undifferentiated cells, they can sometimes differentiate to form new shoots or roots. This plays into woody plant management via "de novo" or adventitious buds later in our discussion.

So, apical meristems extend plant shoots and roots outwards into space during primary growth, creating all the cells the plant needs to function, while lateral meristems then thicken the shoots and roots, strengthening the plant structurally and creating new sapwood to store energy and conduct water, minerals, photosynthates, and hormones throughout the plant.

Wood and Bark: Tissues of the Trunk

For many of us who work with woody plants, we are most concerned about the stem or trunk. The stem is where the bulk of wood lies, forming an energy-dense carbon store that we can cut and dry for fuel or convert into any number of useful craft and building products. Understanding the cross-sectional structure of trees and shrubs helps inform us as woodworkers and tree stewards. Let's take a more detailed look at that structure, examining the components of wood and bark, working our way from the outside in. Follow along with Figure 2.4 to see where and how these parts relate to one another.

Bark: Outward Growth of the Cambium

Bark surrounds woody plants' circumference, serving as the first line of protection against damage from abrasion, weather, fire, and insect pests, bacteria, and fungi. From the outside

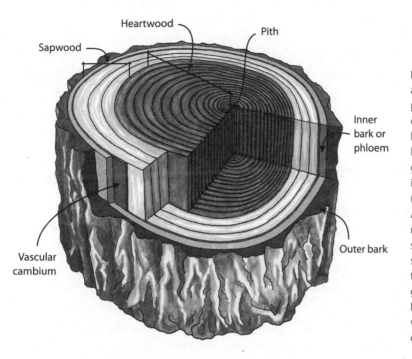

Sapwood
Heartwood
Pith
Inner bark or phloem
Vascular cambium
Outer bark

FIGURE 2.4: Cutaway cross-section of a 21-year-old black locust tree. A few patterns worth noting: the distribution of heartwood and sapwood. Once it has begun forming heartwood, black locust only tends to maintain 2 or 3 growth rings of sapwood. Also note the increased rate of growth in recent years (thicker annual growth ring increments). And last, the tighter growth rings on the right side of the log and off-center pith seem to suggest the tree developed some reaction wood that often occurs on trees growing on hills. The tighter wood grain on the downhill side of the tree helps stabilize it against gravity. Reaction wood can lead to irregular warping and checking during drying.

in, bark is composed of the periderm (which includes cork, **phellogen** or "cork cambium," and in some cases, the phelloderm), cortex, and phloem (or inner bark), which is formed by the outward division of the vascular cambium.

The periderm's outermost cork cells contain waxes and long fatty acids called **suberin**, the compound that makes bark "corky." Very few organisms are capable of breaking down cork, which is impermeable to gases and water. Next, the cork cambium or phellogen is meristematic tissue that behaves similarly to the vascular cambium, dividing outwards to form new cork cells. On some species, the cork cambium also divides inwards to form the phelloderm. Together these three layers comprise woody plants' outer bark or periderm.[6]

As we discussed earlier, the vascular cambium forms phloem (or inner bark) as it divides outwards, and the phloem serves as the transport network for photosynthesized sugars from the leaves, downwards through the stem, to the roots and even the surrounding soil biology.

Though often ignored by modern wood markets, bark has many uses in craft and industry. Up to 40% lignin, it is a tough, durable material. Many cultures have used bark for waterproofing roof membranes, including birch and ash bark in a shingle-type fashion in Scandinavia and North America, respectively. Tannins are a key constituent of bark, making the bark of some species, like oaks, highly valued in the tanning industry. And of course, cork for wine stoppers and all things cork all come from the rotational harvest of living cork oak (*Quercus suber*) tree bark in Spain and Portugal.

The inner bark (phloem) of several species can also be valuable craft material for weaving, cordage, and even edible purposes. In the eastern United States, hickory (inner) bark offers a fantastically durable material for traditional woven chair seats, with reports of seats lasting six decades or more. In other areas, elm has been used for the same purpose. *Tilia* spp. inner bark is a preferred cordage-making material. For these sorts of applications, the best time to harvest stems is during the period of active sap flow in mid-late spring, when bark can be literally peeled from some species like a banana.

As we continue our anatomical journey inwards, we encounter the other product of vascular cambium division—wood.

Wood: Inward Growth of the Cambium

The bulk of woody plants' stems are composed of **wood**. An orderly arrangement of axial (vertical) and radial cells, wood is composed of 40% to 50% cellulose, 20% to 35% hemicellulose (a diverse group of cellulose-like compounds), and 15% to 35% lignin, as well as secondary components including tannins, resins, and oils. Conifers typically have more lignin and less cellulose and hemicellulose than broadleaf species.[7]

Cellulose, the product of solar energy trapped in a molecule made from carbon dioxide and water, is the most abundant natural material in the world.[8] All plants are built of cellulose. Lignin, nature's second most abundant

material, makes stems woody and can only be broken down by a few organisms. Together, these substances provide trees and shrubs with their major survival feature—the remarkable mechanical support that enables them to stretch skyward and optimize vertical space and capture incoming solar energy.[9]

Three Types of Wood Cells: Vessels, Fibers, and Rays

Broadleaf species' wood is composed of three types of specialized cells: water-conducting **vessels** (~30% of total volume), strength-providing small-diameter **fibers** (~50% total volume), and **rays** (~20%). Most of these cells have an elongated vertical form, though their ratio of length to diameter varies widely.

Often easily visible when exposed along a cut surface, vessels provide trees with conduction and support. We call their exposed ends **pores**. Though vessels start life as an end-to-end series of vertical cells no more than 0.04 inches (1 mm) long, their end walls largely disappear as they die. This creates continuous sap-conducting channels that stretch anywhere from an inch or so long (2.5 cm) to the entire length of the tree. While generally fairly large in diameter, vessel size may vary widely, even within the same species.

The long, narrow, tightly bound, closed-ended, thick-walled fiber cells are smallest in diameter and poorly suited for conduction. They impart strength to the wood, forming a matrix that holds the vessels.[10]

Ray cells follow a horizontal axis and radiate outwards from the pith through the cambium and into the phloem. Rays connect xylem and phloem into a large circuit and offer horizontal sap conduction and temporary carbohydrate stores. When trees produce more food than can be immediately used, living ray cells in wood and bark provide a major location for storage. This also occurs during autumn months as trees absorb nitrogen and other minerals from the soon-to-be shed leaves. Rays may possess gum canals packed with resins, latex, and other protective compounds as well as oils that impart unique smell and taste. They also help stabilize wood's cellular structure, minimizing internal crack formation caused by sudden shrinkage or expansion.[11]

Rays resemble flattened ribbons of cells with their thin plane oriented vertically. Ray size varies considerably in broadleaf species, usually stretching many cells in width, though in some broadleaf species (like poplar/aspens, *Populus* spp., and in all conifers), rays may be just a single cell wide. In contrast, white oak's (*Quercus alba*) rays may stretch 4 inches (10 cm) high or more.[12]

Large well-developed rays also have important woodworking implications. Pronounced rays create planes of structural weakness. Green woodworkers and firewood processors can exploit this characteristic, as species with distinct rays often split cleanly and reliably. But by the same token, rays' natural cleavage plane may also encourage checking as the wood dries.

Rings: Earlywood and Latewood

So, as secondary thickening occurs, the cambium divides inwards, forming a ring of new wood around the existing wood cells. These **growth rings** radiate concentrically outwards from the pith, each typically the result of one season's wood growth—at least in temperate zones.[13]

Growth rings depict northern latitudes' annual seasonal cycles of growth and dormancy, their thickness reflecting each season's growing conditions. In parts of the tropics, warmth may support year-round growth, and trees produce wood with no growth rings. In fact, half of all woody species in the Amazon Basin and three-quarters of those in India do not produce growth rings.[14] In areas with irregular precipitation patterns, seasonal rainfall may drive growth ring development. And if fire, flooding, defoliation, drought, frost, etc. interrupt annual growth patterns, some trees may form a second or false growth ring during the same season, marking the point of resumed growth.

Different species' growth rings feature unique cell structures with varying diameters, cell-wall thicknesses, and cell distribution. For many temperate species, wood grown early in the season differs from that grown later. The large, thin-walled, spring-produced **earlywood** vessel cells can carry a large volume of water and stored root energy, enabling plants to grow vigorously during spring and early summer. Following sufficient earlywood development, the plant can invest in producing strong structural wood. These **latewood** cells have thicker walls and a higher density.[15] When plants go dormant in autumn, all growth ceases, ending that season's growth ring development.

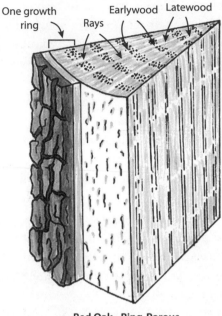

One growth ring   Rays   Earlywood   Latewood

One growth ring

Red Oak - Ring-Porous

Quaking Aspen - Diffuse-Porous

FIGURE 2.5: A look at the difference in growth ring pattern and structure between ring- and diffuse-porous woods (uneven- and even-grained, respectively). The red oak on the left is a great example of a ring-porous hardwood species. Note the large-diameter pores in the earlywood (or spring "produced" wood) compared to the denser grain of the latewood (summer wood). Oak also frequently exhibits prominent medullary rays along the radial surface. Contrast the oak's grain pattern with the diffuse-porous aspen. There's much more consistency throughout the growing season. We'll also explore this characteristic in relation to their annual growth patterns later in the chapter.

In some species, the shift between growth rings' earlywood and latewood is sudden, while in others it is more gradual. This reflects changes in the plant's physiology, which we will get to later. However, the nature of these wood cells plays a key factor determining its usefulness and character for craft and industry.

Ring-porous and Diffuse-porous Woods

Scientists classify broadleaf species as either **ring-porous** or **diffuse-porous** based on the size, number, and distribution of pores and vessels within the wood. Each growth ring formed by a ring-porous species contains two layers with a sharp visible contrast in cell structure between early and latewood. Some ring-porous species include oak, hickory, ash, black locust, elm, chestnut, and catalpa. Diffuse-porous woods like cherry, maple, birch, and beech have many relatively small pores with very little difference in cell size and structure throughout each growth ring.[16]

As usual, we find some grey between these two extremes. A few species, including butternut and black walnut, contain large earlywood pores and smaller latewood pores but no distinct break in the patterning. They are called **semi-ring** or **semi-diffuse porous** woods.

The nature of a species' growth rings also influences its utility. Woodworkers describe diffuse-porous wood as **even-grained**, in contrast to the **uneven-grain** of ring-porous hardwoods. The consistency of diffuse-porous hardwoods' grain pattern makes them excellent for carving and turning. And ring-porous hardwoods' sharp boundary between earlywood and latewood creates distinct planes of weakness that make them particularly well suited to "riven" crafts where raw materials are processed by splitting along the grain. And at the individual level, each tree's growth rings influence strength and structural properties.

Growing conditions affect growth rings' size and structure, and this affects wood strength. Among diffuse-porous hardwoods and conifers, slow growth results in greater overall density and stronger wood. But somewhat counterintuitively, the opposite is true for ring-porous hardwoods. Because slow-grown trees produce a smaller proportion of latewood, their wood is less dense overall. Thus, fast-grown ring-porous hardwoods are stronger and denser than a slow-grown member of the same species. So, this means a fast-grown red oak (a ring-porous species) contains wood that is more dense than a slow-grown red oak, but the opposite is true for diffuse-porous species like sugar maple, where the tight growth rings that develop on slow-grown trees results in denser wood than a fast-grown tree of the same species. And because coppice wood often grows quite quickly, coppicing ring-porous species will tend to produce stronger, more dense wood overall.[17]

Sapwood and Heartwood

Woody plants' youngest xylem rings are the most actively involved in upwards sap flow. We call these tissues **sapwood**. In young trees and shrubs, the entire stem is often sapwood, but as they mature, the entire trunk is no longer necessary to fulfill the branches, twigs, and leaves' sap needs. At this point, the wood near the center of some species' stems begins to form **extractives** in the cell walls, converting sapwood into **heartwood**. Extractives are toxic to decay fungi and often reduce the wood's permeability, helping resist rot. Some species also form bubble-like structures called **tyloses** in their cell cavities. Tyloses impede liquid transmission and also often confer some rot resistance. Tyloses' presence varies from virtually nonexistent (red oaks), to unevenly distributed (chestnut, hickory), to numerous (white oak), to dense (osage orange, black locust).[18] Heartwood often develops a darker color than sapwood, but in some species, there is no distinct color change, though the heartwood is still there. And some species do not appear to form heartwood. Sometimes changes in wood color are instead the result of wounding or fungal decay.

Heartwood contains no living cells, requiring no energetic inputs and providing little to no structural support to the tree.[19] Heartwood often does not follow clean divisions between growth rings, instead meandering across them. Generally, sapwood layers are thicker towards

Table 2.1: Number of sapwood rings of some hardwoods

Latin Name	Common Name	Average Number of Sapwood Rings
Catalpa speciosa	Northern catalpa	1–2
Robinia pseudoacacia	Black locust	2–3
Castanea dentata	American chestnut	3–4
Gleditsia triacanthos	Honey locust	10–12
Juglans nigra	Black walnut	10–20
Fagus grandifolia	American beech	20–30
Acer saccharum	Sugar maple	30–40
Acer saccharinum	Silver maple	40–50
Nyssa sylvatica	Black tupelo	80–100

Adapted from Hoadley, 2000, p. 12.

Table 2.2: Select woody species with good to excellent decay resistance

Bald Cypress
Catalpa
Chestnut
Black Locust
Red Mulberry
Mesquite
Oak spp.
Osage Orange

tree crowns, thinning towards the base.[20] Once trees begin developing heartwood, each incremental increase in sapwood brings an increase in heartwood. When trees suffer severe wounds, they often stall heartwood formation, maintaining more living sapwood to ensure they possess enough energy storage capacity to support new growth.[21]

Different species tend to retain a characteristic number of sapwood rings throughout their mature life. Osage orange (*Maclura pomifera*) and black locust (*Robinia pseudoacacia*) have two of the highest ratios of heartwood to sapwood, with just 1 or 2 and 3 or 4 sapwood growth rings per stem, respectively. In fact, several ring-porous hardwood species maintain just a few sapwood growth rings, which makes sense if you recall that they typically only use the outermost ring to conduct water. Why maintain any more living wood than necessary?[22]

So why do we care so much about heartwood and sapwood? First, due to its lack of extractives, sapwood has little, if any, decay resistance. For outdoor applications, especially those where wood will be buried or in contact with the soil, we particularly value the superior decay resistance of woods with high heartwood to sapwood ratios. Also, although less relevant to coppice poles, heartwood's rich color is often sought after for furniture and flooring applications.

Whew! You've made it! We've examined woody plants' organs, their overall forms, germinal tissues, and the cells, structure, and patterns in wood. With this foundation set, let's explore the way woody plants grow and produce new shoots and wood.

TREE GROWTH, DEVELOPMENT, LIFE SPAN, AND DECAY

The growth of woody plants is intimately linked with their form and structure.[23]

— Stephen Pallardy

Woody plants' growth is so intimately linked with their form and structure that it is nearly impossible to appreciate their architecture without understanding their growth processes. Ironically, these very same growth processes limit their life spans, leading them towards **senescence**. Understanding the energetic dynamics that lead to senescence helps illustrate why coppicing actually prolongs woody plants' lives—if we harvest them properly.

We begin with an overview of woody plants' generalized growth pattern and habits. We'll then discuss seasonal growth patterns, energy dynamics, life span, and plant decay.

Generalized Growth Pattern: Nested Cones

Until now, we've explored numerous facets of woody plant growth, but we've yet to discuss the organization and expansion of the organism's collective whole. We've now got a solid understanding of the tissues and organs that comprise woody plants, the ways they expand and colonize new space, and the structure and patterns of the wood they create. But how does this all manifest in the plant's growth and development?

Kozlowski describes woody plants' growth patterns using the compelling model of "nested cones." Like a branchy expression of a set of Russian dolls, each year woody plants'

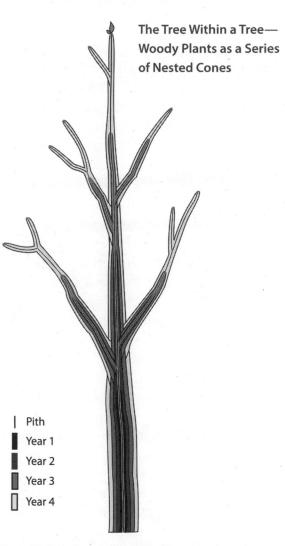

The Tree Within a Tree— Woody Plants as a Series of Nested Cones

| Pith
▮ Year 1
▮ Year 2
▮ Year 3
▯ Year 4

FIGURE 2.6: Woody plants expand outwards and upwards via apical meristems (buds), and they also add a concentric layer of new wood around their entirety each season as the vascular cambium divides and expands. Here we see the expansion of a sapling in height, spread, and thickness. Note that branches always maintain a connection to the pith (save a few exceptions). As the cambium expands, branches are "enclosed" within the wood, becoming knots. Also, note that the branch wood and stem wood aren't actually contiguous. More on that later. Adapted from Kozlowski, 1971.

secondary growth forms a new layer of wood surrounding the entire organism. With this in mind, we can see how their early architecture becomes "frozen in space," as it's enveloped by a newly formed layer of wood each season. Thus, a branch that forms after two years' growth will always maintain its connection to the pith and its original location along the stem. And as long as the plant continues to grow, each year it forms a new layer of wood around that branch, the stem, and the root system. This pattern ensures the annual refreshment of xylem cells for water and mineral transport along with the opportunity to add healthy undamaged sapwood outside older wood layers.

In other words, whatever happens to a woody plant during its early life remains a part of that plant throughout its life cycle (unless it's removed by coppicing). This means that any management activities we undertake will have long-term effects on the health of the stool or pollard. Understanding these implications, what more can we learn about woody plants' growth habit to ensure that our management always works to improve plant health (or at least avoids long-term damage)?

Woody Plant Growth Habits: Above- and Belowground Plant Growth Patterns

Adrian Bell once wrote, "Plants do not grow in a haphazard way but in an organized flexible manner controlled by internal and environmental factors."[24] In this book, we call these patterns their growth habit. Plant habit concerns questions like: How often and in what patterns does the main stem branch, if at all?

What shape does the root system tend to take? Does the plant spread vegetatively by rhizomes or stolons? We'll start by looking at the "organized" parts of plant growth and also try to account for the "flexible" part.

The aspects of habit we focus on here include branching pattern, determinate and indeterminate shoot growth, monopodial and sympodial growth, and root patterns. With a clear understanding of the patterns, we can observe these aboveground growth habits relatively easily. Not so much for belowground habits. While some species follow very consistent growth habits, some exhibit great plasticity and variation. The literature rarely states which species are which.

An individual plant's life history will also affect its habits. Has browse, fire, or ice damaged it? Have humans managed it? Resprout management likely alters woody plants' habit in ways we don't fully understand. We have much to learn about how our management affects plant habits. Nonetheless, appreciating these habits can help us choose which species may be most useful for our needs and also inform several detailed management decisions.

Branching Patterns: Excurrent, Dendritic, and Deliquescent Habits

Woody plants with an **excurrent** habit have a single central stem with small lateral branches. Think spruce trees. This often creates a conical or pyramidal crown. Most conifers have an excurrent habit, as well as some broadleaf species like tulip poplar (*Liriodendron tulipifera*) and catalpa (*Catalpa speciosa*).

Meanwhile, the main stem of most broad-leaf species divides fractally into branches that divide further into smaller branches. This **dendritic** form results in a more rounded, spreading crown, though their trunk may stretch unbranched for some distance.

Plants whose stems divide rapidly and frequently are termed **deliquescent**, meaning the stems quickly diverge into many diffuse, finely divided branches with no main axis or stem. Deliquescent habits occur more frequently among shrubs than trees (e.g., *Hydrangea* spp.). However, there are no hard boundaries between these three habits. Instead, think of excurrent and the deliquescent habits as two ends of a spectrum, with the dendritic habit sandwiched between them.

Excurrent species tend to most easily yield the longest straightest stems. Dendritic woodies may also produce long straight shoots, perhaps even with fewer knots than excurrent species, especially if grown close together. Deliquescent forms probably have less use for craft wood production, though this habit may be great for producing fodder, leafy edibles, medicinals, or ornamental woody cuts. Unfortunately, despite the utility of this information for resprout management, there are few sources that characterize woody species' branching habit.

Woody plants' habit is strongly influenced by growth regulators emanating from a shoot's apical meristem. During the active growth at a twig's tip, hormones suppress the growth of the shoot's axillary buds, a process called **apical**

Woody Plant's Growth Habits—Excurrent, Decurrent/Dendritic, Deliquescent

FIGURE 2.7: Woody plants' habits span a range from excurrent (left)—a strong central leader with smaller lateral branches, to decurrent/dendritic (center) that divide fractally into smaller increments, and deliquescent (right) characterized by frequent, rapid branching and the lack of a primary growth axis.

dominance. The plant's highest shoot regulates the growth of subdominant shoots, affecting its overall shape through **apical control**.[25, 26] Both of these patterns tend to be stronger in excurrent species and weaker in dendritic and deliquescent species especially. We'll explore the implications of apical dominance in greater detail later in the chapter, but keep in mind that the hormones responsible for apical dominance play a crucial role in regulating resprouting following damage to their lead shoot. In other words, apical dominance is a key to unlock woody plants' response to pruning and coppicing.

Furthermore, plant age and growing conditions influence branching pattern. For example, during periods of rapid growth, red alder (*Alnus rubra*) has a strongly excurrent habit, but as it reaches maturity, it becomes "moderately to strongly deliquescent."[27] Woody plants grown in sunny, open field environments tend to branch more frequently, while those in tight quarters will grow longer stems with fewer side branches as they grow upwards to compete for light. To grow straight rods in coppice systems, tight stool spacing therefore makes sense, while maintaining pollards for fodder production at a wider spacing fosters leafier, twiggier, and branchier growth.

To Terminate or Not to Terminate, That's Our Question: Determinate and Indeterminate Shoots

When woody plants' apical meristems grow, they develop into shoots forming a stem, leaves, and flowers, etc., continuing to expand from the shoot's growing tip. As these meristems prepare for dormancy, they form buds that contain the miniaturized stems, leaves, and axillary buds they'll produce when they resume growth. All these preformed structures lie nested together inside the bud just biding their time! We describe plant growth as either **determinate** or **indeterminate** based on the length of the apical meristems' growth period following dormancy.

In shoots with determinate growth, primary shoot growth ends after the bud's preformed structures fully express themselves in a growth flush. If the bud contains eight preformed leaves, apical growth on that shoot ceases once the leaves have unfurled and the internodes have fully elongated. We sometimes also call this "flushed" growth. Occasionally the growth flush ends with the apical meristem turning into a flower, tendril, spine, or other structure. Sometimes the apical meristem simply aborts. For many species, the apical meristem forms a new **terminal bud** with another set of preformed structures for the next growth flush. If so, it goes dormant, awaiting its time and turn. So, a determinate shoot's growth is "determined" by the number of leaf primordia inside the bud that formed during the end of the previous growth flush.

In temperate zones, determinate growth usually occurs during the spring and early summer before ceasing until the following spring. However, some buds may have very short dormant periods, creating two or more growth flushes in a single season. This occurs more frequently in species from warmer climates.

Additionally, the number of preformed structures inside a bud is actually determined by the growing conditions *during the year the bud forms*. So, drought stress in one season may lead to fewer preformed leaves, reducing determinate shoots' growth potential the following year. Also, in tree hay applications, heavily pruning a determinate tree during the growing season may affect the amount of energy available to make preformed leaves in the end-of-year buds, reducing the next season's growth.

In shoots with indeterminate or sustained growth, the apical meristem can grow indefinitely, as long as conditions allow. Apical buds may or may not contain preformed structures, but even after any preformed leaves and internodes fully express themselves, the apical meristem continues to produce new leaves and stem, conditions permitting. As favorable conditions wane, the apical meristem may either abort or go dormant and form a terminal bud.

While hopefully this all seems reasonably clear, things of course get more complex. First, a given species may have both determinate and indeterminate shoots. Terminal shoots may exhibit indeterminate growth while lateral shoots are determinate. They may vary between species within a genus, or even between cultivars of a species.

Interestingly, there's a strong correlation between these growth patterns and the type of wood a species produces. Ring-porous species tend to have determinate (or flushed) growth. This limited pulse of early-season growth correlates with the large earlywood pores that meet the needs of a flush of new leaves. As the

production of new leaves ends and the need for increased sap flow declines, the new wood produced starts to become denser. This makes even more sense given that ring-porous species concentrate water movement through the most recently formed growth ring.

In contrast, species with indeterminate (or sustained) growth would have more of a steady increase in water needs throughout the season as more and more leaves are produced. This correlates with diffuse-porous wood's consistent pore size and the fact that diffuse-porous species spread water uptake across a number of rings.

These two primary growth patterns also relate to other qualities of a species' life history, offering different adaptive advantages under different conditions. Indeterminate plants will often tend to grow more rapidly than determinate plants. When conditions are good, indeterminate species may grow for much or all of a growing season, occupying more space and capturing more available light. Many early- to mid-successional shade-intolerant species have indeterminate growth, which makes sense given their need to rapidly colonize disturbed sites. In fact, the two highest yielding genera in short-rotation forestry biomass systems—willows and poplars—are indeterminate growers. Indeterminate plants therefore may better be able to respond to disturbances by quickly putting on new growth and taking up space.

On the other hand, longer-lived, shade-tolerant late-successional species tend to exhibit determinate growth. Since they stop putting on new growth earlier in the season, they have more opportunity to store the photosynthates produced in late summer and fall for later. They may also use some of that energy to produce lignin and other compounds that build resistance to herbivory or strengthen their tissues for a long life span. Therefore, they may be more resilient than indeterminate plants over the long term.

Indeterminate and determinate shoots will probably respond differently to pruning or harvesting during the growing season. Kays and Canham studied sprout response and root starch storage of four species of North

Table 2.3: Woody species with determinate/ indeterminate shoot growth

Species	Shoot Growth	Branching
American beech	Determinate	Sympodial
Ash	Determinate	Monopodial
Birch	Indeterminate	Sympodial
Black cherry	Determinate	Monopodial
Cottonwood	Indeterminate	Sympodial
Elm	Indeterminate	Sympodial
Hickories	Determinate	Monopodial
Oaks	Determinate	Monopodial
Red maple	Both	Monopodial
Sugar maple	Both	Monopodial
Sweetgum	Both	Monopodial
Sycamore	Determinate	Sympodial
Tulip poplar	Indeterminate	Monopodial
Willow	Indeterminate	Sympodial

Adapted from Franklin and Mercker, 2009, p. 9.

American hardwood saplings in southeastern New York state in relation to timing of cutting, based in part on their growth habits.[28] They cut a number of saplings of each species throughout the year and studied their root starch reserves. They found that determinate species (white ash, *Fraxinus americana*; and black cherry, *Prunus serotina*) cut after August 20 or September 9, respectively, had virtually the same root starch reserves at autumn leaf drop as the uncut controls. In contrast, indeterminate species' (gray birch, *Betula populifolia*; and red maple, *Acer rubrum*) root starch reserves never reached the levels of uncut controls the year after cutting. Gray birch didn't even get close until October 1, whereas by that time red maple only achieved about half the reserves of uncut controls. These species' shoots kept growing until the end of the season, so they had less available excess energy to store. In addition, they were affected by shading far more than the determinate species in terms of total shoot biomass and height three years after cutting.

Based on this study, it appears that sprouting vigor in the two to three years following cutting depends more on starch reserves than growing season photosynthesis, and the ability to stop shoot growth and store starch significantly increases sprout growth, particularly in shadier conditions. But the data also indicates that, with longer intervals between cutting, indeterminate species' growth rates catch up to determinate species.

Determinate and indeterminate growth habits clearly affect woody plants' ability to respond to cutting. This could be critical for plant management, at least in the short run, or for short rotations. Unfortunately, despite its importance, species-by-species data about the determinate or indeterminate nature of shoot growth appears relatively rarely in the literature, so we must often infer it based on wood type, successional role, or field observations.

One Foot or Many Feet: Monopodial and Sympodial Growth

Plant stems or branches can grow in one of two ways: **monopodial** ("one foot") or **sympodial** ("together with feet"). In plants with monopodial growth, the terminal bud maintains apical dominance, forming a central leader that arises from a single base. Shoots on species with monopodial growth end the growing season by producing a new terminal bud that continues shoot growth the following season. These types of species tend to express a pyramid-like shape.

In sympodial growth, shoot elongation usually continues from a lateral bud, once the terminal bud has ceased growth. This forms a new child shoot that arises from the previous parent. Therefore, sympodial plant architecture develops as a collection of many shoots forming one upon another, while monopodial plants tend to continually elongate preexisting shoots or axes. So monopodial growth tends to form excurrent or dendritic growth habits, while sympodial growth leads to deliquescent growth.

And once again, it can get even more complex. Monopodial and sympodial growth patterns may both be either determinate or indeterminate. And some plants even exhibit monopodial *and* sympodial growth, with a monopodial main stem and sympodial lateral branches. For example, red alder's vegetative shoots are mostly monopodial, but the flowering shoots that terminate with a flower are sympodial. These shoots can only extend from lateral buds.[29]

Willows usually grow sympodially. They also happen to be indeterminate, so an apical meristem will grow until the season's end. As autumn arrives, the apical meristem aborts and the current year's shoot tip dies back to a point on the shoot above a lateral bud that hardened off enough to withstand winter. The following spring, that lateral bud becomes the new apex. While it follows nearly the same axis as the parent shoot, it is in fact a child of that parent growing from a different base.

So, we can see that for products requiring long straight shoots, plants with monopodial growth will tend to be best. Species with sympodial growth will have a greater likelihood of crooked branches, but sympodial growth can still grow reasonably straight and can even work well for crafts that require riving or bending.

Root Patterns: Tap-, Heart-, and Flat-rooted

Woody plant root systems largely colonize the upper 1 to 2 feet of soil. In fact, an estimated 60% to 80% of all plant roots exist in this shallow mantle along the Earth's surface. Generally, roots more or less follow one of three main patterns: flat-rooted, heart-rooted, and tap-rooted. We can infer much about these root architectural patterns from their names.

Flat-rooted species like cottonwood, silver maple, birch, and spruce have a lateral spreading form and lack clear primary roots. Heart-rooted species extend several primary roots laterally, which in turn support secondary roots that further branch downwards, creating a heart-like form. Tap-rooted plants send a vigorous primary root-shoot downwards that also supports developing lateral roots. The taproot anchors the plant deep in the soil, helping mine water and minerals from deeper soil horizons. In some cases, tap-rooted plants develop more fibrous root systems as they mature, but continue to benefit from their established taproot. Many nut tree species are tap rooted.[30]

Like most of the habit patterns we've discussed, many species do not follow hard and fast rules and may express aspects of each of these primary forms. While we have limited information on the rooting patterns of many woody species, this knowledge can inform how we place plant species in relationship with others. For example, tap-rooted standard nut trees would seem a good fit in a coppice stand of flat- or heart-rooted species.

Thicketers and Colonizers: Root Suckers and Rhizomes

Some species (poplar/aspen, black locust, sweetgum, plum) colonize space as an expanding population of stems that all derive from the same individual. They often form thickets that extend outwards concentrically from the parent plant. These newly formed vertical stems originate either from root suckers or **rhizomes**. Suckers form as vertical shoots off a plant's root system, often in response to some type of injury or infection. Rhizomes, on the other hand, are specialized underground stems that form nodes and internodes capable of generating new shoots that can form their own root systems. While technically, managing these two different types of vegetative shoots isn't true coppicing, we can utilize an understanding of their growth and distribution to manage them in a very similar way. We'll discuss these traits in greater detail later in this chapter.

The Timing and Pattern of Seasonal Growth

In temperate climates, trees break dormancy in late winter or early spring even before the snow has melted. Growth usually begins as the fine absorbing roots start absorbing water and minerals and mobilizing stored starches. Next, growth shifts to the twig tips as the buds swell. Stored energy in the buds and twigs helps stimulate budbreak and develop new growth. Cambial cell division soon follows, relying mostly on the limited energy reserves from the previous season along with small amounts of newly produced photosynthates. New leaves and shoots also begin to form.[31]

Beginning at the twig tips, a wave of cambial activity ripples down twigs and branches towards the main stem, forming new xylem and phloem, thickening each twig, and continuing downwards along each branch. This wave of cambial expansion then reaches the trunk. It may be another several days or even a few weeks before cambial activity reaches the support roots.[32] Lastly, secondary growth commences along the plant's nonwoody roots.

The speed and pattern of early secondary growth varies between species. In ring-porous species, new wood development spreads downwards through the newly forming growth ring over a matter of days. Low concentrations of the growth hormone auxin in buds developing during early spring stimulates the rapid formation of new wood up to 2 to 3 weeks before bud break.[33] Earlywood develops from the top of the tree downwards, while latewood originates at the base of the tree and moves upwards. As a result, latewood bands are thicker towards tree bases, potentially disappearing completely towards trees' upper reaches.[34]

In diffuse-porous species, secondary growth occurs far more slowly, taking place over a few weeks and as long as 2 months in old trees. When growth begins, diffuse-porous species and conifers immediately produce phloem, making stored food available to growing points. Because existing sapwood is fully capable of conducting ample water until leaves reach full development, conifers and diffuse-porous species do not prioritize new wood growth.[35]

We want to harvest coppice stools and pollarded trees when the materials are best-suited to our needs and when we will cause plants the least harm. Understanding seasonal fluctuations in energy reserves is key to good stewardship. We'll discuss these patterns later in the chapter when we explore the effects of coppicing on woody plants.

The Branch/Trunk Intersection

Branches form as twigs' apical and lateral buds develop and expand. New wood tissues at the base of branches *do not actually enter the trunk*. They turn sharply downwards, forming a collar where they connect to the stem at the branch base. As branch tissues' growth rate slows early in the growing season, trunk tissues begin to grow more rapidly, forming their own collar that overlaps and envelops the branch collar. This overlapping structure creates a more flexible, durable union between trunk and branch. This connection gives branches the necessary resilience to resist wind shear and snow loading.[36]

At the same time, branch development creates a **knot** that extends from the pith to the point the branch emerges from the trunk. Structurally, knots are defects in wood and make machining operations more difficult. Because of this, it can be a very good idea to basal prune young trees to remove low branches and increase the proportion of knot-free stem wood. We'll discuss this practice, called "basal pruning" later on in the book. ☞

Branch and Stem Wood Interface

FIGURE 2.8: Artistic rendering of a longitudinal section of a bolt showing the interface between stem wood and branch wood (after Hartig, 1894, p. 251). We've exaggerated the overlap between the two layers to help illustrate the point. Branch wood (the gray wood that begins at the pith and extends outwards along the branch's axis) is actually surrounded and overlaid by stem wood, creating a strong but flexible union. Also note the collar located at the base of the branch, visible on the bolt's outer face. To make a careful clean pruning cut, follow this plane and avoid cutting into this branch collar.

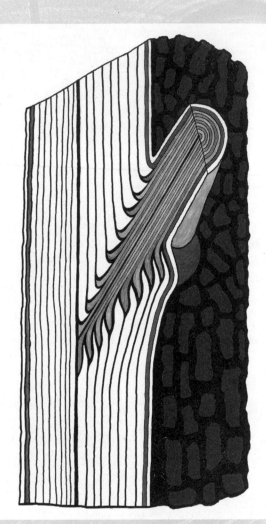

Stages of Knot Formation

FIGURE 2.9: Knots form as woody plants add new stem wood that encircles any persistent branches. On left, a broken off branch stub will continue to be subsumed by stem wood (center), until it becomes enclosed by new wood (right), sealing off the cut surface and thereafter continue to produce knot-free clear wood.

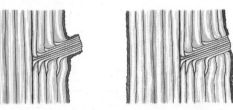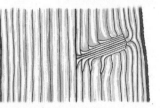

Tree Health and Life Span: An Energetic Balancing Act

The grower of an oak for timber will think about felling it as "mature" at 100 years; after 150 years it becomes a store of capital increasingly at risk from decay; if still standing at 200 years it may be cut down on the plea that it is "dangerous"; but if left alone it may live to the age of 400 or much more.

— Oliver Rackham[37]

Remember—trees evolved to most efficiently meet their needs given the resources available, not to satisfy the needs of foresters, lumberjacks, and craftsmen. As humans, our limited life spans make it remarkably difficult to appreciate time from a tree's perspective.

Unlike animals, woody plants' life cycles are not predetermined. We measure human aging as a ratio of cell breakdown (autolysis) to restoration. When cell restoration equals autolysis, we age slowly, but when autolysis exceeds restoration, the aging process accelerates. A 70-year-old human has experienced the autolysis-restoration process 270 billion times! Trees, on the other hand, cannot complete this process even once.[38]

Each season, woody plants must add a new layer of growth across their entire body through cambial expansion. They must also extend their twigs in order to grow new leaves every year. Annual secondary thickening is the product of the interplay between the crown's photosynthetic productivity and the roots' water- and nutrient-gathering actions. A woody plant's aging ratio compares its **dynamic mass** (the volume of wood capable of energy storage) with the wood that no longer stores energy.[39] This means a tree's health and vigor relates directly to the ratio of crown size to the volume of the trunk, branches, and root system. It needs to balance its "energetic bank account" by maintaining an active crown that's large enough to produce the carbohydrates needed to generate an entirely new layer of growth.

During youth, trees grow in both height and crown expanse. When their growth rate peaks and crown expansion slows dramatically, they enter middle age. But despite the near halt of crown development, the total wood "commitments" the tree must sustain continue to grow. At this point, the tree maintains a balanced energetic bank account, producing enough photosynthates to support each year's growth. With time, increasing commitments and declining crown vigor overburden the plant's ability to meet these energetic needs. Once energy costs exceed energy stored, the branches begin to die back, and the tree enters old age and begins to senesce.[40] Hence, aging is more a function of mass than time (See Figure 2.10).

Rate of Growth Inversely Related to Life Span

Trees that grow rapidly during their early development often reach old age more quickly. Conversely, when tree growth slows due to environmental limitations or pruning, their life span tends to increase because the crown can more easily maintain the slowed stem and root system development. The ancient, weathered, remarkably compact, and very slow-growing bristlecone pines (*Pinus longaeva*) of the

American Southwest offer a living testament. Some are 3,500 years old!

When we coppice or pollard a tree, we significantly reduce its energetic commitments. There's less wood left to encircle with a new layer of growth. This frees up energy to generate and maintain a robust and healthy crown capable of meeting its needs for growth. It essentially "resets" the ratio between the tree's "income" and "commitments," and its "age" now begins at the time it was most recently cut.

According to Oliver Rackham, managed coppice stools "are completely self-renewing and capable of living indefinitely as long as they are not overshadowed by timber trees. An old stool spreads, without loss of vigor, into a ring of living tissue with a hollow center and often an interrupted circumference."[41] In Suffolk, UK, coppice stools in Bradfield Woods up to 18.5 feet (5.6 m) in diameter and at least 1,000 years in age continue to yield good crops of poles.[42] These woodlands have been continuously coppiced since 1252. Assuming a 15-year rotation, these stools would have been cut at least 50 times.[43] So the act of coppicing or pollarding actually serves as a rejuvenating disturbance keeping the plant in a juvenile state and stimulating vigorous new growth.

Woody plants' annual energy cycles may also inform important management decisions. During dormancy, they store energy in their roots, stems, and branches, transferring it to their developing leaves, shoots, roots and stems, and later their seeds in spring and summer.[44] This is one of the reasons why it's best to cut coppice during dormancy, as we may rob the tree of as little energy as possible. While it may be unavoidable if harvesting fodder for livestock during the growing season, at least we're aware of the seasonal effects of harvest timing on energy dynamics.

These energy storage patterns direct woody plants' life processes and ultimately determine their annual growth and longevity. Coppicing and pollarding affects this energy storage. Understanding how woody plants respond to injury and how they've evolved to protect against decay and disease can help us become better stewards.

Hopefully by now you have a solid grasp of the nature of wood and the way woody plants grow, respond to injury, and senesce. So how is it that woody plants have come to develop sprouting responses? Why do they resprout and what function does it serve? How do they sprout, what factors cause them to sprout best, and what does sprouting do to them? What can we learn from these "natural" responses to injury to inform our work? Read on and find out.

SPROUTS AND RESPROUTS

Fortunately for us, the sciences of biology, horticulture, arboriculture, and silviculture have left us a detailed map of the patterns and processes of woody plant growth. But there are still gaps in that knowledge, and they're largely due to our biases as modern plant stewards. Our understanding of the origins of woody resprouts, the factors that influence it, best practices for encouraging strong healthy sprouts, and the reasons why this capability

The Stages of Tree Age and Health and Their Relation to the Coppice Cycle

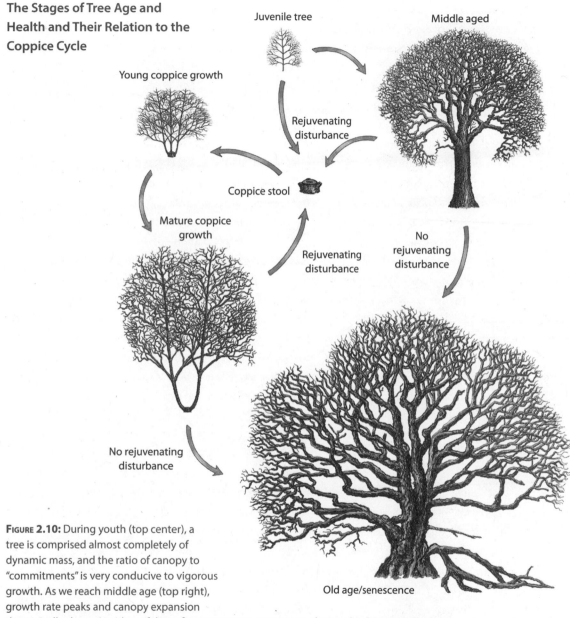

Juvenile tree

Middle aged

Young coppice growth

Rejuvenating disturbance

Coppice stool

Mature coppice growth

No rejuvenating disturbance

Rejuvenating disturbance

No rejuvenating disturbance

Old age/senescence

FIGURE 2.10: During youth (top center), a tree is comprised almost completely of dynamic mass, and the ratio of canopy to "commitments" is very conducive to vigorous growth. As we reach middle age (top right), growth rate peaks and canopy expansion dramatically slows. At either of these first two stages, some type of natural or human-induced disturbance will often rejuvenate aerial growth, effectively resetting it to a youthful growth stage. This is what we do when we coppice a plant. Lacking some form of rejuvenating disturbance, many species will steadily enter old age where the energetic costs of annual stem growth exceed the canopy's photosynthetic output. The tree begins to senesce, and will eventually die.

evolved in the first place have seen far less interest from the scientific community. In the coming pages, we'll explore these questions in detail so as to understand the what, why, when, and how of coppicing. We begin with the "why."

Woody Plants Do Not Heal, They Compartmentalize: Sharing Shigo's Insights

Alex Shigo served as chief scientist and project leader for the USDA Forest Service's "Pioneering Project on Discoloration and Decay in Forest Trees" from 1959 to 1985. A student of biology and plant pathology, he wrote and researched extensively on tree biology, physiology, and protection and defense systems. During his tenure, Shigo used a chain saw to longitudinally dissect over 15,000 trees, studying their growth patterns and learning how they respond to injury and infection.

Until this time, most knowledge of tree biology relied on cross-sections that only convey a partial picture of the way woody plants grow and develop. Shigo sought to understand their physiological processes and responses to injury so people could become better-informed and more conscious stewards. This knowledge should inform the way we prune, coppice, and pollard.

Shigo learned that woody plants do not heal wounds, they compartmentalize them. They evolved as highly compartmentalized organisms with many protection zones.[45] Decay does not develop freely. When injured or infected, their living cambium forms boundaries within the wood that confines rot, resisting defects' spread.[46] Wood cells form the basic compartment. Each growth ring is also a compartment, within which ray cells create even smaller compartments.[47] As a whole, this compartmentalized structure is an extremely effective protection mechanism.

Shigo developed a model he called CODIT, Compartmentalization of Decay in Trees, to describe the ways woody plants deal with injury (see Figure 2.11). His model consists of four metaphorical "walls" that resist decay within the growth rings present when a wound occurs. The first three walls form a continuous protection boundary called the "barrier zone" that chemically resists pathogens' vertical, inward, and lateral spread. The vertical wall is the weakest of the three, so rot tends to spread up and down the stem more than laterally or inwardly. Together, these walls form a protective column within the infected portion of the plant.[48]

The "barrier zone" forms the fourth wall of the CODIT model. It creates a strong chemical boundary separating infected existing wood from healthy, new wood formed after the wound occurrs. Circumferential cracks often form at this structurally weak interface between infected and noninfected growth rings.

So, compartmentalization of injured or infected tissues does not halt decay. It resists the spread of fungi and microorganisms to tissues located within the walled-off compartment.[49] This compartmentalization appears to create the conditions that cause discolored wood, a cosmetic defect that reduces sawlog value.[50] ☞

Compartmentalization requires considerable energy because it walls off sapwood that would otherwise store energy. Woody plants can rebuild this energy storage capacity as they continue to grow, as long as no additional damage occurs, but because defense reactions take energetic priority, recurring wounds weaken trees, increasing their vulnerability.[51]

When we prune, coppice, or pollard a woody plant, we create potential points for infection. Therefore, we should aim to cause as little harm as possible by understanding their biology and physiology. ☞

Pruning Cuts and the Four Walls of CODIT

FIGURE 2.11: This annotated figure from *A New Tree Biology* (Figure 2–32), shows the four walls of CODIT on a red maple bolt along with the effect of a poorly made pruning cut. The white arrow points to the pruning cut that extended well below the branch collar. Compare the extent of the discoloration as a result of the wounds on each side of the bolt. The discolored patch of wood is far larger on the right, despite the wounds occurring at the same point in the tree's life cycle. Removing the collar's natural protection zone with a poor pruning cut led to much more extensive damage.

As for the CODIT walls, pointer 1 indicates wall 1, the weakest wall, which resists pathogens' vertical advance, while pointer 2 shows the second wall, which resists pathogens' inward spread. Wall 2 hasn't developed on the right-hand wound. The discoloration has spread all the way to the pith. Wall 3 is the strongest wall and resists pathogens' lateral spread. Wall 4 (indicated by pointer 4) is the strongest protection zone; also known as the barrier zone. Despite its strength as a protection zone, wall 4 is structurally weak, separating newly formed healthy wood from the infected interior wood. Often cracks forms along the boundary between these two layers of wood. Photo by Alex L. Shigo, courtesy of Shigo and Trees, Associates, shigoandtrees.com.

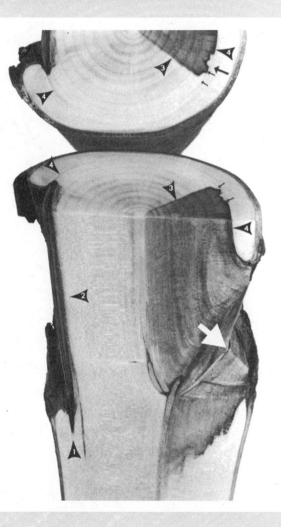

Pruning and the Branch Bark Ridge

Honed over millions of years of evolution, the **branch bark ridge** (BBR) is the boundary created along woody plants' natural protection zone. It's a natural shedding line separating the trunk from the branch. Pruning cuts should preserve the branch bark ridge. A **flush cut** made clean against a tree's stem (see A in Figure 2.12) removes this natural protection zone, inviting fungal invasion and decay. Proper pruning cuts (B and 3) should begin at the base of the branch crotch as close as possible to the outer face of the BBR, extending diagonally to the point where the branch meets the base of the branch collar.[52]

Shigo and associates tested this concept in an 8-year study, pruning 1,200 trees of 12 different species. *Decay rapidly developed both above and*   ☞

FIGURE 2.12: Proper pruning cuts (B and 3) should follow the plane of the branch collar and avoid cutting into it. (A) depicts a flush cut, exposing the stem to drying and infection. For larger branches, use 3 cuts to properly prune a branch and avoid damage to the remaining stem. Begin with an undercut (1) that relieves the tension on the branch's underside some distance from the branch collar. Then cut fully through the branch from above. With the tension and weight no longer on the pruning cut itself, make a clean third and final cut along the branch collar (3).

below pruning cuts made flush along the tree's trunk. Properly pruned branches can form a ring of **callus** surrounding the wound the following growing season, helping the plant form new vascular cambium, capable of producing new tissues that may close the injury.[53]

Note that branch crotches are particularly vulnerable to infection (Figure 2.13). Branch crotches differ from true lateral branching because they form as "codominant stems" with no developed branch bark ridge or branch collar. As the tree matures, these two shoots eventually grow into one another and "include" the bark within the maturing stem. This creates a very vulnerable structural weakness. Also, because bark wrinkles at the junction between branch and trunk, inner bark tissues form a portion of the new surface, easily injured by borer insects.[54] So, if possible, try to avoid the inclusion of branch crotches on coppice stools and pollard heads.

Shigo's insights offer several simple practices to protect and nurture woody plants. Prevent wounds whenever possible, and always try to make proper pruning cuts by preserving the natural defense boundary along the collar.[55] This isn't always possible with coppice stool management because we're often cutting stems between nodes, not along branch stubs, but nevertheless, this knowledge of compartmentalization should help us appreciate how to be better tree stewards.

FIGURE 2.13: Unless one of these two shoots is pruned off, as the young tree increases in girth, these codominant stems will grow into one and another, creating a scenario where included bark between them creates a structural weak point, leaving the tree vulnerable to significant wounding later on in life. Ideally, as we manage growth on coppice stools and pollards, we avoid these types of scenarios and prune trees strategically.

The Adaptive Utility of the Ability to Resprout

Compared to their herbaceous kin, woody plants have long life spans and a large, bulky rigid form. While this allows them to tower above their herbaceous competition and colonize a diverse range of niches, these advantages come with a cost. Woody plants' long life cycle and persistent aerial parts leave them vulnerable to damage from high winds, fires, floods, landslides, ice, snowstorms, and browsing.[56] As a result, they've evolved **life history strategies** to adapt to these types of disturbance.

A species' life history strategy balances the costs and benefits of their investments in growth, survival, and reproduction. These evolutionary adaptations and trade-offs enable species to meet their needs and survive competition, disturbance, and stress so they may bring a new generation into the world. Competition occurs when two or more organisms need the same resource at the same time, in the same space, in the same way, *and* that resource is scarce. Disturbances are discrete events that alter an ecosystem or individuals' physical or relational structures. Stress occurs when an organism's needs are not met, its tolerances are exceeded, its natural functions are prevented, or it's forced to function in a way that it's poorly adapted to.[57] Reproductive strategies are central to a species' life history strategy.

Many woody plants employ a combination of both sexual (seed) and asexual (vegetative) reproduction. Reproduction by seed has many benefits. It enables plants to disperse propagules and expand their distribution, maintain or increase genetic diversity, and adapt to changing conditions. Nevertheless, seed reproduction also has some disadvantages, like the vagaries of pollination, seed dispersal, seed and seedling predation and mortality, and the time required for seedlings to establish and overtop competition. Sexual (seed) reproduction can be very difficult on sites prone to frequent disturbance, like flood plains, and in locations at the limits of a species' physiological tolerances, like heavy shade. Under these circumstances, vegetative reproduction, or resprouting, plays an increasingly important role in woody plants' life cycles.[58]

In general, resprouting species are exceptionally resilient to disturbance and are capable of surviving extended periods with little to no seedling recruitment.[59] They may occupy the same location for hundreds or even thousands of years with minimal change in population size.[60] Sprouts and suckers can also help woody plants colonize unoccupied ground, increase their competitive ability within a plant community, increase plant survival rates in marginal habitat, replace damaged or senescent organs, and restore a plant's entire aboveground form.[61]

As we discussed, resprouting helps balance plants' energy resources, stimulating healthy crown development following damage to dead or senescent limbs. It therefore keeps woody plants in a physiologically youthful state, delaying their aging, shortening their internal transport system, augmenting water and nutrient supplies to the plants' outer reaches, all of which often extend their life span dramatically—effectively doubling it, tripling it, or more.[62, 63] Of course, this increased longevity does come with costs. Namely, the lack of genetic variation that's deployed with each cycle of seedling reproduction.[64]

Disturbance Insurance

At the community level, resource competition and disturbance frequency strongly influence plant architecture and sprouting response. "Successful" species must adapt to effectively colonize a site between disturbance events. Peter Del Tredici describes sprouting as woody plants' "insurance policy" to prepare for future damage.[65]

Investments in seed production support the success of *future* generations. Sprouting species invest to maintain the *current* generation by producing and maintaining a cache of suppressed collar buds, along with a carbohydrate reserve both for maintenance and rapid growth following disturbance. As a result, many sprouting species produce fewer seeds, with lower seedling recruitment and survival, and slower growth rates than non-sprouting species. And their seedlings often take longer to reach reproductive maturity. Their investment in carbohydrate storage enhances survivability at the expense of growth. The cost-benefit of this energetic trade-off ultimately depends on the intensity and frequency of disturbance.[66]

Where disturbance is infrequent, successful plants allocate little energy to storage organs, instead investing energy in growth. These communities tend to rely on seed reproduction. But where disturbance is frequent, sprouting species recover their canopy rapidly, developing a multistemmed form. Species that rely on seed reproduction must grow quickly and set seed earlier than sprouters in order to outcompete them. As long as the next disturbance event occurs before seedlings' taller single-stemmed canopies shade out the lower multistemmed sprouts, resprouting is competitively superior.[67]

Disturbance Severity

Hallé pointed out that a typical coppice stool's architecture largely resembles a tree's canopy. This informed the theory that a resprouting plant's architecture follows a hierarchical framework based on the severity of the disturbance. Different types of disturbance affect different plant tissues (leaves, twigs, branches, and trunk). This leads us to predict that the loss of one tissue type initiates resprouting from the next hierarchical level. For example, leaf damage triggers growth from the leaf axil's axillary bud.[68] When we coppice, pollard, shred, or prune woody plants, we choose the severity of our management to trigger our desired outcome. "Disturbance severity" determines where a given practice lies along the continuum of woody resprout management.

Fire

Sprouting plays a key role in woody species' persistence in fire-prone ecosystems. Here, successful sprouting species allocate more energy resources to roots and their shoots than nonsprouting species. On Australian heathlands, sprouters commonly possess root starch reserves four times greater than nonsprouting

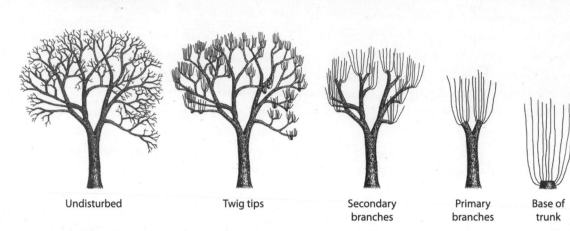

| Undisturbed | Twig tips | Secondary branches | Primary branches | Base of trunk |

Disturbance Severity Continuum

FIGURE **2.14:** A conceptual framework of the extent of disturbance and its effects on sprouting. So perhaps we ought to consider coppicing and pollarding not so much as separate practices but as tools along a continuum. For different products, contexts, forms, and goals, we choose one level of disturbance severity over another. Adapted from Bellingham and Sparrow, 2010.

species, with four to five times higher root-to-shoot ratios, and less than half the biomass of nonsprouters of the same age.[69]

Like other types of disturbance, fire helps rejuvenate sprouting plants, stimulating vigorous new growth from stored energy. Many perennial species in coastal California's chaparral and sage scrub have developed evolutionary adaptations to fire, with some even depending on it for nutrient cycling, sprouting, and/or seed germination. One theory actually suggests that chaparral species have evolved qualities making them particularly flammable (including low silica-free mineral content and high solvent extractives; volatile resins and oils; large volumes of dead biomass; and a dense, continuous multistemmed canopy). This encourages fires to consume all their aerial parts, leaving only buried seeds and belowground tissues to regenerate. During a chaparral fire, soil temperatures reach 500°F to 600° (260° to 315°C) or more, though just 0.5 inches (1.3 cm) below the soil surface temperatures rarely surpass 302°F (150°C), insulating roots from the catastrophic conditions above.[70]

Site Productivity and Sprouting

Site productivity also influences the energetic trade-offs and the rate of regrowth for sprouting and seeding. Sprouting species tend to be more common on less productive sites with frequent disturbance. They regenerate more vigorously and retain this ability longer. On productive sites, single-stemmed seedlings often overtop and shade out multistemmed sprouts. But on unproductive sites, their slower growth and dependence on seed regeneration may impede their success, so species that rely on seed reproduction may be less common.[71]

Site productivity can even affect the reproductive strategies of members of the same species like the myrtle beech (*Nothofagus cunninghamii*) in Australia. On sites with ample moisture and fertility, it commonly reproduces by seed, while on drier sites with low nutrient availability, they often rely on sprouts. The *Eucalyptus* species *Eucalyptus baxteri* and *E. leucoxylon* behave similarly. While they can assume a single-stemmed form reaching 66 to 98 feet (20 to 30 m) tall, on sites with low rainfall and fertility, they may be multistemmed and as small as 3.3 to 6.6 feet (1 to 2 m).[72]

Sprouting Ability and Time

We generally use a binary system (sprout/no sprout) to describe species' sprouting ability, but this really only captures the extremes along an under-researched continuum of sprouting vigor.[73] Though we've just discussed resprouting as a common life history response to severe disturbance, minor disturbances like herbivory, minor wind damage, etc. also stimulate sprouting. Almost all woody plants are capable of resprouting following minor disturbance, but as the severity increases, we find much more variation.[74]

Because of this, it's difficult to quantify a species' relative sprouting response. Further complicating matters, many species' sprouting responses change at different points in their lives. Some species never sprout. Many species' sprouting ability improves with age, plateauing at 20–40 years old. And some are virtually immortal, maintaining a bud bank that enables sprouting at any point in their life cycle.[75] Sprouting during juvenility enhances a species' **recruitment** or new plants' likelihood of successful establishment, while in adulthood

it improves its **persistence** or long-term ability to occupy a site.[76]

A survey of 68 native northeastern North American trees (both angiosperms and gymnosperms) found that all can resprout as seedlings, 84% as saplings, 41% during adulthood, while 25% resprout from root suckers, and 26% from layered branches. While 78% of the angiosperms can resprout as adults or form root suckers or branch layers, only 15% of gymnosperms produce collar sprouts as saplings, 62% form branch layers, and none resprout from the cut stump or produce root suckers as adults.[77]

Seedlings' strong probability of sprouting helps ensure their presence in the forest understory during the early stages of succession. This helps them to fill canopy gaps as they form. In the Missouri Ozarks, Liming and Johnson (1944) found that 77.9% of the seedlings of 7 different tree species were actually resprouts— what biologists call **seedling sprouts**. In Pennsylvania, Ward (1966) found that 58% of all hardwood seedlings in a 53-year-old undisturbed forest were seedling sprouts. In the United States' Central Hardwood Region, Vogt and Cox (1970) found that more than 95% of the trees in second-growth oak stands likely originated from stump sprouts and many 5-to-15-year-old stems had root systems 20 to 30 years old.[78]

So, sprouting is a key life history strategy for many woody plant species. It's an energetic investment that acts as a hedge against disturbance, but it comes at some cost. When we manage for resprouts, we channel these evolutionary adaptations to our advantage, to support a plant's persistence in the landscape.

Next, let's take a step closer and examine the biology of sprouts. What actually occurs to create them? How and where do they originate? What differentiates different types of sprouting responses? And how can we use this knowledge to inform better management?

Beavers: The Original Coppice Woodsmen?

Most of our knowledge about stump sprouting comes from forestry literature on logging. Logging is a unique form of disturbance, relatively new to forest ecosystems. Beaver activity is the only "natural" analog. What first inspired our human ancestors to manage trees for resprouts? Could the beaver have served as an early teacher in the art of coppice "management"? Let's take a brief look at beavers' life cycles and learn how they influence woody plant regeneration.[79]

The earliest fossil record of the beaver (*Castor* spp.) dates 10 to 12 million years ago in Germany. They are believed to have migrated across the Bearing Strait to North America, where fossil teeth dating back 7 million years have been found in Dayville, Oregon.[80] Beavers are the second-largest living rodent in the world. Two species exist today: the North American and Eurasian beavers (*Castor canadensis* and *C. fiber*). Their range spans much of North America and Europe, and extends well into central Asia. Like humans, these keystone species dramatically influence the character and diversity of their ecosystem.

Known for their proclivity to build dams in wetlands and along streams, beavers manage surrounding woodland for food and building materials. They usually inhabit the ponds they create for 5 to 20 years. These herbivores eat the bark of numerous tree species along with sedges, pondweed, and water lilies. Their preferred diet includes willows, aspens, and cottonwood because of their easily digestible, high-protein bark, next favoring oaks, ash, and sugar maple as well as apples, cherries, and other ☞

FIGURE 2.15: Could beavers have inspired humans to leverage woody plants' coppicing response? We'll never know for sure, but here we can see multiple "rotations" of coppice sprouts on a beaver "managed" ash tree in New Haven, Vermont.

members of the rose family. Of less preference are members of the birch family, including musclewood/hornbeam (*Caprinus carolinana*), black and paper birch, as well as gray birch, yellow birch, speckled alder, hophornbeam/ironwood (*Ostrya virginiana*), beech, and red maple. They tend to avoid conifers.[81]

Beavers prefer pole-sized trees 4-to-6 inches (10 to 15 cm) in diameter. Containing a high yield of bark at a size still manageable to harvest and haul, they're beavers' most efficient food source. Beavers "manage" the forest surrounding their habitat, girdling species of low preference and large trees shading out understory growth.[82] Felling hardwood trees often initiates sprouting, yielding several pole-sized stems sometime in the future.

When cutting trees, beavers often leave branch stubs about 1 foot (30 cm) in length, and they never cut back to old wood—the latter being a key practice in traditional coppice management. This appears to help ensure healthy sprouting from buds on the remnant stubs.[83] So, did humans learn to coppice by observing the behaviors of the beaver? We'll probably never know, but it may well have offered them inspiration.

Anatomy and Physiology of Angiosperm Sprouting

While ecologists and evolutionary biologists have long discussed the reasons for sprouting responses, we still know relatively little about them.[84] Let's explore the origin of resprouts, along with their structure, location, longevity, behavior, and integrity. Earlier in the chapter, we examined woody plants' fundamental bud types during undisturbed growth. Now we'll look at the bud types that give rise to new growth following disturbance.

Meristems That Form Sprouts
Dormant Preventitious Buds: Apical Meristems

In 1878, Hartig distinguished two categories of buds that give rise to resprouts—**preventitious** and **adventitious** buds.[85] Most preventitious buds form when shoots first begin to grow and expand. They maintain a continuous trace all the way to the pith. They often contain differentiated preformed structures and remain suppressed until needed. We commonly refer to them as dormant buds although they usually grow enough to keep pace with the stem's secondary thickening. The Society of American Foresters defines a dormant bud as "a bud connected to the pith but which remains undeveloped beneath the bark, and continues to grow throughout the life of a stem. Such buds break through the bark when stimulated as by fire, or when the stem above them is removed."[86]

Preventitious buds are also sometimes referred to as latent or suppressed buds. They may survive for decades or more without sprouting, whereas most axillary buds that fail to sprout die and drop off during the same growing season. Oak and beech preventitious buds may

FIGURE 2.16: I noticed the trace of a preventitious bud's annual expansion on this fencepost while walking in a field in Vermont. The pith is located on the left side and the outer bark on the right. You can see the ripple formed by new wood as it kept pace with and enclosed the bud stele.

Dormant Buds Dividing Fractally

FIGURE 2.17: This conceptual adaptation of a photo from Kozlowski, 1971, Figure 5.11 shows the concept of dormant preventitious buds multiplying fractally, often as a response to a wound or disturbance. Many of these buds lose their viability over time, but some continue to keep pace with cambial growth, remaining capable of sprouting as needed.

persist for centuries, steadily elongating to keep up with trees' annual diameter growth.[87]

Preventitious buds maintain their connection to the pith through a specialized tissue called a bud **stele**. Occasionally these steles or even the dormant buds themselves divide, causing an increase in dormant buds at the node (See Figure 2.17).[88] This suggests that perhaps some well-informed management, especially early on in a woody plant's life cycle, could actually help exponentially increase a plant's bank of preventitious buds, significantly increasing the likelihood of a strong sprouting response following coppicing for generations to come. While this concept has been observed in cross-sections of sprouting species, as of yet it has not been verified in terms of best practices and conformity across species.

In this book, we also call these preventitious buds "disturbance anticipation" or "insurance policy" buds. These types of buds can take a number of different forms including primary axillary buds, accessory buds, shoot germs, meristematic points (which themselves are not buds), rhizome buds, root collar buds, lignotuber buds, or additional root-shoot buds.

For our purposes, we're mainly concerned with the two most common types of preventitious buds, axillary and accessory buds, since they're most relevant to most of us. We examined axillary buds in their undisturbed expression earlier, so we know that they form at young shoots' leaf axils and maintain a direct connection to the pith. In this case, they would remain suppressed but viable, sprouting in response to damage or disturbance.[89]

Figure A1: (Above) Marcus Smart and 10 week old sweet chestnut sprouts (Prickly Nut Wood).

Figure A2: (Top right) 160 year old sweet chestnut stool (Prickly Nut Wood).

Figure A3: (Below) One season's regrowth on 160 year old sweet chestnut stool (Prickly Nut Wood).

Figure A4: (Left) 18 year old sweet chestnut regrowth (Prickly Nut Wood).

Figure A5: (Below) 1 acre cant of sweet chestnut following coppicing (Prickly Nut Wood).

Figure A6: (Top left) High quality sweet chestnut stool (Prickly Nut Wood).

Figure A7: (Top right) Short rotation sweet chestnut coppice, Limoges, France.

Figure A8: (Bottom left) Cross section of a black locust stem clearly showing pith, sapwood, phloem, outer bark. Note the bud stele radiating right from pith to preventious bud.

Figure A9: (Bottom right) Root collar sprouts developing on a young willow stool.

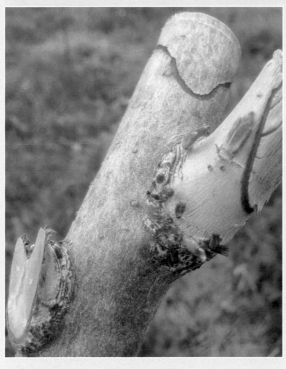

Figure A10: (Top left) Sprouts forming on a corkscrew willow coppice stool—3rd rotation.

Figure A11: (Top right) New sprouts just beginning to emerge.

Figure A12: (Center left) Preventious bud swell and development on a willow cutting.

Figure A13: (Center right) Root collar sprouts developing on an elderberry.

Figure A14: (Bottom right) Adventitious sprouts developing on a seaberry shoot (*Hippophae rhamnoides*).

FIGURE **A15:** (Top left) Adventitious sprouts developing from the cut face of an oak stool, northern Greece.

FIGURE **A16:** (Top right) Root suckers along a patch of sandbar willow (*Salix exigua*), northern California.

FIGURE **A17:** (Bottom left) Root sucker on black locust. Photo credit: Dave Jacke.

FIGURE **A18:** (Bottom right) Hybrid poplar stool (note that new growth has fully enclosed the cut face of the stool).

Figure **A19**: (Top left) Black locust coppice stool one season after cutting as a 2 year old seedling.

Figure **A20**: (Top right) Young black locust coppice stool.

Figure **A21**: (Center left) Callus tissue forming across the cut surface of a young coppice stool.

Figure **A22**: (Bottom right) Ish Shalom with Oregon myrtle sprouts (*Umbellularia californica*) from a large old tree, Coquille, Oregon.

Figure A23: (Top left) Mulberry pollards, Greece.

Figure A24: (Top right) Pollard in Aedipsos, Greece.

Figure A25: (Center left) Pollarded street tree in Athens, Greece.

Figure A26: (Center middle) *Platanus* pollards near Taunton, UK.

Figure A27: (Center right) Young common buckthorn (*Rhamnus cathartica*) pollards New Haven, VT.

Figure A28: (Right) Sprout development on young hybrid poplar pollard knob, New Haven, VT.

Figure **A29:** (Top left) Pollarded head on a hybrid poplar.

Figure **A30:** (Top right) Christopher Peck with massive eucalyptus stool, Sonoma County, CA.

Figure **A31:** (Bottom left) Rick Valley with Oregon ash (*Fraxinus latifolia*) near Dexter, Oregon.

Figure **A32:** (Bottom right) Helen Read with ancient beech pollard (estimated 500 years old) Burnham Beeches, Slough, UK.

FIGURE A33: (Top left) Ancient beech wood pasture, Burnham Beeches, Slough, UK.

FIGURE A34: (Top right) Queen Elizabeth II oak (an ancient pollard that's still alive) near Midhurst, UK.

FIGURE A35: (Center right) Ancient linden (*Tilia cordata*) coppice stool, Czech Republic.

FIGURE A36: (Below) Hedgerows linking the English countryside.
Source: http://fotovoyager.com and istock photo

Accessory buds are axillary buds that subtend (lie below and adjacent to) their adjacent terminal or axillary bud. Usually smaller in size, they often contain preformed differentiated structures and only grow in response to disturbance. Accessory buds can form behind larger axillary buds, near terminal buds, or even within the axils of a larger bud's **bud scales**.[90] See Figure 2.18. These too can form new shoots when the time is right, and some authors suggest that accessory buds are a primary source of the preventitious buds we rely on when we coppice and pollard.

Several factors can trigger preventitious bud sprouting: the death of a shoot's apical meristem, disturbances to nearby trees that lead to increased light reaching the stem, etc. Absent these events, preventitious buds eventually end up buried slightly below the bark. Each year during the week or two preceding bud break, many species' preventitious buds elongate minutely, often changing color, swelling, and later resuming their typical appearance once canopy leaves develop. Some dormant buds appear to add a set of bud scales annually.[91]

We often find the greatest concentrations of preventitious buds at the plant's **root collar**, with concentrations decreasing with increasing height. The preventitious buds closest to a wound grow most actively following damage. By coppicing a tree low to the ground, we increase the likelihood that new sprouts will develop from root collar buds which tend to be more firmly attached to the parent stool. In some cases, when sprouts form at or below the

FIGURE 2.18: Here we see the location of several preventitious buds on a young sugar maple sapling. Typically these very small buds are located at the base of the previous season's bud scar.

soil surface, they can actually even develop their own adventitious root system, helping to stabilize them physically while also increasing their access to water and minerals. **Epicormic buds**, buds that form "upon the trunk," are a type of preventitious bud that forms new branches instead of stems due to their location.[92]

Adventitious Buds: Lateral Meristems

Adventitious buds form "de novo" from undifferentiated lateral meristem (cambium) as a result of stress, damage, or disturbance and a lack of viable preventitious buds.[93] They can immediately form a new shoot or a suppressed, yet viable, bud capable of expanding with secondary thickening and leaving a trace from its point of origin. They do not have a connection with or trace to the pith. Although relatively common on some species' roots, adventitious buds and their shoots are fairly uncommon on stems or branches.[94]

On coppice stools and pollards, adventitious buds typically form from exposed cambium located near a wound, bud, or callus tissue. They often appear along the edge of a stump or pruned branch following cutting.[95] While capable of rapid elongation, adventitious growth may have a short life span. Their sprouts often form a fairly weak mechanical attachment to the parent stool, and high winds or abrasion can easily dislodge them. In some species (especially ash), torn edges along broken branches initiate more adventitious bud formation compared to the relatively even surface made by a saw.[96] This may have implications for management if clean-cut surfaces encourage better-attached adventitious growth.

Five Sprout Origin Types

Resprouts can develop from numerous points on woody plants from both preventitious and adventitious buds. Peter Del Tredici organizes these sprouts into five main categories: *root collar sprouts* at the base of the trunk; *epicormic sprouts* that develop upon the trunk, branches, and twigs; *root sprouts;* and sprouts from specialized organs and stems including *lignotubers;* and *rhizomes*.[97] See Figure 2.19.

As growers, we usually seek to stimulate resprouts that develop vigorously and upright, attach firmly to the stool, avoid decay, and support the longevity of the stool or pollard. In general, sprouts that form from the root collar best meet these needs, but in some cases, the species we're managing may influence how and where new sprouts emerge (i.e., lignotubers, suckers, and rhizomes).

By understanding these various points of origin, we may appreciate their respective strengths and weaknesses, consider the effects of harvest on stool health, and also perhaps learn how to influence the formation of one over the other. We'll start with the most reliable and long-lasting source of new sprouts: the root collar.

Root Collar Sprouts (Preventitious)

Woody plants tend to sprout most effectively from the root collar. While there's no precise definition of the root collar, it's generally considered the zone at or slightly below the soil surface where root and shoot systems converge. The root collar is a specialized organ of regeneration and rejuvenation, storing carbohydrates and supporting preventitious bud growth and proliferation throughout the plant's life.[98]

Biologists consider collar sprouts juvenile, meaning that, upon sprouting, they retain the characteristics of a young individual. As such, collar sprouts have an enhanced ability to form

The Five Main Types of Woody Resprouts

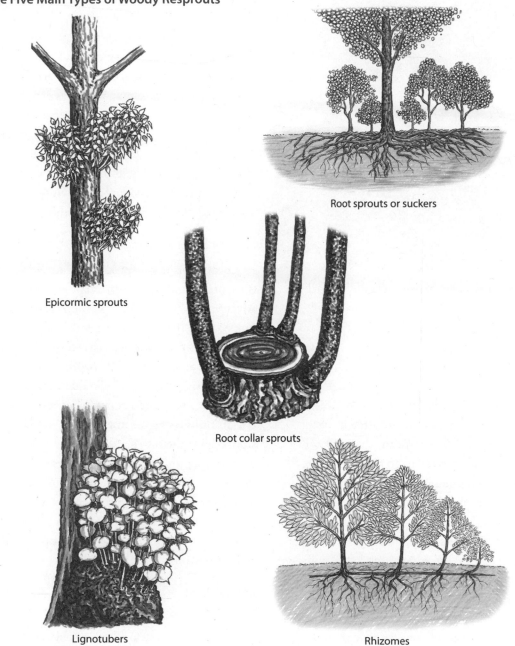

Epicormic sprouts

Root sprouts or suckers

Root collar sprouts

Lignotubers

Rhizomes

FIGURE 2.19: The primary types of woody resprouts we're concerned with.
Adapted from Jenik, 1994, p. 302.

FIGURE 2.20: A dozen or so succulent young stump sprouts emerge from the root collar of a young tulip poplar tree that suffered winter dieback. This transition between the root system and aerial parts often hosts a substantial bank of preventitious buds prepared to sprout in response to injury.

adventitious roots and may exhibit indeterminate growth, retain dead leaves, produce strong vertical growth, and generate large leaves of variable shape.[99] It also means that root collar sprouts generally need to go through all the normal developmental processes that a seedling would before they can become sexually mature, bear flowers, and make seed.

All temperate angiosperms appear to form a collar and develop suppressed latent buds through at least the sapling stage. Conifers generally sprout weakly from the collar and rarely sprout vigorously beyond the sapling stage. The genera *Sequoia* and *Cunninghamia* in the family *Taxodiaceae,* and *Taxus* and *Torreya* in *Taxaceae,* are a few rare exceptions. Most conifers only

form basal sprouts as seedlings, lacking the ability as saplings or adults.[100]

Generally, coppice growth originates from either single buds or bud clusters. Shoots arising from single buds often grow quickly and tend to produce small branches at the base, while bud clusters may generate persistent low branches. As such, single buds are often more desirable than bud clusters. After several years of regrowth, a coppice stool's shoots begin to self-thin, and the survivors are often those that developed initially from single buds.[101]

Epicormic Sprouts (Mostly Preventitious)

Epicormic sprouts—sprouts that develop upon the trunk—may develop from preventitious or adventitious buds and are physiologically identical to buds that form at the collar. The sole difference is their location. "Epicormic" literally means "upon the trunk." If creating a coppice stool, collar sprouts are preferable because they are better located and may form their own adventitious roots. It's rare to cut a coppice stool high enough to leave adequate stem for epicormic sprouts to form. We will often see epicormic sprouts form when we pollard or shred trees. Epicormic sprouts are often undesirable on timber trees as they form new young shoots, creating miniature knots in the outmost layers of wood.

Lignotubers

Some species form specialized organs capable of sprouting. This helps them survive frequent disturbance, namely fire and herbivory. Because they develop belowground, they may form adventitious roots and develop into autonomous plants. Temperate trees produce two main types of specialized organs/stems: lignotubers and rhizomes.[102]

Lignotubers are specialized storage organs adapted to stressful growing conditions that serve two primary functions: they maintain a protected supply of preventitious buds and act as a storage organ for starches and plant nutrients.[103] It's like an exaggerated root collar. The name refers to the organ's woody composition and its morphological and physiological classification as a tuber. Although relatively uncommon in the world of woody plants, lignotubers play a critical role in plant survival post-disturbance, even at the seedling stage.[104]

Built from the same cellular components as stem wood, lignotubers are composed of twice the amount of axial parenchyma (a storage tissue) and about half as many fibers and fiber-tracheids (support tissues). Their considerable girth compensates structurally for their decreased amount of fibrous tissue.[105] Their irregular shape, grain, and large size make it difficult to accurately determine lignotubers' age, but some may reach 500 or more years.[106]

Beadle actually describes the lignotuber as a place of organic compound accumulation, not storage. For example, in *Eucalyptus gummifera*, the nutrient concentration in stems differs little from that of the lignotuber. Altogether though, the lignotuber's size helps it retain a large volume of nutrient reserves.[107]

While most common in subtropical and Mediterranean ecosystems, lignotubers also exist in other climate zones, their broad

FIGURE 2.21: Little leaf linden (*Tilia cordata*), a common street tree in many towns, often produces dozens of sprouts from a substantial lignotuber at the tree's base. Photo credit, Dave Jacke, used by permission.

distribution suggesting it's an ancient and widely adapted response to disturbances including fire, drought, and defoliation.[108] *Tilia americana* (as well as other members of the genus *Tilia*) is the only lignotuber-forming tree native to northeastern North America. *Gingko biloba* and *Sequoia sempervirens* generate shoots from lignotubers.[109] Ninety-five percent or more of all eucalyptus species' seedlings develop a lignotuber within the first year or two following germination.[110]

Lignotubers' size and location affect plants' post-disturbance sprouting. Larger lignotubers deeper in the soil are better poised to survive fires because of their larger bud bank, greater carbohydrate storage, and the soil's insulative protection.[111] Lignotuber carbohydrate storage reaches its lowest levels in summer. Because of this, it's best to coppice in late winter or spring when reserves are high.[112]

Rhizomes

Woody plants' other sprout-forming specialized underground stems are called rhizomes— slender to swollen modified stems that mostly grow belowground. Although not all authors agree, the anatomically similar **stolon** is a slender modified stem more or less above-ground. These terms are sometimes confused and used interchangeably.

As stems, rhizomes possess nodes and internodes, unlike roots. New roots and shoots usually form at the nodes. They may produce new aerial stems some distance from the parent. Drought-adapted southeastern and western North American *Quercus* species (like the gambel oak [*Quercus gambelii*]) frequently produce rhizomes, as does the northeastern *Prunus virginiana*. Rhizome-forming woody species are relatively uncommon, but they do represent another vegetative reproduction strategy.[113]

Suckers: Shoots from Roots

Many temperate early-successional broadleaf species are capable of producing new shoots from root suckers. This includes most species in the alder, seaberry (*Hippophae*), poplar, and elm genera.[114] Other common suckering species include sassafras (*Sassafras albidum*), black locust, paw paw (*Asimina triloba*), prickly ash (*Xanthoxylum americanum*), sumac (*Rhus*

spp.), and American beech.[115] The only two known gymnosperms that produce root suckers are the tropical monkey puzzle (*Araucaria cunninghamii*) and kerapui (*Dacrydium xanthandrum*).[116] Suckering species behave very similarly to rhizomatous species, but they form new sprouts from their root system as opposed to a modified stem.

While suckering species may produce root buds under any condition, they usually suppress sucker development until the trunk has been subjected to some form of trauma. Light is not required to produce root suckers, but suckers do require light to produce vigorous, healthy growth.[117] Shady conditions tend to suppress root sucker development, or at least limit growth to 3.3 feet (1 m) or less, whereas full sun increases the likelihood of sucker development, even without injury. In one closely monitored clear-cut stand of quaking aspen (*Populus tremuloides*), 40,000 suckers per acre (98,840 per ha) formed under full-sun conditions, but at 50% full sun or less, sucker density plummeted to less than 3,000 stems per acre (7,400 per ha).[118]

Early-successional, shade-intolerant suckering species like aspen usually expand in a concentric pattern with the oldest plants in the center and younger plants around the periphery. The colony continues to expand until it encounters competing plants or physical barriers to growth. Eventually, the oldest shoots at the clump's center begin to senesce, and this gradually spreads across the colony.[119] Let's take a closer look at two of the most successful North American suckering species.

Aspen

Quaking aspen is one of the most remarkably expansive clonal woody plant species. In the Rocky Mountains, it may form colonies that cover 100 acres (43 ha) or more with one exceptional individual comprising an estimated 47,000 distinct stems![120] In the moist temperate forests of eastern North America, competition with shade-tolerant vegetation limits quaking aspen colonies to little more than 3.7 acres (1.5 ha).[121] Under favorable conditions, almost any part of an aspen root can sucker. Most aspen clone roots possess thousands of suppressed shoot primordia although parent root depth and diameter affect the success and vigor of aspen suckers. When stems in a clone are cut, suckers emerge from either new or existing meristems on the roots. Suckers that originate from new primordia or meristems develop with far greater vigor than those emerging from existing suppressed buds. In Utah, a survey of post-clear-cut regeneration of a mature aspen stand found 92% of new shoots originated from roots, 7% from root collars, and 1% from tree stumps.[122]

Sweetgum

Common in the southeastern United States, sweetgum (*Liquidambar styraciflua*) frequently forms clonal colonies from root sprouts. In fact, it's not uncommon for a single parent tree 10 to 12 inches in diameter (25 to 30 cm) to generate up to 40 or more suckers from seedling to sapling size.[123] Georgia researchers exposed the shallow root systems of 12 sweetgum trees ranging in age from 30 to 80 years during the early spring months to examine the effect of environmental conditions on adventitious bud development. All exposed roots produced several sprouts the following season, and it appeared as if these suckers rely almost completely on suppressed preventitious buds. And in contrast to aspen suckers, which can form their own root system, sweetgum suckers appear fully reliant on the parent tree's root system.[124]

Regardless of the type and origin of sprouts, sprouting responses stimulate marked physical and physiological changes in woody plants. We conclude this chapter by examining the effects of sprouting on woody plants so we can better understand how our actions influence plant vigor and growth.

Triggers for Sprouting: Apical Dominance

There are still many mysteries we have yet to unlock as to how woody plants manage growth and energy allocation. As best we know, the same apical dominance that drives woody plants' strong vertical growth is the key that unlocks their disturbance response. When a plant's apical meristem is destroyed, these hierarchal organizational relationships are completely reorganized.[125] Internal physiological changes trigger the release and/or formation of preventitious or adventitious buds, causing the rapid elongation of new shoots—sprouting.

The original root system remains alive, and the plant directs new growth to stem replacement. Lateral branching dwindles, and the vast majority of the new growth stretches upwards. Under average conditions, an undisturbed shrub may add 5 inches (12.5 cm) per year to its branch tips, but following fire or other disturbance, it may grow 5 feet (150 cm) in less than 6 months, an astounding 1,200% increase in growth![126]

The apical dominance theory proposes that the downward flow of the hormone **auxin** from tree leaves and buds maintains strong vertical growth and suppresses lateral bud sprouting. These inhibited buds naturally contain higher than necessary auxin concentrations. Damage to the crown interrupts the flow of auxin, triggering dormant bud break and sprout elongation. Some theorize that increased light levels can transport, destroy, or otherwise alter auxin levels, also stimulating new sprout development.[127] Several plant species' roots also exude gibberellins and cytokinins, hormones known to contribute to apical dominance.[128] Within a month after sprouting commences, the production of growth regulators appears to inhibit the sprouting of remaining dormant buds. Any buds that remain viable continue their slow annual growth, keeping pace with the cambium.[129]

Factors Affecting Sprouting Vigor

Sprout vigor depends on a number of factors, including site conditions (soils, climate, and available water and light), plant characteristics (stool age and size, tree species), and management decisions. We can influence several of these with our management and decision-making, but others ought to direct our process. For example, good sites stimulate better sprouting than poor sites. Since improving site quality can be a fairly large undertaking, we might need to carefully identify the most favorable patches or choose species that are best-suited to the current site conditions.[130]

Site Conditions

Light

Light is one of the most important variables we affect with management. Creating a large enough canopy gap to flood a cut stool with light dramatically improves sprout vigor.[131] Although seedling sprouts may regenerate even in a shady forest understory, sprouts that originate from established trees require high light levels to develop strong, healthy new stems.[132] We'll discuss the practical limits and application of gap size in both the Ecology and Design chapters, but remember: *full sun is essential to encourage vigorous healthy resprouts.*

Site/Soil Quality

Soil quality includes organic matter, depth, texture, mineral profile, pH, macro- and micro-nutrients, available moisture, the presence of restrictive layers, biology, toxicity, and land-use history. It plays a pivotal role in the productivity of a site as a whole. We'll discuss the forestry concept of "site index" in chapter 4, but these soil characteristics should determine which species are best-suited for a given site, which practices and patterning are most appropriate for the topography, and what if any amendments or site prep work should be carried out prior to stand creation; and they should help guide our expectations for site productivity.

Table 2.4: Effect of site quality on chestnut sprout growth

Site Quality I - Bottomland Sites II - Average Hardwoods III - "Rocky Ridge"	Year	Average Height Growth	Average Height Growth (meters)	Total Height After 3 Years	Total Height After 3 Years (meters)
I - (Best)	1	6.08'	1.85	11.22'	3.42
	2	3.41'	1.04		
	3	1.73'	0.53		
I - (Average)	1	5.69'	1.73	10.32'	3.15
	2	2.62'	0.80		
	3	2.11'	0.64		
II	1	4.48'	1.37	8.67'	2.64
	2	2.37'	0.72		
	3	1.82'	0.55		
III - (Worst)	1	3.63'	1.11	7.04'	2.15
	2	2.11'	0.64		
	3	1.30'	0.40		

Adapted from Mattoon, 1909, p. 42.

Topographic location offers a partial indicator of site quality. In the case of the holm oak (*Quercus ilex*), a valuable sprouting species blanketing a large band of the Mediterranean basin, topographic location strongly impacts sprout height. Stands on bottomland sites maintained growth rates 1.5 times greater than upland sites, although slope position did not appear to affect sprout diameter. These results suggest that soil nutrients (in this case via topographic location) limit sprouts' vertical growth, while shade and site quality both limit diameter growth.[133]

Plant Species Characteristics
Genetics and Life History Strategies

Less than 1% of all forest seedlings reach reproductive maturity.[134] In a forest stand, different individuals express unique genetic qualities: faster growth, insect or disease resistance, or optimal site adaptations, etc. As we discussed earlier, species' life history strategies determine how they allocate energy and resources towards growth and reproduction. Some species invest more energy in seed production and increased genetic diversity, while others invest in sprouting ability and the persistence of the individual in a niche in space and time. And these unique life history strategies have presumably even been shaped by millennia of human selection and interaction through management.

Shigo found that genetics play a moderate to strong role in shaping trees' ability to form effective defense boundaries. Some individuals form stronger boundaries than others.[135] Thus, individuals that exhibit fast growth are not necessarily good compartmentalizers. As Shigo points out, tree selection based on growth rate rather than effective protection systems may cause us to us to grow defective trees faster.[136]

It's also quite possible that sprouting ability and sprout vigor is also a function of genetics. This means that some individuals will prove better suited for resprout silviculture than others. Although there's little research on this topic, many European sprouting species may have co-evolved with human management over the past several centuries and beyond with humans having selected individual plants that sprout with greatest vigor either intentionally or perhaps unintentionally. If this is the case, we have a lot of catch-up to do on this continent, as little if any work has been done to identify and select plants that sprout reliably and vigorously.

Parent Stem Age/Size

Plant age (and/or diameter) can have a major impact on sprouting responses. In general, younger plants sprout more vigorously than mature ones. As seedlings, sprouting is key to survival in the face of herbivory, deep shade, desiccation, and pathogens. These pressures

diminish for most woody plants as they become more mature.[137]

Every woody species has a unique relationship between the likelihood of resprouting, the optimal size/age of stump at the time of coppicing, and the number of sprouts a stool forms.[138] Most tree species respond best to coppicing if cut before the age of 50. Some broadleaf species like Italian alder (*Alnus cordata*), European beech, birch, bird cherry (*Prunus avium*), and some poplars do not coppice vigorously or only when cut at a relatively young age.[139] European hazel, alder, ash, linden, willow, sweet chestnut, holly, elder, and hawthorn all coppice very strongly, reliably sprouting from the stump. And the European sycamore, hornbeam, oak, elm, aspen, field maple, birch, crabapple, whitebeam, yew, and rowan are also often quite reliable sprouters. Birch generally will not regenerate much beyond the age of 30, whereas hazel, ash, sweet chestnut, and field maple will all respond to cutting much later in life.[140]

Forestry literature suggests that most temperate broadleaf species 2-to-6 inches (5 to 15 cm) in diameter will sprout vigorously, and the majority retain this vigor up to 10-to-12 inches (25 to 30 cm) in diameter. Beyond 12 inches (30 cm) in diameter, sprouting likelihood declines precipitously among non-oak species. While there's no consensus as to why sprouting vigor diminishes with age, it probably relates to dormant buds' limited life span due to genetic, anatomical, and physiological factors.[141]

Beyond 40 years of age, the likelihood of stool mortality increases for most species. Because of the waning viability of many species' dormant buds, it's best to initiate coppicing while still relatively young. Long-managed coppice stools suffering from an extended lapse in management may experience a gradual decline in sprouting response.[142]

In a 1975 study in southwestern Wisconsin and nearby Minnesota, Illinois, and Iowa, 56% of northern red oak (*Quercus rubra*) trees 8-inch (20 cm) dbh (diameter at breast height or 4.5 feet (1.3 m), the standard height to measure tree diameter since stems often flare wider at the base) formed sprouts, while no sprouts formed on trees 26-inch (66 cm) dbh

or larger. Larger stools usually generated more sprouts initially, but after 23 years, all stools averaged about 4 sprouts regardless of parent stump dbh. Sprout height appeared to be more directly affected by site quality and the number of stems per stool.[143] See Figure 2.22 and 2.23.

In a Virginia mixed oak stand, almost all 3-to-4-inch (7 to 10 cm) dbh stumps sprouted following cutting. While 78% of white oak (*Quercus alba*) stumps 12 to 16 inches (30 to 40 cm) in diameter produced sprouts, only 12% of trees 17-to-22-inch (42 to 55 cm)

Percentage of stools with living sprouts

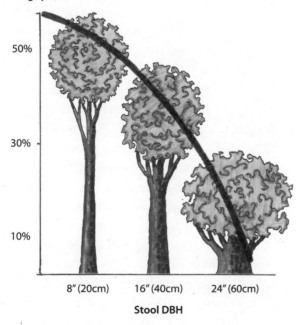

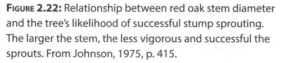

Stool DBH

FIGURE 2.22: Relationship between red oak stem diameter and the tree's likelihood of successful stump sprouting. The larger the stem, the less vigorous and successful the sprouts. From Johnson, 1975, p. 415.

Number of Living Red Oak Sprouts Per Stool Over Time for Three Stump Diameters

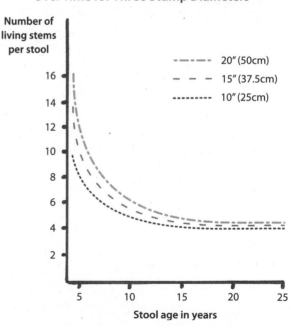

Stool age in years

FIGURE 2.23: Smaller-diameter red oak trees tended to produce slightly fewer sprouts than large trees, but these figures almost completely even out as the stool naturally self-thins with increasing age. From Johnson, 1975, p. 415.

dbh, and 2% of 23-inch+ (58 cm) dbh trees sprouted. In terms of age, white oaks up to 100 years old produced abundant sprouts, though beyond that sprouting declined considerably. In contrast, up to half of the black, chestnut, and scarlet oak stools 101 to 150 years old produced vigorous sprouts.[144]

Stored Energy Reserves

Woody plants' carbohydrate reserves both above and belowground (also known as non-structural carbohydrates, or NSC) fluctuate throughout the season. Stored throughout the plant's living tissues, these sugar and starch reserves are available for woody plants to draw upon as needed when respiration demands exceed energy created by photosynthesis. NSC stores help support plant growth during spring time leaf out and periods of stressful growing conditions, and they are crucial to resprouting following coppicing and pollarding.[145]

Botanists generally agree that NSC reserves peak during late summer/fall, steadily declining after leaf fall and throughout dormancy, but there is still much work to do to thoroughly decipher NSC levels and their contributions to new growth at the organ level, especially taking into account seasonal fluctuations. To date, research seems to suggest that, as shoots begin spring growth, root energy reserves decrease sharply, translocating it to their developing leaves, shoots, roots, and stems. This leaves stools particularly vulnerable to browse or other damage at this point. Plants rebuild these stored NSCs during the growing season so that roots can maintain high carbohydrate stores during

dormancy.[146] At least this is what conventional wisdom seemed to suggest until relatively recently.

According to a study published in 2018 monitoring the organ-level seasonal NSC dynamics of five woody species (red oak, white ash, red maple, paper birch, and white pine) at the Harvard Forest in Petersham, MA, things aren't quite as clear-cut. Researchers collected monthly samples from the leaves, branches, stemwood, and roots of 24 trees, measuring their NSC concentrations and multiplying this by each organ's estimated biomass to determine projected whole tree NSC storage. Their results confirmed some accepted theories and called others into question.[147]

Firstly, the two ring-porous species (red oak and white ash) had the greatest NSC pools at the whole tree level.[148] This supports our earlier discussion on determinate and indeterminate species' energy reserves when cut at different times of year. Gray birch and red maple (indeterminate growth and diffuse-porous) had the longest window of low-energy reserves. In contrast, ash (determinate growth and ring-porous), with its flushed cycle of growth formed largely from stored reserves, completed much of its growth fairly early in the growing season. Because of this, the species had a much shorter window of vulnerability.[149] Because indeterminate species use the current year's phytosynthates to generate new shoot growth well into the growing season, they are much more vulnerable to harvest during the growing season.[150] This should be especially interesting to readers considering tree hay harvest when choosing their preferred species and optimizing harvest timing. See Figure 2.24.

While *whole tree* NSC levels did correspond with existing theories, peaking in late summer/fall and declining thereafter, the seasonal balance of storage at the *individual organ level* followed a far more dynamic pattern. Branches were actually the largest NSC storage organ (30% to 40%), followed by roots (25% to 35%), stemwood, and leaves. Surprisingly, root NSC pools actually remained relatively stable throughout the year, whereas the branches showed an early season decline in NSC, presumably due to their close proximity to buds as plants began to leaf out in early spring. The relative stability of root NSC levels led the authors to theorize that they serve as a bit more of a long-term, catastrophic insurance policy.[151] Just the type of disturbance that coppicing causes!

So, it's clear that resprout silviculture relies on these stored NSC reserves to stimulate vigorous new growth following harvest, but we still have plenty of questions to answer before fully understanding how this plays out at the whole tree level. And what's more, these patterns also vary between species. We know that dormant-cut trees sprout more vigorously than those cut during the growing season. But while roots clearly play a major role in post-coppice resprouting, it now appears that they aren't as significant an energy storage as we once thought. Until we learn more, we can rest assured that root NSC stores seem to be sufficient to power healthy new regrowth as they have done for millennia and beyond.

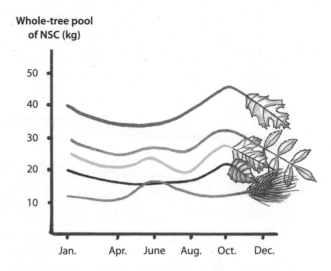

Whole-tree pool of NSC (kg)

Figure 2.24: Seasonal changes in nonstructural carbohydrate reserves for five tree species in Massachusetts. From top: red oak, white ash, red maple, paper birch, white pine. Note that the broadleaf species peak at the end of the growing season in October and generally bottom out during spring months. Adapted from Furze et al., 2018, p. 1470.

Management Factors
Plant Density

Like most management factors, we must constantly choose between balanced trade-offs, plant density being a big one. Traditional coppice stands are often managed quite densely with canopies (and also root systems) interwoven with one another. We discuss these layouts in more detail in chapters 4 and 6, but dense spacing tends to encourage straighter upright growth with minimal branching. This of course may come at some cost to the growth of the individual sprout or stool, but on the whole, it ensures dense regrowth at the stand level and helps develop quality sprouts useful for crafts and other value-added needs.

Cutting Time

We just briefly examined harvest timing in relation to plant energy reserves, but once again, timing our harvest with dormancy is the best strategy for long-term stool and pollard management. Another benefit of dormant season harvest is that sprouts begin their new growth with the cool temperatures and often ample moisture of spring months and then have the entire growing season to mature and harden off before the cold temperatures of fall and winter set in.

Cutting Height

As we've learned from our in-depth overview of bud types and sprout origin, coppicing plants as low as reasonably possible helps ensure we trigger preventitious bud development from the root collar. This increases the likelihood that shoots grow straight and upright and may also lead to them forming their own adventitious root system, better anchoring in the soil. It's unclear if cutting height has an effect on the actual vigor of sprout development, but it certainly has implications for the long-term stool health and vitality.

Cutting Frequency

Harvest frequency is usually determined by the size of the material needed. In general, longer rotations give the plant more opportunities to recharge stored energy reserves and also replenish site fertility through annual leaf litter deposit and decay. It doesn't appear that rotation length has a huge impact on sprout vigor in the short term. That is, new sprouts don't seem to grow any faster or healthier if they're cut on longer or shorter rotations. That said, particularly short-rotation coppice systems (1 to 3 years) may in many cases exhaust stools, requiring the stand to be replanted after just a couple of decades. Because of this, coppice rotations usually span at least a few years and often range from 3 to 15, providing plants with adequate time to recharge their essential energy stores. We'll discuss the effects of rotation length at the community level in chapter 3.

Cutting Geometry

Whenever possible, it's best to make sloping cuts oriented away from the center of the stool to minimize water collection and associated decay. New cuts must always be made in "new wood" so as not to cut into the tissues remaining from a previous harvest. Some anecdotal experience suggests that the orientation of cut faces can have an impact on sprout vigor. Apparently, in Mediterranean climates, pollard faces oriented northwards can help limit excessive drying, while south-facing cuts in Scandinavia increase access to light and help reduce the likelihood of decay. The geometry of the stool or pollard will also ultimately inform the best angle and orientation, but because cutting geometry is a decision we ultimately have to make, it's good to consider the implications of each option.

Effects of Coppicing on Woody Plants

Coppicing and pollarding alter the plant's remaining tissues in four primary ways:

- It exposes cut faces of bark, cambium, and wood to microorganisms.

FIGURE 2.25: A massive centuries-old linden (*Tilia cordata*) coppice stool, at Děvín Wood, Czech Republic.

- It causes cut surfaces to dry and changes moisture and gas flows in its tissues.
- It reduces the volume of foliage and sapwood and decreases whole plant carbohydrate stores.
- It disrupts hormonal growth-coordinating signals.[152]

Reducing Energetic Commitments Lengthens Woody Plant Life Spans

We discussed this theme in greater detail earlier, but essentially, coppicing or pollarding significantly reduces a plant's energetic commitments. There's less wood left to envelop with a new layer of growth. This frees up energy to generate and support a robust and healthy crown capable of photosynthesizing sugars to support new growth. Coppicing essentially resets the ratio between the tree's income and commitments, "keeping it young" and stimulating vigorous new growth.

Photosynthesis Rates

Coppicing may also induce changes in leaf photosynthesis rates. In a Dodgeville, Wisconsin, red oak (*Quercus rubra*) forest, resprouts maintained higher daily leaf photosynthesis maxima and averages throughout the growing season than uncut stems. This supports other trials that found improved photosynthetic performance from new and residual leaves on damaged or defoliated woody plants.[153]

Additionally, coppice stools' smaller canopy (compared to uncut trees) alters the balance between leaves and roots. Assuming a particular root system can support a particular canopy gas exchange threshold, the reduced canopy

Year 1 and 2 Growth Rates for Chestnut, Red Maple, Ash, Red, and Chestnut Oak Resprouts

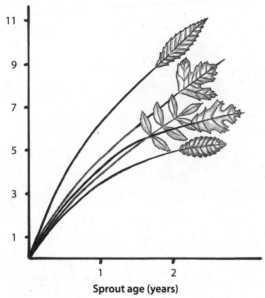

FIGURE 2.26: Average height of resprouts from five species after one and two years' growth in Connecticut (1909). From top: American chestnut, red maple, ash, red oak, chestnut oak. Adapted from Mattoon, 1909, p. 45.

Table 2.5: Effect of light and shade on chestnut sprout growth

	Partial Shade	Full Sun
Year 1	5.1'	5.6'
Year 2	2.4'	4.0'
Year 3	1.5'	2.9'
Total	9.0'	12.5'

Adapted from Mattoon, 1909, p. 41.

following coppicing may stimulate an increase in total canopy productivity. In Dodgeville, while coppicing caused a 33% reduction in tree canopy size, average sprout photosynthetic rates increased 35% over controls.[154]

Growth Rate

Several variables affect sprout growth rates: the size and age of the parent tree, the season of cutting, light regime, site quality, and topographic position. Most research suggests that, at least during early development, coppice shoots outgrow seedlings of the same species.

In Australia, Drake et al. (2009) compared the growth rate of *Eucalyptus globulus* seedlings and coppice sprouts. Coppice stools showed accelerated rates of biomass production and **leaf area index** (LAI). While seedlings initially rely on the carbohydrate reserves stored in their seed, coppice stools enjoy the benefit of existing root energy stores.[155]

Trees' lateral root extension develops proportional to stem diameter, increasing as the square of the diameter increases. This means a 2-inch (5 cm) diameter tree's roots ramify nearly 4 times the area of a 1-inch (2.5 cm) tree. Ultimately, sprouts from larger stumps have access to more moisture and nutrients, enabling them to grow larger faster. And larger stools tend to generate more sprouts because their size enables them to form more dormant buds.[156]

Sprout growth usually shows a rapid increase in height during the first few seasons but quickly slows as annual growth causes an increase in diameter. In Connecticut, Mattoon monitored chestnut sprout growth patterns and found young sprouts rapidly increased in height, outcompeting neighboring vegetation and ensuring their access to light. During later years, vertical growth slows while shoots rapidly increase in cross-sectional areas. Under favorable conditions, sprouts can maintain an annual diameter increase of 1 inch (2.5 cm) during the first 8 to 15 years![157]

Sprout Origin and Influence on Decay

Two main factors appear to affect sprout integrity and longevity: the height of the sprout's origin and the stool's diameter. For reasons we've discussed earlier, sprouts originating from the root collar are more prone to develop into healthy, high-quality stems. Sprouts that form aboveground fully rely on the parent stem and root system for water and nutrients and are more susceptible to disease and decay affecting the parent stump (especially on large stools).[158]

Dead branches, sprouts, and stubs at the base of coppice stools may create vulnerable infection points that lead to decay. Also, as the stool's cambium grows, it may engulf an old sprout or create included bark as the space between the new sprout is subsumed by stool growth. Included bark formation also often occurs between stems with a sharp, V-shaped fork. As the two branches grow in diameter, they eventually converge on one another and the bark becomes subsumed or "included" within the wood. This creates an especially weak core within the growing stem, prone to severe breakage as a result of stress or loading and often leading to severe splitting along this plane.

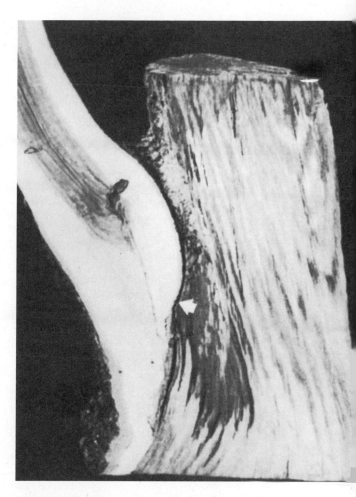

FIGURE 2.27: An 18-year-old red oak sprout on a much older coppice stool. Note that the sprout is actually separate from the original stump/stool. The white arrow points to the interface between the original stool and the new wood formed by the sprout. Here and along this union, the bark of the sprout and the stool intersect, forming a weakened union between the two. Shigo points out that this weak point can be minimized if the sprout can quickly grow outwards and away from the stool or the stool decays quickly. This is also another good reason to make coppice cuts low on a stool so there is less of a chance of included bark forming a weakly attached sprout union. Photo by Alex L. Shigo, courtesy of Shigo and Trees, Associates; shigoandtrees.com. Originally printed in A New Tree Biology, Figure 37–14, p. 497.

To avoid the development of included bark on coppice stools, sprouts must either grow outwards and away from the stool (although less than desirable for long straight poles), or the stool should be cut as low to the ground as possible so the sprout cannot become included.[159]

Genetic Diversity

Because stools regenerate from an established root system, coppice stands do not benefit from the dispersal of new genetic material that occurs in high forest stands of seed origin. The impact of these repeated generations of vegetative reproduction on forest genetics is a common concern regarding the long-term landscape-scale effects of coppicing. Despite this, in chapter 3, we'll discuss how, at least in the case of rebollo oak (*Quercus pyrenaica*), it appears that stands of sprout origin may actually contain more genetic diversity than we might otherwise assume.

CONCLUSION

Woody plants are complex organisms, with an orderly, structured anatomy, an elaborate and elegant set of growth processes, an ingenious strategy to limit the effects of wounds and other damage, and a diverse set of adaptive responses to stimulate new growth. While we need not understand these features and patterns in their entirety to be good tree stewards, the better we are at anticipating plants' responses to our management, the healthier, more vigorous and long-lasting our relationship. After all, the plant is always right—right?

Next, we broaden the scope of our discussion from the individual plant to the ecosystem it's a part of. When we tend our woodlands, we redirect available resources to the individuals and growth forms we deem most desirable. In chapter 3, we examine the many ways resprout silviculture impacts ecosystems at the community level, including changes in habitat quality and type, water and energy resources, and soil nutrients.

Chapter 3:
Ecology of Coppice Systems

The tree was like a social community center. It provided shelter, food, and reproductive sites for many other living things.

—Alex Shigo[1]

In chapter 2, we examined woody plants' physical characteristics and life processes along with the ways resprouting affects their form, longevity, energy reserves, vigor, and life history. But how does sprouting and sprout-based management affect whole populations, communities, and ecosystems?

Clearly resprout silviculture practices cause dramatic changes in forest stand structure, composition, energy cycling, and the character of the habitat they provide. No single technique meets all management goals. There will always be trade-offs. So, what aspects of forest ecology does resprout silviculture enhance, and what compromises does it require? In chapter 3, we aim to answer these questions and more.

THE ECONOMY OF FOREST ECOSYSTEMS

Forests blanket more than 30% of the land on Earth. Home to 80% of all plant biomass, they are responsible for 75% of our planet's **gross primary production**. Forested ecosystems are one of the key foundations supporting life on Earth.

Forests' composition varies widely depending on latitude, precipitation, elevation, soils, subsurface hydrology, previous management/mismanagement, geologic history, and a host of other factors. Broadly speaking, we can organize global forests into three primary groups—tropical, temperate, and boreal—largely based on their latitude. While resprout silviculture is a useful tool for all types of forest ecosystems, this book focuses on the forests and wooded and non-wooded landscapes of the temperate world. Because of this, we will generally use the terms "forest" and "woodland" synonymously, despite the fact that, in some parts of the world, they actually describe uniquely different types of tree-covered landscapes.

In chapter 1, we explored the origins of the term "forest," pointing out that it initially described land set aside for hunting, usually reserved for royalty, with little direct connection to any presence of tree cover. The term has obviously evolved considerably over the past several centuries.

As an ecosystem, a forest is most definitely more than just an assemblage of trees. A forest is a network of relationships formed between living organisms and the landscape's nonliving elements and features. These relationships revolve around the capture, conversion, storage, exchange, and release of energy, biomass, water, and mineral nutrients. Forests are a composition of several identifiable

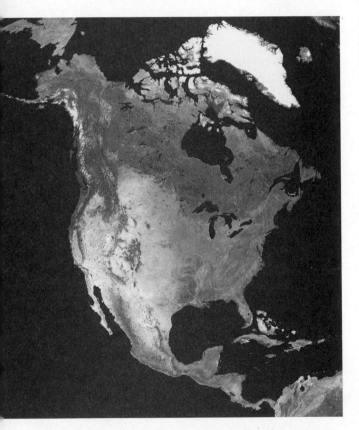

FIGURE 3.1: Even 5000+ miles above the Earth's surface, we can clearly see the distribution of forest cover across temperate North America. Image by wikiimages.org, accessed April 20, 2021.

vegetative layers (canopy, understory, ground cover, etc.) of woody and herbaceous species along with myriad other life forms, including insects, microbes, birds, mammals, amphibians, reptiles, and fungi who either live off the products of photosynthesis or one another.

Forest character and productivity depends on climate, moisture, sunlight, nutrients, resource cycling, predation, fire, and other processes. Broadly speaking, we can divide the forested landscapes of the US and Canada based on their relative locations east or west of the Mississippi River. Gymnosperms (conifers) tend to dominate western forests, with most forest cover relegated to high-elevation mountainous terrain in the dryland interior along with pockets of remarkable biodiversity in the rain-soaked coastal margins. To the east, where precipitation tends to be much more abundant and reliably distributed, we find an especially diverse mix of angiosperms (broadleaf species) and gymnosperms (conifers), with virtually all landscapes rapidly returning to forest cover following land clearance or other forms of disturbance. While forest productivity is a direct result of these regional-scale patterns and processes, dramatic site-specific variations also play a huge role in stand richness and diversity.

As we examine forests, we place particular emphasis on the living elements and processes that guide their growth and resource cycles. But death and decay are also essential to a healthy forest. Mature forests experience a 1% to 2% average death rate resulting from disease, storms, fire, and animal damage. The biological agents that weaken or kill woody plants also act as selection agents, eliminating weak or stressed individuals while creating openings for the next generation.[2]

Of course, as we consider the potential resprout silviculture offers us, our interests extend beyond just forested ecosystems, into the savannah, shrublands, prairies, riparian zones, and even deserts. With proper care, trees can grow in just about any climate, but that doesn't mean we humans can engineer a true forest ecosystem. Coppice agroforestry seeks to ally ourselves with woody plants' resprouting capabilities to steer the evolution of community-scale ecosystem processes towards our own production or conservation goals. The outcomes of our decisions help shape the landscape's character, with cascading results that affect all the other members of the ecosystem. To be a good manager, we must first understand the whole system we're participating in. And there's perhaps no better place to start than at the scale of the community.

NATURAL COMMUNITIES AND FOREST COVER TYPES

As you get to know the forested ecosystems in your region, learning to identify plant *communities* is in many ways more important than the individual species. (But of course, you can't identify communities without already knowing the species that comprise them.) These species assemblages tend to thrive on sites with particular soils, hydrology, aspect, and elevation. And at the same time, the plants' physical characteristics (plant type, size, rooting pattern), moisture and fertility requirements, shade tolerance, and

successional status influence the character of the community. We call these assemblages **natural communities**, and foresters often refer to them as **forest cover types** or forest stands.

In identifying a forest stand, we look for enough consistency in the dominant species, structure (age, arrangement, size), and ecological relationships to develop broad management recommendations for the stand as a whole.[3] The Society of American Foresters (SAF) identifies forest cover types based on the dominant woody species. We describe different cover types using the names of the one to three predominant species like "Hemlock-Sitka Spruce" or "Oak-Hickory."

Foresters use **basal area** as a relative measure of stand dominance. For a much more in-depth overview of basal area, skip to chapter 6. To earn a listing as part of a forest cover type, a species must comprise at least 20% of a stand's total basal area. Stands with 80% or more of the same species are considered "pure stocked." In the United States, many publications that feature silvical descriptions of tree species (i.e., *Silvics of North America*) include references to the SAF cover types where they can be found.[4]

The SAF recognizes 145 cover types in American forests, 90 in the eastern states and 55 in the west. In Canada, provincial governments own much of the country's forestland, and they generally monitor timber production using inventory classifications established by province. Practically speaking, distinguishing between two different cover types can be quite challenging as stand boundaries, or "ecotones," often transition over broad areas.[5]

Climate (namely annual precipitation and minimum temperatures) plays a major role in determining a forest's species composition. In mountainous areas, elevation often directs

Table 3.1: US Forest Service's renewable resources evaluation group forest types

Eastern	Western
White-Red-Jack Pine	Douglas fir
Spruce-Fir	Hemlock-Sitka Spruce
Longleaf-Slash Pine	Ponderosa Pine
Loblolly-Shortleaf Pine	Western White Pine
Oak-Pine	Lodgepole Pine
Oak-Hickory	Larch
Oak-Gum-Cypress	Fir-Spruce
Elm-Ash-Cottonwood	Redwood
Maple-Beech-Birch	Noncommercial
Aspen-Birch	Hardwoods

Adapted from Eyre, 1980, p. 3.

forest community distribution. Increasing elevation often comes with cooler temperatures and increased precipitation. We find some of clearest expressions of these ecosystem gradients in the Rocky Mountain region and the American Southwest where elevation and aspect (orientation in relation to the sun) have dramatic influences on microclimates and plant communities.

In several states, field naturalists and ecologists have taken forest cover types one step further, identifying their region's prominent natural communities using a field guide format. Often offering far more specificity than SAF cover types, these resources include lists of dominant and commonly associated plant and animal species in a given community type, along with associated climatic, geologic, topographic, and hydrologic details. Where available, they're an invaluable tool for bioregionally appropriate site design.

For those who live in the Northeast, The Nature Conservancy's Northeast Habitat Map (https://maps.tnc.org/nehabitatmap/) allows you to click on a particular map tile within the

Figure 3.2: Here on Basalt Mountain in Basalt, Colorado, USA, we see the stunning effects of aspect and elevation on plant communities. Low elevations host virtually no woody plant life, that is except for the quaking aspen stands that follow stream courses. Above this, around 6000-7000' elevation, we see members of the Pinyon-Juniper complex with scattered patches of scrub oak. And at higher elevations and along north-facing slopes, new communities emerge that are adapted to cooler temps and higher humidity.

MESIC CLAYPLAIN FOREST

**DISTRIBUTION/
ABUNDANCE**

This community is known primarily from the Champlain Valley of Vermont, New York, and Québec. A similar clayplain ecosystem occurs on clay soils adjacent to the lower Great Lakes.

THE CLAYPLAIN FOREST ECOSYSTEM

The clayplain forest is best considered an ecosystem composed of several closely-related forest types occurring together as a mosaic on the glacial lacustrine and marine soils of the Champlain Valley. Two natural community types of the clayplain forest ecosystem are uplands (Mesic Clayplain Forest and Sand-Over-Clay Forest) and two are wetlands (Wet Clayplain Forest and Wet Sand-Over-Clay Forest).

The clayplain forest ecosystem covered nearly 220,000 acres of the post-glacial lake and marine plain of the Champlain Valley prior to European settlement. The predominately clay soils of these forests formed from sediments deposited in the Champlain Valley during and following the Pleistocene glaciation, both when the valley was flooded by a large freshwater lake, and later when salt water invaded the basin from the north. Moisture in these soils varies considerably with soil texture, topographic position, and slight elevation differences. In localized areas, lenses of sand and silt lie over the clay and result in variations in the forest communities.

These deep, fertile, stone-free soils have been prized for agriculture, and the majority of the clayplain forest has been cleared since the time of European settlement. The most well-drained areas of the clayplain were preferentially cleared for agriculture and the clayplain forests remnants that are left are generally on the moister sites, although they typically contain a mosaic of wet and mesic areas. It is presumed that the clayplain was nearly 100 percent forested prior to European settlement, but today only 12 percent of the clay soil in the southern Champlain Valley supports forest (Lapin 2003). Because of forest conversion, all four clayplain forest natural community types are rare in Vermont.

FIGURE 3.3: The first page of a detailed profile for one of nearly 100 natural communities identified in Vermont. Reprinted from *Wetland, Woodland, Wildland,* courtesy of Thompson, Sorenson, Zaino, 2000/2019, used by permission. Artwork by Libby Davidson.

database's range and instantly receive information on the type of natural community that ought to be there. Note that the range of data included on this site extends as far south as Virginia and as far west as West Virginia, Pennsylvania, and New York.

So, although the identifying characteristics of forest communities are comprised of and determined by individual plants, we know that the whole is greater than the sum of the parts. How do our management decisions influence the ecology and biodiversity of these communities?

EFFECTS OF COPPICE ON COMMUNITY ECOLOGY AND BIODIVERSITY

Resprout silviculture transforms forest ecosystems structurally and systemically. In most traditional coppice woodlands, the trees are all more or less the same age, and unless the rotations are exceptionally long (say beyond 25 to 30 years), they usually only express the early stages of forest growth and development. The rotational management between cants creates canopy gaps at a far faster rate than you'd find in an unmanaged high forest. Because of this, coppice stands support populations of flora and fauna adapted to these short rotations and periodic intervals of full sun. At the same time, they often lack the complex structure and deadwood resources needed by species adapted to mature forest habitats.

So it's a trade-off. Coppicing creates unique habitat, forming novel ecosystems that support the needs of some species while limiting the potential of others. Let's examine the implications of each of these qualities in detail.

Key Differences Between Coppice and High Forest

British ecologist G.P. Buckley describes coppice woodlands as "an intimate mosaic of small management units of disturbed, and relatively open, young woodland at close intervals of growth." Coppice woodlands differ from high forest stands in numerous ways. They have unique disturbance regimes, structural forms, gap formation rates, frequency of full-sun conditions, quantities of deadwood and mature trees, microclimates, species diversity, and water and nutrient cycles.[6]

Here are some specific differences:

- Coppice stands maintain an uncommonly high amount of open space, forming gaps at a rate of 5% to 10% per year, compared to 0.5% to 2% in "natural" forests. This favors early-successional, light-demanding species, whereas high forest succession trends towards more shade-tolerant canopy species.
- Coppice stands consist of small patches of even-aged growth roughly equivalent to a gap created by the death of 1 to 10 canopy trees.
- Large mature trees are relatively uncommon. Standard trees often lie along woodland margins, and in Europe, they frequently comprise just one or two genera (most commonly oak, *Quercus robur* and *Q. petraea*).
- Snags and large downed woody debris—crucial habitat for birds, mammals, fungi, and invertebrates—are uncommon in coppice stands.
- Woodland rides offer permanent open habitats rarely found in high forest.
- In many ancient European copses, woodsmen often transformed the sites' hydrology, straightening existing streambeds, and draining soils with ditches.
- Coppice stands often exist as isolated patches in the long term reducing total species diversity.[7]

As ancient human intervention caused woodlands to transition from "natural" uneven-aged stands to even-aged coppice, their populations of flora and fauna shifted as well. Organisms dependent on deadwood declined due to its relative absence, while species adapted to young growth, unwooded habitats, and forest gaps increased. Shade-tolerant herbaceous plant species likely suffered little, due to their capability to survive under both dense canopy cover and coppice woods' shifting light

Table 3.2: Environmental characteristics of coppice stands compared to unmanaged woodland

Site Characteristic	Coppice	Unmanaged Woods
Light patterns	Consistently shifting based on rotation, species, spacing, and growth rate	Uneven with high light intervals requiring the opening of large gaps
Soil and air temperatures	Following harvest, soil warms much earlier in spring, reaching higher overall summer temperatures Air temperatures significantly hotter in full sun	More even transitions throughout the year Slow to warm in spring and slower cooling in autumn
Soil nutrients	Potential depletion of phosphorus and other nutrients depending on rotation length, season of harvest, and types of materials removed	Steady accumulation of surface and soil organic matter
Soil disturbance	Largely depends on harvest timing but if done when soils are wet, can cause major damage to soil structure	Aside from harvest and extraction operations, little to no compaction from management

Adapted from Buckley, 1992, p. 117.

regimes. Because coppice systems support both shade-tolerant and intolerant species, some ecologists suggest that coppice systems' diverse habitats may actually support more species than equivalent areas of natural woodland.[8]

In the coming pages, we'll examine coppice woodlands' key habitat characteristics in greater depth in order to appreciate those effects we might consider beneficial, and learn how we might moderate, disperse, mitigate, or otherwise avoid any undesirable effects.

Forest Growth Stages

Forest stands possess unique qualities at different points in their growth cycle. Some ecologists call these stages *young, thicket, pole,* and *veteran* (See Figure 3.4). The young stage begins after a disturbance event and continues until it reaches canopy closure, the point where tree crowns converge and shade the understory. The thicket stage then begins, continuing as the canopy ascends and lower branches begin to die. Eventually, the stand starts to thin and canopy gaps form, marking the beginning of the pole stage, leading to the rapid growth of surviving trees. When stem growth eclipses the point of maximum average annual canopy increase, the stand enters the veteran stage. As canopy growth slows, the understory vegetation races to fill in gaps as they open.[9]

The point where a stand reaches canopy closure marks a radically important shift in habitat. Most coppice stands only experience the young and thicket stages. Under particularly short rotations of 2 or 3 years, the canopy may actually only just begin to close. Generally, the

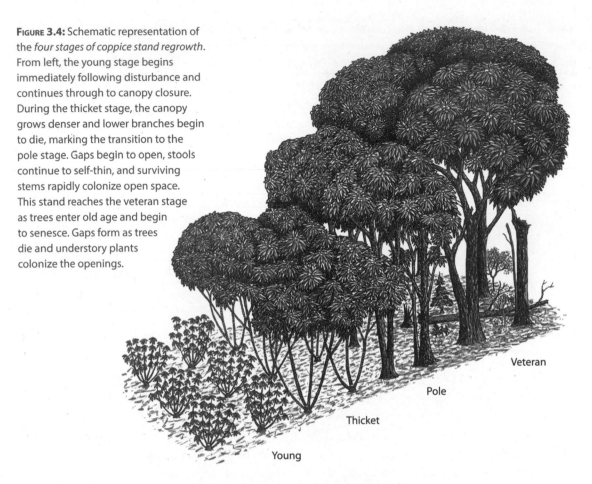

FIGURE 3.4: Schematic representation of the *four stages of coppice stand regrowth.* From left, the young stage begins immediately following disturbance and continues through to canopy closure. During the thicket stage, the canopy grows denser and lower branches begin to die, marking the transition to the pole stage. Gaps begin to open, stools continue to self-thin, and surviving stems rapidly colonize open space. This stand reaches the veteran stage as trees enter old age and begin to senesce. Gaps form as trees die and understory plants colonize the openings.

Veteran

Pole

Thicket

Young

denser the stand, the faster it passes through the early growth stages.[10]

The rate of canopy closure plays a major role in determining herbaceous ground flora composition. In British woodlands, plants like bramble (*Rubus fruticosus*) and bracken (*Pteridium aquilinum*) often dominate large canopy gaps following major disturbance, out-competing shade-tolerant understory herbs.[11] Canopy gaps reset successional processes and set the stage for the next expression of the stand.

Forest Gap Formation

Disturbance events form canopy gaps that transform a forest's structure by reorganizing plant communities and enabling new individuals and species to establish.[12] A gap's size, solar orientation, and relationship to the surrounding forest influence the scale of these changes.

To quantify gap size, we compare the ratio of gap diameter to the average height of adjacent trees. For example, a 100-foot-wide gap surrounded by 80-foot-tall trees creates the same relative opening as a 50-foot-wide gap

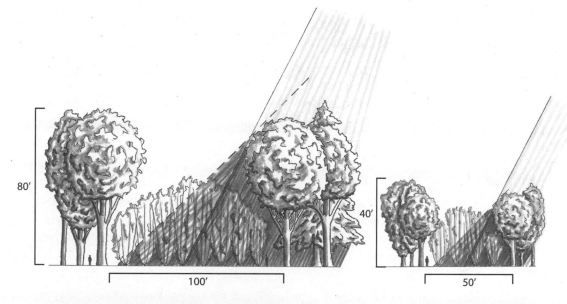

Ratio of Surrounding Tree Height to Gap Size

Figure 3.5: The scale and impact of a forest gap is a function of the gap's width along with the height of the surrounding vegetation. Here, two different-sized gaps create the same relative conditions due to the proportional relationship between gap diameter and the surrounding canopy. Assuming we're in the northern hemisphere, south is to the right, and we see the sun angle and resultant shade modeled at both the summer solstice and the equinoxes at 45 degrees latitude.

surrounded by a 40-foot canopy (See Figure 3.5). The use of a ratio compensates for these variations in scale and indicates the relative amount of available incoming light.[13]

Coppice management and natural canopy gap formation share several similarities, namely, the changes in forest floor microclimate. Gap creation causes a wider daytime fluctuation in air and surface temperatures and increases air movement, soil moisture (due to reduced transpiration), nutrient availability, and available light. Usually, these changes improve plant growth conditions, but gaps also create some drawbacks. Without a forest canopy's insulative

effects, nighttime temperatures drop, increasing the risk of frost damage and possibly a shorter growing season. Gaps may also alter soil hydrology, increasing erosion, runoff, evaporation, and more frequent soil water stress.[14] By understanding these impacts, we can work to accentuate the positive while mitigating the disadvantages.

Coppice-created gaps differ from natural gap formation in both frequency and distribution. **Gap formation rate** measures a stand's percentage of total land area annually affected by disturbance. It typically ranges from 0.5% to 2% per year in most forest ecosystems, translating into a "canopy turnover rate" of 50 to

200 years. In other words, natural forest gap creation would completely replace the forest canopy over the course of 50 to 200 years.

By comparison, coppice systems are usually cut on roughly 5-to-20-year cycles, a turnover rate ten times faster than that of natural forests. The frequency, speed, and regularity of this disturbance cycle eliminates the unpredictability or "stochasticity" of a natural forest stand.[15] On a positive note, the rapid and consistent recolonization of the canopy by resprouts can help support a dynamic yet stable understory population.[16]

Gap Size

In southern Appalachia, Emory University researchers studied the effects of gap size on the ensuing early-successional community. They created four different size gaps (0.04, 0.2, 1, and 4.9 acres [0.016, 0.08, 0.4, and 2.0 ha]) and measured productivity, species diversity, soil and air temperature, and solar radiation. The largest gaps had the highest solar radiation and soil and air temperatures, increasing from the gap's edge to its center. In the largest gap, increased light caused a 300% to 400% increase in net primary productivity (NPP) over the smallest gap. Root and stump sprouts represented the largest contribution to this total aboveground NPP in all gaps. By the second year after cutting, vegetation biomass reached just 0.7% to 2.6% of levels found in undisturbed forests, while aboveground NPP reached 17% to 58% of forest levels.[17]

They found greater plant species richness in the largest gaps, and the species composition of the gaps changed dramatically post-disturbance. Tulip poplar (*Liriodendron tulipifera*) remained a major component before and after logging, though once-dominant oaks and hickories contributed little to the sprout population. Instead, understory species and less common canopy species including black locust, Carolina silverbell, red maple, flowering dogwood, and magnolia (*Robinia pseudoacacia, Halesia carolina, Acer rubrum, Cornus florida,* and *Magnolia fraseri*) sprouted most prolifically.[18] So it appears that, from a biological diversity and productivity point of view, larger gaps are better, opening the stand to more light and creating a more favorable edge to interior ratio.

Genetic Diversity

Because resprout silviculture relies on vegetative reproduction to populate forest stands following harvest, some biologists worry about the effects of repeated generations of vegetative regeneration on forest genetics. These stands do not benefit from the dispersal of new genetic material that occurs in high forest stands of seed origin. While this is certainly a legitimate long-term concern, it appears that this issue may not be as clear-cut as it might initially seem.

A 2008 Spanish study examined genetic diversity in rebollo oak (*Quercus pyrenaica*) coppice stands. Managed for centuries on 12-to-20-year cycles for fuelwood, charcoal, cattle fodder, and bark for the leather industry, the rebollo oak sprouts rapidly in response to the region's frequent ground fires and human disturbance.[19]

Foresters had assumed these stands were composed of large clonal colonies of just a few vigorous individuals, but their research actually found high densities of many small stools resulting in a high level of genetic diversity and a large number of unique genotypes. Despite at least 15 coppice rotations since 1750, they found low levels of clonal propagation, suggesting the presence of many young seedlings with diverse genetics around the time the stand was first coppiced. Subsequent coppicing stimulated intense competition between individuals creating small clones, helping actually maintain genetic diversity.[20] So, perhaps by establishing a new coppice stand following a season of vigorous natural (seed-based) regeneration, we may be able to build a system from a diverse, locally adapted genetic pool.

Effects of Disturbance on Ground Layer Populations

Ancient European coppice stands commonly support richly diverse communities of flora and fauna—especially those species adapted to the early stages of forest growth. Continuity in management, predictability of disturbance, and coppice management's varying light levels all contribute to this diversity.

Forest ecosystems' patterns of light and shade drive their ecological processes and species composition. Following disturbance, most ground flora species flower and seed more prolifically, regardless of shade tolerance. As the canopy closes, flowering and seed production become irregular and less frequent, and shade-intolerant species begin to disappear.[21]

This cyclical light regime is responsible for England's spectacular spring woodland flower displays.

Woodland ground flora includes two primary species groups: *early- to mid-successional, light-demanding (shade-intolerant) plants* whose populations largely rely on their persistence in the seed bank, and *late-successional, shade-tolerant plants* that have adapted to varying levels of canopy closure and to low-light periods by relying on energy storage organs like bulbs and corms. Disturbance helps reorganize ground flora populations, creating a window for new species' establishment.[22]

Many species' seed may remain dormant for long periods between fellings. Early-successional species commonly dominate seed bank populations, while late-successional species' seeds tend to have more limited viability.[23] After about 50 years, the viability of most seeds in the forest seed bank begins an accelerated decline, but wood-spurge (*Euphorbia amygdaloides*) seed can remain viable for a remarkable 125 years![24] Creating temporary canopy gaps increases the likelihood that rare species that rely on seed production can reproduce and establish themselves on new sites.[25]

Frequently disturbed ecosystems often possess larger seed banks with species strongly correlating to those already present, compared to infrequently disturbed ecosystems.[26] Some of these seed bank plants (the St. John's worts, *Hypericum* spp.; fig wort, *Scrophularia nodosa*; wood spurge, *Euphorbia amygdaloides*; and ragged robin, *Lychnis flos-cuculi*) usually appear during the first or second year after coppicing, persisting for 2 to 5 years depending on stool vigor and density.[27] A few relatively shade-tolerant European perennial herbs include wood anemone (*Anemone nemorosa*), primrose (*Primula vulgaris*), bluebell (*Hyacinthoides non-scripta*), dog's mercury (*Merculialis perennis*), and ivy (*Hedera helix*).

In most circumstances, coppice stands reach canopy closure so rapidly after felling that the intervals of high-light conditions do not significantly inhibit shade-demanding species. (Oliver Rackham assessed plant population shifts in five coppice stands using woodland records dating back to the 18th century, and discovered minimal disappearance of shade-tolerant herbaceous species.[28]) Of additional note, during these shady phases, species that spread rhizomatously tend to disperse more effectively than those that rely on seed.[29]

Table 3.3

Adoxa moschatellina
Calamagrostis canescens
Campanula latifolia
Carex pendula
Conopodium majus
Narcissus pseudonarcissus
Orchis mascula
Paris quadrifolia
Primula vulgaris
Ranunculus auricomis
Veronica montana
Viola reichenbachiana

Select European perennial herbs likely to benefit from coppice management. Adapted from Buckley, 1992, p. 142.

The "intermediate disturbance hypothesis" proposes that disturbance events of *moderate* frequency and intensity cause the greatest species diversity.[30] In the absence of disturbance, forest communities trend towards homogeneity, favoring the species best-adapted to the site. A 29-year study in an undisturbed North American forest found a sevenfold increase in populations of the species already most abundant and the decline or disappearance of rare ground layer species.[31] At the landscape scale, concentrated disturbance events like coppicing often help promote diversity.[32]

Large swaths of European woodlands have been coppiced for hundreds, if not thousands, of years, leading to the evolution of specialized ecosystems adapted to coppice management's cyclical rotations. Many coppice stands are actually long-managed remnants of original old-growth forests and include species well-adapted to the site's unique characteristics. Ancient coppice woodlands' disproportionate presence of rare indigenous plant and invertebrate species with limited powers of dispersal and colonization is a living testament to this history.[33] These species have either persisted in place from the original old-growth stand or the stands have been managed long enough to provide rare and relatively immobile plants time to colonize it.[34]

So, we see that traditional coppice systems' centuries-long management patterns have created diverse, shifting ground flora populations that respond to varying light regimes. At the patch scale, these patterns constantly change, but at the landscape scale, they maintain relative stability, helping support enhanced biological diversity. In a similar way, each patch's microclimate follows a steadily shifting pattern that also influences wildlife and vegetation populations.

Growth Stages and Microclimate

A **microclimate** includes the localized set of climate conditions affecting a small area (solar radiation, shade, wind, precipitation). Microclimates exist at varying scales. When at

The Tree Community: Woody Plants, Mycorrhizal Associations, and More

A tree without fungi would be like a stage without actors.

— Alan Rayner[35]

Ninety-five percent of green plants depend on at least one fungal relationship to survive. All **mycorrhizal** (literally "fungal root"—symbiotic fungi that provide plant roots with nutrients and water in exchange for plant-photosynthesized sugars) relationships are beneficial and aid the growth and development of both species. Fungal populations are so widespread that they may comprise as much as 55% of a forest's biomass.[36]

Mycorrhizae's extensive fungal network can increase woody plant roots' absorptive surface area by 700% to 1,000%, improving drought resistance and nutrient uptake and contributing to the plant's overall health and vitality.[37] Phosphorus is the most common nutrient transferred to roots via mycorrhizal association, especially in soils otherwise deficient. In exchange, many plants transfer 20% or more of the total carbon they assimilate to their fungal partners.[38]

The forest's mycelial network creates "communication channels" that allow mature plants to support nearby seedling growth. It may also help reduce interspecies competition and improve nutrient uptake among interconnected individuals.[39]

A German study investigated ectomycorrhiza (EM) formation in poplar and willow (*Populus nigra x maximowiczii* and *Salix viminalis*) short-rotation coppice managed on 3- and 6-year rotations. EM are the most common mutualistic poplar and willow symbionts. Generally, EM colonized poplar fine root tips more frequently than willow. Interestingly, the 3-year poplar rotation hosted substantially larger EM concentrations than the 6-year rotation. The researchers concluded that shorter, more frequent harvests promote mycorrhizae association with poplars, theorizing the plants send signals to nearby colonizing mycorrhizae to meet their increased demand for mineral nutrients.[40]

Trees also maintain essential relationships with decay-causing, **saprophytic** fungi, the only organisms capable of breaking down lignin. Saprophytic fungi drive forest ecosystems' biomass and nutrient cycles. Fungal decay and hollowing play a part in woody plants' natural aging process, and they also support increased biodiversity as food and/or habitat.[41] This interconnected network mycelium create and saprophytic fungi's constant cycling of decaying deadwood are potent expressions of the forest as a "whole system."

the same stage of growth and comprised of the same species, coppice and high forest stands will have similar microclimates. Ultimately, it's the patch-scale rotation length that drives microclimatic changes. And at the landscape scale, these unique structural patterns create a patchwork of microclimates. While high forest generally has a homogenous microclimate across large expanses, the mosaic of managed coppice stands that express all age classes at any given point in time have a huge potential for microclimatic variation.[42]

Besides aspect (the site's orientation in relationship to the sun), the primary factor that influences a site's solar and thermal radiation, temperature, humidity, and wind is canopy density or **canopy cover**. Dense, thick canopies better insulate the understory from fluctuations in temperature, radiation, and humidity and reduce wind velocity. A study monitoring air temperatures in four chestnut coppice stands of varying ages found average air temperatures in stands yet to reach canopy closure 3.6° to 5.4°F (2° to 3°C) higher in late spring and early summer and 1.8° to 3.6°F (1° to 2°C) higher during the rest of the year than those under a closed canopy.[43] Patches that include standard trees are better sheltered from the effects of hard frosts and drying winds.[44]

Solar radiation has a major effect on the vigor of a forest's herb layer. **PAR, or photosynthetically active radiation** (400–700 nm wavelengths), and the R/FR ratio, or the ratio of red to far red radiation, are the two primary types of radiation scientists measure.

Essentially, think of PAR as light *quantity* and the R/FR ratio as light *quality*.[45]

Researchers monitored 12 coppice plots over 18 months in three south England locations, tracking variations in PAR and the R/FR ratio in various sweet chestnut (*Castanea sativa*) and willow coppice stands. In summer, the dense canopy caused low PAR levels, and even in winter months just 50% of incident PAR reached the understory. Interestingly, the open canopy and increasing solar altitude in early spring actually showed the highest recorded PAR levels, enabling more light to penetrate tree crowns just beginning to break bud.[46]

Most woodland plants grow and flower at PAR levels of 5 E/m^2/day, while levels of 1 E/m^2/day allow shade-tolerant species to survive, though few will flower. In pre-canopy closure chestnut stands, the understory received more than 5 E/m^2/day PAR in spring and summer. But after reaching canopy closure (usually the third year of regrowth), just a few percent of incident PAR penetrated the fully developed but leafless canopy. And in the case of willow, fast growth and dense spacing caused stands to reach canopy closure in the first year, transmitting PAR levels of 5 E/m^2/day for just a few months in early spring during the second year of growth. Even in winter months, dense leafless stems intercepted 55% to 75% of incident PAR.[47]

Management periodicity has a huge effect on coppice stands' herbaceous diversity. Herbaceous plant populations often decline with lengthening management cycles so that even shade-tolerant species begin to disappear.

Our coppice management density and rotation length determine each growth stage's interval along with the timing of the transitions. If we understand the needs of the flora and fauna we want to encourage, we may be able to choose management strategies to maintain their optimal growth stage and microclimate.[48]

And even individual trees can create uniquely important microclimates. Ancient trees can provide essential habitat for specialized flora and fauna including lichens, invertebrates, hole-nesting birds and bats.[49] Old pollard boles' morphology, physiology, and longevity support substantial bryophyte (moss and liverwort species) diversity, especially when compared with unmanaged trees. Their diverse substrates (buttresses and crowns, exposed roots, bark fissures, epicormic cankers, decaying ancient wood, and the smooth young bark on young poles) create diverse substrates for them to grow on. Rough surfaces and an open crown collect and store rainwater, while sunlight penetrates deeply into the crown.[50]

And while this patchwork of canopy cover leads to varied and diverse habitats, some of Europe's most valuable habitats can be found along the access ways that transect managed woods.

Effects of Access Rides and Woodland Roadways

Over the past 300 years, **woodland rides**, or wood product access and extraction routes, began to emerge in Europe, as an expansion of the existing network of footpaths and small tracks. These networks often comprised intermingled grassy strips, herbaceous plant communities, and shallow marshlands, connecting grasslands to adjacent coppice stands.[51] While roads fragment forest ecosystems, they also create diverse new habitats for woodland flora and fauna. Coppice woodlands often lie along the edges of town margins, connecting village centers with riparian habitats, with networks of hedgerows linking otherwise discontinuous coppice cants.[52] Forest roads'

Table 3.4: Light, nutrient, and disturbance levels of four habitats adjacent to woodland rides

	Disturbance Level	Nutrients	Light
Young Stands Roadway	High	High	High
Mature Stands Roadway	Medium or Medium-High	High	Medium-High
Young Stands Interior	Medium or Medium-Low	Mid	Medium-Low
Mature Stands Interior	Low	Low	Low

A comparison of key ecological characteristics of four different habitat types along woodland rides and interiors in young and mature French lowland oak forests. Adapted from Avon et al., 2010.

ecological effects extend well beyond the spatial extent of the road itself, with edges between ecosystems blurring with increasing distance.[53]

French researchers examined the effects of forest road distance on understory diversity in 20 young and mature sessile oak (*Quercus petraea*) stands with long management histories (Table 3.4). Both stand age and distance from the roadway markedly influenced total species richness. They found the greatest species diversity in young stands and along the road margins.[54] Beyond 16.4 feet (5 m) from the road edge, total species richness declined considerably with more or less homogenous diversity further into the forest interior. Available light and soil nutrients strongly influenced this species richness. Forest herbs favored the road verge and embankment in the adult stands, preferring the forest interior in young stands.[55]

While stand structure, light regimes, and the age when stands reach canopy closure have a strong effect on understory flora, they also influence a woodland's avian population.

Coppice and Bird Species

Bird species' abundance and diversity often directly relates to forest structural complexity.[56] British research concludes that a woodland's regrowth stage plays a major role in the bird species present, breaking them into three broad phases: *establishment*, including the years immediately following coppicing; *canopy closure*; and *maturation*, or the years following canopy closure.[57] So as coppice stand structure changes from exceedingly open, to low, dense

vegetation, to a full ascending closed canopy, different bird species occupy the varied niches.

Spatial variation in shrub layer density may be the single-most important factor determining bird distribution. The most abundant bird populations exist in woods with well-developed shrub layers. Migrant bird species prefer the more open habitat of early-successional stages, while resident species tend to prefer mature forest and other structurally complex habitats. As the shrub layer disappears, so do its abundant nesting and foraging niches, and migrant bird populations tend to relocate to other sites.[58]

European coppice woodlands support high densities of songbird species.[59] Often occupying compact, distinct territories, songbird habitat preferences closely match the scale and structure of individual coppice cants.[60] In Britain, the tree pipit (*Anthus trivialis*), dunnock (*Prunella modularis*), whitethroat (*Sylvia communis*), and yellowhammer (*Emberiza citrinella*) prefer the establishment phase, while the canopy closure phase supports increasing numbers of wrens (*Troglodytes troglodytes*), robins (*Erithacus rubecula*), nightingales (*Luscinia megarhynchos*), blackbirds (*Turdus merula*), blackcap (*Sylvia atricapilla*), chiffchaff, several warblers, and bullfinch (*Pyrrhula pyrrhula*). Robin and tits (*Paridae* spp.) dominate stands during the maturation phase.[61]

Coppice stands can also support extraordinary populations of breeding migrant birds. These populations often peak during the establishment and canopy closure phases. In the UK, turtledove (*Streptopelia turtur*), tree

Table 3.5: Songbird densities and resident to migrant bird species ratios in different coppiced habitats in the UK

Habitat Type	Breeding Pairs per Hectare	Ratio of Migrant to Resident Species
Basket Willow Coppice	15.0	0.66
Short Rotation Willow Coppice	8.6	0.71
Traditional Coppice Woodland	7.6	0.64
Unmanaged Willow	6.0	0.54
Unmanaged Traditional Coppice	5.0	0.12

This table demonstrates the high value of managed and pre-canopy closure coppice to bird species. We also tend to see higher bird population densities in shorter rotation systems. Adapted from Sage and Robertson, 1994, p. 44.

pipit, nightingale (*Luscinia luscinia*), white-throat, garden warbler, and willow warbler (all of which winter south of the Sahara), and the blackcap and chiffchaff (which winters around the Mediterranean) often take up seasonal residence in coppice stands.[62] The modern loss of dense scrub habitats and thick coppice regrowth has led to the severe decline of populations of one of Britain's best-loved birds, the nightingale. Though migratory, nightingales tend to only return to coppice stands with a long continuity of management and are reluctant to colonize coppice that has been restored after many years of neglect.[63]

Management can affect bird habitat quality via rotation length, stool density, woody species present, and cant size. Rotation length governs the frequency and duration of pre-canopy closure habitat, the most valuable growth periods for birds.[64] High stool density tends to provide the best-quality habitat. Patchy stands with scattered stools may never reach canopy closure, creating poor habitat for breeding migrant birds.

Different tree species support different bird species, offering unique cover and feed opportunities.[65] Cant size also plays an important role, as most migrant species require a minimum breeding territory of 1.25 acres (0.5 ha). Therefore, stands managed for bird habitat should include cants this size or larger. To maintain habitat continuity, growers can harvest coppice cants adjacent to one another to encourage a gradual transition in stand structure.[66]

In 1987, British researchers conducted bird counts in alder, ash, birch, hazel, and oak coppice at Bradfield Woods, Suffolk, UK, where documented coppice management dates back to at least 1252. They identified 43 breeding species in total, 26 of which were songbirds. They found the lowest total songbird density in very young (<3 years) and old (>11 years) coppice stands, with no species in the oldest stands. Total species peaked between years 3 and 7, with the greatest densities in 5-year-old stands. Stands with a low canopy at the middle stages of growth and nearing closure had the greatest bird species diversity.[67]

In another British study in a 54-acre (22 ha) sweet chestnut (*Castanea sativa*) copse, resident and migrant bird species populations steadily increased during the establishment phase, maintaining high, stable densities during the first 4 years of regrowth. After the eighth season, species richness dropped almost immediately as many stands approached canopy closure.[68]

Similarly, a five-year study at Ham Street Woods in Kent, UK, in the early 1970s found that migrant bird species, and species with narrow habitat amplitudes, strongly preferred the early stages of coppice growth with moderate stool densities, whereas resident bird species tended to prefer more mature woodland habitats. This preference for young coppice appears consistent throughout much of England and Europe, with migrant species usually only using old coppice when bordered

by adjoining young coppice. The results of these studies appear to suggest that coppice managed for diverse bird species conservation should maintain as large an area as possible in pre-canopy closure coppice.[69]

Coppice stands also create excellent habitat for pheasant populations who favor woodland cover during the winter months and find breeding territory in open crop fields during spring and summer. The quality of this habitat depends on both the shelter provided by shrubby cover and a high ratio of edge to internal area, exactly the type of habitat that young managed coppice growth provides. In fact, pheasant hunt success points to significantly higher densities in woodlands with well-developed shrub layers. Coppice stand structure during the first few years after cutting is ideal, peaking at 3 years for shrubby species like hazel and 6 for the slow-growing hornbeam. As stands age, pheasant populations appear to level off. For optimal pheasant management, maintain small cants on a short rotation, designed to maximize edge-to-area ratios using compact or irregularly shaped plots, woodland rides, and boundaries between stands of different age classes.[70] See Figure 3.6.

There's some dissent as to the quality of bird habitat provided by alder, hybrid willow, and poplar short-rotation coppice. Some European research indicates they offer little support for bird populations of interest to most conservation organizations. Rapidly reaching canopy closure due to their fast growth and high density, short-rotation coppice often supports low insect populations and provides little food

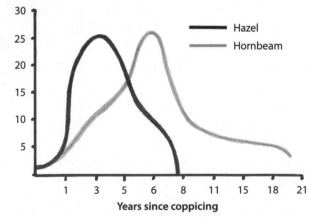

FIGURE 3.6: Attractiveness of different age classes of hazel and hornbeam coppice to pheasants. Adapted from Buckley, 1992, p. 197.

for many bird species. Nevertheless, managed osier coppice (*Salix viminalis*) in Attenborough, UK, has the greatest breeding bird concentrations of any British scrub habitat. The *Salix* genus supports more arthropod species than any other native British tree, and willow's short rotation ensures that ⅓ of the managed area will typically be cut in any given year, creating a large amount of internal edge, and attracting a rich arthropod food supply for insectivorous birds. Willow coppice at Leighton Moss in Lancashire attracted 2 to 3 times as many songbirds as an unmanaged stand.[71]

Standard trees support arboreal bird species including hole-nesters who prefer their

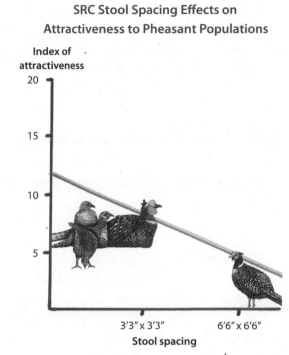

SRC Stool Spacing Effects on Attractiveness to Pheasant Populations

FIGURE 3.7: Higher-density short-rotation coppice (SRC) stands provide more attractive habitat for pheasant populations. Adapted from Sage and Robertson, 1994, p. 45.

Table 3.6: Effects of planting density, edge-to-interior ratios, and available groundwater on vegetation and bird populations

	Vegetative Effects	**Effects on Bird Populations**
Ratio of Coppice Edge to Interior	Sites with a higher ratio of edge to interior create more interfaces between coppice patches and other ecosystems, causing varied microclimates and species mixes and densities	Supports larger woodland bird species like the song thrush, missile thrush, and woodpigeon in addition to ground and leaf foraging bird species
Planting Density	High density stands only usually result in one year of pre-canopy closure conditions. This causes less available understory vegetation and open space	Undesirable for many species that prefer young woods and shrublands like wren and whitethroat
Available Groundwater	More groundwater = wetter soils and perhaps different dominant vegetation	Assuming suitable understory vegetation, wetter sites support moist-preferring species like bluethroat and reed bunting

Adapted from Londo et al., 2005, p. 2.

structure and height for feeding and nesting sites. They also offer song posts and foraging sites for insectivorous bird species. But a 1989 study in Ham Street Woods found higher standard densities tend to dissuade migrant bird species populations, finding larger populations in young coppice with few standards (<10 per acre or 25 per hectare) than similarly aged coppice growth with many standards (>20 per acre or 50 per hectare).[72]

Mature diverse hedgerows also offer many farmland birds quality nest sites, especially those bound by unmanaged vegetation that yields insects and seeds. This allows birds to forage and feed young in the relative safety of the hedgerow. As with coppice stands, different bird species prefer hedges with different structural forms. Wrens, robins, hedge sparrows, pheasants, and other game birds prefer hedges with a thick base whose cover allows them to safely scratch for insects and find shelter from winter cold. Songbirds like the blackbird and thrush forage at the base of hedges, nest in the middle regions, and sing from the top.[73]

Coppice and Small Mammal Populations

Small mammal populations also benefit from the increase in edge habitat and diversity of food and cover available following forest gap formation. According to Carey and Johnson, the two most important habitat components for small mammals are understory vegetation and coarse woody debris or deadwood. Coppice management can provide these features, often by the end of the second year, with abundant ground cover 3-to-6 feet (1 to 2 m) in height,

a more balanced microclimate, and increased populations of herbaceous vegetation and invertebrates.[74]

Most small mammals' (including shrews, voles, and mice) life span rarely exceeds one year. Their populations have a high turnover rate. In quality habitat, animals travel shorter distances over a more concentrated range, enabling higher population densities. It appears that several small cants (1.25 to 2.5 acres or 0.5 to 1 hectare) offer better habitat than a few large ones, while hedgerows, woodland rides, and other edge habitats further increase structural diversity and facilitate travel between habitats.[75]

Despite seasonal fluctuations in food supply, small mammal populations typically peak during autumn and early winter, declining during winter on into spring. One-to-7-year-old coppice stands support the highest small mammal diversity. A study at Suffolk's Bradfield Woods found the greatest diversity and highest concentrations in 3-year-old coppice with populations declining thereafter.[76]

If managed on long-enough rotations, chestnut and hazelnut coppice can also support small mammals via their mast (nuts). While bearing age depends on site quality and species, hazelnut and sweet chestnut roughly take 4 to 5 and 4 to 7 years, respectively (in the UK) to produce seed (although chestnut may not produce good seed crops until 20 years after cutting). Standard trees (ideally 10 to 20 per acre or 25 to 50 per ha) create even more structural diversity and increase the available food supply via flowers, seeds, and invertebrate populations. Squirrels particularly gravitate to cants that include standard trees, especially if they're mast-producing species like oak and beech.[77]

Of course, many small mammals can create big problems for coppice growers. In British hazel coppice, squirrels can consume as much as 81% of the nut crop. Gray squirrels also strip bark off smooth-barked coppice poles (including sycamore, hornbeam, and sweet chestnut), rendering them unsaleable.[78] Rodents girdle the stems of young sprouts as they feed on the nutrient-dense cambium below, while rabbits browse down the tender young shoot tips. Needless to say, improved habitat for small mammals is a pretty uncommon goal for most growers, and in many cases, an unfortunate unintended management byproduct.

But English coppice stands that support small mammals do create valuable hunting grounds for predatory species, including fox (*Vulpes vulpes*), stoat (*Mustela erminea*), weasel (*M. nivalis*), mink (*M. vison*), polecat (*M. putorius*), pine marten (*Martes martes*), and wild cat (*Felis catus*). In a similar way, they also can support woodland bat populations by hosting diverse invertebrate communities. Managed pollards and ancient trees also offer valuable bat habitat for summer nursery roosts and potential protected winter hibernacula in cavities formed by woodpecker holes, lightning damage, hollows, and cracks.[79]

In addition to supporting numerous small mammals and their larger mammalian predators, coppice (somewhat regrettably) creates high-quality cover and browse for some of the most ubiquitous mammals in the suburban and peri-urban landscapes of today—deer.

Coppice and Deer

While it's pretty unlikely that you're aiming to create coppice to support deer populations, it does provide excellent deer browse and cover. During the first few months of regrowth, coppice yields little available forage, but that changes with time. Once stands reach canopy closure, deer enjoy the large quantities of woody and herbaceous browse available from late spring through early summer and on into winter. In late summer, autumn, and winter, they prefer more mature stands that lack a developed herbaceous understory. During the 5 to 10 years following cutting, large populations may achieve high reproductive rates and body weight gains.[80]

Of course, deer often dramatically impact successful coppice regeneration. As they browse, they affect the stand's vegetative structure, potentially impacting small mammals, invertebrates, and birds—especially species that nest in low vegetation and bramble. In an experiment comparing the growth of protected (fenced) vs. unprotected hazel coppice, deer browsed more than 80% of unprotected shoots on coppice stools, with the longest reaching just 4 inches (10 cm) height the October after felling, whereas protected sprouts grew 4.9 to 6.6 feet (1.5–2 m) tall. While careful design of coppice cant layout can help to limit browse damage, fencing is the only guaranteed long-term sprout protection strategy.[81] We'll explore strategies to minimize deer's impact on coppice stands in chapter 7.

Despite the damage they can cause, deer do play an important role in forest ecosystems, and moderate levels of browse can actually benefit forest dynamics.[82] In some regions, foresters

Case Study: NRCS-funded White Pine and Aspen Patch Cut

During the summer of 2017, we received a Conservation Stewardship Program grant from the Natural Resource Conservation Service (NRCS) to help fund a patch cut in our 40-acre woodlot under Conservation Practice 666: Forest Stand Improvement. With close to 8 acres of low-quality 60-year-old pasture pines and quaking aspen, very little understory growth and virtually no marketable value, we identified a 1-acre patch that fit the key practice parameters. It had low populations of opportunistic (invasive) species like Japanese honeysuckle and common buckhorn, a minimum of one acre in size, and was located outside nearby riparian buffers.

We were interested in harvesting building materials from on-site, and many of these pines were 14-to-20-inch dbh and close to 100 feet tall. Although they were covered in branches all the way to the ground, we'd only intended to use the materials for our own on-site construction projects, so we weren't too concerned about quality. It was clear that creating a targeted disturbance in the woods would produce much more diverse habitat in an otherwise uniform, low-quality stand that would benefit birds, mammals, and insect species and stimulate rapid regeneration.

Although we were required to do a fair bit of prep work, removing all buckthorn and honeysuckle plants before commencing cutting across the patch cut area and within a 50-foot buffer surrounding it, we were allowed to keep all materials and were given a yearly payment, spread over the course of 5 years to incentivize the work.

The primary aim of the patch cut was to increase and enhance food sources and habitat for diverse wildlife populations by initiating the growth of thousands of young new woody stems per acre. This forest architecture creates excellent brood rearing conditions for the American woodcock and ruffled grouse, chestnut-sided warbler, turkey, sparrow, and other species who value early-successional habitats offering thick cover. Additionally, insect biomass often explodes in these zones of young growth yielding valuable protein-rich food for young birds.

The copious woody browse that develops post-harvest also benefits deer, moose, bobcat, snowshoe hare, and turkey. We were allowed to retain up to 10% canopy cover including any valuable seed trees, mast-producing tree species, as well as potential northern long-eared bat roost trees like shagbark hickory (*Carya ovata*). The plan also called for us to retain as much woody debris as possible to protect the soil, retain moisture, and decompose and feed soil biology.

We completed the cut over the course of about 5 weeks that winter using a chain saw and 50-horsepower tractor, selecting and sorting sawlogs into stacks, building brush piles, and leaving low-quality logs on the forest floor to decompose. We also girdled 8 low-quality mature trees to create standing dead snags to serve as roosts for birds and provide habitat for small mammals.

While the ground was still frozen, we hired a local logger to bring out his tractor and forwarding trailer to carefully remove the sawlogs and stack them ☞

FIGURE 3.8: Two growing seasons following the completion of a 1-acre patch cut. The small grove of quaking aspen in the original stand has expanded to occupy close to ⅓ of the total area, with growth dwarfing the next most vigorous woody species. Whereas pine and oak seedlings are 8 to 10 inches (20 to 25 cm) tall on average, most aspen sprouts are more like 6 to 8 feet tall (1.8 to 2.4 m) after just 2 seasons. It is rapidly filling in with healthy new growth, although large swaths of the patch are still relatively open with young seedlings still several years from reaching a closed canopy.

neatly at our log landing for milling that spring and summer. The 1-acre cut yielded close to 14,000 board feet of lumber and 3 cords of firewood.

Within the first season following the patch cut, the small patch of aspen had exploded, creating thousands of suckering shoots that were 6 feet tall by the end of the summer. The white pine seedlings were much slower to take off but, after several seasons, are just finally beginning to colonize the space. While the few small seed trees we were able to retain haven't made a noticeable contribution to the stand's population, we have a rejuvenated, vibrant young stand now occupying a formerly dark, dense, low-grade patch of woods, creating an oasis of sunlit habitat for diverse wildlife populations. And we received financial support to carry it out. As the poplars continue to grow and mature, I plan to selectively thin them, using the poles for substrate to grow oyster mushrooms.

even seek to support and encourage deer populations both for ecosystem health and to maintain healthy population dynamics. In this case, coppicing is a great tool to provide deer with abundant, vigorous, supple young browse.

Coppice and Invertebrates

In forested ecosystems, invertebrate diversity tends to increase with age due to increases in structural diversity. Complex high forests with diverse populations of large trees can support more invertebrate species than herbaceous vegetation by an order of magnitude. Often lacking in deadwood and structurally simple, coppice systems once again tend to favor invertebrate species adapted to the early stages of forest growth.[83]

Mature trees contain myriad niches capable of supporting far more invertebrate species than seedlings. And yet, young trees and shrubs can still support considerable invertebrate diversity during the first few years of growth. In as little as 2 to 3 years, diverse coppice growth may host over 100 species of macrolepidoptera (larger butterflies and moths), while a 10-to-15-year cycle should provide habitat for a wide diversity of invertebrates, especially if it includes standard trees.[84]

Many invertebrate species associate closely with only one or just a few woody species. This means more diverse woodlands usually support greater invertebrate diversity. Some ecologists suggest coppice-with-standards systems' increased structural diversity creates optimal invertebrate habitat at standard densities of 6 per acre (15 per hectare) or fewer.[85]

Coppice woodlands' sheltered microclimate accelerates invertebrate development during their early life stages. August ground temperatures in first-year coppice stands may be as much as 18°F (10°C) higher than in closed canopy stands and 5.4°F (3°C) warmer at 2 to 3 years age. This protection, along with higher overall temperatures, can help increase ground and herb-dwelling invertebrate diversity and populations.

Young coppice growth's tender, nutrient-rich leaves and shoots also offer valuable food for phytophagus (plant-eating) insects. Coppice stools' young leaves grow particularly large, sustaining high rates of photosynthesis. They also tend to be more nutrient-rich than mature leaves, containing high concentrations of water and available nitrogen. This young growth provides an abundant food source for developing insects.[86]

Swiss researchers monitored the effects of renewed coppice-with-standards management in an abandoned woodland on the diversity and abundance of four beetle families (*Buprestidae, Cerambycidae, Lucanidae,* and phytophagous *Scarabaeidae*). Stumps and branch heaps offered habitat well-suited to beetle larvae development. Coppice stands' dynamic light regime particularly benefitted thermophilous (heat-loving) beetles, and with adequate deadwood, coppice-with-standards plots may even recreate ecotones like those that occur as ancient trees fall and create gaps in primeval forests.[87]

Coppice stands also create quality habitat for many moth species. Of the 800 species known in the UK, ⅔ regularly utilize woodland habitats. Scientists in southern England sampled night flying micro- and macro-moth populations in coppice stands ranging from 1 to 20 years regrowth and found that different species preferred the varied conditions present in three general growth stages: post felling (years 1 to 4), pre-canopy closure (5 to 9), and post-canopy closure.[88]

Moth species' larvae associated with the earliest growth stage primarily feed on herbaceous vegetation like bracken, rushes, grasses, and sedges, while over half the species occupying the middle growth stage feed on trees and shrubs, tending to utilize open woodland and scrub habitats. The greatest number of moth species and individual moths resided in the post-canopy closure cants. They included specialist species whose larval food includes decaying leaves and lichen, several of which are particularly scarce and threatened. As a whole, these shifts in moth species distribution with increasing cant age suggests moths may colonize new habitats as they become available.[89]

Some of the most active advocates for British coppice habitat conservation have taken on the plight of moths' fellow lepidopterans, butterflies.

Butterflies

In the UK, populations of butterflies that breed in early-successional woods and grasslands have declined precipitously over the past century and a half. Conservation organizations largely attribute this to a decline in coppice management, which has been reduced by 82% in just the past 50 years alone. Like many insect species, butterflies must complete their life cycle each year, making them particularly vulnerable to habitat

loss. Some of these populations require regular access to newly cut coppice cants. Even patches as small as 2.5 to 5 acres (1 to 2 hectares) can support a viable population.[90]

Nearly ¾ of the 59 butterfly species found in British woods breed in habitat found near or within woodlands. Some of these species include the silver-washed fritillary (*Argynnis paphia*), pearl-bordered fritillary (*Clossiana euphrosyne*), Duke of Burgundy (*Hamearis lucina*), wood white (*Leptidea sinapis*), heath fritillary (*Mellicta athalia*), purple emperor (*Apatura iris*), and white admiral (*Limenitis camilla*).[91]

Just two British butterflies breed in the dappled shade of woodland interiors, and none utilize environments with dense shade. Most butterfly larvae feed on herbs and grasses in the open, sunny zones along glades, rides, grasslands, and other open woodland habitats. Therefore, these declines in butterfly populations results from the lack of new gap creation, not the disappearance of woodlands. Without fresh-cut coppice, woods become too shady and dense to support diverse butterfly populations.[92]

Most butterfly species' developing larvae require a warm microclimate and food plants that need full to part sun. Because of this, their populations increase rapidly following coppicing, peaking some time during the first 3 years of regrowth. Once stands reach canopy closure, populations rapidly decline.[93]

Standard trees also influence butterfly breeding territory, casting heavy shade on larval food plants. Stands with 25% or less canopy cover support the highest densities of woodland butterfly populations, while coverage of 60% or more is unsuitable for virtually all ground-dwelling species.[94]

Individual butterfly colonies tend to have limited powers of dispersal. Most species form distinct colonies and rarely stray far from the sites where they emerge. In Blean Woods (UK), most butterfly populations colonized new clearings within 985 feet (300 m) of the original colony during the first year after coppicing, but when the distance to the next cant increased to 1,970 feet (600 m), colonization was at best delayed and often did not occur. Researchers estimate that butterflies will probably not colonize new clearings more than 3,280 feet (1 km) from a colony, suggesting that small fritillary butterfly species require a regular supply of new clearings for breeding, and these clearings must maintain a high degree of connectivity to facilitate movement from one patch to the next. Planning harvest rotations that shift between adjacent stands serves butterfly populations well.[95]

Short cutting rotations favor most butterfly species as long as the post-canopy closure phase lasts long enough to shade out the most vigorous light-demanding ground flora (brambles, etc.) while helping support less competitive butterfly food plants like violets (*Viola* spp.). Wide access rides that help connect colonies with newly cut cants can support butterfly populations in stands with longer rotations. These rides are ideally up to 100 to 130 feet wide (30 to 40 m) and oriented east-west to minimize shade. Up to ⅔ of this space may be occupied by managed coppice cut on a shorter interval.[96]

Until now, much of our discussion on coppice woodlands' habitat has revolved around species that thrive during the early stages of forest growth, when stands are bathed in sunlight and support abundant herbaceous vegetation. But at the other end of the forest's life cycle, deadwood resources support an equally vital community of organisms.

Deadwood

Dead and dying wood is an essential resource in healthy forest ecosystems. Deadwood supports as much as one-fifth of all woodland wildlife, providing habitat and food both directly and indirectly. Holes in standing dead trees provide roosts for mammals and birds, while deadwood furnishes woodpecker feeding sites. Coarse woody debris offers protective cover for both predators and prey, bridges streambeds, slows the flow of water and organic matter down slopes, and forms structures in streambeds that create pools. Fallen trees support new seedling establishment, fuel the mycelial web connecting individual members of the forest, and store and cycle nutrients, offering a slow release of organic matter. With as much as twenty times more water-storage capacity than an equivalent volume of mineral soil, deadwood also stores and filters water.[97]

While coppicing provides excellent habitat for flora and fauna adapted to the early stages of forest succession, it often lacks significant deadwood resources. In many cases, only ancient pollards, standards, and veteran trees provide any significant forms of dead or dying wood. In Europe, these trees were often located along wood boundaries.[98] Because of this, coppice management has presumably led to a decline in invertebrate populations whose life cycles rely on deadwood.

The small-diameter brush and deadwood created in most coppice systems decomposes quickly. Coarse woody debris of at least 3-inch (7.5 cm) diameter or larger offers a much more valuable ecological resource. Beyond 8-inch (20 cm) diameter, deadwood's decay rate slows rapidly as diameter increases.

Because coppice stands often only yield minor deadwood resources, growers looking to increase quantities of large available deadwood might allow cants to grow on a longer rotation, permit individual trees to senesce and decay, retain old individual pollards and coppice stools, girdle living trees to create new snags, build deadwood piles in the woods, and leave larger materials in the cant to decay.[99] Even low-value materials will help support fungi, invertebrates, and other wildlife, though because small-diameter material decays quickly and at more or less the same rate, it's short lived. To accommodate this, large volumes of small brushwood and poles laid out in a long, thin pile creates a better mimic of natural deadwood than a large compact cordwood-type stack. A sinuous pattern also creates superior invertebrate habitat.[100]

Management for deadwood may even occur at the scale of the individual cant. Simply allowing a few trees to reach maturity and senesce helps diversify stand structure and provides canopy deadwood and rotting trunks. Many foresters suggest a target of 1 or 2 overmature trees per acre (4 or 5 per hectare) in high forests. In coppice stands, just a single tree per acre could suffice. Locate these trees along cant edges to minimize the shade they cast on the underwood. By selecting snags of both fast- and slow-growing species, you can help create more deadwood continuity. Mature pollards often develop rotten heartwood that offers consistent deadwood throughout their life cycle. Locating them along internal boundaries instead of cant perimeters can help further spread these resources more widely throughout the stand.[101]

Deadwood provides something of a slow-release fertilizer that retains nutrients held in woody debris, cycling it through living biology before returning to the forest floor. While the same nutrient cycling processes play out in both coppice and high forest systems, coppice management's short rotations accelerate these patterns. How do these nutrient dynamics play out, and what are the long-term implications of the harvest of coppice-grown wood?

Nutrient Dynamics in Coppice Woods

It stands to reason that any extractive management system will cause a net export of nutrients from a site, leading to a long-term decline in soil fertility. Intensive clear-cutting can impact soil bulk density and porosity, stimulate soil acidification, deplete organic matter and other nutrient reserves, and lead to a loss in plant biomass and species diversity.[102] Nevertheless, considerable historical evidence suggests that some forms of long-rotation coppice management could perhaps continue almost indefinitely. British research suggests long-rotation coppicing does not cause any significant soil nutrient depletion.[103] Similarly, soils in Central Spain managed as chestnut (*Castanea sativa*) coppice of varying intensity show no evidence of soil decline or textural changes.[104] So how can this be?

In temperate deciduous species, the highest concentrations of mineral elements are found in the leaves and bark, followed by branches, roots, and finally, the trunk. A tree's foliage may possess upwards of 30% of its total stored nutrients. Woody plants translocate nutrients throughout the growing season as a conservation measure. Early in the growing season, foliage rapidly accumulates nutrients as it quickly expands. As dormancy approaches, woody plants withdraw nitrogen (N), phosphorus (P), and potassium (K) before abscission (leaf shedding), translocating these nutrients to their stem (though some evidence suggests trees only retranslocate N and P to the stem, directing K to their roots). In this way, retranslocation may supply *50% to 69% of all nutrients required for new growth* (according to Coles, forest stands return 60% to 80% of annual nutrient uptake to the soil via leaf litter).[105] And British ecologist F.B. Goldsmith maintains that many elements probably return to forest ecosystems via aerosols and precipitation and that phosphorus is likely the only element reduced by the system.[106]

So, this cyclical redistribution of nutrients has significant implications for fertility in coppice stands. Harvesting poles during dormancy ensures leaf litter can replenish forest soils. And leaving brushwood in the cant to decay further contributes to these nutrient cycles.

Tropical shifting cultivation systems, aka "slash-and-burn" agriculture, are in many ways similar to temperate Europe's traditional coppice systems. Short-rotation shifting cultivation systems generally lead to deficiencies for most nutrients, but soil and phytomass nutrient pools respond differently to short-rotation management. While we commonly see a large reduction in nutrients stored in aboveground biomass, soils usually show no net nutrient depletion as a result of the return of nutrient-rich ash deposits from post-clear-cut burns.[107]

Several studies point to increased rates of leaf litter decomposition after clear-cutting, presumably due to an increase in soil moisture, temperature, and microbial activity, as well as greater exposure to wind, rain, and sun. And the regenerating coppice stand's rapid march towards canopy closure helps reduce erosion, while increased transpiration rates reduce the volume of water available to leach nutrients into deeper soil layers.[108]

Soil compaction during harvesting and extraction is one of the primary causes of forest soil degradation, reducing water infiltration, soil pore-size distribution and volume, and hydraulic conductivity, while increasing runoff and waterlogging. Harvesting and extraction operations may also cause mixing of soil horizons, puddling, and rutting.[109] In 1976, Dickerson found that most soils recover from compaction within 14 to 20 years, which in the case of oak coppice could be about half of a typical rotation.[110] We can avoid these impacts by conducting harvesting and extraction operations when soils are frozen or at the very least dry and less compaction-prone. Dormant season harvests also allow the nutrient-rich leaf litter to decompose in place.

Historically, virtually all woody material would have been collected and harvested with even the smallest material used for kindling, leaving little biomass to replenish soil fertility. Some woodsmen use the term "tired" to describe coppice whose yields have begun to decline following management. In the UK, the estimated annual removal of an average of 0.9 to 1.8 tons of biomass per acre per year (2 to 4 MT/ha/yr) over many centuries may have caused a net loss of soil nutrients (namely phosphorus). Oliver Rackham theorizes that this nutrient export led to declines in coppice yields during the Middle Ages.[111]

A German study compared mineral soil, aboveground phytomass, and organic layer nutrient pools between silver birch (*Betula pendula*) coppice and 140-year-old European beech high forests under similar conditions. The coppice stand had a higher pH (4.24 versus 3.85) and significantly greater topsoil concentrations of total carbon and nitrogen than the high forest stand. In addition to greater concentrations of calcium (Ca), potassium (K), and magnesium (Mg), at a depth of up to 2 inches (5 cm) in the coppice stand's soils, plant nutrients were more available than in the beech high forest.[112]

While coppice stands had four times less aboveground phytomass and considerably less organic matter as leaf litter than high forests, their higher nutrient concentrations and pH suggest a more hospitable nutrient reservoir in the organic soil layers. Additionally, the coppice stands' larger pools of total nitrogen and exchangeable Ca, K, and Mg, and smaller carbon-to-nitrogen ratio in the mineral soil (up to 8 inches, 20 cm deep), further suggested that coppicing improves nutrient availability while increasing soil biological activity. They concluded that coppicing does not appear to lead to soil impoverishment despite more frequent nutrient export via phytomass and nutrient removal.[113]

In Austria, the Institute of Forest Ecology collected and compared similar soil data from beech-dominated (*Fagus sylvatica*) high forest and oak-dominated (*Quercus petraea*) coppice. Coppice stands showed slightly higher nitrogen and soil carbon content than the high forest. Subsoil (2 to 16 inches, 5 to 40 cm) concentrations of these nutrients did not differ. Coppice forests' topsoil horizon (up to 2 inches, 5 cm) had significantly higher carbon and nitrogen concentrations than the high forest stand, with higher soil pH than the high forest at all measured depths. They concluded that coppice stands' higher carbon and nitrogen pools probably result from the erosion of organic matter from the steep terrain of the high forest to the flat landform occupied by the coppice, and the coppice stand maintains higher rates of branch residue and leaf litter decomposition. They believed that lower concentrations of carbon and nitrogen in coppice forest soils at 2-to-4-inch (5-to-10 cm) depth were caused by considerable historical biomass removal.[114]

Nutrient Dynamics in Short-rotation Coppice Systems

So, although it appears that longer coppice rotations may not have a dramatic effect on soil fertility, this is not necessarily true for short-rotation systems. Some worry that frequent harvest of woody biomass from these intensive systems will lead to a net loss in soil nutrients, steadily eroding site quality over time. Even though nutrient turnover from living biomass generally occurs in less than one year, forest stands require a similar amount of nutrients to support new growth each and every season. Because of this, few forest soils can satisfy short-rotation forestry systems' intense nutrient demand. As a result, productive short-rotation forests tend to require prime soils with adequate water and fertility.[115]

Several studies on short-rotation forestry systems that utilize whole-tree harvesting find that these systems will dramatically reduce available supplies of Mg, Ca, K, and P in acid soils.[116] Whole-tree harvest only yields 30% more biomass, but it increases nitrogen removal by 84% and potassium removal by 52%. To better balance yield and nutrient export, data suggests leaving all materials smaller than 2.75 inches (7 cm) to decay on-site.[117]

In the late 1970s, researchers along the Mississippi River floodplain monitored the nutrient dynamics of short-rotation American sycamore (*Platanus occidentalis*) plantations harvested on 2-to-4-year cycles. They found that harvest removed 20 to 145 kg/ha of N, P, K, Ca, and Mg, as well as small quantities of manganese (Mn), zinc (Zn), iron (Fe), and copper (Cu). For all nutrients, sycamore branches contained greater nutrient concentrations than stems. Their model suggested that natural nitrogen gains more or less balanced

Table 3.7: American sycamore dry matter production at various spacing and disturbance chronologies

Spacing	Disturbance Chronology	Stem (t/ha)	Branches (t/ha)	Total Plant (t/ha)
2'x5' (0.6x1.5 m)	4 cuts annually for 4 years (16 total)	15.8	3.7	19.5
	2 cuts over 4 years (every other year)	28.2	3.1	31.3
	1 cut after 4 years	30.5	1.6	32.1
4'x5' (1.2x1.5 m)	4 cuts annually for 4 years (16 total)	12.3	2.2	14.5
	2 cuts over 4 years (every other year)	19.9	3.2	23.1
	1 cut after 4 years	27.4	1.4	28.8

In this study, sycamore dry matter production peaked at the denser stand spacing with longer intervals between harvest. In this case, annual harvests demonstrated the lowest total plant biomass productivity. Adapted from Blackmon, 1979, p. 2.

out the amount extracted, though for P and K, losses exceeded gains. When the rotation was increased from 1 year to both 2 years and 4 years, nutrient removal increased by 50% to 60% at the closer spacing and 70% to 90% at the wider spacing.[118]

When considering the sustainability of these operations, it's better to examine a site's net loss or gain of nutrients over time than nutrient drain via biomass removal. Natural processes including weathering, precipitation, mineralization, and other sources all contribute

Table 3.8: Total macronutrient removal (in branches and stems) during a 4-year period at two different plant spacings

Spacing	N lbs/ac (kg/ha)	P lbs/ac (kg/ha)	K lbs/ac (kg/ha)
2'x5' (0.6x1.5 m) Spacing			
Available Soil Reserves (same for both spacings)	56 (63)	414 (464)	1024 (1148)
Expected 4 year gains from weathering, mineralization, precipitation, etc. (same for both spacings)	120 (134)	8 (9)	40 (45)
Harvest Loss			
Once a year for 4 years	81 (91)	17 (19)	52 (58)
Every 2 years for 4 years	128 (144)	28 (31)	82 (92)
Once after 4 years	130 (146)	29 (32)	84 (94)
Gain Loss Index Year 1	1.47	0.47	0.78
Gain Loss Index Year 2	0.93	0.29	0.49
Gain Loss Index Year 4	0.92	0.29	0.48
4'x5' (1.2x1.5 m) Spacing			
Harvest Loss			
Once a year for 4 years	65 (73)	16 (18)	40 (45)
Every 2 years for 4 years	104 (117)	26 (29)	64 (72)
Once after 4 years	128 (144)	32 (36)	80 (90)
Gain Loss Index Year 1	1.84	0.50	1.0
Gain Loss Index Year 2	1.14	0.31	0.62
Gain Loss Index Year 4	0.93	0.25	0.50

Adapted from Blackmon, 1979, p. 3.

to a site's nutrient pool. For example, in the case of nitrogen, gains sometimes exceeded losses, with the most significant gains occurring during annual rotations.[119]

This nutrient drain shows how vulnerable intensive systems can be on some soils. For example, the study's author points out that, in the case of phosphorus, as few as two rotations with no fertility input can deplete soil reserves on sites with 35.7 to 44.6 lb/ac (40 to 50 kg/ha). And this study was conducted during dormancy (post leaf fall). Whole-tree harvests that remove foliage and woody biomass could cause an even greater nutrient drain.[120]

In New York state, Hector Adegbidi examined the nutrient contributions of leaf litter produced by a one-year rotation willow and hybrid poplar bioenergy plantation. Figures varied depending on the variety and fertilization treatment used. Table 3.9 compares the

return of N, P, K, Ca, and Mg to the soil in leaf litter with the amount removed in harvesting operations. After 7 years of short-rotation management, results showed little noticeable effect on soil properties. Plots fertilized using ammonium nitrate (NH_4NO_3) showed a decrease in pH.[121]

In France, a 15-year study collected soil nutrient data, yields, and potential nutrient export from 19 coppice stands managed on cycles of 20+ and 7+ years, respectively. They used this data to create valuable tables for forest managers, enabling them to convert biomass production data into nutrient values for different harvest intensities.[122]

Using genetically improved, routinely fertilized poplar clones, the short-rotation systems produced considerably more biomass than the extensively managed stands but also possessed significantly higher contents of all nutrients. The most productive chestnut stand yielded approximately 530 cubic feet (4.15 cords or 15 m³) dry matter, just over half the most productive poplar stands. Whole-tree harvest as compared to removal of only those poles with a top diameter of 2.75 inches (7 cm) removed 55% more dry matter in extensively managed stands and 70% in intensively managed stands, causing double the nutrient export. Branch and top removal caused the most dramatic nutrient drain. Though highly productive, short-rotation stands immobilized far more nutrients than extensively managed stands, in the long term requiring regular fertilization to maintain productivity.[123] See Table 3.10. These studies highlight the value

Table 3.9

Nutrient	Litter Returns to Soil lb/ac (kg/ha)	Amount Removed by Harvest lb/ac (kg/ha)
N	1071–5532 (1200–6200)	26.8–62.5 (30–70)
P	20.5–115 (23–129)	3.6–8.9 (4–10)
K	1.8–9.8 (2–11)	12.5–35.7 (14–40)
Ca	12.5–98.1 (14–110)	17–52.6 (19–59)
Mg	24–135 (27–151)	2.7–4.5 (3–5)

Nutrients returned to soil in leaf litter as compared to nutrient volume removed by annual harvests in New York state willow/hybrid poplar short-rotation coppice plantations. Adapted from Adegbidi, 1994.

Table 3.10

	Dry Matter	N	P	K	Ca	Mg
Chestnut	54.1%	102.4%	150.0%	185.7%	94.5%	110.0%
Mixed broadleaf	56.2%	133.3%	150.0%	114.3%	100.0%	66.7%
Poplar	71.0%	151.3%	121.7%	87.7%	134.2%	105.3%

This table indicates the differences in dry matter and nutrient removal between whole-tree harvest and stems greater than 2.75 inches. In all cases, the increase is substantial, suggesting that leaving leaves, twigs, and small-diameter branches in a coppice stand helps return significant nutrients to woodland soils. Adapted from Ranger and Nys, 1996, p. 98.

of leaving small-diameter materials in the woodland to decompose and help replenish soil fertility whenever possible.

To allay concerns about the nutrient dynamics and fertilization needs of biomass crops, short-rotation alder coppice may provide a more ecologically sustainable option. Many alder species show considerable promise as short-rotation forestry species. Along with their symbiotic root associates *Frankia* spp. bacteria, alders help increase soil N, C, and available P content, while improving microbial biomass and activity. In fact, alder stands may need little if any fertilization at all.

In Estonia, the grey alder (*Alnus incana*) has attracted attention for its potential as a bioenergy crop. Grey alder maintains high productivity on mineral and organic soils, and its capacity for symbiotic nitrogen fixation enables it to meet much of its annual nitrogen demand, setting it apart from other short-rotation forestry crops.[124]

In a study on abandoned agricultural land in eastern Estonia, the test plot received no fertilization or weed control and had an initial stool density of 6,377 per acre (15,750/ha). Soil nitrogen and carbon pools increased on the test plot from 0.11% to 0.14% and 1.4% to 1.7%, respectively. The alders annually fixed 135 pounds N per acre (151.5 kg N/ha), about 74% of their annual demand. Upon reaching canopy closure after 10 years, the resulting decrease in understory biomass led to a lower annual nitrogen demand. Between 5 to 10 years of age, the alders' nitrogen demand increased by a factor of 1.5—mainly the result of woody biomass production—while nitrogen demands in the foliage stabilized after 7 years. Seventy percent of the trees' aboveground nitrogen demand accumulated in the leaves, and by age 10, the stand generated 84.8 lb/acre/yr (95 kg/ha/yr) N via the leaf litter. The stand's nodule biomass remained stable at both years 5 and 10, leading researchers to predict that by the age of 5, an alder stand on abandoned farmland will have formed ample nodule biomass to fix sufficient nitrogen to support the stand's needs.[125]

Mediterranean Systems

To this point, the bulk of our nutrient dynamics case studies have focused on cold temperate ecosystems. How does coppicing affect nutrient pools in Mediterranean climates? One study in southwest Spain's Extremadura region examined the impact of sweet chestnut (*Castanea sativa*) coppice management on forest soils. Averaging a 50-year rotation during the past 200 years, stand management also included intermediate thinning treatments to remove overtopping competitors. This thinning left 2 to 4 sprouts per stool in 1 or 2 diameter classes: 6 to 8 inches and 8 to 10 inches (15 to 20 and 20 to 25 cm). The researchers measured soil qualities from plots clear-cut in 1957, 1967, 1982, and 1993 in order to develop a chronological sequence for comparison.[126]

Over the course of the 50-year rotation, phosphorus, iron, and pH showed no significant change, while organic matter and general nutrient availability declined for 15 to 30 years after harvest. After clear-cutting, pH showed a short-term increase, likely due to more rapid slash decomposition as a result of increased solar radiation; pH later decreased as the stands matured. Minimal canopy cover during the first few years after cutting caused an increase in soil water movement vertically and laterally, increasing the potential for surface horizon leaching losses of organic matter, potassium, and nitrogen. Nitrogen and soil organic matter took 30 years to recover to pre-harvest levels, while potassium required 40 years.[127]

Although these chestnut stands reach their maximum annual growth increment of 1.83 cords per acre per year (16.37 m³/ha/yr) after about 20 years, this study's results suggest soils need 40 to 50 years to fully recover. Under the region's stressful growing conditions, it appears that soil conservation should take precedence over maximum economic output.[128]

While just a single study can only teach us so much about coppice system nutrient dynamics in more brittle ecosystems, we can once again infer that longer rotations appear to help replenish soil nutrients and help mitigate concerns about the long-term depletion of soil fertility both in temperate climates where water is well distributed and abundant and in drylands where it's often deficient.

At the global scale, centuries of deforestation, soil loss, and unchecked industrial growth have created the conditions for a dramatically reorganized carbon cycle. Can resprout silviculture provide a tool to help channel and redirect this imbalance, restoring soil and ecosystem health, while at the same time meeting our material needs?

Coppice and Carbon

The global carbon cycle is directly and indirectly responsible for the exchange and flow of organic matter and other essential nutrients, the concentration of atmospheric CO_2, global temperature and precipitation patterns, and much more.[129] Woody plants sequester carbon at rates unlike just about any other living organisms. Resprout silviculture systems and other forms of perennial agriculture can and arguably must play an important role in rebalancing the global carbon cycle. But how do coppice stands impact soil and atmospheric carbon? While there's still a general dearth of data on the subject, we do have at least a few studies to draw from.

In *The Carbon Farming Solution*, Eric Toensmeier writes that agroforestry woodlots in Puerto Rico and Kerala, India, have sequestered up to a remarkable 24.9 tons/acre/year (55.8 MT/ha/yr) and 10.7 tons/acre/year (23.9 MT/ha/yr), respectively, in soil and aboveground biomass combined. Different short-rotation coppice stands have been shown to sequester 1.9 and 3.1 tons/acre/year (4.3 and 7.0 MT/ha/yr), respectively. Toensmeier rates various management systems based on their carbon sequestration efficacy with 0.4 to 2.2 tons/acre/year (1 to 5 MT/ha/yr) considered medium, 2.2 to 4.5 tons/acre/year (5 to 10 MT/ha/yr) high, 4.5 to 8.9 tons/acre/year (10 to 20 MT/ha/yr) very high, and 8.9+ tons/acre/year (20+ MT/ha/yr) extremely high. Even the most unproductive coppice-based systems on challenging soils in difficult climates are capable of moderate sequestration, with many systems demonstrating incredible potential.[130]

In Belgium, researchers used 10-year computer model simulations to compare carbon sequestration projections for short-rotation poplar coppice and an oak-beech plantation. The model assumed each system began with the same starting conditions on agricultural crop land (100 MT/carbon/ha). The simulated poplar plantation used 10-inch (25 cm) cuttings planted in early spring at a density of 4,050 per acre (10,000/ha), cut at

Table 3.11: Carbon sequestration of various types of coppice stands

	SOC/n AGB	T/acre/yr	MT/ha/yr
Puerto Rico		24.9	55.8
Kerala		10.7	23.9
SRC poplar, USA	SOC	2.4	5.4
SRC willow, USA	SOC	1.9	4.3
SRC poplar, US, age 8	SOC	3.0	6.7
SRC Black Locust, Germany	SOC	3.1	7.0
Belgium, 10 year simulation - poplar SRC	AGB	2.77	6.2
Belgium, 10 year simulation oak-beech plantation	AGB	1.07	2.4
Spain Sweet Chestnut coppice	AGB	2.34	5.25
Spain Sweet Chestnut coppice annual excess		2.1	4.7
Iran - *Quercus macranthera* coppice	Both	0.66	1.49
Total C-Seq per 150 year rotation			
Belgium simulation - **Total** after 150 Years - poplar SRC	Both	72.3	162
Belgium simulation - **Total** after 150 Years - oak-beech plantation	Both	144.5	324

SOC = Soil organic carbon; AGB = Aboveground biomass.
Sources: Deckmyn et al., 2004, Lancho et al., 2004, Kafaky et al., 2009, Toensmeier, 2016.

the end of the first season, and then coppiced on a 3-year rotation until replanting 25 years later. The simulations included annual fertilization of 54 pounds N/acre/yr (60 kg N/ha/yr) with harvests occurring after leaf fall. The oak-beech plantation was spaced 6.6 feet by 6.6 feet (2 m × 2 m) for a total density of 1,012 trees/acre (2,500 trees/ha), managed on a 150-year rotation.[131]

The simulated oak-beech stand maintained a stable yet low net primary production of 1.07 tons/carbon/acre/year (2.4 MT/carbon/ha/year) after 150 years, while the short-rotation poplar plantation sustained 2.77 tons/carbon/acre/year (6.2 MT/carbon/ha/year). Remember though that the poplar plantation is producing biomass fuel feedstock that quickly returns this sequestered carbon to the atmosphere, whereas most of the larger-diameter materials produced by the oak-beech stand would become value-added products with a long useful life, locking up the carbon in a stable solid form.[132]

Both systems increased soil carbon pools, though the increase was smaller for short-rotation coppice due to regular soil disturbance every 25 years. While soil carbon pools continued to grow in the oak-beech forest, the rate slowed over time. After 150 years, the mixed woodland had a total carbon pool (living biomass, wood products, and soil) of 144.5 tons/carbon/acre/year (324 MT/carbon/ha/year), while the figure was just 72.3 tons/carbon/acre/year (162 MT/carbon/ha/year) in the poplar stand.[133]

In Spain, researchers quantified the carbon sequestered by sweet chestnut (Castanea sativa) coppice in the Sierra de Gata Mountains in the country's central-western region. Managed on a 25-year rotation, these stands average

0.12 inches (0.3 cm) annual diameter increase and an annual sequestration of 2.34 tons/acre/yr (5.25 MT/ha/yr), with an average annual excess of 2.1 tons/acre/yr (4.7 MT/ha). These relatively low figures reflect the region's acidic, nutrient-poor soils. But despite this, chestnut coppice in the region fixes as much as three times more carbon than the estimated rates for nearby oak (Quercus pyrenaica) coppice.[134]

In Iran, challenging growing conditions lead to even further reduced growth and carbon sequestration rate in oak (Quercus macranthera) coppice. Covering over 12,355,000 acres (5 million ha) throughout the region, coppiced oak stems average just 0.09 inches (0.23 cm) in diameter growth and 6.15 inches (15.6 cm) in height annually for an average annual increase of 0.705 tons biomass per acre (1.58 MT/ha). These stands sequester 0.66 tons/carbon/acre/yr (1.49 MT/carbon/ha/yr).[135]

These select studies illustrate some of the potential for resprout silviculture stands to serve as a tool to sequester carbon—especially when the materials are used for long-lasting products like crafts or construction—basically applications where the wood isn't burnt as fuel or allowed to rapidly decay.

The imbalance in atmospheric carbon and the resulting climatic instability appears to also be contributing to major shifts in the distribution of perhaps the most important resource for life on Earth: the global water cycle. How do resprout silviculture systems interact with water's movement in soils and between other members of forest ecosystems, and how might they respond to increasing irregularities in the water cycle?

Soil-Water Dynamics

Annual average precipitation and its distribution largely determines a region's potential to support agriculture and forest growth. Over the course of millennia, woody plants have developed adaptive strategies to optimize the use of available moisture. In arid regions, some species develop remarkably deep root systems like the roots of trees in some acacia forests that may reach 50 feet (15 m) deep and rarely experience water stress despite modest annual precipitation. In non-brittle temperate ecosystems, abundant water resources allow many species to form much shallower root systems. But despite higher and more evenly distributed annual precipitation, trees growing in these climates may actually, counterintuitively, suffer from water stress more frequently.[136] Choosing climatically adapted species managed at appropriate spacings is one way to help limit these hydrologic stresses.

Soil texture has a huge influence on the moisture available to plants. Light soils drain freely and also readily give up available soil-water to plant roots. In contrast, heavy clay soils' enormous surface area tightly adsorbs water molecules, making it difficult for plant roots to access. Therefore, soil type dictates how and when soil moisture becomes available to growing plants.[137]

Research exploring coppice systems' belowground hydrology indicates a rise in the water table following cutting. With increasing age and density, soil-water content decreases, causing an increase in the percentage of observed water stress days. These fluctuations in available soil water also impact herbaceous vegetation. Trials at Swanton Great Wood, UK, found soil-water levels were higher under 3-to-5-year-old coppice than in 9-to-11-year-old stands, and the difference was even greater during dry years.[138]

In another British study, an increase in coppice stool density from 873 to 1,220 per acre (2,157 to 3,014 per ha or 7 × 7 foot spacing to 6 × 6 foot) doubled the percentage of stress days during the drier summer months of March through October from 11.8% to 23.5%, leading to the hypothesis that younger coppice at lower densities is less prone to water stress.[139] Therefore, the implications of these two studies suggests that lower density short coppice rotations of 5 to 10 years may help reduce soil-water deficits and water stress, while longer higher density rotations of 10 or more years may increase water stress. This also has implications for ground flora as shorter rotations could support improved clonal growth and more homogenous stands of flowering perennials. Longer intervals, on the other hand, may stimulate rhizome-forming perennials to branch and lead to greater overall understory patchiness, also favoring shade-tolerant understory species with deep root systems.[140]

In Australia, comparisons between *Eucalyptus globulus* seedlings and coppice sprouts found that, while coppice sprouts grew more rapidly, they used more water and drew on deeper stored soil-water reserves. Seedlings only tapped into the top 2.95 feet (0.9 m) of the soil profile, while coppice stools reached depths of at least 14.8 feet (4.5 m)! Coppice sprouts' increased demand for soil water may increase competition, leading to a reduction in growth rate. But because stools appear capable of drawing on deep soil water, they may better withstand extended dry periods. Several studies conducted on Eucalyptus species in the northern hemisphere have also found that seedlings show lower drought tolerance than coppice stools.[141] We see many similar if not more dramatic patterns in short-rotation coppice systems.

Swedish research on irrigated short-rotation willow coppice in the late 1980s and 1990s found that they use considerably more water than most tree species and agricultural crops. Simulation models of short-rotation willow (*Salix viminalis*) coppice found the highest mean transpiration rate during the month of July averaging 0.17 inches per day (4.4 mm/day) over 4 years with the highest daily rate at 0.31 to 0.35 inches per day (8–9 mm/day). This may be concerning to some growers because this intense water demand could lead to a reduction in aquifer recharge and river flows, especially in drier regions.[142]

During the summer of 1994, a group of British researchers in Swanbourne, England, monitored transpiration by two stands of unirrigated short-rotation poplar coppice with different leaf size, number, and morphology (Beaupré - *P. trichocarpha* Torr. & A. Gray x *P. deltoides* Bartr. ex Marsh. - a common commercial variety, and Dorschkamp - *P. deltoides* x *P. nigra*).[143]

Beaupré clones have twice the leaf area of Dorschkamp, which clearly impacted transpiration during June and July when soil water supplies were plentiful. As the season progressed, during late July and August, the difference in transpiration rates between the two species narrowed as Dorschkamp's transpiration rate increased, despite the increasing soil water deficit. By August of the second growing season, Beaupré had significantly outperformed Dorschkamp, reaching mean heights of 9.2 and 6.6 feet (2.8 m and 2.0 m), respectively. The results led researchers to conclude that these high early-summer transpiration rates could have an adverse effect on soil water resources.[144]

Like most of the characteristics we've discussed in this chapter, we once again see a trade-off when it comes to water use efficiency. Coppice stools often appear better adapted to dry conditions than unmanaged trees due to their extensive root systems and their capacity to draw on deeper soil-water reserves. But this can also lead to higher rates of evapotranspiration and declines in soil-water resources. These complexities remind us that our land management decisions need to respond to the unique realities of the region where we apply them. And this is true for all aspects of resprout silviculture as a whole system that leads to the creation of novel emergent ecosystems. So, after examining the effects of coppice on plant and animal communities, microclimates, and soil and water resources, how does coppice stack up as a silvicultural tool?

"A CONSERVATION PANACEA?"

Coppicing is widely regarded as a conservation panacea, resulting in high structural diversity and species diversity, but it can also be viewed as producing a dwarf, uniform, even-aged, biologically impoverished, nutrient depleted woodland. Which view is correct?

— FB Goldsmith[145]

Despite the unique ecological qualities we find in coppice systems, like all management systems, it ultimately favors some elements while excluding others. No ideal management system exists—at least in the sense that it's capable of concurrently supporting all possible goals. Coppicing transforms the character and quality of forest ecosystems, maintaining a patchy landscape mosaic characterized by frequent and regular gap formation, short intervals between full-sun and canopy closure conditions, rapid growth, large populations of both shade- and sun-loving herbaceous ground flora, quality habitat for bird species that dwell in woodland thickets, invertebrates adapted to edge ecologies and early-successional habitats, and small mammals.

If done well, it can perhaps offer a balance of woody production with modest soil nutrient drain while also helping sequester carbon. What's more, coppicing produces small-diameter wood on a consistent, relatively short rotation, allowing growers to concentrate and intensify management that could perhaps help to free up more land for minimal or non-intervention forestry.

While coppice management does have its drawbacks at the ecosystem level (namely the relative lack of deadwood and other mature trees, stand-scale structural simplicity, and intensive patch-scale disturbances), we may approach system design and management by embracing these drawbacks, working diligently to address and improve on these shortcomings so our systems support as broad and diverse a range of goals as possible. This is one of the keys to developing designs for resprout silviculture techniques at the systems level. In chapter 4, we'll discuss how all of these individual parts we've discussed so far fit together as complex and interrelated wholes, how these systems have been expressed over time, and what new insights and opportunities might guide our adaptation of these techniques in the coming decades and beyond.

Chapter 4:

Woodland Management Systems

Much of the existing literature on coppicing and resprout silviculture places far more emphasis on the **practices** themselves than the larger **system** they're a part of. In this chapter, we'll expand on this existing knowledge and understanding with an eye towards clarity, organization, and practicality.

How might an understanding of the individual parts of a system collectively inform the creation of uniquely site-specific silvicultural systems? Well, the first step is to identify what those overarching characteristics are—uncover the essential core of resprout silviculture systems. With that done, we see that as we move from the micro to the macro, each of these individual characteristics affects and contributes to the whole. We can also see how a well-designed system must integrate each of these characteristics in order to function well.

This chapter marks an important transition in our exploration of resprout silviculture. It's a synthesis of all of the themes we've considered to this point. And as we assemble each of these pieces, we soon find ourselves crossing the threshold from assessment and interpretation into design and application. So as we explore "what it is that makes a system a system," we'll begin with a framework that integrates those key factors that contribute to the whole: the species, economics, disturbance severity, disturbance chronology, and landscape pattern that they express and relate to.

We begin this chapter with an overview of the bits and pieces that comprise our systems framework. With that groundwork laid, we'll explore how these characteristics together shape resprout silviculture systems both traditional and modern. We'll also compare and contrast more conventional silvicultural systems both to better understand how resprout silviculture fits within the larger field of forest management and to equip ourselves with more tools to help steer forest succession. And finally, we discuss how these insights can help refine our skills as designers.

SYSTEMS FRAMEWORK

Our systems framework organizes the fundamental characteristics of managed sprout-based woodlands into their most basic components, from detailed specifics to broad pattern. In a natural ecosystem this may include: Species + Disturbance Severity + Disturbance Chronology + Landscape Pattern = Vegetation Architecture/Habitat Structure; while a designed

Table 4.1: Resprout silviculture systems framework

An overview of the fundamental characteristics of resprout silviculture systems and the specific attributes that determine their population, purposes, structure, rotation, and physical expression. Created by Dave Jacke with input from Mark Krawczyk.

Species Characteristics	Economy - Products/ Purposes	Disturbance Severity/ Practices/ Cut Location	Disturbance Chronology/ Management Timing/ Age Structure	Landscape Pattern
Ecology	Leaves/Flowers/ Fruit	Above ground	Episodic/Random/ Unplanned	From Smallest to Largest
Habitats	Fodder	Leaves	Storms	
Tolerances/ preferences	Medicine	Browsing	Harvesting as neccessary	Nuclei
Functions/niches	Food	Stripping	Draw-felling pollards/stools	
Form	Green Manure	Buds	Clear-felling pollards/stools	Random Distribution
Tree/Shrub	Twigs	Browsing, winter kill	High Forest: Selection harvest	Drifts/Scatters
Standard/Multi-stem	Basketry	Pinching/ rubbing	Strategic/Directed Sequences	Clumps/Masses
Individual/Clonal	Kindling	Twigs	First Coppice/Pollard harvest	Lines/Strips/Belts, Rows
Root Pattern	Garden products	Browsing, winter kill	Training pollards to shape	
Habit/Behavior	Ramial wood	Shredding/ chipping	Disturbances to direct succession	Arrays
Evergreen/deciduous	Fodder	Shearing	Cycles	Patches
Sprout type/ physiology	Branches	Branches	Natural fire intervals	Corridors (1 row)
Terminal/axillary buds	Crafts	Heavy browsing, winter kill	Rotations (Frequency/Length)	
Adventitious buds	Furniture	Storm, ice, rubbing (antlers)	Even-Aged (1 layer)	Fields
Dormant buds	Wattle	Pruning	Pollards: Clear-felling patches	Mosaics/Patchworks

Table 4.1: *Continued*

Species Characteristics	Economy - Products/	Disturbance Severity/ Practices/ Cut Location	Disturbance Chronology/ Management Timing/ Age Structure	Landscape Pattern
Basal/Root Collar buds	Stakes	Lopping	Copses: Clear-felling cants	Webs
Lignotuber	Biomass	Stems/Individuals	High Forest: Clear-cuts	
Rhizomes	Stems	Beaver	Multiple-Aged (2+ ages/layers)	In Pattern Order
Root suckers	Polewood	Fire, flood, herbivory	Coppice with Standards	Nuclei
Cut tolerance	Greenwood products	Storm damage, breakage	Coppice with Reserves	Lines/Strips/Belts/Rows
Overall resilience	Building materials	Thinning	High Forest: Shelterwood	Corridors (1 row)
Sprout number, vigor, longevity	Fuelwood	Draw-felling rods	Uneven Aged (multiple layers)	Alleys (multiple parallel lines)
Sprout growth rate	Mushroom logs	Clear-felling stools	Draw-felling pollards/stools	Networks (intersecting lines)
Wood Qualities	Bark	Communities/ habitats	Clear-felling pollards/stools	Webs
Strength	Tanning	Fire, flood, land-slide, tornado, etc.	High Forest: Selection Management	Patches
Durability	Twine/Rope	Clear-felling patches or cants	High Forest: Patch Cut	Drifts/Scatters
Hardness	Medicine	Clear cutting whole stands or forests		Clumps/Masses
Workability	Adjacent Products			Arrays/Blocks/Cants
Plant chemistry/ composition	Pasture/hay below	Below-ground disturbances		Mosaics/ Patchworks
Nutritional value	Arable production below	Surface		Fields
Medicinal value	Conservation use below	Shallow		
Chemical value		Moderate		
Allelopathy		Deep		

ecosystem is comprised of: Species + Purposes/ Products + Disturbance Severity/Practice(s) + Disturbance Chronology/Management Timing/ Age Structure + Landscape Pattern = Production System. It is in the amalgam of all of these factors that whole resprout silviculture *systems* emerge. And these whole systems in turn reflect the resources and constraints of local and regional culture, economics, and ecosystems. Let's examine each in detail. Note that from here on, throughout the text, we'll use icons to indicate when our discussion involves different aspects of each of these characteristics.

Species

The building blocks of a resprout silviculture system lie in the species that call it home. When managing or transforming existing woodland into resprout silviculture, the species composition directly reflects the quality and history of the site. It expresses the way selection pressures due to soils, hydrology, aspect, successional stage, microclimate, and past land use have refined the community to those species and individuals best-adapted to current site conditions. Of course, this is not necessarily the case for systems designed from scratch, as species are often chosen for their economic utility rather than their ecological suitability.

Woody species' fundamental traits including form, habit, tolerances, and type of sprouting behavior all influence their growth rate, stature, suitability, and the ways they occupy space. What type of sprouts develop? How and where do they emerge from the plant? How rapidly do they grow? What harvest frequency and

intensity will they tolerate? These physiological characteristics determine a species' suitability for management.

At the same time, the properties of each species' wood, leaves, flowers, buds, and sap determine their economic utility and direct the fiscal motivations for sustained management. So, while individual species serve as the actors in our successional drama, it's the way they relate to one another and the wider economy that brings relevance to the part they play in the system.

$ Economy

The practical demands of each unique product and the regional demands for goods and materials further refine and direct the nature resprout silviculture systems.

The economic context driving system management answers key questions about the values and needs you aim to address. What is your intended use of the polewood, rods, tree hay, or other materials? What part or parts do you need to harvest? How much processing or skilled labor are required to convert this material into a salable form? Are you looking to attain greater self-reliance or develop products for retail or wholesale markets? Do you recognize an existing demand for these products or materials, or will you need to educate potential clients? These answers help clarify the specifics, scale, and sequence of your resprout silviculture operation.

Disturbance Severity

Disturbance severity describes the location and intensity of material harvest. In most cases, this disturbance occurs aboveground, although it may occur anywhere from the root collar to the twig tips, buds, or leaves. A function of the type of materials needed for a given product or use, disturbance severity is often used to describe the management system itself. Plants cut at ground level become coppice stools, while those managed by harvesting branches or twigs cut above the reach of livestock become pollards. But the act of coppicing, pollarding, shredding, or hedgelaying aren't the actual systems— they're a consequence of the type of materials required to meet a particular set of needs.

We discussed this concept in greater detail at the individual plant level in chapter 2. Disturbance severity influences many characteristics of the system, including harvest frequency, light intensity, spacing, accessibility, and potential for integration with other elements like grazing livestock or row cropping. It's even possible that within the management goals of a larger system, an individual plant may be coppiced and pollarded at different points in its life cycle.

⏱ Disturbance Chronology

Disturbance chronology adds the element of time to the emerging system. It addresses how and when planned disturbance resets successional processes and initiates resprouting. Although unlikely in most modern management, random/stochastic disturbances may stimulate sprouting in response to unplanned ecological events like flooding, fire, windthrow, etc. These types of events may have inspired prehistoric humans to use intentional, strategic disturbance to stimulate sprouting. And evolutionarily speaking, it would seem that resprouting in woody plants developed in response to stochastic disturbance events.

Most resprout silviculture systems hinge on relatively consistent rotations or regular management cycles. But even within these fairly rigid planned rotations, we often find a certain degree of irregularity due to unforeseen realities in any given year. Fluctuating market demand, prolonged drought, family illness, or pest and disease outbreaks all may introduce variability in an otherwise regular harvest cycle. But in the aggregate, most management rotations follow a consistent average.

Rotations express both the frequency of disturbance over time and the length of the interval between disturbance events. Most resprout silviculture management creates cyclical patches of even-aged stands where the vast majority of the stems belong to a single age class.

In coppice-only systems, this amounts to clear-cutting small patches usually ranging from ⅓ to a few acres in size (0.13+ ha). The key to the scale and frequency of the disturbance is ensuring enough light may reach the stools and stimulate vigorous new growth. In more conventional high forest management, foresters use clear-cuts to create even-aged stands.

As we discussed in chapter 1, many historical resprout silviculture systems contained multiage classes. Coppice with standards is the most widespread multiage system. Here, coppice stools comprise the understory, interspersed with widely spaced single trees grown for lumber and mast on much longer rotations. And in some cases, coppice stands were managed to contain multiage classes within the same stand. "Selection coppice" or "coppice with reserves" involved either the selective harvest of individual poles from a stool or multiage classes of coppice stools dispersed throughout a cant. Based on our knowledge of stump sprouts' light needs, it seems that this type of management would have been complex to sustain, but it would create a far more diverse stand structure while still permitting routine polewood harvests. Because it maintains continuous canopy cover, this type of selection management would seem to favor shade-tolerant over light-demanding species.

These varied forms of disturbance chronology determine the structural composition of a forest stand at the patch scale. If we expand our view even more broadly, we examine the configuration or pattern the systems express across the landscape. This is the next layer of system expression.

🌀 Landscape Pattern

Resprout silviculture can be practiced on a scale as compact as an individual managed stool or as expansive as hundreds of acres of managed woodlot or silvopasture. This overall pattern affects many aspects of system management, including the nature and vitality of sprouting, the character and quality of habitat, the light regime within the stand, the patterns and practicalities of access, and also reflects the grower's primary goals.

Once again, ensuring adequate light levels for good regeneration is crucial to a system's pattern on the landscape. As long as it doesn't experience too much competition from neighboring vegetation, buildings, etc., even a single stool may achieve high productivity. As scale

increases, the patterns become more complex. The two main forms larger-scale systems assume include patches (contiguous areas of a 0.25 acre (0.10 ha) or more, cut in a single season) and alleys (belts or rows with laneways in between for other purposes, like grazing, row crops, understory production), but there exist any number of possible variations. And in the case of wild-managed systems that haven't been intentionally planted, we find innumerable shades of gray.

When succession proceeds uninterrupted, individual species' distribution patterns determine the way they colonize space. For example, species known to vigorously sucker will tend to trend towards a thicket, whereas those whose seeds are distributed by wind may follow a more scattered and diffuse configuration.

Coppice systems have usually been managed as even-aged compartments at the (patch) scale, and mosaics or patchworks at the landscape scale. That is, if we were to observe them at ground level, they appear as fairly contiguous tracts of similarly aged sprouts. But if we broaden our view using aerial imagery, we see a far more varied and diverse arrangement of stands at different phases of regrowth.

Pollard-based silvopasture systems tend to follow a wider range of patterns depending on their origin and the land manager's goals. When planted, they often take on a fairly regularly spaced configuration with rows or alleys, offering the simplest layout for fencing and protection from browse. But when created from existing woodland, pollard distribution may be far more random and irregular, assuming more of a savannah-like structure. This may pose greater management challenges if it's necessary to maintain fencelines, but brings a dynamic structure and aesthetic that can be extremely difficult to replicate.

There are no rules to the patterns resprout silviculture systems take. There is plenty of room for creativity and experimentation. The general trend points to patch and clump-type arrangements when polewood is the primary product, and alleys or rows when paired with other forms of agriculture like grazing or alley cropping. Just remember the pattern/s you choose have significant effects on the intensity, distribution, accessibility, and diversity of the overall system. And ultimately, this characteristic becomes the composite expression of the whole being managed. Together, all of these factors—the species, economic context, disturbance severity, disturbance chronology, and landscape pattern—bring temporal and spatial form to the unique systems that emerge in each individual design. So, what does the sum of each of these characteristics look like in practice?

RESPROUT SILVICULTURE SYSTEMS BOTH HISTORIC AND MODERN

Let's take a look at the broad range of resprout silviculture systems from intensive to extensive, consider how they express each aspect of our systems framework, and explore examples both historic and contemporary.

Adding Trees to Fields

We begin with systems that often integrate well into other forms of agriculture: hedges and hedgerows, windbreaks, riparian buffers, a few variations on pollard and pasture-type alley cropping, and savannah systems. In most cases, this amounts to adding, protecting, or enhancing the presence of woody plants in fields.

Hedgerows

Hedgerows or hedges are dense rows or belts of trees and shrubs that partition fields and property boundaries while yielding wood products and providing numerous ecosystem services. Widespread throughout the countryside in many parts of the world, hedgerows act as living fences, valuable wildlife corridors, shelterbelts providing wind protection and shade for crops and livestock, erosion control, and productive forest strips. Hedgerows often include a diverse mix of woody species and provide varied habitat for understory flora and fauna—especially invertebrates, birds, and small mammals. Hedgerow management may also yield numerous forest products, including polewood, firewood, fodder, and kindling.

In landscapes long-impacted by human agricultural development, hedgerows provide linear corridors of diverse habitat. Hedgerows' origin often reflects the agricultural history of the local landscape. Northern Ireland's sparse woodland and centuries of intensive land use have left a staggering 60% of all broadleaved trees relegated to hedgerows![1] In the UK in the 1980s, hedgerow trees provided 20% of all domestically produced hardwood.[2] In the US, hedgerows often emerged of their own accord, along fencelines and old stone walls where woody plants found protection from grazing animals and mowing. Fence posts, wire, and

FIGURE 4.1: A recently laid hedge at the Weald and Downland Museum, West Sussex, UK. Note the standard trees retained as part of the hedgerow.

rocky crevices also created perches and pockets for birds and small mammals to enhance seed dispersal and germination. While they serve similar functions, these divergent origins (planted vs. wild) tend to strongly impact how hedgerows fit within the systems framework.

Planted hedgerows usually include species chosen for specific functions, including thorniness, dense growth, nutrition and palatability to livestock, rapid establishment, tolerance of site-specific conditions, and potential economic outputs. Wild hedgerows, however, reflect the diversity of the local seed bank, the whims and preferences of dispersal agents, and they tend to favor fast-growing, early-successional species that can tolerate herbaceous competition and full-sun conditions.

Hawthorns (namely *Crataegus laevigata* and *Crataegus monogyna*) have long reigned as the most popular hedgerow species in Europe. Multifunctional, mixed-species hedgerows planted in the 18th and 19th centuries also included fruiting shrubs like damson plum (*Prunus domestica* var. *institia*), gooseberry (*Ribes* spp.), blackthorn (*Prunus spinosa*), and apples (*Malus* spp.).[3] In the northern Italian plains, farmers managed elm (*Ulmus campestris*), common osier willow (*Salix viminalis*), and mulberry (*Morus alba*) hedges for firewood.[4]

A managed hedgerow's disturbance severity and chronology may run the gamut from the complete harvest of the entire row to a hands-off approach that allows uninterrupted succession. Hedgerow function is the key driver of management intensity and sequencing.

Wild hedgerows may see occasional harvest of individual trees for firewood, poles, or sawlogs or the wildcrafted harvest of fruits, flowers, and fodder as available or needed. Often this may take the shape of more random or unplanned intervention where material needs or natural disturbances inform tree selection and felling. In the US, wild hedgerows rarely have a history of resprout silviculture-type management or an intentional management rotation or sequence.

Planted hedgerows may also experience a broad range of disturbance intensity and chronology, but in many cases have been established with the goal of multifunctional management on relatively short harvest cycles. Take, for instance, the "laid" hedges of the British Isles that are subjected to severe disturbance every 7 to 15 years or so. Hedgelaying results in the thinning and stool-level harvest or intentional wounding of stems, often along entire stretches of the hedgerow. And between management cycles, these hedgerows may experience varying degrees of intentional disturbance, from occasional browse of leaves, buds, and twigs by livestock or wildlife to the lopping of branches for cut and carry fodder, kindling, biomass, or craft materials. These hedgerows are often even-aged, although some growers choose to keep individual standard trees to diversify the structure and yield in time.

Almost by definition, hedgerows assume a pattern of belts or lines, usually containing no more than a few trees along their width and often just one or two rows wide. Within-row spacing is usually remarkably

Figure 4.2: A countryside mosaic of fields bounded by hedgerows is a sight to behold. Offering diverse habitat, habitat connectivity, wind protection, and wood and non-timber products, living hedges are invaluable features in rural landscapes. Photo Source: istockphoto.com.

dense—sometimes as close as 6 inches (15 cm) between individuals, though 12 to 18 inches (30 to 45 cm) tends to be a bit more common. Spacing is informed by hedgerow function. Those serving as stockproof barriers will require much tighter spacing than one that aims to simply create shade and perhaps yield fruit, fodder, and wood products.

Because hedgerows frequently delineate property boundaries, they also may reflect generations-old patterns of field partitioning and assume rather straight lines. In other cases, hedgerows may follow the topography of the landscape, along the contour or even slightly off-contour, perhaps following the **Keyline** pattern of land development.

At the landscape scale, especially in areas with a rich history of managed hedgerows, they become networks or even webs of intersecting lines. And at this scale, we begin to see a blurring of lines between systems, where hedgerows may almost assume the structure of alley crop-silvopasture, bringing us back full circle to the species selected, the products desired, and the goals of the grower.

Windbreaks

We can use resprout silviculture techniques to enhance the functionality and structure of windbreak plantings. Many resources already exist on the fundamentals of windbreak design, so we will focus on the ways coppicing can serve as a useful tool for windbreak management.

For year-round protection, windbreaks really require a significant proportion of coniferous species, but in a multirow stepped planting, coppicing can maintain low, dense cover along the edges of the installation, while yielding useful material on a multiyear rotation. In these instances, productive, fast-growing hardy species with minimal establishment costs often make the most sense. Willow and poplar species are some of the best for this application. Others include dogwood species, elderberry, hazelnut, lilac, serviceberry, buffaloberry, viburnum species, wild plum, crab apple, wild cherry, aronia, hawthorn, elm, honey locust, basswood, silver maple, oak species, and black walnut.

The severity of management really depends on the density of the windbreak, the number of rows it contains, the desired/required height, and the type of materials most useful to the land manager. The lowest row on the windward side could be coppiced at ground level. If planting rows are staggered in an offset pattern, you might choose to alternate between stools to maintain wind protection while keeping plant height manageable. In situations where it's desirable to keep plants from growing too tall to avoid casting shade on plantings or buildings on the leeward side, pollarding could also offer a great option.

To my knowledge, there are no established precedents or design guidelines for resprout silviculture-managed windbreaks. This means we're co-creating the state of the art, but allow the systems framework to guide your decision, based on an informed understanding of traditional windbreak design principles.

Riparian Buffers

We call belts of woody vegetation flanking rivers, streams, or other bodies of water "riparian buffers." These vital ecosystems have long been subject to the pressures of agricultural expansion along fertile floodplains. Because of this, these sensitive areas are often denuded of most, if not all, the woody vegetation that they have co-evolved with. Their disappearance often brings increased stream bank erosion and undercutting, over-nutrification of surface waters due to reduced infiltration of nutrient-laden agricultural runoff, increased water temperatures causing stress to aquatic ecosystems, and the interruption of important wildlife corridors. Resprout silviculture can offer farmers and land managers a useful tool to retain or enhance these qualities while also obtaining a yield from these valuable zones.

Once again, there's little established precedent to integrate resprout silviculture rotations within riparian buffers. From a pattern perspective, they usually form belts flanking either side of a water body. Wider is better, and the steeper the landscape, the more important the breadth of the buffer. A 50-foot (15 m) buffer is the minimum recommended on the flattest of landscape, while grades of 20% ideally require buffers of at least 130 feet (40 m). See Figure 4.2.

Many of the species native to these ecosystems readily sprout, some quite vigorously. Because riparian buffers often serve as conservation measures, planting stock that can be cheaply and easily propagated like

Table 4.2: Minimum riparian buffer strip width for various slopes

Slope %	Buffer Width
0	50' (15 m)
10	90' (27 m)
20	130' (40 m)
30	170' (52 m)
40	210' (64 m)
50	250' (76 m)
60	290' (88 m)
70	330' (100 m)

Adapted from Pavlov, 2016, p. 27.

dogwoods, willows, poplars, and elderberry are especially common. Frequently, riparian buffer installation finds funding and support from conservation programs like those administered by the Natural Resources Conservation Service. As a result, the economics tend to focus less on the utility of the species and more on their suitability to the site. But when managed using resprout silviculture, their potential to yield supplemental fuel, building poles, fruit, fodder, and biomass could very well contribute to the resilience of farm and homestead operations.

Wood Pasture and Wooded Meadows (Coppice and Pollard Meadows)

Found throughout the European continent including the Mediterranean, Scandanavia, much of Western Europe, the Balkans, the British Isles, and likely the African continent, the Middle East, Asia, Australia, and North America, wood pasture and wooded meadows present two variations on silvopastoral systems: tree crops integrated within pasture or hayfield. These systems benefit from the synergies that occur when woody plants are paired with livestock and wildlife. These benefits include nutrient cycling, multistoried production, partitioning of light and shade, wind amelioration, and the stacking of yields and financial returns over time. In **wood pasture** systems, livestock graze below an evolving canopy of woody plants. **Wooded meadows** also include a sparse woody canopy, but the understory is managed for hay production. Here we'll call wooded meadows with coppice and pollard

Table 4.3: European wood pasture systems

Predominant Tree Species	Animals	Management and Products
Quercus petrea, Q. robur	Cattle, sheep	Coppicing, branch lopping for fodder, bark for tanning
Corylus avellana, Populus tremula, Fraxinus excelsior, Quercus robur, Tilia cordata	Cattle	Pollarding, coppicing, hay, shredding, cultivated fields
Betula pubescens, Fraxinus excelsior, Picea abies, Quercus robur	Cattle, sheep	Coppicing, branch lopping for fodder
Fagus sylvatica, Quercus petrea, Q. robur, Carpinus betulus	Cattle, pigs, sheep, deer, horses	Pollarding, branch lopping for fodder, shredding
Quercus robur, Q. petraea, Q. pyrenaica, Carpinus betulus, Pinus sylvestris	Sheep, cattle, horses	Pollarding, shredding, beekeeping
Quercus pubescens, Q. petraea, Q. frainetto, Q. cerris, Castanea sativa	Sheep, cattle, pigs	Pollarding, shredding, acorn collecting
Quercus robur, Ulmus spp., *Fraxinus excelsior, F. angustifolia*	Cattle, pigs, horses	Pollarding, shredding, hay
Quercus pubescens, Q. frainetto, Q. cerris, Carpinus orientalis, C. betulus, Juniperus oxycedrus	Cattle, goats, sheep	Coppicing, pollarding, branch lopping for fodder, bark for tanning
Quercus ithaburensis, Q. pubescens, Q. frainetto, Castanea sativa	Cattle, pigs, sheep	Cultivated fields, branch lopping for fodder, pollarding, acorn collecting
Quercus coccifera, Cupressus sempervirens, Acer sempervirens	Sheep, goats	Branch lopping for fodder, pollarding, charcoal, beekeeping

Adapted from Bergmeier et al., 2010.

stools coppice meadows and pollard meadows, respectively.

Wood pastures need not include coppice stools or pollards, although they frequently do contain widely spaced pollards so animals may graze freely below without damaging the growing sprouts above.[5] In the UK, much of the modern literature on wood pastures focuses on the preservation and restoration of ancient wood pastures.

Many localized variations of wood pasture evolved across the European continent. Kratt (*krattskogar*) is oak-dominated (*Quercus petraea, Q. robur*) deciduous coppiced woodland in northern Central Europe and southern Scandinavia. In Germany, pollards and shredded oaks, beech, hornbeam, and other broadleaf species interspersed throughout pasture is known as hudewald. In Mediterranean regions, the coppicing and/or burning of evergreen holm or holly oak (*Quercus ilex*) creates dense sclerophyllous broadleaved or ericaceous scrub or macchia (*makija, maquis*), similar to the Spanish matorral.[6]

Pollards in wood pasture and pollard meadows yield polewood or livestock fodder. In Europe, almost all deciduous tree species (and some conifers) have been used for fodder.[7] In Scandinavia, pollards usually included wych elm (*Ulmus glabra*), ash, *Betula pendula, B. pubescens, Salix caprea,* and less commonly rowan, poplar, alder, linden, and oaks (*Sorbus acuparia, Populus tremula, Alnus glutinosa, Tilia cordata, Quercus robur,* and *Q. petraea*).[8] Many considered wych elm (the species with the highest

nutritional value of all European woody plants throughout the growing season) and ash the best-quality fodder trees. Farmers often planted them along roadsides, streams, and property boundaries.[9]

Ash and wych elm were used to feed cattle and occasionally pigs. Shepherds favored leaves from birch and alder species. In silk-producing regions, mulberry trees are pollarded to feed silkworms.

Fodder from pollard shoots harvested during summer months provides an important nutritional supplement for graziers and shepherds in many parts of the world, though the management techniques, preferred species, and forage utilized varies by locale.[10] Norwegian farmers believed tree fodder offered a first-rate feed, providing an important nutritional complement to hay, while helping maintain animal health.[11]

Silvopasture systems' harvest frequency depends on the products they yield. Fodder rotations on pollarded trees range from 2 to 7 years and polewood from 10 to 15. Norwegian rotations usually spanned 4-to-7-year cycles, taking care to ensure branches didn't grow too large between cuttings, making harvest difficult.[12] Pollards appear most productive when managed on 5-year rotations. This gives trees ample time to recover from the previous harvest. Farmers commonly divided fields of fodder pollards into zones with regular harvest rotations and clear usage rights like coppice cants.[13] This created similar intermittent, patchy structural diversity at the landscape scale. Some scientists consider wood pasture systems an "ecosystem complex"—meaning that they occupy a spatial level somewhere between that of ecosystem and landscape.[14]

Pollard height depends on either the type of livestock or the length of the bole (trunk) most useful to builders and craftspeople. Generally, this amounts to 4 to 6 feet (1.2 to 1.8 m), though in some cases, pollard boles may stretch as high as 18 feet (5.5 m).

Silvopastures may express several different spatial patterns. Alley-style silvopasture features rows of trees and shrubs between alternating pasture belts. And silvopastures may also take on more of a savannah or park-like character with scattered woody plants throughout. In planted systems, rows, belts, alleys, patches, and clumps create the simplest arrangements to maintain and protect from browse. When carved out by thinning existing oldfields and woodland, wood pastures may take on a more random patchwork pattern, responding to the topography of the land, the location of existing valuable trees, optimal density for balanced cover, along with any number of other factors.

Tree density in wood pasture systems can range widely, though wood pastures typically contain more open than shaded ground.[15] While very general, this supports the typical prescription for cool temperate silvopasture that stands possess no more than 50% canopy cover (30% in warm temperate climates) to support a healthy herbaceous understory. Also, remember that to ensure long-term system viability, you'll need to retain some seedlings as replacement trees.

In Estonia, wood pastures are often quite dense, averaging roughly 149 trees/acre (368/ha). Compare this to 81/acre (200/ha) in central Sweden, 11.3/acre (28/ha) in young Spanish stands, 6.47/acre (16/ha) in old Spanish stands, and just 3.24/acre (8/ha) in Romania. Some researchers theorize Estonia's high tree density is a result of their origin as closed canopy forests that were gradually converted to silvopasture without much effort to consciously thin the stand.[16]

Silvopasture systems can help graziers balance forage yields throughout the season. In the temperate northeastern United States, cool-season forage species commonly dominate pastures. As summer temperatures increase, these species decline in growth rate and nutrition. Silvopasture systems' partial shade can offer grazing livestock more diverse forage and a more favorable microclimate for cool-season species to maintain more even growth.

Research in Beaver, West Virginia, by the USDA's Agricultural Research Service found that some cool-season grass species grown in the light shade of silvopasture woodlands have a higher nutritional content and greater palatability than the same plants grown in full sun.[17] Holechek et al. (1981) found forage produced under canopy cover possessed higher crude protein content and greater in vitro organic matter digestibility than open grass forage. More digestible forage contains more fiber and a greater digestible energy content. Increased protein makes more nitrogen and peptides available to rumen microflora, stimulating stock growth, making more protein available to the animal. Research by Frost and McDougald (1989) found herbaceous production under scattered oak canopies to reach 115% to 200% of that of open grassland due to improved physical and chemical soil properties and more moderate soil temperatures. This effect becomes even more pronounced in areas that receive less than 20 inches (50 cm) of annual precipitation.[18]

Silvopasture systems offer valuable protection for livestock, providing both shade and wind protection. In warm climates, studies have shown that shade improves animal performance, helping ameliorate heat stress. Lower wind speeds reduce animal stress, while improving overall health and increasing feeding efficiency. In Canada, research indicates that cattle raised on open range during winter months require an additional 20% feed to compensate for energy losses due to exposure to wind and cold. Simply providing livestock with adequate wind protection may reduce the direct effects of cold temperatures by half or more.[19]

Wooded Meadows

A wooded meadow is a sparse forest stand with a regularly mowed herbaceous layer. It's maintained by annual mowing and the shaping of its tree and shrub layer."[20] According to entomologist Helen Read, wooded meadows were managed for midsummer hay harvest, and sometimes the regrowth was grazed following the cut.[21] Like wood pastures, wooded meadows may include standard trees, coppiced trees (coppice meadows), and/or pollards (pollard meadows).

The introduction of the scythe influenced the spread of wooded meadows. This innovation in mowing technology facilitated the harvest and storage of large volumes of winter fodder. Wooded meadows were once especially common throughout Estonia, central and southern Sweden (namely Gotland), and southern Finland (particularly the Åland Islands). Their distribution peaked during the late 19th and early 20th centuries, when they covered nearly 18% of Estonia (2,099,500 acres, or 850,000 hectares).[22]

Today wooded meadows are renowned for their exceptional species richness. On the calcareous soils of western Estonia and Saaremaa, some wooded meadows contain more than 50 species of vascular plants per square meter (50 per 10.75 ft²), peaking at a staggering 76 species per square meter (76 per 10.75 ft²)![23, 24]

Pollard meadows like those found in southwest Finland, southern Sweden, western Norway, Spain, northern Italy, south east Europe, and the UK produced both hay and tree fodder.[25] Some modern wooded meadows of Nato Island in the Åland Islands of southwest Finland include European ash and black alder (*Alnus glutinosa*) pollards interspersed between multistemmed alder trees. The village of Ribaritsa, in Bulgaria's Balkan Mountains, hosts orchard meadows bordered by hazel, hornbeam, ash, and linden pollards.[26] It was not uncommon to simultaneously cultivate and harvest hazelnuts, crab apples, and other fruits in these dynamic ecosystems. And of course, these meadows were also grazed during spring, summer, and fall.[27]

Estonia's best wooded meadows have between 20% and 40% canopy cover in the tree layer and 10% in the shrub layer.[28] George Peterken describes remnants of pollard meadows in Sweden, Slovakia, and Croatia as a "sparse scatter of lopped trees mixed with a few large timber trees which collectively covered no more than 10–20% of the ground."[29]

In contrast, Peterken writes that "coppice meadows were densely and fairly evenly stocked with trees and shrubs, so the cover was more-or-less complete at the end of the rotation, and larger open spaces appear to have been unusual."[30] It feels hard to reconcile this high level of stem density with our knowledge of the relationship between canopy cover and understory production. But perhaps due to spatial limitations and the desire to maximize the utilization of all "cultivated" land, the first few seasons following coppicing would have yielded valuable grazing paddocks with the added benefit of livestock cycling nutrients and helping maintain each plot's fertility.

Woodland-based Systems

Until now, nearly all of the systems we've discussed have maintained a fairly open canopy, straddling the lines between agriculture and silviculture, while incorporating aspects of both. The systems we explore next place a much stronger emphasis on wood products and other woody plant yields. These systems often rapidly advance towards a closed canopy. We'll begin with the quintessential expression of resprout silviculture—rotational copses.

Rotational Copses

Rotational copses or coppice woodlands, are dense stands of woody plants, cut close to ground level, and often managed using patch-scale clear-cuts. We break them into two categories—short- and long-rotation—based on the interval between harvests.

The species in coppice systems should be well-suited to the site qualities and economy of the woodland and surrounding community. As we already know, just about any broadleaf species will coppice—but will it yield a useful product in a reasonable time frame given the resources and constraints of the site? Of course, species selection depends on how the system is established. Planted systems will meet the grower's production goals but may not pair well with certain aspects of site quality. Systems crafted from existing forest stands tend to include species well-suited to the site but not necessarily aligned with the grower's economic goals and needs. Most traditional coppice stands either evolved as polycultures, reflecting the species present in the original forest stand, or as planted monocultures. It's fairly uncommon to find coppice stands that include more than three primary woody species.

By definition, coppice systems involve severe disturbance—the entire removal of a stool's aerial parts. But this intervention may

Table 4.4: Common coppice system types and terminology

System	Species	Products/$	Disturbance Severity	Disturbance Chronology	Landscape Pattern
Simple Coppice (short rotation)	Often fast growing (willow, poplar, sycamore, alder, black locust) or useful for crafts (willow, hazel)	Biomass, fodder, woven crafts	Cut to ground level to stimulate basal sprouting	Usually even-aged patch cuts with just one vegetation layer, cut on cycles of less than 10 years	Usually managed with rotational patch cuts, but can be managed at the scale of the individual plant, in rows/belts, etc.
Simple Coppice (long rotation)	Depends on products, site conditions - often monocultures but can be polycultures of up to 3 woody species	Widely varied	Usually cut to ground level to stimulate basal sprouting	Usually even-aged patch cuts with just one vegetation layer, cut on cycles of 10+ years with cycles exceeding 30 years rare	Usually managed with rotational patch cuts, but can be managed at the scale of the individual plant, in rows/belts, etc.
Coppice with Standards	Varies as above; standards generally chosen for mast production and timber value	Coppice wood for misc. uses and lumber/mast for wildlife and livestock	Usually cut to ground level to stimulate basal sprouting, standards cut to ground level	Usually even-aged patch cuts with cycles of 10+ years; standards cut on 60+ year rotations or 3–8 coppice rotations	Usually rotational patch cuts

Adapted from Evans, 1992, p. 19.

vary in the years between cuts, with the selective harvest or draw-felling of individual stems from a stool, the pruning or topping of poles to create pollard heads, or the harvest of buds, leaves, flowers, or even fruit. Despite these potential alterations, true coppice systems ultimately return to the complete harvest of the plant's aerial parts at the start of the next management cycle.

Coppice stands are often managed at a relatively small scale and involve clear-cuts of between roughly 0.3 and 3 acres (0.13–1.2 ha), though in some parts of the world, harvests may span dozens or even hundreds of acres. Small-scale clear-cuts employ intense patch-scale intervention. In these fairly simplistic ecosystems, any given patch offers modest habitat diversity at best. But if we examine the system at the landscape scale, we see considerable diversity *between* patches. If, for example, we were to manage 10 acres of woodland on a 10-year rotation, we would find a significant degree of between-habitat diversity, with each patch displaying 1 year's incremental development at any given point. As a whole, these systems offer stable, yet steadily shifting, habitat for species that thrive at particular successional stages.

Depending on the length of the rotation and the species involved, coppice stands are densely stocked, with individuals spaced as close as 2 feet to 8 feet (0.6 to 2.4 m) from one another (680 to 11,000 per acre, or 1,740 to 27,800 per hectare). Both short- and long-rotation coppice systems maintain an even-aged stand structure. While it's possible to develop more complex systems with two or even three age classes managed on staggered rotations (for example, hazel cut on a 7-year rotation and oak cut every 21 years, or 3 hazel rotations) generally, in a given stand, coppice stools are all the same age and occupy a single layer of vegetative growth. Coppice stands may also contain some seed trees that help renew the stand and maintain genetic diversity.[31]

Because short- and long-rotation coppice systems each possess unique qualities, we'll explore them individually in greater detail.

Short-rotation Coppice

Short-rotation coppice stands are managed on cycles of 10 years or less, though the specific interval varies between sources.[32]

$ These systems yield small-diameter materials suitable for crafts, including baskets, yurts, woven fencing, artist's charcoal, stakes, rustic furniture, garden products, erosion control materials, livestock fodder, compost and mulch, propagation stock, woody cut stems, and most commonly, biomass for heat and energy production. The term "short-rotation forestry" (SRF) is another name for the intensive short-rotation coppice systems that produce biomass for the energy sector. While SRF systems also rely on coppice regeneration, they often differ from traditional short-rotation coppice systems used to produce craft, fodder, and fuel in their scale and intensity.

Short-rotation Coppice for Biomass

The relatively modern development of short-rotation forestry (also known as short-rotation woody crops, short-rotation intensive culture, short-rotation coppice, intensive culture of forest crops, intensive plantation culture, biomass and/or bioenergy plantation culture, biofuels feedstock production system, mini-rotation forestry, short-rotation fiber production system, mini-rotation forestry, silage sycamore, wood grass, etc.) descends directly from the ancient traditions of short-rotation coppice management. Tree breeding work during the early 20th century inspired the development of the "silage sycamore" concept in the southeastern United States during the mid-1960s. Integrating traditional coppice management, plantation willow culture, Eurasian poplar agroforestry systems, and modern advances in tree breeding, the concept blurs the lines between forestry and agriculture. In Sweden, many short-rotation foresters actually come from a farming, not forestry, background.[33]

🧬 The most commonly utilized short-rotation forestry species include willow, poplar, sycamore (*Platanus occidentalis*), pawlonia (*Pawlonia* spp.), and eucalyptus (*Eucalyptus* spp.)—especially willow and poplar due to their rapid growth rates and easy, reliable propagation using hardwood cuttings. Fast-growing and tolerant of frequent cutting (2-to 5-year cycles), these species yield an abundance of biomass and can provide farmers with an economically viable land management system. To optimize yields, SRF systems require productive soils and adequate moisture. Waterlogged, compacted shallow soils with low fertility will not likely support the level of growth necessary to justify the monetary investment required for system establishment.

Alder species, black locust, silver maple (*Accer saccharinum*), sycamore, and sweetgum also show some promise for SRF applications though they are currently relatively uncommon.[34] Researchers in the US, UK, and Scandinavia have maintained decades-long trials of farm and industrial-scale copses, managed using existing agricultural equipment.

💲 Economically, SRF systems produce materials with limited markets due to its small-diameter and low ratio of wood to bark. These systems usually require the harvest of relatively large patches, resulting in less structural diversity and less diversity between-habitats than smaller-scale systems. The sites with the best potential for SRF are also often some of the most productive agricultural sites, so they must compete with food production for their landbase.[35]

A number of folks at SUNY-ESF, Syracuse, NY, have done fantastic work exploring the state of the art, and fleshing out the opportunities for these systems to make economic and ecological sense in temperate North America. Their data suggests average yields from the first rotation of willow biomass of 4.5 oven dry tons/acre/year (10 oven dry MT/ha/yr). This means that the first true harvest should produce 13 oven dry tons/acre (30 oven dry MT/ha/yr) after 3 years' growth. Because subsequent harvests will benefit from the coppiced regrowth's established roots, yields may increase by 30% to 40% to an estimated 6 oven dry tons/acre/year (13.6 oven dry MT/ha/yr).[36] British trials found an average yield of 3 to 5 oven dry tons/acre/year (7 to 12 oven dry MT/ha/yr). This amounts to mean annual growth increments of 353 to 1,059 cubic feet or 2.75 to 8.3 cords (10 to 30 cubic meters) of wood per hectare per year, varying considerably depending on species/variety, site characteristics, climate, and management.[37]

🕐 🪵 Like modern row crop agriculture, short-rotation forestry systems maintain high plant densities to maximize solar energy conversion and biomass yield while outcompeting herbaceous competition.[38] They take two primary forms: high-density, short-rotation plantings (2,025 to 8,100 per acre, or 5,000 to 20,000 per ha) cut on 1-to-5-year rotations, and lower-density longer rotations (405 to 1,000 per acre, or 1,000 to 2,500 per ha) cut on 8-to-12-year cycles. The high-density short-rotation systems function well for biomass production and phytoremediation and filtration applications where material quality is of little importance. The lower-density longer rotation systems offer a better alternative when the market requires a higher ratio of wood to bark and a more versatile larger-diameter raw material.[39]

Of course, these intensive, industrial-scale systems also have drawbacks, several of which we discussed in chapter 3. Short-rotation stands often have a relatively short life span (an average of 25 to 30 years) thereafter

Understanding Plant Spacing

Throughout this book, we talk quite a bit about plant spacing, so let's explore a few different ways we typically express these patterns. At the patch or stand scale, we often note the number of trees, stems, or stools per acre. If you're not used to thinking in these terms, a little math can be a very helpful ally so that you can reduce this number to the area occupied by a single plant.

A common density for managed hazel stands in the UK is roughly 600 stools per acre. An acre is a nice round 43,560 square feet. So, if we divide that figure by 600, we get 72.6. That means each stool covers roughly 72.6 square feet. Assuming even spacing in all directions, we can take the square root of 72.6, and we get approximately 8.5 feet. This means each stool is roughly 8.5 feet from another stool in any direction.

The other way to discuss spacing is based on the "between-row" and "within-row" distances between plants. In many cases in this book, these will be our primary units of measure. While in many coppice systems, plants are laid out in a grid that's more or less equidistant (meaning the between- and within-row spacing is the same), in silvopasture systems, more complex agroforestry plantings or custom coppice-type systems, we may choose to have different within- and between-row spacings.

Let's assume it's an alley cropping silvopasture system with pollarded trees spaced 15 feet apart within row and the rows 50 feet apart. (Note that we base this decision on tree species, disturbance severity, the type of cropping or grazing that occurs in the alleys between tree rows, the orientation of the field, and the scale of the planting, among other factors.) Once we know these two figures, we can then measure our planting area to determine how many trees there will be per row, how many rows per patch, and how many trees per acre. By multiplying the between- and within-row spacing, we determine the area each tree occupies (in this case, the trees won't necessarily span the full 50-foot row width, but the equation works nevertheless). So, 50' x 15' = 750 ft^2/ tree. By dividing 43,560 ft^2 by 750 ft^2, we discover that this spacing amounts to roughly 58 trees per acre. This basic information helps make our installations and plant material requirements that much more tangible. See, math really is your friend!

requiring clearing, stump removal, and replanting. Also, the large volume of cuttings required leads to high establishment costs and higher up-front risks for growers (although this would certainly be the case for any large-scale agroforestry installation). Because they often involve the harvest of the entire tree/ stem, these systems can also cause considerable nutrient drain. To compensate for this, future harvests usually require fertilization. It's essential to control herbaceous weed competition during the early years of establishment (again often also the case for longer rotation agroforestry systems), but at these scales, growers usually resort to repeated cultivation and/or multiple herbicide applications. And because many SRF plantations feature monocultural plantings or just a few varieties of one or two species, insects and disease may become problematic.[40]

While SRF systems may offer a valuable complement to other types of renewable energy systems (and a biological alternative to petroleum and nuclear energy sources), the details lie outside the scope of this book.

Short-rotation Coppice for Cottage Industries

Short-rotation coppice systems may also generate top-notch materials for artisanal products, especially just about any and all woven crafts. Willow and hazel are two of the most frequently utilized short-rotation coppice species. Basket makers have long realized the benefits of the management and use of coppice shoots for their products. Because they rely on an abundance of straight stems of consistent diameter, supple young coppice shoots provide an ideal raw material.

Like SRF systems, these installations generally feature high-density plantings, with basket willows as tight as 1 foot by 1 foot (30 cm × 30 cm). They're usually cut every 1 to 3 or so years. Technically, based on our characterization here, a number of other coppice-generated products fall within this short-rotation category including woody cut stems for the floral industry, live stakes and **fascines,** propagation stock for the nursery trade, artist's charcoal, and a number of other value-added products. We'll explore these products in greater detail in chapter 5 along with the nuances of these systems in chapter 6.

Long-rotation Coppice

In this book, we consider coppice stands managed on cycles of 10 years or longer "long-rotation." (Note that some sources use 20 years as the threshold between short- and long-rotations.) Long-rotation coppice yields polewood between 1.5 to over 8 inches (3.8 to 20 cm) in diameter and 20 to 60 feet (6 to 18.3 m) in length used for fuel, charcoal, fencing materials, mushroom substrate, building material, furniture, and much more.

During the past three centuries, many coppice stands have been developed and maintained as monocultural plantations due to the efficiency of management and harvest, the high commercial value of select species (chestnut, hazel), and the management simplicity that

Cross-section of Common Resprout Siviculture Systems

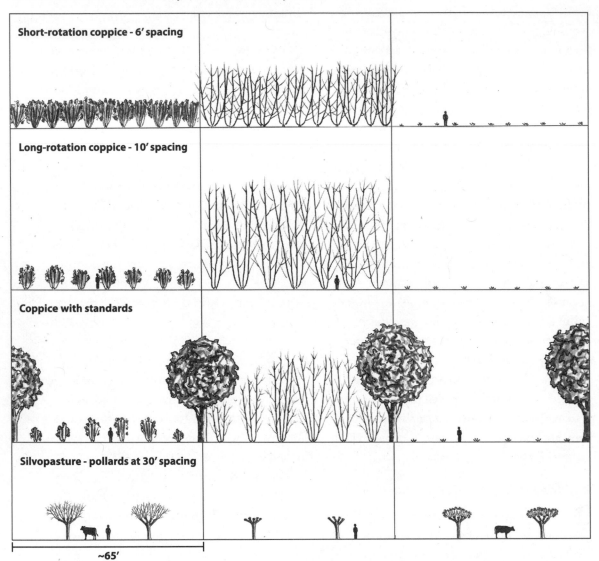

Short-rotation coppice - 6' spacing

Long-rotation coppice - 10' spacing

Coppice with standards

Silvopasture - pollards at 30' spacing

~65'

FIGURE 4.3: An overview of four of the classic forms of resprout silviculture management.
Columns are roughly 65 feet wide. The top sequence is a short-rotation stand at roughly 6-foot
spacing, harvested after 3 to 5 years. Second from the top, a long-rotation stand with 10-foot
spacing between stools, harvested when 30 to 40 feet tall after 10 to 15 years. Coppice with
standards follows with similar spacing to the long-rotation system (note the shading effects
on underwood growth). The bottom frame depicts a simple silvopasture system with belts of
pollards spaced 30 feet apart, pollarded at roughly 10-foot height.

results from the lack of inter-species competition due to differing growth habits, etc.

🕐 Like most forms of coppice management, long-rotation coppice stands have an even-aged structure with rotations as long as 60 years, though generally not much more than 20 to 30 years. While not all species will resprout after such a long interval, in some parts of Europe, oak, beech, and chestnut coppice are managed to produce small-diameter sawlogs. In this case, they often prune the stools to just one or two of the highest-quality stems. In more typical long-rotation coppice management (10-to-30-year cycles), stools are allowed to self-prune because the costs and time required for thinning often do not outweigh the benefits.

Ecologically speaking, in terms of nutrient cycling, the longer the coppice rotation interval, the more balanced the system. Because longer rotations allow more time for autumn leaf fall to build soil fertility and organic matter, stands maintain more stable fertility and nutrient cycles than short-rotation systems where declining soil fertility is a concern.[41]

📗 Long-rotation coppice systems' landscape pattern follows no established precedent. Stands are also generally dense, often ranging from 6.5 feet by 6.5 feet (2 m × 2 m) to 10 feet by 10 feet (3 m × 3 m), depending on the species, rotation length, and equipment used for management and harvest. When established by planting, these systems often assume some variation on a grid-type pattern, although it can be a simple north-south/east-west layout or something more topographically informed like contour or Keyline. However, when coppice stands are developed from existing forest, the pattern tends to be far more irregular, a function of the layout and spacing of the tree cover present at the time of coppice, with gaps later filled in by planting.

Coppice with Standards

Coppice-with-standards systems maintain a two-age stand, with coppice stools below a canopy of standard trees (single-stemmed trees left to grow to maturity). This architecture offers more diverse yields temporally and materially. The standard trees provide a long-rotation timber yield and often also produce mast (nuts, seeds, and fruits) that provides additional fodder for wildlife and livestock.

📗 In England during the reign of Henry VIII, copses were required to contain at least 10 standard trees per hectare (4 per acre). At the time, large-diameter trees were a rare commodity, and it was imperative to maintain a steady supply of sawlogs to support the shipbuilding needs of the navy. Historically, oak was the most common standard tree, comprising up to 90% of all British standards.[42] Ash and beech were also common, while in latter years, chestnut enjoyed some favor. Woodsmen usually retained standards for 60-to-100-year cycles, spanning 3 to 8 coppice rotations.[43, 44]

As a general rule, foresters aim for 10% to 15% of total stand canopy cover (the proportion of the canopy they comprise) devoted to standard trees, though some recommend as much as 20% to 30%. This density seeks to optimize standard production while minimizing the shade effects on coppice growth. Standard-tree species that have an open canopy, cast a dappled shade, and are late to leaf out, help enhance underwood productivity by minimizing competition for light. Ideally, woodland ground flora completes as much as ⅔ of their annual photosynthesis before the standard canopy fully leafs out.[45]

To maintain a consistent population, coppice with standards patches often retained two age classes of standards so replacements were available as mature trees were felled. In so doing, woodsmen selected and graded coppice regrowth, identifying **tellers** as future replacement standards.[46] To convert recently cut or existing coppice to a coppice-with-standards structure, some growers select a single sprout from a well-sited stool. This process, called **singling** produces standard trees known as **maidens**. Otherwise, growers plant seedlings to fill in gaps left following standard harvest.[47]

So, when compared with simple coppice systems, coppice with standards yields a wider diversity of forest products and materials (short and long term, polewood, and timber), a more diverse structure supporting a broader diversity of wildlife, and the opportunity to incorporate fruit and nut-producing species. It's a more structurally integrated and diverse system. But it requires planning on a multigenerational level, and causes reductions in coppice growth at the expense of the developing standard canopies.

Coppice with Reserves and Other Obscure Variations

Although uncommon, there are several references made in the literature to systems known as coppice with reserves, but clear definitions are difficult to come by. It appears there's some confusion as to the difference between coppice-with-standards and coppice with reserves. Some sources describe coppice with reserves as a two-aged system with 8 to 12 standard trees per acre (20 to 30/ha) to yield seed for forest regeneration and a future timber yield. This is essentially what most sources refer to as coppice with standards.[48]

Zlatanov and Lexer (2009) describe coppice with standards as a combination coppice/high forest system, with a multistoried, uneven-aged structure. In their definition, the canopy retains several age classes that reflect different phases in the coppice rotation. According to this definition, a coppice- with-standards system retains replacement standard trees as reserves to succeed mature standards once harvested.[49] It seems that these descriptions are confused and that a coppice with reserves system would actually possess more architectural diversity to serve as "reserves" to replace standards as they're harvested or senesce. We won't place too much emphasis on these nuanced differences since the overall concept is the same—a multistoried system producing timber and wood.

In parts of Southeastern Europe and Italy, we find scattered descriptions of "selection coppice forests" where people **draw-felled** poles of the largest-diameter classes while also removing low-quality, small-diameter sprouts.

On sites with challenging climates and steep slopes, this can be an excellent alternative to patch cut-type harvest since it maintains permanent soil cover, helping prevent erosion.[50] While uncommon, selection coppice offers a gentler option for wood products harvest, maintaining continuous canopy cover with more frequent harvests. Details on system management are scant, though, and it's unclear how to best rejuvenate the stand without creating a fairly large gap.

Another variation called "polewood coppice" involves both sprout- and seed-based stand regeneration. It's something of a hybrid between coppice management and a **shelterwood** system, with natural seedling reproduction occurring beneath the shelter of retained seed trees and sprouting coppice stools. Harvest rotations may span up to 80 years. In practice, foresters apply two cuts, the first of which, like a shelterwood cut, removes 30% to 40% of the stand's total wood volume. Once good regeneration has established (generally about 5 to 10 years later), they harvest the rest of the stand.[51]

Occasionally coppice stands were allowed to grow beyond their normal rotation length, becoming "stored coppice." The term apparently implies that materials won't be harvested until there's appropriate market demand.[52] With species that produce polewood suitable for fuel, building poles, fence posts, etc., this added flexibility may help growers earn higher returns for the materials. But in some cases, storing coppice may prove inappropriate, namely, when managing species used in craft work (i.e., hazel) that may not be useful as an added-value

material beyond a certain diameter class.[53] Stored coppice does not appear to require any significant changes in stand structure, disturbance severity or landscape pattern.

While these past few systems add some flexibility to more traditional coppice systems, they still focus almost exclusively on wood production and silvicultural management. Next, we examine several highly complex agroforestry systems integrating sprout production, pasture management, and annual cropping.

Most of the systems we've discussed so far involve a few key yields spread out over fairly wide-ranging cycles. Coppice-only systems may include modest grazing or understory harvest during the narrow window following resprout emergence but before canopy closure. Coppice with standards vastly lengthens the systems' two-story time horizon, while some of the variations we just discussed create more complex and flexible polewood harvest schemes. Wood pasture and wooded meadows add the seasonal management of the herbaceous understory. But we find some of the most dynamic, interwoven, time and space-stacked expressions of multifunctional land use in several complex "systems-of-systems."

Agro-Silvopastoral Systems: Hackwald, Hauberg, Alnoculture

"Hackwald" describes a German mixed system of oak coppice for bark along with field crops. Because it requires more widely spaced stools limiting per acre bark yields, it's not necessarily the best system for bark production. Yet, regular soil cultivation and burning appear to actually benefit bark production. Following each felling, soil is collected from around the stools and mounded into beds to support the grain crop.[54]

From the 15th through the 19th centuries in northwestern Germany's Siegerland region, people cooperatively managed a shifting mosaic of land use that included coppice, planned burns, grain production, a fallow period, and an interval of wood pasture. Known as the Hauberg cycle, the entire rotation spanned 18 to 22 years and yielded many diverse products. Wood from the birch-oak coppice was converted to charcoal to smelt ore. Oak bark was harvested for tanning. Remnant brush was burnt and the ash returned to the woodland soils. During the first 1 to 2 years following coppicing, people sowed oats (*Avena*), rye (*Secale*), and wheat (*Fagopyrum*). During the next 5 to 7 years, they left the land go fallow, during the course of which they harvested broom (*Cytisus scoparius*) whose germination was enhanced by fire. Once the coppice sprouts stretched beyond browse height, the patch was grazed as wood pasture until the closing canopy led to diminishing forage yields. The stand was then left to mature until the coppice sprouts reached the desired size for charcoal, and the cycle began anew. Apparently, examples of this system have been reestablished in parts of the Siegerland region.[55, 56]

During a similar time frame, more than 500 miles south, we find another regionally unique expression of a complex agro-silvopastoral system. In Italy's Aveto Valley, map and manuscript sources belonging to the ruling Doria Pamphili family during the 16th to 18th centuries describe complex systems management of what have today become known as "alnocultura" (alnoculture, as in *Alnus* or alder). The cycle began by clear-felling alder coppice stools every 5 years, stripping the sod, and burning the biomass. They then cultivated rye on the cleared land for 1 to 5 years, later grazing sheep and goats until the coppiced alder reached harvestable size. Shepherds presumably would have prevented livestock from damaging cropland and young coppice stools (or at least minimized the damage). The alder coppice rotation length varied from 3 to 10 years, averaging 5, depending on the time required for the shoots to reach a useful size.[57]

The alnoculture system demonstrates a complex agroecosystem that cycles nutrients through multiple processes. Alder leaves and roots accumulate and release nutrients both into and on top of the surrounding soil through leaf drop and root dieback; livestock consume pasture grasses and woody browse, returning them to their soil in their manures; and colonies of *Frankia* spp. nitrogen-fixing bacteria living on the alder roots also fix atmospheric nitrogen. Despite the alder coppice's short rotation, these nitrogen-fixing bacteria are actually most productive during the first 5 years of plant regrowth.[58] Together, all of these processes maintained a steadily shifting pattern of highly productive land use.

As modern forestry practices grew widespread during the late 19th century, alnoculture disappeared either because it was outlawed or otherwise perceived as inferior. During the same period, the conversion of land to private

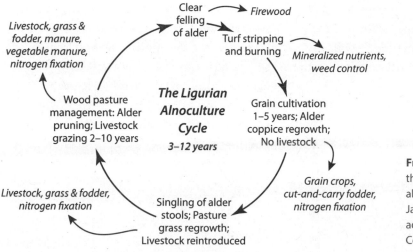

Clear felling of alder → Firewood

Turf stripping and burning → Mineralized nutrients, weed control

Livestock, grass & fodder, manure, vegetable manure, nitrogen fixation

Wood pasture management: Alder pruning; Livestock grazing 2–10 years

The Ligurian Alnoculture Cycle

3–12 years

Grain cultivation 1–5 years; Alder coppice regrowth; No livestock

Grain crops, cut-and-carry fodder, nitrogen fixation

Livestock, grass & fodder, nitrogen fixation

Singling of alder stools; Pasture grass regrowth; Livestock reintroduced

FIGURE 4.4: A conceptual overview of the phases and yields of the Ligurian alnoculture cycle. Graphic by Dave Jacke, 2011, used by permission, adapted from Bertolotto and Cevasco, 2000.

property and the erosion of "the commons," along with land abandonment and rural depopulation, all contributed to the disappearance of these complex, interwoven management practices.[59]

Unfortunately, there's little detail on the spacing and landscape patterning of alnoculture and Hauberg cycle systems. To allow for more than a year or two of grain cultivation, it would seem that the stools must have been relatively widely spaced, or the stump sprouts' rapid growth would have quickly shaded out annual plants. And equally so to allow adequate light to support any significant pasture production. Were the stools in relatively widely spaced rows, scattered patches, or even more random and irregular configurations? We may never know for certain.

So, many variations on resprout silviculture have arisen during the past few millennia, and some of these practices continue to evolve. This next set of practices, although not complete "systems" per se, utilize vegetative reproduction to produce rootstock, seedlings, fruit and nuts, holiday ornamentation, and more.

OTHER TYPES OF RESPROUT SILVICULTURE
Shredding

Rare in most of the modern world, shredding involves the regular harvest of a tree's side branches and top. Like coppicing and pollarding, this stimulates sprout formation from both preventitious and adventitious buds. After 2 to 3 years, these sprouts are used for livestock fodder, small-diameter fuel, and craft material. Because the tree's main stem continues to grow, shredding also yields a harvestable pole on a longer time horizon.

In *The Woodland Way*, Ben Law describes widespread shredded woodlands he witnessed during a visit to the village of Vukaj in northern Albania. In late summer, villagers harvest branchwood from oak trees on 2-to-3-year cycles, saving dry leaf fodder for goats and cattle, and using larger branches for cooking fuel. They manage these shredded woods by annually harvesting individual cants, felling the

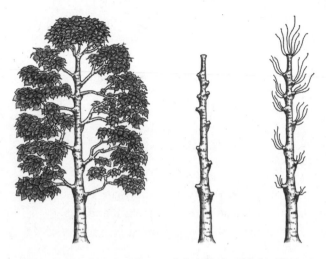

FIGURE 4.5: A shredded poplar tree fully leafed out (left), after removing side branches and topping (center) and with the first year's dormant regrowth (right).

trunks once they reach a size useful for building construction.[60]

Compared to coppicing and pollarding, shredding is especially labor-intensive (and in some cases even dangerous), requiring one to strip a tree bare to the stem by either scaling it or using a pole saw at ground level. For this reason, it may not be an economically viable management technique for many today. Nonetheless, it is another variation on resprout silviculture that some readers may find useful.

Stooling

Stooling, or stool layering, is a horticultural technique that yields rooted clones from a coppice stool. To stool a woody plant, cut a well-established seedling (or an existing coppice stool) within 6 inches (15 cm) of ground level during dormancy. When the shoots reach about 6 inches in height, firmly mound up about 3 inches (7.5 cm) of fine soil, compost, or

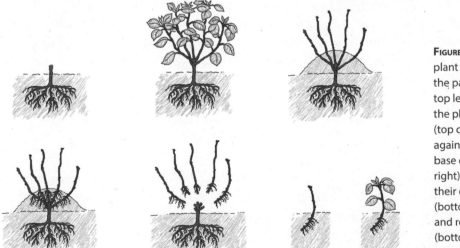

FIGURE 4.6: Stooling a woody plant to yield rooted clones of the parent plant. Clockwise from top left: After cutting/coppicing, the plant generates new shoots (top center), soil is mounded up against the stem, covering the base of the young sprouts (top right). Sprouts begin to form their own adventitious roots (bottom left), soil is removed, and rooted sprouts are cut free (bottom center). Each rooted cutting can then be transplanted.

organic matter around the base of the sprouts. As the growing season proceeds, continue to mound the base with soil to a height of up to 6 inches. The buried portion of the shoot should start to form roots as the sprouts grow.

In autumn, peel away the mounded soil and inspect the shoots for adventitious root formation. Under favorable conditions, rooted shoots may be ready for transplant as early as October or November of the same year. Carefully remove well-rooted shoots and pot them up or transplant them directly. Leave weaker shoots attached to the stool for another year to develop a stronger root system. Growers can use this technique to propagate seedlings to fill in gaps in a coppice stand, while nurserymen also use the technique to clone plant stock—especially rootstock for fruiting species. Commercial growers recommend 3-foot (90 cm) spacing between rows and 1-foot (30 cm) spacing between stools in each row.[61]

Own-root Fruit Trees

Most fruit trees planted today are grafted. This means young shoots from varieties with known desirable characteristics ("cultivars," or cultivated varieties) like flavorful fruit, staggered flowering and ripening times, hardiness and storability are carefully spliced onto a rootstock variety that controls the mature size of the tree (dwarf, semi-dwarf…standard) and is well-suited to particular soil conditions. Grafting has become so common in modern fruit production that many believe grafted trees are the only way to produce known varieties of certain species. But a resprout silviculture

variant offers a possible alternative management technique.

While literary evidence describes the practice of producing own-root (OR) fruit trees as early as 1807, British breeder and nurseryman Hugh Ermen is largely responsible for its modern revival. To produce OR fruit trees, growers root cuttings of known varieties and plant these OR trees as one would a grafted tree. Research suggests OR fruit trees may produce healthier growth, with improved nutrient uptake capability, better fruit set, improved pest and disease resistance, and more reliable storage.

The primary drawback of OR fruit trees lies in the lack of control of the mature tree size. Because of this, British grower Phil Corbett began coppicing fruit trees, effectively "pruning" the tree to ground level to encourage low, accessible, vigorous regrowth.

Taking the technique one step further, Corbett developed the concept of the coppice orchard, planting north-south oriented rows of OR apples, plums, pears, hazelnuts, and nitrogen-fixing woody plants. When the orchard's canopy closes, Corbett coppices an entire row, planting annual vegetables in the understory during the ensuing years of full sun. The following year, he coppices another row, but instead of cutting the row adjacent to the previous year, he skips ahead so that the increased solar exposure stimulates the development of more fruit buds. This "alternate row" coppicing system maximizes the edge within the orchard and makes optimal use of the interval between fruiting for coppiced fruit trees.[62]

Though many questions remain regarding the nuance and long-term application of coppice OR orchards, Corbett's work helps illustrate how such an approach could revolutionize fruit production. It's been difficult to find detailed explanations of Corbett's system along with data on its performance over time, but it does provide some excellent inspiration for the ever-evolving use of these age-old practices.

A bit closer to home (for those of us in North America), the development of woody agriculture practices brings a similar vision of coppicing integrated into large-scale agroforestry in order to perennialize staple food crop production while replacing the corn-soy monocultures covering vast swaths of the interior plains states. A pioneer in this evolving industry, Minnesota-based Badgersett Research Corporation and Farm have been developing this vision for over 30 years.

Building their agroforestry system on chestnuts and hazelnuts, both known for their strong coppicing responses (along with hickory and pecans), the farm has been breeding their own varieties adapted to local site conditions, productive yields, and suitability for this type of management. When I visited the farm back in 2013, they were about to start their fifth breeding cycle for hazelnuts, their second for chestnuts, and their first for hickory. Their system employs a 10-year coppice cycle to eliminate the need for laborious manual pruning, at the same time revitalizing the plants and stimulating new vegetative growth. Nuts can be harvested mechanically using overhead harvesters, while sheep and horses graze herbaceous forage between the rows and pigs are

allowed to feed on any unharvested nuts. You can read more about their work in their 2015 book *Growing Hybrid Hazelnuts*. Their work has helped stimulate and inspire much innovation in the world of woody agriculture throughout the midwestern states.[63]

And beyond food crops, resprout silviculture also provides a tool to reimagine the way commercial growers manage an iconic woody crop—Christmas trees.

Stump Culture or Coppiced Christmas Trees

Contrary to the commonly held belief that just a handful of coniferous species resprout when cut, a practice called **stump culture** enables growers to produce cyclical crops of Christmas trees and conifer boughs from the same root system. Using this method, growers allow planted trees to reach a height 3 to 4 feet (0.9 to 1.2 m) taller than normal. Once the first round of trees is ready for harvest, they cut the "tree," leaving several whorls of branches (or about 30% to 40% of the total foliage) at the base. This triggers the stump to send up new shoots along the stem and branches. During the following season or two, growers select the sprouts they wish to retain, pruning off the rest. Some have produced a new Christmas tree from the same stump every 5 years for 60 years or more.[64] If you'd like more information on the subject, Emmet Van Driesche of the Pieropan Christmas Tree Farm published a book in 2019 called *Carving Out a Living on the Land* that details his farm's exploits producing stump culture Christmas trees in Massachusetts.

And in parts of British Columbia, Canada, some growers have taken the stump culture practice a step further, also cultivating a productive understory. In the dry East Kootenay region, native grasses commonly colonize the ground layer beneath Christmas trees. Once stump culture shoots have adequately developed, they bring in livestock to graze the herbaceous understory, fertilizing the trees while also helping reduce fuel loads and fire potential during dry summer months.

In the humid western reaches of British Columbia, including Vancouver Island, some growers cultivate marketable floral crops in both the shady understory and scattered sunny patches. Some of the shade-tolerant species include salal (*Gaultheria shallon*), bear-grass (*Xerophyllum tenax*), falsebox (*Pachistima myrsintes*), sword fern (*Polystichum munitum*), deer fern (*Blechnum spicant*), and evergreen huckleberry (*Vaccinium ovatum*). Sunnier patches may also support Scotch broom (*Cytisus scoparius*) and Oregon grape (*Mahonia nervosa*).[65]

So, with stump culture, we see that the "rules" of sprouting may be more malleable than once thought. Resprout silviculture offers us a diverse and powerful management tool. But please remember that this tool kit is part of a much larger cache of management practices. Resprout silviculture systems are not always the most appropriate strategy for all ecosystems. In many cases, healthy, diverse high forest stands deserve their own forms of ecological management.

In parts of Mediterranean and southeastern Europe where coppice systems continue to

FIGURE 4.7: A pruned balsam fir stump that has been managed to produce both Christmas trees and greens for wreaths and garlands for 60 years at Pieropan Farm, Ashfield, MA. Note the two upright shoots that remain on the topmost branches in the center of the photos. These are likely to be selected as future Christmas trees in another 5 to 8 years. Photo courtesy of Emmet Van Driesche, used by permission.

dominate forested landscapes today,[66] contemporary foresters struggle with an ecological and economic dilemma. With 25% to 50% or more of their forested landscapes under coppice management and few market opportunities for coppice wood other than fuel, many are working from the opposite direction in order to transform centuries-old copses into uneven-aged high forest.

Concerned with declining productivity, the relative lack of mast for wildlife, low market value, and the sheer inappropriateness of large-scale clear-cuts on steep forested mountain slopes, foresters have found it difficult to successfully transition coppice regeneration into high forests of seed origin. In these circumstances, we see how scale, context, and economics suggest a diverse approach to forestry.

In other words, resprout silviculture systems have many benefits, but they also have limitations. So how does resprout silviculture fit within the broader array of silviculture systems? How are high forests typically managed? What other options do we have when it comes to managing our woodlands? And ultimately, what lessons can more conventional silviculture teach us about how to design and transform resprout silviculture systems? We conclude this chapter

with an overview of the fundamental principles and systems that underpin modern silviculture.

SILVICULTURAL FOUNDATIONS OF HIGH FOREST MANAGEMENT

 The European term "high forest" describes woodland managed for single-stemmed trees yielding timber. The term is used to contrast with "low forest" or coppice systems. While high forest management can take numerous forms from clear-cutting to selection management, the overall goal is to produce high-quality large-diameter timber for conversion into a wide range of products, including veneer, **sawlogs**, fuelwood, pulp, and biomass, with sawlogs—tree stems converted into

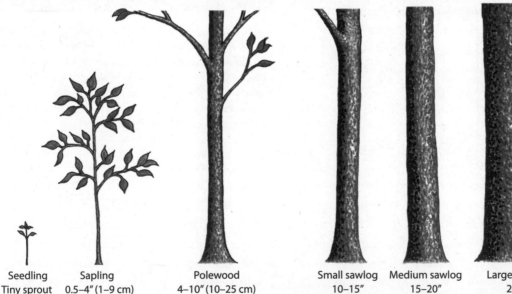

| Seedling | Sapling | Polewood | Small sawlog | Medium sawlog | Large sawlog |
| Tiny sprout | 0.5–4" (1–9 cm) | 4–10" (10–25 cm) | 10–15" (26–37 cm) | 15–20" (38–49 cm) | 20"+ (50 cm+) |

Figure 4.8: Foresters describe trees based on their age class. Somewhat counterintuitive, "age class" refers more to a tree's diameter at breast height than its age. Age class categories include seedling; sapling; polewood; and small, medium, and large sawlog.

lumber—being the most common high-value use. And beyond wood, high forest management may also yield wildlife habitat, non-timber forest products (maple syrup, fungi, understory herbs), recreational value, and a wealth of ecosystem services including erosion control, water retention, soil creation, and carbon sequestration.

High forest management yields lumber products very effectively. As we discussed towards the end of chapter 1, widespread improvements in transportation infrastructure

and materials processing, along with the standardization of building materials, helped high forest management become the industry standard, displacing complex resprout silviculture systems in a matter of decades.

Several foundational ecological assumptions guide the art and science of **silviculture** (the growing and cultivation of trees) and the application of management prescriptions. First and foremost, a forest's potential annual growth on a given acre is theoretically fixed. That is, each site's soils, aspect, available water, and nutrients

Table 4.5: A random sampling of site indices and MAI for different types of forest stands in different ecosystems

Forest Location	Community Type	Soil Type	Site Index	Mean Annual Increment
Central Maine	American Beech	Elliottsville-Monson complex, 3 to 15 percent slopes, very stony	55'	29 ft³/ac/yr
Mifflin, PA	Black Cherry	Laidig extremely stony loam, 8 to 25 percent slopes	80'	57 ft³/ac/yr
Clay, Alabama	Loblolly Pine	Clymer association, steep	70'	Unavailable
Forest County, Wisconsin	Sugar Maple	Argonne-Sarona sandy loams, 6 to 15 percent slopes, very stony	63'	42 ft³/ac/yr
Coos County, Oregon	Douglas Fir	Preacher-Bohannon loams, 3 to 30 percent slopes	126'	186 ft³/ac/yr

A random sample of forest data from different parts of the US reflecting varied climates and soil types, derived from the USDA's Web Soil Survey "Soil Data Explorer." On the site where this data is available, look under the Vegetative Productivity drop-down menu where you'll see at least two potential metrics for forest productivity (site index and cubic feet per acre per year). Note that this data is not available for all sites. Largely a function of climate and soil type, these figures are unique to the particular tree cover indicated. And remember, site index suggests the average height of these trees at 50 years age (assuming they haven't been planted and maintained when it can be as short as 25 years). The mean annual increment tells us roughly how much new wood in cubic feet a stand can generate annually per acre. Data source: USDA Web Soil Survey.

together dictate the land's productive capacity. Foresters characterize landscapes using a **site index** that indicates a stand's relative productive potential. A stand's site index describes trees' projected height at a given age (usually 50 years). Therefore, trees in a forest stand with a site index of 60 should reach 60 feet by age 50.

This concept of fixed stand potential implies that *management cannot actually increase growth in a forest stand, but it can direct growth potential towards a predetermined goal,* often high-quality sawlogs or volumes of fuelwood. We do this by concentrating and redirecting the annual growth in the most desirable species and individuals. In this sense, management decisions transfer annual growth potential from the trees selected for harvest to those trees chosen for retention. This underlying assumption forms the foundation of forest management.[67]

To ensure sustained forest productivity, foresters develop management plans that aim to achieve a stand's "maximum sustainable yield." They determine this figure based on the quality of the site and the **mean annual increment** (MAI), the average volume of wood trees produce in a given year on a site. The related figure **current annual increment** (CAI) measures woody plants' annual volume increase. Because CAI varies significantly with age, MAI offers an a more consistent basis for comparison between stands.

The **allowable annual cut** (or AAC, the accrued growth among marketable surviving trees and new trees reaching marketable size) describes the volume of wood one could theoretically harvest without depleting the forest's wood resource. Think of it like a bank account. The CAI represents the interest rate, and the AAC amounts to the net annual interest earned. Theoretically, one could spend the interest each year without depleting the principal. This figure also represents the **net community productivity**, the additional energy plants store beyond the communities' maintenance needs. Theoretically, as long as forestry operations only harvest the net community productivity, they do not excessively mine resources from the forest stand.[68] Theoretically.

The practice and philosophy of ecoforestry takes a particularly holistic view of forest ecosystem management, striving to "maintain and restore full functioning, natural forest ecosystems in perpetuity, while harvesting forest goods on a sustainable basis."[69] Ecoforesters' management practices aim to mimic natural forests' patterns and processes, enhancing their abundance while providing a sustainable yield. The following three primary principles guide ecoforestry management:

- All actions address *forest needs first.*
- Trees removed to meet human needs are selected to also help improve overall ecosystem health.
- All forest product removal and management activities preserve the structural and ecological functions of existing forest components.[70]

Whichever management practices we choose, when we intervene to direct and concentrate forest productivity, we reshape stands to generate useful products while supporting healthy ecosystem functions. Holistic conscious land management emerges from a broad knowledge base and an openness that enables us to choose and apply the appropriate tool to achieve our desired goal. Resprout silviculture won't be our best option in all circumstances. It's one of many management tools available to us. Next we take a brief look at a suite of common forest management systems to add diversity and context to our silvicultural tool kit.

Even- vs. Uneven-Aged Forests

Forests fundamentally express one of two forms based on their temporal and physical structure: **even-** and **uneven-age**. In an even-aged forest, canopy trees all comprise a common age class. As such, even-aged forests result from some stand-level form of disturbance—a landslide, fire, blowdown, ice storm, tornado, hurricane, flood, or clear-cut. At one point in their

evolution, essentially all forests will express an even-aged form as they reorganize following disturbance.[71] Because of their relative lack of age diversity, most even-aged forests not influenced by human intervention are comprised of early-successional species that thrive in high-light conditions. This structure begins to change as canopy gaps open and shade-tolerant species gradually colonize the canopy.

Uneven-aged forest stands exhibit significant structural diversity and a range of age and size classes. This structure typically indicates a longer interval since the last disturbance event; thus, they tend to reflect mid- to late-succession forest ecosystems and may include a higher proportion of shade-tolerant species. Uneven-aged forests may also result from smaller-scale patch-level disturbance events that transform forest structure at the patch scale.

Even- and uneven-aged stands develop in "natural" (those not influenced by human

Even- Vs. Uneven-Aged Forests

FIGURE 4.9: In an even-aged stand (left), canopy trees are more or less all members of the same age class. The uneven-aged stand (right) has a more irregular structure with gaps having formed, mature standing dead trees, and seedlings and polewood-sized trees colonizing the open space.

intervention) and managed forests. They each have strengths and weaknesses, and these structural forms can be applied to meet specific management goals. Most forest management systems maintain an even-aged structure, although the intensity and frequency of intervention varies. When comparing systems, remember that scale plays a huge role in their relative suitability. While many of us have seen the devastating effects of large-scale clear-cuts on steep mountain slopes, on a small scale, patch cuts can mimic natural disturbance events that can actually have a rejuvenating effect on forest ecosystems, creating diverse habitat at the patch scale.

In its traditional form, coppicing is an even-aged forestry system. Contiguous clear-cut patches create disturbance events that expose a patch to high-light conditions, stimulating rapid vegetative reproduction from stools' established root systems. At an appropriate *scale*, these clear-cuts can achieve positive results while minimizing the drawbacks of this disturbance.

Although ecological high forest management yields diverse valuable products, we will only discuss it here briefly. But once again,

Table 4.6: Pros/cons and qualities/characteristics of even- and uneven-aged forest management

	Pros	Cons
Even-Aged Management	Much simpler management and harvest than uneven-aged systems	Less diversity in vertical structure, species mixes, and tree age and size
Favors shade intolerant species	May provide all stages of forest succession, especially early stages	Can cause habitat fragmentation that strains the needs of some wildlife
		Often lacking in deadwood resources and mature trees
		Less aesthetically appealing to many
Uneven-Aged Management	Potentially high levels of structural, species, age, and size diversity	May require more extensive access
Favors shade tolerant species	Often requires much smaller scale disturbance	Sometimes lacking in "between habitat" diversity since much of the stand is at the same successional stage
	Has a more dynamic aesthetic that many find more appealing	Management and harvest is more complex and often therefore expensive
	Often contains deadwood, mature trees, and snags	

Adapted from Mark Hatfield, National Turkey Wildlife Federation, nwtf.org/conservation/article/forest-management-why-how-what high forest management must play a foundational role in an ecological future. Resprout silviculture and high forest management are complementary, each capable of producing useful products for specific needs under particular circumstances. And a basic understanding of high forest management can also provide us with valuable techniques for converting high forest stands to coppice, a theme we'll explore in more detail in chapter 7.

High Forest Management Systems

Our suite of high forest management techniques either initiate a new forest stand (**regeneration**) or modify the species, spacing, quality, age, or size of trees in an existing stand (**intermediate**). All techniques alter the competition between the trees on a site, removing undesirable trees while giving competitive advantage to the most valuable individuals and species. Our management decisions primarily involve choosing which species and age classes to favor based on their suitability to the site, the current quality of the stand, and its perceived future economic value (often predicted by analyzing past trends).[72]

In most cases in the humid temperate forests of the northern and eastern United States, high forest management relies on seed for regeneration. (In contrast to the Pacific Northwest and parts of the Southeast, where faster growth rates tip the cost-benefit scales

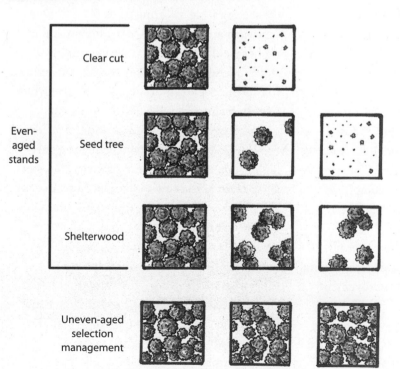

FIGURE 4.10: An overview of the primary forms of silvicultural management for high forest stands. Each patch is roughly 100 feet by 100 feet. The top three practices all create and maintain even-aged stands. Clear-cuts remove all trees larger than 2-inch dbh. Seed tree cuts remove all but 1 to 10 well-formed trees of valuable species per acre. Shelterwood cuts use 2 or 3 management interventions to incrementally transform a stand to a new even-aged community. Selection management involves the strategic removal of individual trees in order to preserve continuity in canopy cover, maintaining an uneven-aged stand.

to replanting following clear-cutting or other silvicultural treatments to ensure a well-stocked stand of desirable species). Herein lies one of the primary distinctions between high forest and coppice systems. While vegetative reproduction from coppice sprouts produces fast, voluminous growth and maintains a continuous, intact root system, seed reproduction ensures that the stand perpetuates genetic diversity, helping the evolving forest adapt to changing soil, moisture, cultural, and climatic conditions. This succession of woody plant genetics ensures forests' long-term adaptability and allows foresters to continually reshape the character of forests as they apply treatments that retain the best-quality individuals while removing those with little future promise. We start our examination of high forest systems with a look at a range of intermediate treatment options.

Intermediate Treatments

Intermediate treatments improve the future value of young regenerating forest stands. They maintain moderate levels of species diversity, favoring the most valuable species. They usually feature a well-developed canopy comprised of 2 or 3 species of more or less the same age and a shade-tolerant understory of young woody plants and herbaceous flora. Because intermediate treatments usually occur in young developing stands, most stand improvement efforts yield products with little commercial value. As such, they're often considered a form of **pre-commercial intervention**, which is usually the most significant drawback to intermediate treatments. The treatment costs time and money, but their benefits may not be realized for several decades. But at the same time, well-timed interventions restructure a young forest stand, concentrating growth in the best-quality species and stems, helping ensure a higher economic return in the future and a healthier better-stocked forest. So intermediate treatments must be seen as an investment in the future value of the woodland.

There are several different types of intermediate treatments including **release cutting**, **improvement cutting**, thinning, and pruning. Let's look at each of these briefly.

Release Cutting

Release cuts thin a young forest stand during the sapling stage, favoring the most desirable species in order to produce the highest-value timber yield at maturity. After identifying the stand's highest-quality trees, foresters examine tree form, spacing, and species to inform which trees to retain. Cleaning and weeding cuts remove trees of low value and poor form. To encourage straight growth protected against high winds and ice and snow damage, foresters try to maintain adequate stand density. Usually any trees removed during these pre-commercial cuts are left to decay on the forest floor.[73]

In a **liberation** cut, foresters remove trees older than those of the favored species that otherwise restrict understory growth. Major storm damage or previous high-grading may both create the need for a liberation cut. These stems may be either harvested and sold, left to decay, or girdled and left upright to slowly die and decompose. Girdling reduces under-story damage while creating valuable habitat and requires neither felling nor extraction. Economics ultimately drive this decision.[74]

Improvement Cuts and Thinnings

Improvement cuts and thinnings reduce the total number of stems in an unmanaged or overcrowded stand's canopy once they've reached pole size or larger. These treatments usually occur after the stand begins to show a declining rate of diameter growth. The intent is to increase the growth rate of the remaining desirable trees and harvest suppressed trees before they're lost. If suppressed for a long enough period, many tree species will not rebound with increased growth.[75]

Improvement cuts focus on the removal of undesirable species while thinnings strive to optimize tree spacing, rather than species or stem quality. Foresters often prescribe frequent light thinnings during stand development to maintain balanced growing conditions for the remaining trees.[76]

Pruning

And in some cases, foresters will recommend pruning individual pole-sized trees once they can clearly identify high-quality crop trees. This increases the sawlog's value by removing branches to heights of 12 to 24 feet or more (3.7 to 7.3 m) using a telescoping pole saw. Because, as we know, branches ultimately form knots in lumber, basal pruning produces high-er-value sawlogs. It's best to prune trees once they've reached 4 to 6 inches in diameter (10 to 15 cm) and before they've reached 8 inches (20 cm). Remember though that generally fewer than 100 trees per acre (247 trees per hectare) will survive long enough to yield a harvestable sawlog, so don't invest time pruning any more trees than are likely to offer a return on the investment.[77]

Regeneration Methods

As a forest stand matures, trees' annual growth rate eventually slows. When the annual growth rate drops below previous years' annual average, foresters consider the stand "biologically mature." They may then choose to prescribe regeneration treatments that modify light levels to accommodate desirable species and initiate a transition in stand age, composition, and structure.

Like intermediate treatments, most regeneration treatments maintain an even-aged stand with just a few dominant canopy species. The rotation length depends on the species and their value, but they often span 60 to 80 years. Regenerating forest stands should already possess some **advanced regeneration** (established seedlings and saplings) to fill in following harvest.

Allowing a forest stand to naturally regenerate from seed of nearby trees maintains genetic diversity, fostering resistance to pests and disease and resilience to shifting soil fertility and climatic conditions. Ideally regeneration treatments coincide with the seed set of remaining trees to ensure the germination and growth of a healthy population of seedlings. Logging operations may actually help stimulate regeneration as tree felling and extraction scarify the leaf litter, helping germinating seedlings' roots make contact with bare mineral soil.[78]

High forest regeneration methods take five primary forms: **clear-cutting, seed tree cutting, shelterwood, selection,** and **patch cuts**. These techniques mainly differ in the light, moisture, and temperature conditions they create. All but the last two treatments create an even-aged stand, while selection and patch cuts maintain an uneven-aged stand that includes multiple age classes.[79]

Clear-cutting

A clear-cut completely removes the forest over-story (technically all trees larger than 2- inches (5 cm) dbh). Clear-cuts can be an appropriate treatment in a few scenarios—to release established advanced regeneration of a desirable species or to regenerate shade-intolerant species. To regenerate shade-tolerant species, it's important that a reasonable amount of seedlings and saplings at least 2 to 4 feet (30 to 60 cm) tall are already present. Lacking adequate advanced regeneration, clear-cutting is a risky prospect, especially on sites where there are pressures from opportunistic or invasive species. For shade-intolerant species, a successful clear-cut must open a large enough area to ensure that shade cast by surrounding trees doesn't compromise seedling growth.[80]

When discussing clear-cutting, it's important to consider the scale and context of the intervention. At a small scale on moderate slopes, clear-cuts may mimic a natural disturbance like a blowdown, ice storm, fire, or flood. However, industrial-scale clear-cuts, especially on steep slopes, dramatically impact landscapes, bisecting habitats, causing soil erosion and voluminous runoff, while liquidating forest resources.

Clear-cutting's main advantage lies in its relative efficiency on a large scale. By removing all trees in a forest stand, identification, felling, and extraction operations are much simpler.

Seed Tree Cutting

Seed tree cuts remove all but 1 to 10 vigorous, high-quality trees of seed-bearing age per acre, (0.4 to 4 per hectare) leaving them to produce seed to aid in stand regeneration. Whereas clear-cutting largely relies on seed and seedlings already present or dispersed from nearby stands, seed tree cuts retain the stand's best individuals to steer the next round of forest succession. In this way, seed tree cuts may restore or redirect a stand's species composition. On sites with shallow or extremely wet soils, seed tree cuts may be risky due to the potential for blowdowns during high wind events. In areas where stands sufficiently regenerate following clear-cutting, seed tree cuts are uncommon because clear-cuts require less time and expense.[81]

Shelterwood Cuts

Best-suited to species with medium to high shade tolerance, shelterwood cuts remove the forest overstory incrementally while stimulating advanced regeneration. Shelterwood cuts usually involve at least three harvests or interventions. The canopy is only completely removed once there's sufficient advanced regeneration of the desired species.

During the first cut (preparatory), the canopy is thinned by 20% to 40%, targeting undesirable species and low-quality trees. This leaves the highest-value individuals to continue to disseminate seed. Gaps of 19 to 33 feet (6 to 10 m) between canopies of remnant seed trees, permit enough light to penetrate and ensure strong regeneration.[82] These canopy trees also help nurture young seedling growth by shading out competing weeds while protecting them from wind and frost damage.

In subsequent cuts, foresters continue to winnow down the canopy to only the best-quality trees. The second, or "seed," cut is timed to coincide with a good seed year, further aiding germination by opening the stand to more light. This cut sometimes occurs during the summer months to scarify the soil surface and assist germination. If logging operations are poorly timed or the seed set is insufficient, seedlings may be planted or seed broadcast to compensate.

Once advanced regeneration of the appropriate density, volume, and size is established, the final cut or cuts remove all remaining canopy trees. At this point, seedlings should be at least 2 to 4 feet (60 to 120 cm) tall to ensure they're strong enough to survive logging. In areas with harsh winters, snow cover may offer seedlings adequate protection during winter logging.[83]

In a two-cut shelterwood treatment, loggers skip the preparatory cut and remove 30% to 60% of the canopy during a good seed year.

They make the second (final) cut once sufficient regeneration has established.

Shelterwood cuts do present some challenges. Because at least one cut must occur during a good seed year, which can be difficult to predict, the timing may not coincide with a favorable economic climate. The system also requires careful harvesting operations to avoid damaging seedlings or retained overstory trees. It also demands steady monitoring and interventions over a fairly long period, perhaps up to 20 years or more.[84]

Selection Method

Unique among regeneration treatments, selection management maintains an uneven-aged stand with continuous canopy cover. Selection management mimics and accelerates natural succession by harvesting trees in the oldest age class to encourage understory regeneration, while also reshaping the forest understory using intermediate thinning treatments. Selection management optimizes the growth potential of all age classes in a stand so that several mature high-quality trees and several fuelwood trees may be harvested on a relatively short cycle from a given acre. Selection-managed stands possess a diverse, multilayered structure, including individuals from a wide range of age classes.

Selection forestry does not involve rotations. Timber harvests steadily regenerate the understory while simultaneously carrying out timber stand improvement. Selection management cutting plans prescribe optimal basal area targets based on the forest cover type and site-specific qualities. These stand density targets provide a metric to guide the selection of trees for removal.[85]

Best-suited to shade-tolerant species that can respond even at a relatively mature age, selection forestry aims to maintain optimal stocking rates in each diameter class. Individual trees are selected and removed to improve overall stand quality. Trees suffering from disease, demonstrating poor form, of a relatively large diameter, or otherwise compromising the growth of other quality stems nearby may all be candidates for removal.

The selection method creates small stem-scale disturbances, producing a continuously evolving stand at the landscape scale that offers a highly stable structure and continuous canopy cover. This high structural diversity and steadily shifting dynamic are some of selection management's strongest attributes: it maintains fully forested landscapes capable of supporting diverse wildlife populations and uncompromised ecosystem functions.

Selection forestry's primary drawback lies in the perceived efficiency of management. Harvesting single trees in a forest stand makes felling and extraction operations more complex. There are more opportunities for trees to get hung up as they're felled, while access and log removal are more constricted. As such, it is well-suited to low-impact extraction techniques (draft animals, tractor and forwarding trailer, etc.). Selection forestry also requires more careful and informed forest inventory and management prescriptions that may challenge the skills and training of some landowners and loggers.

Patch Cuts or Group Selection

More heavy-handed than selection forestry, yet less extensive than a clear-cut, patch cuts open sizable canopy gaps while maintaining a mosaic of forest blocks very similar to traditional coppice systems. Something of a hybrid between even- and uneven-aged management systems, patch cuts employ intense patch-level disturbance to reshape forest ecosystems on a broader scale than selection forestry.

While patch cuts have a greater impact on forest cover than selection management, they also create openings that receive considerably more light, supporting natural or artificial regeneration of light-demanding species. To help ensure this, patch cut gap diameter should be at least 1.5 to 2 times the height of the trees

FIGURE 4.11: Stacks of eastern white pine sawlogs in a one-acre patch cut, creating diverse habitat for wildlife in an otherwise even-aged forest stand in New Haven, VT.

surrounding the gap. Gaps themselves are only a few acres in size.[86] Seedling growth rates vary within patch cuts depending on trees' location within the opening. Those in the center tend to grow rapidly, while individuals along the edges may suffer from excess shade.[87]

In summary, all of the high forest systems we've discussed can be applied with care and good intention to create and maintain healthy forest ecosystems that provide valuable ecosystem functions while also yielding diverse wood products. And as we'll discuss in chapter 7, we may choose to use some of these techniques to aid in converting high forest stands to coppice woodlands. As we continue along our journey, we'll soon begin to make the transition from analysis to design. And with this understanding of the various components that comprise systems, along with a suite of management tools to employ as appropriate, we become more and more equipped to make confident well-informed decisions.

PUTTING IT ALL TOGETHER: CHOOSING THE RIGHT SYSTEM FOR YOUR LANDSCAPE AND NEEDS

We modern land managers have a remarkable diversity of forest management strategies at our disposal. Having evolved over hundreds of years, each offers opportunities to intervene in forest succession and shift stand species composition, reduce competition, reshape stand structure, rejuvenate the stand, maintain desirable habitat conditions and ecosystem functions, improve genetics, and harvest wood products that meet our needs.

Familiarize yourself with these systems, learn their strengths and weaknesses, and carefully analyze which are most appropriate for your woodland and your goals. What are your needs? What resource base do you have access to? What species naturally grow there? What will it take to establish the system/s that best meets these requirements? What does the landscape tell you it needs?

Silviculture leverages intentional targeted disturbance to achieve social, economic, and ecological goals. As we establish and maintain forestry systems, we're responsible for choosing and applying the tools that offer the most appropriate long-term benefits for the forest and its human and nonhuman inhabitants. Recall our discussion on the ethics of coppice agroforestry in the introduction. How can your efforts help regenerate and revitalize damaged ecosystems while supporting healthy ecosystem functions?

Resprout silviculture systems provide us with a rapid, powerful intervention strategy to intensify wood production using the wonders of vegetative reproduction. We might choose to create landscapes with a diverse multistoried architecture that includes understory plants, standard trees, and complex rotations or integrate grazing animals into a mosaic of land use patterns. Resprout silviculture systems are not appropriate in all circumstances, but our understanding and appreciation of their fundamental principles can help us better steward the wooded ecosystems we engage with.

Hopefully this systems-level discussion has left you feeling energized and excited, pondering the benefits and drawbacks of a wide range of silvicultural practices. In chapter 5, we shift our focus to a deep dive into perhaps one of the most exciting aspects of resprout silviculture—the potential to add value to poles and rods to create myriad products and craft our own livelihoods and cottage industries around our woodlot management. Here's where we learn about traditional and modern uses for coppiced material and imagine how these practices become relevant to our lives and local and regional economies. So let's explore this vision of the modern polewood economy!

Chapter 5:
Coppice Economics and Products

It is said that money does not grow on trees…. The economic time scale of forestry is so vast and unique that to many investors and in terms of ordinary economic theory it really is not economic at all…. It takes a certain kind of ambivalence to keep the economics of forestry in perspective. The decision to practice forestry is usually a matter of ethics, politics, and social concern for posterity but not basically one of conventional economics.

— Mark S. Ashton and Matthew J. Kelty,
The Practice of Silviculture: Applied Forest Ecology (2018)[1]

Coppicing evolved to meet clear and immediate societal needs, and it persisted because it was highly effective. Some of these needs were essential: fuel, fencing, fodder. Others formed the foundation of highly developed production networks, fueling the craft economy. These unique coppice economies reflect each woodland's structure and management strategies. Systems that produce useful yields receive care and attention.

We strive to design coppice systems that don't just meet the needs of our families and communities but also benefit our larger bioregions socially, ecologically, and economically. We can design highly functional agricultural ecosystems, but if they aren't economically sustainable, they will tend to fail in the long term.

We discover one of agroforestry's largest hurdles when we compare its profitability to that of conventional land management. Here, the metrics we choose determines the outcome of our analysis. A truly comprehensive model must account for the ways agroforestry reduces landowner risk, helping stabilize income through diversification, and provides essential ecosystem services whose value grows increasingly clear.[2]

In an industrial forestry model, "efficiency" describes only the cost-benefit of wood production compared to the costs of doing business (fuel, labor, infrastructure, etc.). It pays little attention to the social and ecological costs affecting the entire system (soil loss, habitat fragmentation, flooding, pollution).[3] We have numerous economic tools to guide agroforestry planning at our disposal, including enterprise planning (estimates of profitability), risk assessment models (projected event probabilities and their influence on profitability), and nonmarket valuation models (estimated values for environmental goods and services like water quality, carbon sequestration, and erosion prevention).[4] But regardless of the tools we use to plan, for our enterprise to be economically viable, we must identify and connect with a willing market.

MARKETS

As we develop resilient local economies employing ecological land-use practices, we must remember that several elements comprise a whole economy. A strong woodland-based economy demands a reliable supply of quality materials, skilled and passionate craftspeople, and a sound market for their produce.[5] A gap in any one of these areas creates bottlenecks that strain the whole. Many readers presumably plan to involve themselves in the first two parts of this economic triad, but even the most well-intentioned producer often skips the hard work of market research. Taking the time to analyze market demand for your products will pay long-term dividends, helping ensure that your early investments find a customer base eager to support your work.

Market demand is one of the most important factors in developing a viable agroforestry enterprise. Markets frequently act as the primary constraint limiting expanded production.[6] There are no modern mass-market demands for greenwood products here in North America as there were for 19th-century European products like hurdles, hop poles, fencing, besom brooms,

The Key to a Healthy Woodland Craft Economy: Markets, Craftsmen, Materials

FIGURE 5.1: A strong craft economy requires quality raw materials, access to markets and reliable demand, and people with the skills, vision, and drive to add value. Balancing these three characteristics creates numerous opportunities to produce value-added products and support viable land-based livelihoods.

hoops, bobbins, and turnery. Back then, industry and agriculture were coppice workers' largest, most reliable customers.

In modern Britain, most coppice products are marketed to the "tourism, leisure, environment, and heritage" sectors. Demand for garden products, wattle hurdles, barbecue charcoal, custom furniture, plant trellises, and pea and bean poles are many coppice workers' most important and reliable markets.[7]

Diversification is key to the viability of today's coppice enterprise. Most modern coppice craftsmen build and sell 3 to 7 different products, while some produce more than 20.[8] In 2004, ⅔ of all products made by British coppice workers included rustic fencing, charcoal, firewood, wattle hurdles, and chestnut pale fencing. And in many parts of the UK, conservation-based coppicing to create early successional woodland drove as much as 25% to 30% of all coppice management.[9]

Perhaps we have a great product idea that lacks an existing market. What to do? Start by considering two primary options—develop the market and create demand before building production capacity; or build capacity to produce the product, anticipating the future growth of the market.[10]

One strategy focuses on production first, believing market demand will follow. This carries significant risk, especially for wood products that take years to reach harvestable size. In the alternative, producers begin with a thorough market analysis, identifying threats and opportunities before developing production capacity. This enables producers to better anticipate future successes and challenges. Either way, economic sustainability ultimately relies on market demand.[11]

So which strategy is better? Well, like most things, it depends, and it's most often a bit of a both-and scenario. It's wise to start developing your business after doing a bit of market research, testing the waters with your new products, identifying local competition, and getting a good handle on your likelihood for success. But at the same time, these systems take many years to mature, so it's hard not to start implementing while you're immersed in market analysis. And like we said earlier, more often than not, developing a diverse line of valuable products builds a far more resilient enterprise.

In the traditional woodland economy, woodsmen were also craftsmen. And most of these entrepreneurs took whatever opportunities they could to add value to the raw materials they harvested. Added-value products offer much more earning power than raw material sales, encouraging diversification and customization. The potential to add value to coppice-generated materials spans a wide spectrum. When considering these opportunities, imagine where a product fits along a conceptual continuum of added value. This tool helps illustrate and articulate how raw material quality, volume, and invested labor influence a product's monetary value.

THE VALUE-ADDED CONTINUUM

The "value-added continuum" compares wood products based on the time, energy, and skill

The Value-added Continuum

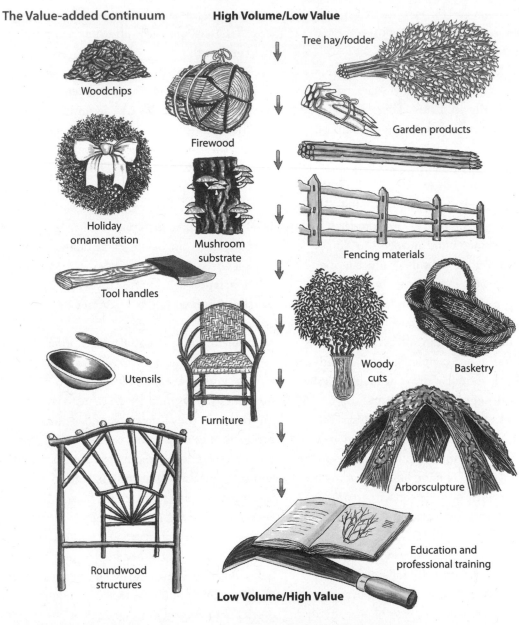

High Volume/Low Value

Woodchips

Firewood

Holiday ornamentation

Mushroom substrate

Tool handles

Utensils

Furniture

Roundwood structures

Tree hay/fodder

Garden products

Fencing materials

Woody cuts

Basketry

Arborsculpture

Education and professional training

Low Volume/High Value

FIGURE 5.2: With creativity and access to local or regional markets, it's possible to turn the same raw materials into any number of useful products. The key lies in creatively adding value to those materials in order to make best use of them. The continuum begins at the top of the illustration, from low-value materials that require minimal processing, to higher-value return products that tend to become more and more specialized.

required to transform a raw material into useful product. It directly affects the materials' market value and each enterprise's economic viability. To paraphrase Ben Law, "You can make a living off 1,000 acres or just one—it all depends on how you add value." It's so obvious, it almost goes without saying—exploring creative ways to add value to rods and poles can vastly increase earning potential, creating craft-based livelihoods that sustain ecological land management.[12]

In a research paper published in 2000, small-diameter timber values reportedly averaged $26 to $60 per green ton in the midwestern USA. But this was only partly true. While firewood and woodchips sold for roughly $26 to $40/green ton, fence posts were worth $140/green ton, lumber $140 to $160/green ton, and small poles $200/green ton.[13]

At one end of our value-added continuum, we find low-value materials that might otherwise be considered waste. Bundles of twigs and brush used for firing bread ovens and mitigating stream bank erosion both utilize materials with no market value. Often these materials may comprise as much as half of all wood harvested in a coppice cant by volume, and usually it would be chipped, burnt, or left in the forest as slash. While not useless, it's a low-value, high-volume product, providing a return measured in dollars per ton. But with the right tools, skills, and market access, they can become a financial opportunity.

Moving along the continuum, we can add some value to raw polewood by cross-cutting, splitting, and seasoning as fuelwood. Meeting a clear need for many rural folks well into the future, fuelwood also provides considerable

Table 5.1: A few potential products made using different diameter rods and polewood

0.5–1.5" (1.25–3.75 cm)	1.5–3" (3.75–7.5 cm)	3–5" (7.5–12.5 cm)	5–8" (12.5–20 cm)
Ramial Woodchips	Ramial Woodchips	Woodchips	Woodchips
Wattle Fencing	Fuelwood	Fuelwood	Fuelwood
Basketry	Charcoal	Charcoal	Charcoal
Artist's Charcoal	Garden Products	Mushroom Bolts	Mushroom Bolts
Besom Broom Heads	Holiday Ornamentation	Tool Handles	Tool Handles
Arborsculpture	Utensils	Rustic Furniture	Rustic Furniture
Rustic Furniture	Rustic Furniture	Green Woodworking	Green Woodworking
Woody Cuts	Decorative Fencing	Polewood Construction (Small)	Polewood Construction (Large)
Fodder			
Garden Products			

potential to innovate through creative marketing. In some communities, CSA-style (Community Supported Agriculture) fuelwood models invite customers to pay a premium for wood they know has been harvested and processed in an ecologically responsible way.

But why stop with fuelwood when we can make charcoal? Now we can convert low-grade biomass into a higher-value fuel or soil amendment. While locally crafted artisanal barbecue charcoal may be a tough sell in some regions, the demand for quality artist's charcoal already exists and could provide an excellent return for a small volume of quality product.

Log-grown edible mushrooms are another practical product with an established market and reliable production system. Still making use of relatively low-grade (curvy, irregular)

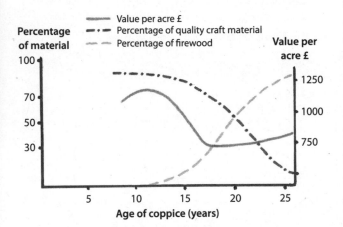

FIGURE 5.3: This graph depicts how increasing age affects material's utility and relative value assuming a mixed species coppice with at least 40% hazel. Because out-of-rotation hazel is too large for woven crafts, it is often only useful as firewood. And so we see value peak around age 12 or so, dropping until year 17, and then slowly increasing thereafter as fuelwood. Adapted from Tabor, 1994, p. 149.

small-diameter wood, growers may sell inoculated logs or develop their own home or production-scale operation, marketing their produce to health food stores, restaurants, or direct to customers. Shiitake and oyster mushrooms may fetch as much as $15 per pound retail and a single 6-inch-by-40-inch log may produce a pound or more of fresh mushrooms per year.

At this point along the continuum, we start to shift from the modest conversion of relatively low-quality material towards products that require a greater degree of craftsmanship and serve more specialized needs. Although these products will tend to generate far greater returns, they usually require much higher-quality materials, skill, and investment, rendering them more demanding from a production standpoint. While this dichotomy in materials quality and processing creates a bit of a dilemma, it also drives growers to develop operations that make economical use of *all* materials—both high- and low-grade.

Prioritize straight, even, regular poles for those products that demand it, and make best use of whatever remains. We'll explore each product in detail later in the chapter, but as a general trend, as we progress further along the spectrum, products shift from utilitarian to artistic. This is not to say that a product can't reflect both functionality and artistic expression, but that the markets with the highest return often have little connection to meeting immediate needs. Practically speaking, one's ability to access any of these markets relates directly to the economic climate of one's region and the scope and skill of one's marketing and outreach.

BUSINESS DEVELOPMENT

To explore a material's product potential, start by identifying its complete range of useful possibilities. From there, we can examine how to best utilize it, matching quality with use, while considering new and creative value-adding opportunities.

As we explored in chapter 1, traditional coppice workers usually specialized in one or two products that they produced during spring and summer. While this model appears to have been functional then, today this single-product approach could be vastly improved.

Successful modern coppice operations typically produce a wide range of products in order to optimize their use of harvested materials and buffer themselves against market unpredictability. The economy of coppice both influences and is influenced by several fundamental factors, namely, the species being coppiced, their size and maturity at harvest, and the needs of the neighboring economic markets. Thus, coppicing is highly site-specific. The products one chooses to produce relate directly to the materials available, their personal interests and passions, and their knowledge of the economic climate and opportunities surrounding them. To expand the relevance and demand for specialty products, we need to find ways to innovate and intensify production.

Always look to utilize resources for their highest possible practical use. While this "highest value use" is subjective, the concept reminds us to be both conscious and flexible when exploring our options. Our ultimate goal lies in creating a cascade of products along the value-added continuum. Ben Law's sweet chestnut copse in West Sussex, England, is an excellent example of a thoroughly integrated, economically viable coppice system.

Ben manages 100 acres (40.5 ha) of woodland that includes160-year-old sweet chestnut coppice; hazel, ash, and field maple coppice with oak standards; and a larch plantation. Ben produces fence materials, tree stakes, walking sticks, yurt kits, garden structures, firewood, charcoal, roundwood timber frame structures, rustic furniture, faggots, woodchips, shiitake mushrooms, and more. He has designed his management to adapt to the demands of his customers, adjusting his annual management plan to meet market demand. For example, if he finds considerable demand for yurt kits before beginning the winter harvest, he can choose to cut a cant that meets that demand rather than strictly adhering to a predetermined rotation. This flexibility provides much greater security—he knows well in advance that he has customers lined up willing to purchase his products.

His broad product line also allows him to make far more efficient use of harvested material. If cutting a stand of 20-year-old chestnut, for example, he can grade materials, prioritizing long straight pole sections for building and fencing, while setting short curved sections aside for mushroom production, fuelwood, charcoal, or rustic furniture. The slash can be chipped and returned to the woods, composted, sold as mulch, converted into fascines, etc. Short, straight small-diameter pole sections may make good yurt poles, tree stakes, or posts. This spectrum of potential optimizes Ben's utilization of coppice wood, enabling him to nimbly navigate each year's orders.

Similarly, a chestnut shake (hand-split roofing shingle) factory in central France maintains chestnut coppice to meet their material needs. Usually more than half of all harvested raw materials are unsuitable for shakes—but these leftovers aren't waste. They process the larger waste logs into cordwood, smaller sections into kindling, and shavings into high-quality firestarter. They also convert some of their scraps into charcoal. In this way, they manage a far more diversified operation that attracts a broader customer base and provides a more stable income stream. An economically viable coppice system will likely take a very different form in your locale, but these flexible, resilient, integrated systems provide excellent models.

The remainder of this chapter offers a more in-depth exploration of the wide range of

Pricing Your Product

The Industrial Revolution brought major shifts in the value and availability of energy and labor. Traditional cultures relied on abundant cheap labor with minimal external energy sources. Today, throughout the "developed" world, cheap energy abounds, while high labor costs relegate most craft work to the likes of luxury items accessible to a narrow demographic. Once you've identified your audience, how do you ensure your products are accessible to them and you earn a healthy wage for your efforts?

Finding the right price point for your products can be a major challenge for many craftspeople. Building a business is difficult enough—determining a fair market price can be exceedingly daunting. British green woodworker and master bowl and spoon carver Robin Wood points out that value (and price) always depends on context. How do your products stand out? How do you tell a compelling story about your work? How do you present your work to the public, and how and where do they discover it? These factors and more make the difference between clients finding your work ordinary versus extraordinary.[14]

Perhaps the most straightforward approach to choosing a fair price for your products is through a cost of production model. To do this, you need to determine your enterprise's fixed and variable costs. Fixed costs include such items as shop space, maintenance, utilities, tool and machinery upkeep, employees, insurance, and marketing. These costs exist regardless of how many items you produce. They form the foundation of all it takes to keep your business afloat.

Next, project the variable costs associated with the actual manufacturing of your product. This should include wood costs/value, other essential materials (glue, fasteners, seat material if making a chair, etc.), any shipping or packing materials, labels, and of course, *your labor*. It's a good general rule to double your estimate of the time required to make a product in order to more accurately project your true costs of production since often at least 50% of your ☞

working time is spent talking with customers, developing marketing materials, sourcing craft materials, traveling to and setting up at craft fairs, accounting, etc.

Once you've compiled all of this information, you can determine the actual costs of producing a single item or a batch. This signifies the lowest possible price that you can afford to make a product. This potent assemblage of information also gives you flexibility to adapt your pricing for wholesale orders or perhaps sliding scale pricing tiers. And it helps you identify bottlenecks in your production and opportunities to reduce costs by streamlining an operation, choosing a different supplier, or perhaps purchasing materials in larger quantities. While this process may not be the fun part of your production, taking the time to do this analysis is essential to creating a business that is truly sustainable. And in so doing, it provides a reality check

to help ensure that these processes and costs actually align themselves with your larger overall values and life goals.

The words of an Appalachian shingle maker help contextualize some of these quality and value-based decisions:

> If I had a portable shingle mill and a gasoline engine, and gasoline to run it, and a helper, a horse to draw it about, feed for my horse, a place to keep my mill when idle, some attachments for other kinds of work, and a little money for keeping up repairs I could make as many shingles in a day as I can by rivin' in a week… but when it comes to lastin', a roof of properly hand rived boards will last 4 times as long as them cut with a saw.[15]

Table 5.2: Projected costs of hazel coppice rental and harvest in Southern England (11-year rotation) ca 1990 and US today

	1990 per Hectare Estimate (UK)	2021 per Acre Estimate (US)
Volunteer person-days for coppicing and extraction	275 days	110 days
Skilled woodsman (10x volunteer speed)	27.5 days	11 days
Minimum daily coppice worker pay	£40	$160 ($20/hr)
Labor costs (27.5 x £40)	£1100	$1760
Tools, fuel, chain saw	£100	$350
Rental costs - £12/ha/yr	£132	$100/acre/yr = $1100
Break even costs	£1332	$3210

Adapted from Buckley, 1992, p. 304.

products one may make using woody resprouts. Whenever possible, for each product we'll explore the processes involved in production; the most important wood qualities; the best-suited species; the skill, labor, and quantity of materials required; and the current market value for the finished product.

Today we inherit a remarkable trove of wood products and craftsmanship. This blessing bestowed upon us by our ancestors expresses the unique character of local cultures, species, needs, resources, and skills. In creating our own craft-based livelihoods, we connect ourselves to these traditions and continue to build on their legacy.

To learn something about crafts and craftsmen is to learn about the history of the race. Each craft is a rich repository of many years of practical experimentation and knowledge by men and women whose very lives were shaped and enhanced by the work of their hands....They also demonstrate, with beauty and precision, how generations of creativity went into developing and refining hundreds of regional variations that are only now blending, losing their identities....Of course many variations developed depending on local conditions: many an English countryman made a profession of coppicing because he could easily sell the forest products within the local community, but a settler in Ohio, with no such market, produced entirely for himself and his family.

— John Seymour[16]

Figure 5.4: Bodger at work, Mr. Silas Saunders, ca 1940s. A bodger using a drawknife to shave riven wood into parts for green wood crafts. He's sitting atop a shaving horse—an ingenious foot-operated vice that helps the woodworker grip the piece as they shape it using hand tools. Just one of the many craft legacies passed on to us by our ancestors. Image used with permission from the Museum of English Rural Life, University of Reading.

WORKING ALONG THE VALUE-ADDED CONTINUUM FROM LOW TO HIGH

Woodchips: Biofuel, Soil-improving Mulches, Woodchip-Clay

Desirable wood properties/qualities—Depends on application: mulch, species that readily decay (or provide long-lasting cover); woodchip-clay, less dense woods offer a better R-value; fuel, denser is better, though they generally grow more slowly

Most promising species—Depends, but often fast-growing species like willow, poplar, alder, sycamore, silver maple, black locust

Special tools or machinery required—Chipper, truck, trailer

Materials volume required—Varies depending on end use and need but often large volumes

Skill level—Low

Price range—Low, free to $20+ cubic yard

Biofuel

Chipped and shredded pole and branchwood rate low along the value-added continuum. Woody biomass for heat and electricity production only makes good economic sense at a relatively large scale. But as woody biomass markets grow, it increasingly becomes possible for landowners to generate some income from low-grade wood products.[17] High establishment and raw material transportation costs along with the need for relatively specialized harvesting and processing equipment are a few of woody biomass production's greatest barriers to adoption.[18] But considering that Swedish short-rotation *Salix viminalis* (basket willow) and *S. dasyclados* plantations have yielded as much as 14.2 tons/acre/yr (35 MT/ha/yr), we can begin to understand why these systems have garnered so much interest over the past few decades.

Quite a bit of solid research and development has been invested in creating tools for growers to establish commercial-scale short-rotation coppice (SRC) systems. Visit appendix 1 for a list of recommended resources. Here we'll also explore a few other fairly common ways to add value to raw woodchips.

Mulch

Various forms of woodchip (or bark) mulch play a crucial role in the commercial landscape industry for which there is a near-continuous demand. Many landscapers use softwood *bark* mulch to conserve moisture, protect soil, impede weed growth, and create a clean finished look. But many bark mulches decompose slowly, lasting for several seasons before needing to be refreshed, doing little to build soil fertility and biology.

Woodchips, on the other hand, offer a fantastic feedstock for fungi and soil microbes.

In the northeastern United States, bark mulches sell for $30 to $50 per cubic yard ($40 to $65 per cubic meter) in and of itself a modest return on investment at best but one that could help supplement income generated from more lucrative value-added products.

Ramial Wood

Ramial wood, young woody tissue less than 2.75 inches (7 cm) in diameter, quickly decomposes creating rich soil and supporting biologically active topsoil. Coppicing generates large volumes of this valuable material. The Ramial Chipped Wood technique (RCW) developed by Professor Lemieux at Quebec's Laval University stimulates natural nutrient cycles and benefits many types of microorganisms during the first 6 months following application. Fungal populations may swell to 10 times their typical number for as much as 1 to 2 years, while microarthropods graze on their hyphae, and their excreta feeds bacteria who make nitrogen and other nutrients available to nearby plant roots.[19]

The benefits of RCW have been studied since the 1980s. In Belgium, RCW applications increased soil organic matter by 1% in just 10 years as compared to 50 years for manure, while in Canada, RCW application led to a 300% increase in strawberry biomass and 30% dry matter increase for potatoes.[20]

The best RCW comes from dense hardwood species with a high lignin content like beech, oak, and maple (conifers should not exceed 20% of total RCW biomass). Early-successional species like poplar and birch will not generate the same effects. RCW is applied mechanically with a manure spreader or by hand with a rake. Researchers recommend applying a 0.6-inch (15 mm) layer at a rate of 80 cubic yards per acre (150 m³/ha), ideally incorporating it into the soil to a depth of 4 inches (10 cm) with subsequent applications of 40 cubic yards per acre (75 m³/ha) every three years.[21]

Figure 5.5: Mixed with planer shavings and sawdust and coated with clay slip, woodchips can be used to create well-insulated wall systems for cold-climate buildings. Here thin wooden lath strips retain the woodchip-clay mix within the cavities formed by a double-stud wall.

Woodchips for Natural Building

Woodchips can also form an insulative wall system called "woodchip-clay." This system uses a mix of woodchips, wood shavings, and sawdust coated in a liquid clay slip (water-clay suspension) as infill between wall-framing members. Poured into stud bays and loosely compacted, it's held in place by thin wood lath strips that are usually finished with a clay or lime-based plaster.

Woodchip-clay's R-value (insulative value) appears to vary between 1.5 and 1.7 per inch wall thickness. (The higher the number, the greater the insulation value.) Although not exceptional insulation in a single-stud wall system, builders often compensate by creating double-stud cavity walls 8 to 18 inches (20 to 45 cm) thick to achieve much higher performance standards.

Fodder

Desirable wood properties/qualities—Readily digestible leaf material with high nutritional value

Most promising species—White mulberry, oak spp., black locust, elm, mesquite (note that some species do have toxicity concerns so please take care and thoroughly research any woody fodder species before feeding to livestock)

Special tools or machinery required—Loppers, billhook, polesaw, drying racks, dry storage

Material volume required—Depends on needs. Can range from supplemental feed to potentially a dried and stored winter feed

Skill level—Moderate. Harvesting materials is relatively straightforward, but training/creating pollards or managed fodder banks takes some degree of knowledge and expertise

Price range—Presumably low. No real existing commercial market. Better suited to small-scale livestock operations striving to achieve winter forage subsistence, although there may be possibilities to scale these systems up.

Both as a dried, stored, dormant season livestock feed and a nutritional supplement during the growing season, tree hay offers a great

FIGURE 5.6: A bundle of freshly cut fodder from a mix of woody plant species en route to supplement the diets of a small flock of sheep.

complement to herbaceous forage. Consider that a single mature aspen tree may produce roughly the same volume and quality of forage as a round hay bale.[22] In Europe, elm and ash are said to produce the best leaf fodder, while maples, alders, birch, hornbeam, European walnut (*Juglas regia*), mulberry, poplar, oak, willow (especially *S. alba* and *S. caprea*), rowan, and linden are also commonly pollarded for tree hay. In Mediterranean regions, graziers also pollard (and/or shred) *Melia azedarach*, white mulberry, olive (*Olea europaea*), oak species, *Platanus orientalis, Tamarix* spp., *Populus nigra,* and numerous eucalyptus species.[23] While the specifics certainly vary considerably based on species, etc., Karl Gayer's 1896 *Manual of Forestry* reports that 150 kilograms of leaf fodder (330 lb) (not including branches) is considered equivalent to 100 kilograms (220 lb) of average quality hay.[24]

When considering the nutritive value of a feed, two of the most common metrics include crude protein and dry matter digestibility. Crude protein is a measure of the amount of nitrogen in a forage. This is important because it is the building block for amino acids that form proteins. At insufficient levels, the bacteria in ruminants' guts cannot adequately process forage. Keep in mind though that crude protein is really more of a qualitative measure of forage value and not a measure of its available energy.

Digestible dry matter or dry matter digestibility provides a measure of the digestible proportion of a forage. It's often measured both in vitro (in a test tube/laboratory setting) and in vivo (within a living animal). The primary challenge in quantifying these qualities lies in their seasonal variation and that the digestibility varies by animal species and breed. Generally, both of these figures peak with the first flush of

Table 5.3: Fodder species with relatively high crude protein content

Genus	Species	Common Name	Average Reported % CP
Albizia	julibrissin	mimosa	19.3
Alnus	incana	European gray alder	20.1
Betula	allegheniensis	yellow birch	23.5
Elaeagnus	angustifolia	Russian olive	18.0
Hippophae	rhamnoides	sea buckthorn	19.7
Morus	alba	white mulberry	20.5
Populus	tremuloides	quaking aspen	13.8
Robinia	pseudoacacia	black locust	19.8
Salix	caprea	goat willow	18.8

A brief selection of common fodder species with moderate to high averaged reported crude protein content. Turn to appendix 2 for a much more comprehensive list.

spring growth, steadily declining as the season progresses, varying between species. See Figure 5.7. What's more, fodder volume per tree steadily increases later in the growing season. This means spring harvests optimize forage value but result in minimal yields, while late-season harvests provide a larger supply of lower-value feedstock. So, timing of fodder harvest involves a series of complex management decisions.

Among 10 North American broadleaf species, black locust and honey locust forage had the highest nitrogen availability and digestibility for ruminants. In a 2002–2003 Arkansas study, black locust fertilized with 535 pounds P per acre (600 kg P per hectare) averaged 1.56 tons/dry matter/acre (3.5 MT/dry matter/ha) at a density of 6,072 trees/acre (15,000 trees/ha), reaching peak yields of 4,729 lb/acre (5,300 kg/ha) in August. Crude protein and in vitro dry matter digestibility levels generally decreased as the season progressed, but were still sufficient to maintain beef cattle. Foliar crude protein decreased from 239 to 170 g/kg during the growing season. Digestibility decreased between June and September, later increasing in October (663, 534, and 557 g/kg, respectively). While many pollarded species require a fallow period every few years to reduce carbohydrate reserve depletion in stem and root tissues, black locust is known for its ability to tolerate repeated annual pollarding.[25]

Similar comparisons between black locust and mimosa (*Albizia julibrissin*) found leaf yields of 1,695 and 1,427 pounds of leaves per acre (1,900 and 1,600 kg per ha), respectively, at densities of 4,980 per acre (12,300/ha) or

Effect of Harvest Date on Crude Protein and Dry Matter Digestibility of Black Locust Fodder

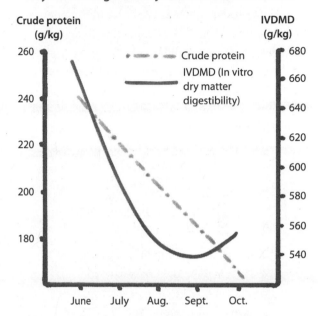

FIGURE 5.7: In general, the quality of black locust forage declines as the growing season progresses. Here, crude protein content shows a steady decline between June and October, while in vitro dry matter digestibility follows a similar trajectory between June and late August only to rebound slightly from summer's end until October. Adapted from Burner, Pote, and Ares, 2005, p. 212.

3 feet (0.9 m) in-row and between-row spacing. Both species should provide sufficient nitrogen for growing goats and beef cows. While goats preferred black locust fodder, growing goats would require protein supplementation, whereas mimosa fodder could meet and even exceed growing cattle and goats' nutritional nitrogen requirements.[26]

And blanketing as much as 84 million acres (34 million ha) across the American Southwest, mesquite species (*Prosopis* spp.) are another promising livestock feed. Honey mesquite

(*Prosopis glandulosa*) pods offer a valuable food source for humans and animals, and the leaves' crude protein, gross energy, and fiber values are similar to mature alfalfa. Note though that allelochemicals in the leaves (flavanoids, alkaloids, and nonprotein amino acids) may cause flavor aversions and some digestive drawbacks.[27]

Fuelwood

Desirable wood properties/qualities—Density, fast growth

Most promising species—Osage-orange, black locust, oak, maple, alder species

Special tools or machinery required—Depends on scale: chain saw, splitting maul, truck/trailer, hydraulic splitter, firewood processor

Materials volume required—Depends on need/demand

Skill level—Low to moderate

Price range—$100 to $300+/cord ($28 to $83 m^3) depending on species and whether it's green or air dried

Over ⅓ of the global population uses fuelwood for cooking and heating today. In the developing world, 80% to 90% of all wood resources are used for fuel, with over half devoted to cooking. In much of Asia, Africa, and Latin America, fuelwood for cooking and heating is a major living expense and continues to grow in scarcity with increasing population and development. In Nepal, many rural residents must travel up to a full day to collect ample wood, whereas just a generation earlier, they could find sufficient wood within an hour's radius of most homes.[28]

For many readers, fuelwood may be the most practical yield coppicing provides. Small-diameter coppice-grown fuelwood is easy to handle and transport, requiring less splitting but more cross cutting.[29] Although it has a steady market value that will presumably only increase with time, the value of developing self-sufficiency likely greatly exceeds the income it can generate unless you invest considerably in equipment to streamline the process (tractor, logging winch, firewood processor, etc.).

In the United States, we sell fuelwood by the cord, a pile 4 feet by 4 feet by 8 feet or 128 cubic feet (~3.6 m^3) (although gaps between logs reduces the average cord to only about 80–95 cubic feet of solid wood). Depending on species and moisture content, a cord's weight may vary between 1 and 2 tons.[30] However, not all fuelwood is created equal. Wood density directly affects its heat value. Denser woods possess more heat-generating capacity. And wood's moisture content strongly influences its combustion efficiency.[31]

Burning green (or unseasoned) wood may take as much as 20% of wood's total energy to evaporate the latent moisture, equivalent to wasting one out of every 5 year's growth! (And what's more, it also leads to creosote buildup, increasing the risk of a chimney fire.) This means *properly seasoning wood yields more energy per unit time than the natural rate of tree growth.*[32] At a minimum, season green firewood 1 year before burning, ideally 2.

Volume of Wood Equal to 30 mBTUs for Three Coppicing Species

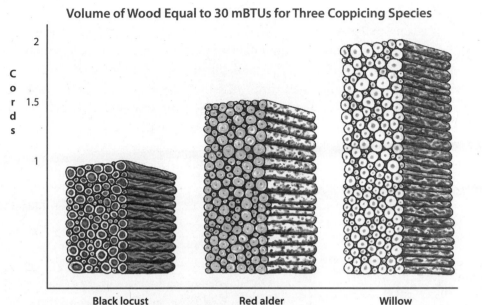

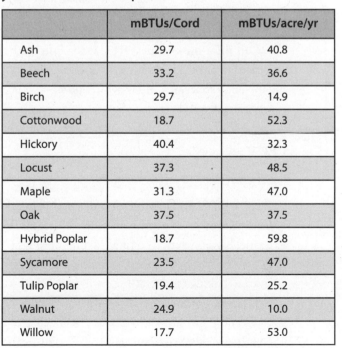

FIGURE 5.8:
A comparison of the fuelwood values of black locust, red alder, and willow spp. One cord of black locust contains nearly two times the energy of an equal volume of willow.

Table 5.4: Energy value and total annual per acre energy yield of different tree species

	mBTUs/Cord	mBTUs/acre/yr
Ash	29.7	40.8
Beech	33.2	36.6
Birch	29.7	14.9
Cottonwood	18.7	52.3
Hickory	40.4	32.3
Locust	37.3	48.5
Maple	31.3	47.0
Oak	37.5	37.5
Hybrid Poplar	18.7	59.8
Sycamore	23.5	47.0
Tulip Poplar	19.4	25.2
Walnut	24.9	10.0
Willow	17.7	53.0

Despite the low heat value of many fast-growing tree species, they are actually among some of the most productive per unit area due to their prolific, rapid growth. Adapted from Beattie et al., 1993, p. 166.

While yields vary considerably based on species, climate, and site quality, many readers in non-brittle ecosystems can generally anticipate a mean annual increment (MAI) of 0.5 to 1 cord per acre per year (0.7 to 1.5 m³/ha/yr)—about the same as the sustained annual increment from well-managed high forest in the northeastern US. Therefore, 7 acres (2.83 ha) managed on a 7-year rotation should yield 3.5 to 7 cords (12.5 to 25.5 m³) per acre at harvest. In Britain, 40-year-old coppice

Table 5.5

Log dbh Inches (cm)	Average Number of "trees" to Make a Cord
5 (12.5)	46
6 (15)	21
7 (17.5)	15
8 (20)	10
9 (22.5)	8
10 (25)	6
12 (30)	4
14 (35)	3
16 (40)	2

This table gives a general idea as to the number of trees of various diameters required to yield one cord of wood. It doesn't state tree length, but keep in mind that the usable length of trees will vary considerably, and larger-diameter trees will tend to generate longer useful lengths. Length is built into this general figure so that the smallest diameter trees are assumed to yield roughly 25 feet (7.6 m) useful length while the number increases to 45 feet (13.7 m) or more for larger-diameter stems. Adapted from Beattie et al., 1993, p. 43.

growth should yield 28 to 33.6 tons (25 to 30 tonnes) of fuelwood per acre as well as several hundred poles. Assuming about 3,000 pounds per cord, this amounts to roughly 18.7 to 22.4 cords (67.8 to 81.2 m³) per acre (plus several hundred poles) or right around 0.5 cord/acre year (0.733 m³/ha/yr).[33]

Often we look to hardy fast-growing pioneers capable of quickly colonizing deforested landscapes as some of the most promising fuelwood species. Fast-growing Brazilian eucalyptus plantations may be harvested in as little as 7 to 8 years after planting, with subsequent coppice harvests on rotations of 5 to 6 years.[34] In the temperate world, black locust is remarkable. A super-dense, rot-resistant wood that grows quickly and sprouts readily.

And, of course, sustainable fuelwood production is just one piece of an energy-efficient future. We should always begin with conservation since the cheapest BTU is the one you never burn. Weatherization, insulation, and efficiency strategies, along with high-efficiency wood heat systems must all parallel renewable fuelwood production so we can meet our needs on as little land as possible.

Faggots

European bakers once commonly used tightly bound bundles of brushwood called "faggots" to fuel bread ovens. The small-diameter twigs generate a hot fire, quickly bringing a masonry oven to baking temperature. Farmers also buried lines of these brush bundles to create free-draining channels to dry out fields, like a woody version of drainage tile.[35] Faggots generally measure 3 feet (90 cm) in length and 2 feet (60 cm) in circumference, a measurement standardized in 1474.[36]

Although uncommon today, creative marketing and access to specialty outlets could create unique sales opportunities. In the ski town of Warren, VT, I've seen a local crafts-person selling beautifully bundled faggots as kindling for tourists enjoying a ski vacation. Although a hard sell in many communities, the producer's attractive packaging in an area with a lively tourism industry allows them to turn a few dozen twigs into a small cottage industry.

Charcoal

Desirable wood properties/qualities— Density; Depends on the end use

Most Promising Species—Hedge and sugar maple (*Acer campestre* and *A. saccharum*), hickory species, beech species, red alder, white and black willow (*Salix alba* and *S. nigra*)

Special tools or machinery required— Some type of kiln or strategy to control oxygen intake during pyrolysis

Materials volume required—Varies based on kiln design. Ranging from a cookie tin's worth for artist's charcoal to several cords for larger kilns

Skill level—Moderate. Novices may effectively pyrolyze wood, but honing the craft requires skill

Time to produce—A few hours to 48 or more depending on kiln size and design

FIGURE 5.9: This device cinches together a tight bundle of brushwood to produce a high-intensity fire, often used to fuel masonry ovens. Photo courtesy of William Scott Bode, used by permission. Photo taken outside the Wood Museum (Le Musée du Bois) at the Aubonne Valley Arboretum (L'Arboretum du Vallon de l'Aubonne), Aubonne, Switzerland.

Price Range—Varies depending on location/market: 6.6-pound (3 kg) bag in the UK costs £6; $10 to $25/bag in Canada (15.5 lb or 7 kg); $1.30 to $1.85/lb in US for mass-produced hardwood charcoal

Humans have probably been producing charcoal since we first began smelting metals at least 5,000 years ago.[37] From ancient times through the Industrial Revolution, charcoal fueled iron, steel, glass, and gunpowder industries. It's also used in soap making, penicillin, plastics, paints, sugar and nonferrous metals refinement, and tar making and is one of the most effective natural filtration materials.[38]

It's hard to overstate charcoal's importance to early industrial cultures, but until the widespread development of coke ovens in the 18th century, steel and iron industries depended entirely on charcoal fuel. In 1282, 900 British charcoal burners worked in the King's woodlands in the Forest of Dean alone.[39]

Charcoal production or "the destructive distillation of wood," is a highly controlled pyrolysis process. By heating wood in oxygen-starved conditions, the process drives off water, volatile gases, tars, and other chemical constituents, resulting in a porous, pure form of carbon, a near-ideal fuel source. Because half or more of wood's weight is water, pyrolysis reduces its weight and volume between one-half and two-thirds. Charcoal sustains consistent combustion temperatures of over 1,832°F (1,000°C), yielding no smoke and little ash, and generating about twice the heat energy of wood.[40]

Wood used to make charcoal should be seasoned for at least 3 months. The greener the wood, the longer the pyrolysis process and the lower the yield as additional energy is required to drive off excess moisture.[41]

There are two primary types of charcoal burning technologies: kilns and retort systems. Kilns, like the portable ring kiln, generate heat inside the chamber where charring occurs, allowing limited air to enter while burning off some of the wood to drive pyrolysis. Unfortunately, ring-style charcoal kilns release large volumes of unburnt particulate into the air.

In retort systems, an external burn chamber supplies the heat that drives pyrolysis. Wood gases comprise 30% to 40% of wood's total energy. Retort kilns harness this energy, increasing the conversion efficiency of wood into charcoal, while releasing far fewer emissions than traditional kilns.[42]

Different wood species produce charcoal with unique characteristics. The barbecue market prefers charcoal that stays intact during transport. In the UK, ash, beech, birch, oak, hornbeam, alder, and hazel all yield quality barbecue charcoal.[43]

While it's unlikely we'll see charcoal emerge as a viable industrial fuel any time soon,

FIGURE 5.10: A freshly lit 6-foot-diameter steel charcoal ring-kiln in a coppice stand.

charcoal has reemerged as a local alternative to barbecue briquettes, providing consumers with an opportunity to support land-based livelihoods and enjoy a superior product. And while a much smaller market than barbecue, artist's charcoal is a particularly high-value product requiring little more than twigs for raw materials. Willow, ash, and hornbeam are said to produce some of the highest-quality artist's charcoal.[44]

Biochar

Renewed interest in widespread patches of highly fertile *terra preta* ("black earth") in the Amazon River Basin has stimulated a major revival in pyrolysis and its benefits to climate stability and soil health. While there's no consensus on exactly how or why ancient Amazonians created terra preta, these fertile pockets of deep, rich tropical rainforest soils are generally accepted to be massive midden piles, the remnants of well-developed ancient human civilizations.

Charcoal and other charred biomass is a key ingredient of terra preta—essentially a matrix of charcoal and ceramic pottery shards that improve drainage, retain soil moisture, create habitat for soil biology, and provide a media with a high cation exchange capacity, helping prevent nutrient leaching.

The art of creating and applying biochar lies well beyond the scope of this book, but it's something that's seen a lot of interest and innovation over the past decade plus. You can essentially pyrolyze just about any organic matter—wood just happens to be a great feedstock. What makes it biochar is the fact this charred material is charged with fertility, either by mixing with animal bedding, decaying organic matter, urine, etc. before it's applied to the soil. It's usually lightly tilled into agricultural soils. And because charcoal is a highly stable form of carbon storage, biomass that's charred and added to soil can provide a long-term carbon sink, helping contribute towards reversing high atmospheric carbon levels.

Carbon Farming

Carbon farming includes a suite of land management practices focused on combating climate change by developing stable stores of carbon in soils and perennial plants. Carbon farmers strive to optimize living and organic ground cover and cycle it through a healthy soil food web, intensively managed rotational grazing systems, and agroforestry systems while concurrently building viable agricultural enterprises. Woody plants are some of carbon farmers' most effective tools, and coppiced trees' permanent roots and vigorous growth make them especially useful.

Most forest-based carbon sequestration strategies require trees to remain standing to provide long-term carbon storage. While logs continue to store carbon even after harvest, their end use dictates how and when that carbon is released. Fuelwood is a short-term storage, crafts much longer storage, while biochar may sequester carbon for centuries. Resprout silviculture techniques can form the foundation of cut-and-come-again carbon-negative biomass production.

Garden Products

Desirable wood properties/qualities— Straight growth, flexibility, durability

Most promising species—Ash, hazel, black locust, chestnut, willow, whatever is convenient

Special tools or machinery required—None

Materials volume required—Depends on need but can range widely

Skill level—Low to medium

Price range—From 25 to 50 cents each to several dollars or $25+ for more involved products (bean poles, £11 for a bundle of 10; flower stakes, £4 for a bundle of 10; pea sticks, £8 for a bundle of 20; tree stakes, 3 to 6 for £2.50 to £4, $1.50 to $2 per 5-foot oak hardwood stake

Young quality coppice growth can yield an assortment of useful garden products like pea sticks, bean poles, and tree and flower stakes. Growers usually expect these products to last just a few years at most, so rot-resistant species aren't necessary. In fact, many species will have only just begun producing heartwood by the time they're ready for harvest. These products include the following.

Tree stakes—measuring at least 1 inch (2.5 cm) in diameter at the tip (and no larger than 2.5 inches, 6.25 cm) and usually 4 to 6 feet (1.2 to 1.8 m) in length, often with a pointed end, tree stakes should ideally extend at least 18 inches into the soil. Larger bolts (log sections) can also be riven into several pieces. A 4-inch (10 cm) diameter bolt can yield four stakes.

Flower stakes—Four-foot (1.2 m) rods at least 1 inch (2.5 cm) in diameter at the tip with a pointed butt end.

Bean poles—These trellis poles should be straight and at least 7 feet long (2.1 m) and 1 to 1.5 inches (2.5 to 3.75 cm) in diameter at the butt end. Gardeners often build a vertical framework using 5 pairs of poles tied together by horizontal rods spanning their length.

Pea sticks—Flat, well-branched, fan-shaped branch tips 4 to 5 feet long (1.2 to 1.5 cm) with 1 foot (30 cm) of straight length at the bottom (inserted firmly in the soil) provide structural

FIGURE 5.11: Raised bed edging, garden partitions, plant stakes, and more are but a few of the many wooden garden products coppicing can produce.

support for pea plants. Hazel, elm, birch, and linden are some of the best species.

Progs—These support arching tree and shrub branches and top-heavy flowers with a Y-shaped fork at the branch tip. They're usually 4 feet long (1.2 m) or more and 1.5 to 3 inches (2.8 to 7.6 cm) in diameter, but their size may vary depending on the proportions of the branch they support.

Hop poles—These are 3 inches (7.6 cm) in diameter at the tip and 16 to 20 feet (4.8 to 6 m) in length, usually made from ash, linden, or chestnut. Each hop plant requires 2 poles, with hop yards requiring roughly 2,000 poles to the acre (800/ha).[45]

Garden hoops—Long, thin, flexible shoots can make very useful hoops to support agricultural row cover over plants, protecting them from cold and pests.

Holiday Ornamentation

Desirable wood properties/qualities—Evergreen foliage, colorful stems, berries

Most promising species—Holly, fir, spruce, red osier dogwood, winterberry

Special tools or machinery required—Minimal: loppers, mowing equipment, shears

Materials volume required—As little as just a few sprigs

Skill level—Moderate

Price range—$1 to $5 for small garlands/swag, $10 to $50+ for wreaths, $25 to $150 for Christmas trees

While just a month or so of the year, the Christmas holiday is responsible for a massive proportion of annual retail sales. Holiday wreaths, garlands, and Christmas trees meet a guaranteed yearly demand, and although very few conifer species coppice in the true sense, resprout silviculture offers several options to meet these product needs. Note that the pitch pine (*Pinus rigida*), coast redwood (*Sequoia semperviens*), Mexican pinyon pine (*Pinus cembroides*), and one-seeded juniper (*Juniperus monosperma*) do truly coppice.

Many holly species (*Ilex* spp.) do coppice. Their pointed, glossy, thick foliage and deep red berries make lovely wreaths and garland material, woven onto a frame made from flexible young deciduous species' shoots.

Stump Culture

And as we discussed in Chapter 4, many coniferous trees will resprout if managed using stump culture. By allowing Christmas trees to grow 3 to 4 feet (90 to 120 cm) taller than their typical size before harvesting, growers can then leave a remnant "skirt" of branches (ideally about 30% to 40% of the tree's foliage) to continue photosynthesizing and supporting new growth. The larger the skirt, the faster the tree regenerates. The remaining stump develops new leaders along existing branches that can be pruned to shape after a year or two. Stump culture allows growers to reduce production costs, minimize herbicide use, and tap into trees' natural resilience.

Reportedly just about any conifer species will respond to stump culture. California growers use it on spruce, sequoia (*Sequoia sempervirens*), and Monterey pine (*Pinus radiata*). In British Columbia, some stump culture-managed Christmas trees have produced a new tree every 5 years for over 60 years.[46]

Stump culture management also offers growers an opportunity to maintain a herbaceous and/or woody understory that yields floral, medicinal, and agricultural products. Pacific Northwest growers cultivate valuable shade-tolerant floral crops like salal (*Gaultheria shallon*), bear-grass (*Xerophyllum tenax*), falsebox (*Pachistima myrsintes*), sword fern (*Polystichum munitum*), deer fern (*Blechnum spicant*), and evergreen huckleberry (*Vaccinium ovatum*). More widely spaced stands may support floral products like Scotch broom (*Cytisus scoparius*, which some consider to be highly invasive) and Oregon grape (*Mahonia nervosa*).[47]

Case Study: Coppicing Christmas Trees Using Stump Culture in Massachusetts' Pioneer Valley

Name: Emmet Van Driesche; Christmas tree farmer, spoon carver, author

Location: Ashfield, MA

Ecoregion: 58f Vermont Piedmont

Kopeen-Geiger Climate Zone: Dfb—Humid Continental Mild Summer, Wet All Year

Average annual precipitation/distribution + relative brittleness: 48.88 inches (124 cm), evenly distributed with 3.2 to 4.6 inches (8.1–11.7 cm) per month; very non-brittle

Hardiness zone: 5a

Elevation: roughly 1,200 feet (366 m)

Site aspect: generally south/south west, highly variable slope, some flat and some 35 to 40 degrees

Soil types: sandy loam-ish soil with lenses of clay somewhat ledgy in some areas ☞

FIGURE 5.12: Pieropan Christmas Tree Farm, Ashfield, MA. Photo courtesy of Emmet Van Driesche.

Project Description

For the past 13 years, Emmet Van Driesche and his family have managed 10 acres (4 ha) of balsam fir in Ashfield, MA, for Christmas trees, wreaths, and greens using stump culture. He and his wife first began this journey at the age of 25 while renting a house from innovative Christmas tree grower Albert Pieropan as he inched towards retirement. Pieropan convinced the young couple to take on the enterprise, sharing his knowledge and experience with them. Soon thereafter, they were off and running.

Pieropan originally began the farm in 1953, inspired by local grower Linwood Lesure who's credited with "discovering" stump culture. To understand this history, it's important to note that, prior to the 1950s, the market for Christmas trees was very different than it is today. During the early phases of Lesure's career in the 1930s, most Christmas trees were harvested from naturally grown forest trees of the right size and shape. Lesure noticed that some trees were prone to sprout and regrow following harvest, and he began to key into the details of these observations, thereby developing this coppice-type management technique.

The practice gained some steam during the 1930s, 40s, and 50s, especially in British Columbia and the Pacific Northwest where stump culture-grown trees proved far more resistant to periods of dry conditions than seedlings. But despite knowledge of the relative simplicity of this technique, as the market demand for Christmas trees grew and milk prices plummeted in the 1950s, land once used for pastures gradually gave way to the small-scale Christmas tree plantations that are much more common in modern times.

Pieropan gradually established his Christmas tree farm on rocky, undulating cow pasture, incrementally transplanting volunteer balsam fir seedlings he collected from colder pockets along the Massachusetts hills during his commute to work. Despite the relative success of Pieropan's operation, stump culture is an anomaly in the region, and there are just a few similar operations known in the northeast today.

By the time Al Pieropan passed on the farm to Van Driesche, it was largely overgrown, becoming more and more difficult to salvage. But the foundation was there, and over time, they've steadily restored the operation and built a successful, diversified small business.

Van Driesche's operation consists of two 5-acre (2 ha) plots, one of which is a you-cut grove that as a whole is too inefficient and inconvenient to manage for retail/wholesale production. The other 5 acres consists of a series of hand-cut groves. Today they produce 500 to 600 wreaths per season and about 500 Christmas trees, selecting mature sprouts that have reached the appropriate size often on clusters of 3 or 4 stumps. The farm's overall layout is rather irregular. Trees aren't planted in rows. They don't do any major mowing or path maintenance or fertilization. After all, it's really marginal agricultural land at best.

While the farm is almost exclusively devoted to balsam fir production, it appears that just about any conifer species should respond to stump ☞

culture management. Remember that this technique differs from coppicing in that it's essential to leave a minimum of 2 and ideally 3 "whorls" or layers of branches along the lowermost section of the tree after harvest to ensure the plant can continue to photosynthesize and stimulate healthy new growth.

💲 In addition to wholesale, retail, and you-cut trees, the greenery produced by this management system is tremendously important to the business's viability, ultimately making it competitive with conventional tree farming. They make about 60% of their money from Christmas trees, 30% from wreaths, and 10% from the sale of greens to garden centers and other local wreath-making outlets. According to Emmet, roughly 50% of his Christmas tree customers will also buy a wreath if well-made and affordably priced, so having multiple products at different price points really helps diversify and support their operations' sustainability. While yields have been variable during their tenure on the farm, at this point, they expect to net between $25,000 and $28,000 per year, grossing roughly $34,000, with $12,000 of this total in wreath sales and $2,000 from greens. 👉

FIGURE 5.13: Bundles of balsam fir greens at Pieropan Christmas Tree Farm, Ashfield, MA. A pickup load of greens cut from managed stumps for wreath making. Photo courtesy of Emmet Van Driesche.

🕐 Seasonally, harvest work rapidly picks up following the second hard frost of the season, usually in late October or early November. Van Driesche highlighted the importance of waiting till after this point in the season since greens or trees cut earlier are prone to rapidly losing their needles, whereas materials cut later often retain the bulk of their needles until January.

From early November to Thanksgiving, he harvests greens, later transitioning to Christmas trees until just before the Christmas holiday. During this narrow window, he works seven days a week.

🪓 As mentioned earlier, stump culture is similar to coppicing but yet also quite different. If starting with seedlings, most growers wait until trees have reached roughly 8 to 10 feet (2.4 to 3 m) before making the preliminary cut. Be sure never to take more than 50% of the leaf-bearing branches. One of the great things about stump culture is that it's exceedingly easy to transition an existing operation. All you need to do is allow the trees to grow a bit taller and make your cut a bit higher off the ground.

Most trees respond to this cut with vigorous sprouts the following season—usually far more than are desirable or useful—so it's important to do some early pruning and reduce them to just a few of the most dominant. Sprouts will put on between 4 to 12 inches (10 to 30 cm) of new growth in a season. In a well-managed operation, it's possible to manage multiple-age classes of sprouts on a single stump: one mature sprout that's nearing harvest after 7 to 8 years' growth, a 3-to-4-year-old sprout set to replace it, and several young sprouts that can be selected for retention as the more

mature sprouts are harvested. In practice, this takes a lot of observation and skill to do well, but it means that you may be able to shorten the tree production cycle and optimize the use of space in a stand.

Emmet doesn't particularly concern himself with the type or origin of sprouts though he did mention that sometimes branches will turn upwards following harvest, sending their tips skyward to compete to become the new leader. He said these are often quite irregularly shaped at their base, needing to grow taller before they're useful, so he tends to prune them off. The most vigorous sprouts tend to be those that emerge towards the top of the branch where the main heading cut was made.

His primary tool kit consists of a set of pruners and a pruning saw made by the Japanese company ARS. They are high-quality tools priced affordably. Whereas bowsaws are common on many Christmas tree farms, he only uses them for his you-cut customers due to their low cost and ease of blade replacement.

Modern Christmas tree farms typically feature relatively dense spacings of 6 feet by 6 feet (1.8 by 1.8 m) or 6 feet by 8 feet (1.8 by 2.4 m). This is far too tight for stump culture production since the skirt of branches surrounding the stumps will continue to grow in diameter as they mature. Emmet recommends spacings closer to 15 feet (4.6 m) for anyone considering establishing a new planting, and aiming for more of a labyrinth of paths rather than perfectly straight rows. He was keen to point out the importance of good, safe convenient access, noting that increasing distance from tree to truck rapidly diminishes operational efficiency. Keep in mind that herbaceous 🖙

competition, especially from brambles, can be considerable during the first 10 to 20 years. It may well be necessary to invest energy in controlling this competitive growth initially, but once well established, relatively little understory management is necessary. This resulting diversity may actually help create a balanced ecosystem with little to no significant pest or disease issues.

As stumps get bigger, the growth rate gradually increases. At the same time, the stump tends to get taller as each successive tree is cut, so keep this in mind when making initial stumping cuts and later managing for future trees. Skirts of branches should be cut back every 3 to 4 years, this material yielding valuable wreath-making greens mentioned earlier. When harvesting boughs, make the cuts at branch junctions so that the branches remain alive and increase in complexity. Don't cut branches back to the trunk, as they will not regrow.

Emmet also shared a few important insights as to helping ensure an operation's success. He warned that often folks new to the process tend to want to retain too much growth, leaving the stumps far too crowded. It's better to thin appropriately, and only keep one possible new tree despite there being a second viable option. Also, he has a mobile 7-foot-by-12-foot (2.1 to 3.7 m) workshop that is parked down by the road at a central location, where he can comfortably tie wreaths in a warm, protected space while also engaging with you-cut customers as they arrive.

Overall, despite the tight production window, stump culture has provided him and his family with a reliable, profitable land-based livelihood and a model enterprise that could be readily replicated in many parts of temperate North America. You can learn more about the practice and his farm in his book *Carving Out a Living on the Land*.

Culinary and Medicinal Mushrooms

Desirable wood properties/qualities—Density (More cellulose = more food for fungi). Avoid rot-resistant woods

Most promising species—Oak, maple, alder species

Special tools or machinery required—Mycelium-inoculated substrate, drill/angle grinder, bits, inoculation tool, wax/sealant

Materials volume required—Depends on scale: 3 to 8-inch (7.5–20 cm) diameter ideal; one 5.5-pound bag of colonized sawdust is enough for 20 to 30 three-foot (0.9 m) log sections

Skill level—Low to moderate

Price range—$10 to $15 per pound ($22 to $33/kg) retail for shiitake and oyster mushrooms

While we can't eat wood, fungi can—and culinary and medicinal mushrooms can be an especially promising coppice wood-based enterprise. Hardwood logs are a fantastic substrate for shiitake, oyster, reishi, maiitake, chicken of the woods, lion's mane, and turkey tail mushrooms among others.

By inoculating 3-to-4-foot-long (90 to 120 cm) log bolts 3 to 8 inches (7.5 to 20 cm) in diameter with a myceliated medium, a single log may yield 3 to 5 pounds (1.4 to 2.3 kg) of mushrooms over its 3-to-5-year lifetime, with locally grown shiitake and oyster mushrooms fetching $10 to $15 per pound ($22 to $33/kg) retail.

Freshly cut logs from healthy hardwood trees should ideally be inoculated within 2 months of harvest to avoid colonization by wild fungi. In cold climates, late winter is often the best window for felling and inoculation. Ideal log sizes require a 12-to-25-year coppice rotation, depending on species and diameter. These small-diameter logs optimize the ratio of sapwood to log volume.

The inoculation process is simple. You begin by literally riddling each log with 12-millimeter holes in staggered rows at roughly 4-to-6-inch (10-to-15 cm) intervals around and along the log's length and circumference. More holes help accelerate colonization, but it also consumes more spawn. To dramatically speed up the drilling process, use an angle grinder and drill bit mounting adapter.

Once the holes are drilled, pack them full with an inoculated substrate, usually either sawdust or hardwood dowels. While hardwood dowels are quick and easy and require no specialized tools, they are a lot more expensive than sawdust spawn if you're planning to do more than a couple of dozen logs. For sawdust

FIGURE 5.14: A flush of shiitake mushrooms on inoculated red maple logs.

Table 5.6: Top wood species for log-grown shiitake and oyster mushrooms

Shiitake	Oyster
Oak species	Aspen
Sugar Maple	Cottonwood
Hophornbeam	Willow
American Beech	Box Elder (*Acer negundo*)
Alder	Hackberry (*Celtis occidentalis*)
Musclewood (*Carpinus carolinana*)	Mulberry
Sweetgum	Tulip poplar
Red Maple	

Table 5.7: Sample shiitake enterprise plan

	Year 1	Year 2	Year 3	Year 4	Year 5
Production Information					
Logs Inoculated	200	200	200	200	200
Fruiting Logs	0	195	385	580	675
Anticipated per Log Shiitake Yield (lbs)	1	1	1	1	1
Total Shiitake Yield (lbs)		195	385	580	675
Income					
Retail Sales (lbs)	0	90	175	220	240
Wholesale Sales (lbs)	0	90	175	280	300
Price/lb (retail)	$0	$15	$15	$15	$15
Price/lb (wholesale)	$0	$12	$12	$12	$12
Dried Mushroom Sales (oz)	0	30	70	160	270
Price per Ounce Dry	$0	$8	$8	$8	$8
Total Income	$0	$2670	$5285	$7940	$9360
Expenses					
Variable Costs					
Log Harvest Labor ($10/hr)	$180	$180	$180	$180	$180
Shiitake Spawn ($25/5.5 lb bag)	$200	$200	$200	$200	$200
Wax/Sealant	$50	$50	$50	$50	$50
Inoculation Labor ($10/hr)	$500	$500	$500	$500	$500
Misc Supplies - Log Labels, etc	$50	$50	$50	$50	$50

This plan lays out the key details, income, and expense opportunities for a modest log-grown shiitake business. Adapted from UVM's Center for Sustainable Agriculture's 2013 "Best Management Practices for Log-based Shiitake Cultivation in the Northeastern United States." While most line items should carry through to most enterprises, there may be elements missing for your specific context. The numbers given here are rough estimates based on my own experience with small-scale shiitake enterprise. Be sure to update and/or research these numbers as best you can before building out your own business plan to ensure they are appropriate to your unique context.

Table 5.7: *Continued*

	Year 1	Year 2	Year 3	Year 4	Year 5
Fuel	$35	$35	$35	$35	$35
Shocking/Harvest Labor ($10/hr)	$0	$150	$290	$425	$535
Processing/Packaging ($10/hr)	$0	$100	$185	$245	$285
Fixed Costs					
Inoculation Tools	$150				
Angle Grinder/Bits	$100	$20	$20	$20	$20
Irrigation/Soaking Supplies	$150	$125	$100	$50	$35
Refrigeration	$25	$35	$45	$45	$45
Maintenance and Repairs	$100	$100	$100	$125	$125
Insurance	$0	$200	$250	$250	$250
Accounting	$75	$100	$125	$150	$150
Loan Payments					
Marketing/Advertising	$50	$125	$150	$150	$150
Total Expenses	$1665	$1970	$2280	$2475	$2610
Net Profit or Loss	-$1665	$700	$3005	$5465	$6750

spawn, most growers use a hand-operated inoculation tool, dabbing it into the colonized sawdust, lining it up over each hole, and depressing the plunger to pack the hole full.

Once full, seal each hole with wax to avoid colonization by competitor fungi, prevent the mycelium and log from drying out, and protect it from wildlife. To do this, the wax needs to be heated to a liquid consistency. A single wax dip using a dauber or foam brush should cover several holes.

Once inoculated, store logs in a protected shady location with convenient access and ideally near a water source. A coniferous canopy provides the best cover. The mycelium gradually spreads throughout the wood, taking 9 to 18 months to completely colonize the log. Because it can be difficult to know exactly when logs will fruit, and their harvest window is relatively narrow, ideally site them in a regularly visited area.

Colonized logs continue to yield flushes of mushrooms for 3 to 5 years. The number of flushes a log produces relates directly to the wood species' density, with heavier woods yielding more flushes and a greater overall yield. Since it takes the same energy, time, and material to inoculate logs regardless of density, if possible, choose oak or hard maple to ensure a better return on investment.

One unique characteristic of some varieties of shiitake is that they can be "force fruited" by shocking them in a cold-water soak for 24 to 36 hours. This triggers them to begin fruiting, and you can generally count on a healthy flush of mushrooms within 7 to 14 days, depending on conditions. Logs managed this way require 8 weeks of rest between fruiting, but this offers growers far more predictability and control in order to spread out harvests more evenly across the growing season.

And another great thing about mushrooms is that any excess can be dried and stored, rehydrating them before cooking. Generally, ½ pound of fresh mushrooms yields about 1 ounce dry. And another income-generating opportunity lies in selling fresh-cut log bolts and/or inoculated logs.

Mushroom log culture provides coppice operations with a valuable complement to other value-added products. Because they don't require straight logs, growers can high-grade harvested polewood, using short lengths from irregular logs for mushrooms.

Food/Medicine/Wine

Desirable wood properties/qualities—Edible/medicinal leaves, shoots, roots, inner bark, fruits, flowers

Most promising species—Linden/basswood, elderberry, mulberry, hazelnut, see appendix 2 for a more in-depth species list

Special tools or machinery required—Depends on application, none for most food products, but medicine and wine may require additional materials

Materials volume required—Depends on scale but can range from a single plant to a commercial enterprise

Skill level—Low to moderate

Price range—Varies widely

Woody agriculture pioneer J. Russell Smith once wrote, "The crop yielding tree offers the best medium for extending agriculture to hills, to steep places, to rocky places and to the lands where rainfall is deficient."[48] In addition to diverse ecosystem services, numerous woody plants also produce fruit, seed, and leaf yields that dwarf the productivity of many annual crops on a per acre basis.

Cereal (wheat, rice, barley, oats, etc.) yields in the US between 2006 and 2010 range from 5,300 to 6,185 pounds per acre (6,000 to 7,000 kg/ha).[49] By comparison, Ontario apple yields between 1995 and 2005 averaged 17,000 to 24,000 pounds per acre (19,046 to 26,889 kg/ha),[50] and the pod-bearing honey locust (*Gleditsia triacanthos*) may yield 66 pounds (30 kg) of quality livestock fodder per tree at 5 years age.[51]

Many woody species have edible or medicinal parts that can be coppiced to improve vigor and harvestability. Young basswood and linden (*Tilia* spp.) leaves make a tender salad green. Coppicing basswood and linden keeps leaf production at a convenient height and encourages strong, healthy growth. *Tilia* species' flowers have a wonderful fragrance and can be harvested, dried, and stored for tea. They're also fantastic fodder for bees and other pollinators.

Gingko biloba also responds well to coppice management, yielding high-value medicinal leaves. In the Pacific Northwest, the bark of Cascara sagrada (*Rhamnus purshiana*) trees can be used as a natural laxative. When coppiced, they can produce stems with harvestable bark just a couple of years after cutting.[52]

Coppice-with-standards systems can also integrate overstory nut- and/or fruit-producing standard trees within polewood producing copses. Walnuts, pecans, hickories, butternuts, almonds, oaks, chestnuts, pears, apples, and more could all help augment and diversify coppice system yields for home use, resale, or value-adding. Additionally, standard fruit and nut trees are a fantastic option along sunny woodland rides or as living cant boundaries where they benefit from easy access and increased sun exposure, especially during the years following coppicing. You could even design your layout to favor fruiting species by planning more frequent harvests of cants due south of fruit avenues.[53]

For some fruit- and nut-producing species, coppicing can also serve as a form of intensive pruning, helping rejuvenate the plant while eliminating dead and diseased wood. Coppiced chestnut in the UK tends not to respond well to this type of management, producing only a small crop of nuts by 10 years of age, although stools located along a cant's southern edge tend to be more productive.[54]

And as we briefly discussed in chapter 4, coppicing also holds promise as a large-scale pruning strategy for woody agriculture systems. Once again, at Minnesota's Badgersett Research Farm, they've been monitoring the growth of crosses of *Corylus avellana* x *C. americana* x *C. cornuta* (Common/European hazel, American hazelnut, beaked hazelnut) for close to 40 years, finding that these hybrids demonstrate considerable potential as both a specialty and staple crop with significant cold hardiness. This versatile species could provide a productive, multipurpose crop for windbreaks, living snowfences, riparian buffers, and alley cropping.[55]

And beyond growing food, coppice systems have generated materials used to control and enhance food production since the early days of agricultural production.

Fencing

Desirable wood properties/qualities—Durability, straight grain, riveability

Most promising species—Black locust, osage orange, redwood, oak, chestnut

Special tools or machinery required—Wedges, sledgehammer, froe, drawknife, drill, chain saw

Materials volume required—Can be considerable but depends on fence style and length

Skill level—Moderate to high

Price range—Ranges widely depending on fence style and materials chosen

Next to fuelwood and charcoal, communities' needs for reliable fencing were once one of the major drivers of coppice management.[56] Fences enclose fields, define boundaries, keep livestock and wildlife both in and out, and provide a line of defense.

Wooden fences take a range of forms. By the 1870s, various forms of split rail fencing had become the predominant North American fence style. At that point, fences stretched an estimated length of 4 million miles across the country! The split rail, or post and rail, fence encloses space without the need for fasteners, using mortised posts to support fence rails (horizontal members), often 10 feet (3 m) in

FIGURE 5.15: A section of riven chestnut post and rail fence.

length. Fence builders create post mortises by removing a pocket of wood using a drill bit and a chisel and mallet, forming a cavity that accepts the rail's tapered tenon.

A mortise-free split rail variation uses two parallel posts about 8 inches (20 cm) apart, connected by short wooden horizontal nailers that hold the rails in place. This style requires twice as many posts along with metal fasteners to hold the nailers in place, but it doesn't require joinery, and makes rail replacement simple. Post and rail fences usually include one to three rails and stand about 4 feet (1.2 m) high. Two rails can somewhat effectively enclose large livestock like cattle, while small stock need at least three. And in reality, if your fence absolutely must be stockproof, there's no substitute for wire fencing.

The development of barbed wire in the mid-19th century revolutionized fence construction, enabling landowners to enclose fields in areas lacking sufficient wood resources.[57] Modern high tensile wire allows fence installers to tightly stretch wire lengths and span straight stretches of up to 1,000 feet (305 m). Once it's strained, fallen tree branches or aggressive animals will not easily stretch it. To sustain this enormous tension, it's essential that fence corners be well-braced. Although more expensive than mild steel wire, high tensile wire requires fewer support posts.

Materials and Selection

Regardless of style, most wooden fences call for similar wood qualities. Because posts are partially buried, durable rot-resistant woods are always best. Remember that rot-resistant species' durability is limited to their heart-wood—sapwood will often decay quite readily, even in otherwise rot-resistant species. Black locust, white oak, osage-orange, chestnut, and redwood are a few coppicing species whose wood is especially long-lasting. Prioritize these species for applications where materials come in contact with the soil, i.e., posts as opposed to rails.

"Riveability," the ability to reliably split along the grain, is another valuable fence wood quality. Species that rive well can yield twice as many pieces as those that don't. Ring-porous hardwoods with straight, even grain and minimal knots are some of the best riving woods. Keep in mind that while most members of a species possess similar strength and working characteristics, there can also be considerable variation. For example, people commonly describe black locust as having a dense irregular grain, but in my experience, it can often have a beautifully straight even-grain pattern, splitting remarkably reliably. Genetics, site quality, and other factors may all influence these characteristics.

Fence posts often measure 3 to 6 inches (7.5 to 15 cm) in diameter and 4 to 12 feet (1.2 to 3.6 m) in length. While smaller posts may be used whole, bolts 4 to 5 inches (10 to 12.5 cm) across can often be split in two, and larger poles may even yield four posts. Fence rails are usually either split or sawn from 10-foot (3 m) logs, 6 to 10 inches (15 to 25 cm) in diameter. If large enough, it's possible to quarter the log, yielding four pie-shaped rails.

In terms of strength and durability, riven wood far outperforms sawn stock. Because the splitting process leaves wood fibers fully intact, riving offers an excellent option for materials processing, usually creating a longer-lasting fence. We'll discuss these qualities in more detail later in this chapter.

When selecting riving stock for fence posts or rails, observe bark patterns to predict the

Table 5.8: Rot-resistant species also well-suited to fencing applications

Genus	Species	Common name
Castanea	dentata	American chestnut
Catalpa	bignonoides	southern catalpa
Catalpa	speciosa	northern catalpa
Diospyros	virginiana	common persimmon
Juglans	nigra	black walnut
Maclura	pomifera	osage-orange
Quercus	alba	white oak
Quercus	bicolor	swamp white oak
Quercus	gambelii	Gambel oak
Quercus	garryana	Oregon white oak
Quercus	macrocarpa	bur oak
Quercus	montana	mountain chestnut oak
Quercus	stellata	post oak
Robinia	pseudoacacia	black locust
Sassafras	albidum	sassafras
Sequoia	sempervirens	redwood
Taxodium	distichum	bald cypress
Umbellularia	californica	California laurel

character of the underlying wood. Save poles with spiraling bark and frequent knots for uses that don't require riving—they will be much more difficult to split evenly. Straight 10-foot (3 m) poles with even grain are a high-value material. Treat them like it!

Wattle Fencing

Desirable wood properties/qualities— Straightness, flexibility, durability

Most promising species—Hazel, willow

Special tools or machinery required— Minimal; many wattle hurdle makers use a jig for the uprights and a spar hook to aid in splitting weavers in half

Materials volume required—High

Skill level—Moderate to high

Price range—Starting at £3 per square foot; there is little wattle fence production that I'm aware of in the US or Canada, so I've been unable to find a price equivalent here, but it's a relatively high-value product.

If longevity isn't a pressing need and you don't have a huge area to enclose, wattle fencing is a beautiful option. Wattle is a woven framework of sticks. Wattle fences use upright posts spaced 12 to 18 inches (30 to 45 cm) apart, and horizontal weavers or **withies** about 0.75 to 1.5 inches (1.9 to 3.8 cm) in diameter and at least 3 feet (90 cm) long (the longer, the better).

While wattle fence construction isn't difficult, it requires considerable time and materials. To build a wattle fence, pound 1 to 1.5 inch

FIGURE 5.16: Wattle fencing.

(2.5 to 3.8 cm) diameter uprights in the ground at least 12 inches (30 cm) deep. To extend the fence's life span, rive out uprights from more mature rot-resistant log sections since small-diameter rods won't have developed any heartwood at that size.

You could also create living fence posts using cuttings from vigorous species like willow. Live willow stakes will easily root when left in water for a few weeks or even just inserted directly into moist ground. Once established, they can be shredded, pollarded, or coppiced to control their size and generate additional weavers for repairs and expansion.

Wattle fences' horizontal weavers are traditionally made from straight, flexible, fast-growing species. Hazel is the most common, while willow, elm, maple, ash, and even birch and linden may be used. Withies must flex and bend when laid between uprights, forming a strong, solid barrier.

Wattle fences consume a massive amount of high-quality material. Assuming 18-inch (45 cm) spacing between uprights and 8-foot-long (2.4 m) weavers averaging 0.5 inches (1.25 cm) in diameter (to accommodate for taper along their length), a 3-foot-high (90 cm) fence section 10 feet (3 m) long requires 7-4.5-foot-long (1.37 m) uprights and 90 weavers.

Wattle construction makes it easy to understand why coppice became such an important traditional woody production system. They often need repair or replacement after 5 to 10 years. While too expensive to serve as functional field enclosures, wattle fences create decorative garden and landscape partitions,

trellises for vining plants, sound barriers, privacy screens, and raised bed edging. Their aesthetic beauty makes them a timeless coppice product.

Wattle Hurdles

Desirable wood properties/qualities— Long, straight, flexible

Most promising species—Hazel, willow (6-to-8-year rotation)

Special tools or machinery required— Billhook, pruners, loppers, hatchet

Materials volume required—For a 6-foot-by-6-foot wattle panel ~40 to 50 0.75-to-1.25-inch hazel rods (usually about 7-year growth); 45 splitter rods, 14 weavers, and 6 zale rods[58]

Skill level—Medium to high

Time to produce—Roughly 2 or 3 6-foot-by-6-foot hurdles/day (with high-quality stock, skilled hurdlers can make up to 4 per day)[59]

Price range—Averages £40 to £60 in the UK.[60] Little available data in US/Canada

Hurdles are portable fence panels—kind of like old-school Electronet fencing. Taking several forms, British hurdles made from coppice wood were once produced by the thousands. Farmers used hurdles to confine livestock during spring lambing and manage grazing rotations, moving livestock between paddocks by adding three new sides to an existing edge and creating a new fenced-in enclosure.[61]

During their heyday, large farms may have used as many as 200 hurdles and 50 lambing pens, requiring replacement every 5 to 6 years. There are two main styles of hurdle: wattle hurdles built with flexible, small-diameter rods

and gate hurdles made using riven stock from larger-diameter poles.[62]

Wattle hurdles are a woven framework of uprights and horizontals that create a strong, dense fence panel. Hazel's long, thin, straight growth habit and uncanny ability to twist and bend 180 degrees without snapping make it the most common wattle hurdle-making material. That said, just about any species that produces shoots of similar quality will work. Good-quality rods should be straight, unbranched, and 1.5 inches (3.8 cm) in diameter or less.[63] Cheap, mass-produced wattle hurdles may last just 2 to 3 years, while quality hurdles should last at least 7 to 8 years and up to 10 to 15.[64]

To build a wattle hurdle, craftsmen first set the uprights (zales) into holes drilled in a log form. They then begin the "weft" (horizontal weave), working in 0.75-to-1-inch (1.8 to 2.5 cm) rods in an alternating over-under

Figure 5.17: Sheep grazing inside a gate-hurdle enclosed paddock in Eastbury, UK. Photo courtesy of the Museum of English Rural Life, University of Reading.

Figure 5.18: Gate hurdles in foreground and continuous wattle fence in background. Weald & Downland Museum, Chichester, UK.

pattern. Occasionally, hurdle makers bend and twist hazel rods 180 degrees at the panel ends, weaving them back towards the hurdle's center. This twisting action causes the fibers to intertwine like cordage and adds strength and stability.

Whenever possible, hurdle makers rive rods in two. This requires considerable skill and is often done using a billhook, opening up a split at one end and, using one hand to work the billhook back and forth, tightly holding the rod with the other, helping control the split. If the split begins to "run out," the hurdler applies downward force towards the thicker half to recenter it.

Due to the difficulty sourcing sufficient high-quality rods to make wattle hurdles, British coppice craftsman Colin Simpson developed the Westmoreland panel. Just 4 feet (1.2 m) wide, the uprights are affixed to permanent mortised top and bottom rails, holding them firmly in place. This eliminates the need to weave the horizontals back into the panel's center.[65]

A skilled wattle hurdle maker building 12 to 15 6-foot-by-6-foot hurdles per week could gross up to £25,000 annually ($38,300 USD, April 2013) assuming each sells for £45 ($69 USD). To maintain this level of production, a hurdle maker would require 3.5 to 5 acres (1.4 to 2 ha) of quality hazel coppice each year. This means each craftsman would require 28 to 35 acres (11 to 14 ha) to support their craft over a 7-year rotation.[66]

Today, cheap manufactured fence panels and relatively high labor costs make wattle hurdles all but impractical for their original purpose, but their textured form and aesthetic beauty make them attractive garden trellis panels. As many as 25,000 wattle hurdles and screens are sold annually in the UK.

Gate Hurdles

Desirable wood properties/qualities— Riveability, straight growth

Most promising species—Chestnut, ash, oak

Special tools or machinery required— Froe, drawknife, drill and bits, chisel

Materials volume required—Roughly 15 riven pieces about 2 to 3 inches wide by 4-to-6-feet long for a single hurdle

Skill level—Moderate

Time to produce—1.5 hours each

Price range—2'6" × 3'3" costs £65; 3'3" × 3'3" cost £75

Gate (or Sussex) hurdles are joined using rough mortise and tenon joinery and metal fasteners. They consist of several horizontal rails mortised into two outer uprights, stabilized with diagonal braces. Uprights should be straight and up to 3 inches (7.5 cm) in diameter. Rails only need be 1.5 to 2 inches (3.8 to 5 cm) and may bow slightly.

Different regions developed hurdle designs suited to their needs. The primary difference lies in the number of rails and the brace locations. East Anglian hurdles stretch 6 feet long by 4 feet high (1.8 m by 1.2 m) with 6 horizontal rails. Kentish hurdles have 5 rails and measure 8 feet long by 4 feet high (2.4 m by 1.2 m).[67] The three bottom rails are often spaced more tightly to keep sheep from putting their heads through. Shepherds pin adjacent hurdle sections together using wooden pegs knocked through corresponding holes on the end uprights.

Hurdle makers drill and clean out mortises in the uprights and shape the rails' tenons with a drawknife, fixing the joint with nails. Some gate hurdle builders forego the mortise and tenon joint, simply face nailing the horizontals to the uprights. Although weaker, they still serve their intended use. A skilled craftsman can build one in an hour and a half.

While too cumbersome and costly for modern graziers, gate hurdles can still create beautiful, multifunctional modular fence systems in gardens and landscape installations.

Other Types of Fences

There are many other creative fencing styles that make use of small-diameter wood. One common style in parts of Eastern Europe features a sturdy post and rail framework that supports a vertical array of saplings and brushwood, creating a dense wall-like partition.

Paling fences are another lightweight design often made using coppiced wood. First conceived in 1905, paling fences continue to meet the construction industry's demand for durable, light, portable, reusable, semi-self-supporting temporary fencing. Palings about 1.5 inches wide by 0.75 inches (3.8 by 1.9 cm) thick and 3 to 5 feet (0.9 to 1.5 m) in length are bound together using two or three strands of twisted wire. This clever portable system can be rolled up for easy transport, and if suspended 2 inches (5 cm) or more off the ground and made from a rot-resistant wood, they may last 20 years or more.[68]

The best paling material comes from easily riven woods (usually ring-porous hardwoods) cut on a 10-to-14-year cycle. A single acre may produce up to 25,000 pales, enough to build a mile (2.2 km) of fencing.[69] Varying from 2 to 6 feet tall (0.6 to 1.8 m), individual palings

Figure 5.19: Paling fences often feature split staves of wood bound together using 2 or 3 runs of twisted wire, allowing them to be fashioned into a lightweight roll that one can quickly and easily erect, take down, and move. While this particular fence is permanent, it utilizes riven paling-like pickets that also demonstrate a similar concept of a functional green wood fence design. Photo courtesy of Dave Jacke, taken at Colonial Williamsburg, Williamsburg, VA, used by permission.

are spaced 1 to 5 inches (2.5 to 12.5 cm) apart, depending on the use.[70]

Various forms of fences have shaped the built environment for centuries, and many of these styles have evolved around the creative use of small-diameter coppiced material. While this list is by no means exhaustive, hopefully it provides enough historical context to either replicate or reinterpret these patterns for our own unique context.

Shelterbelts/Windbreaks/Soundbreaks

Desirable wood properties/qualities—Fast growth, dense form, wind tolerant

Most promising species—Conifers, poplar, willow

Special tools or machinery required—None: most of the work lies in planting; perhaps a mechanical tree planter for particularly large installations

Materials volume required—Depends on plant spacing but can be as much as 4 or 5 plants per linear meter (39")

Skill level—Low to moderate

Price range—Difficult to price - costs include site prep, plants, installation, maintenance

Shelterbelts—vegetated buffers that protect crops, livestock, and buildings from wind, pollution, drifting snow, noise, and odors—can serve as multipurpose plantings managed using coppicing and pollarding. Many technical resources already exist on shelterbelt design, but they rarely discuss the potential to use sprouting as a tool to rejuvenate the hedge and yield useful materials.

Many fast-growing species like willow and poplar can provide rapid, inexpensive protection from strong winds. For species easily propagated by cuttings, it can be as simple as inserting one-year-old hardwood cuttings directly into the soil at your chosen spacing. Depending on site quality and climate, in just 2 to 3 years, these plantings may reach 9.8 to 13.1 feet (3 to 4 m) in height.[71]

In parts of New York State, growers are using hybrid willow to create living snow fences to protect roadways from winter drifts and harvesting the biomass on semi-regular cycles for fuel, pulp, and/or animal bedding.[72]

In parts of Germany and the UK, a few companies have developed "living wall" systems that serve as sound barriers, helping shelter communities along major roadways. Based in Ontario, Canada, the Living Wall Inc. earned a patent for its living noise barrier design.

A 2005 Quebec study explored the cost-benefit of a living sound barrier along Highway 116 in a residential stretch in the city of Saint Bruno. The 98-foot-long (30 m) wall used 2,000 *Salix viminalis* stems, at least 11.5 feet (3.5 m) long and 1.6 to 2.3 inches (4 to 6 cm) in diameter, installed vertically in two 3.3 feet (1 m) deep trenches spaced 4 feet (1.2 m) apart. Held firmly in place by wood uprights and steel rods, they fastened a thin geotextile fabric over the entire inside length of the wall, backfilling the trench with sandy soil, resulting in a wall 8.2 feet (2.5 m) high and 4 feet (1.2 m) wide.[73]

Just 8 weeks after the April installation, the wall was completely green, and by the end of the first season, branches stretched 6.6 feet (2 m), forming a wall 15 feet (4.5 m) high. The following season, the wall resumed growth, showing no signs of freeze damage or stress from road salt. While they did not measure this wall system's acoustic properties, German tests on a similar design found it reduces sound as effectively as just about any other type of acoustic barrier (concrete, metal, etc.).[74]

These types of purpose-built living barriers are really just variations on hedgerow design.

Hedges

Desirable wood properties/qualities— Browse-resilient, fast growth, thorny, multifunctional, edible/medicinal fruits/leaves, etc.

Most promising species—Hawthorn, osage-orange, mesquite, black locust, willow

Special tools or machinery required— Not necessarily; perhaps a Yorkshire billhook or long-handled slasher for hedgelaying and brush management

Materials volume required—Plants spaced about every 15 to 18 inches (35 to 45 cm) for stock resistant hedges and wider spacing or other purposes

Skill level—Moderate to high

Price range—Hedgelaying: £5 to £10 per meter (underwoodcrafts.co.uk)

Figure 5.20: A short section of freshly laid hedge. Note the pleaching cuts made at the base of the saplings.

Hedgerows can function as enclosures that keep livestock in, barriers excluding wildlife, shelterbelts, wildlife corridors, and living field partitions. Multispecies hedgerows may generate small-diameter fuelwood, polewood for crafts, nuts and fruit, medicine, livestock protection, erosion control, and habitat, all while creating a breathtakingly beautiful landscape mosaic. Because managed hedgerows require semi-regular maintenance (thinning and reworking on 10-to-15-year cycles), they also provide recurring employment for specialized workers.

While standard fencing begins to decay once it's installed, hedgerows continuously grow and mature, yielding diverse products while becoming increasingly complex. Hedgerows create varied habitat for birds and insects along farm field edges, enhancing pollination, also offering natural checks on insect pest populations.[75] For these reasons, hedgerows provide an invaluable solution to enclosure needs in a low-energy future. We'll examine hedgerow design, establishment, and management in more detail in chapters 6, 7, and 8.

Specialized Traditional Crafts
Besom Broom Making

Desirable wood properties/qualities—Fine, soft, brushy, wear-resistant branches

Most promising species—Birch; handles from any strong, straight wood

Special tools or machinery required—Broom making horse—a device used to clamp birch tops tightly before binding

Materials volume required—One 3-foot-long bundle of twigs 10 to 12 inches (25 to 30 cm) in diameter; 1 handle

Skill level—Moderate

Time to produce—15 to 20 minutes for a skilled worker

Price range—$15 to $100 (£5 to £11 in the UK, averaging £7 to £10)[76]

The art of "besom" broom making has been practiced in the UK since the early medieval period.[77] Besom brooms feature a head made from the brushy tops of birch saplings, bound tightly and held in place by a thin greenwood lashing.

For a broom to last, season raw materials for several months. The best material comes from birch trees at least 7 years old. Sort this

brushy material into two grades based on length and coarseness with the long, coarse material forming a stiff core and the shorter, smoother shoots on the outside.[78] Broom heads are formed from a 10-to-12-inch (25 to 30 cm) diameter bundle of twigs about 3 feet (90 cm) long. A 3.5-foot-long (106 cm) handle is inserted into the broom head and held in place by two greenwood lashings. A skilled besom maker can produce as many as 36 in a day![79]

Although besom brooms are often still sold today at craft and antique shops for their rustic aesthetic, they're also quite useful for cleaning sidewalks and roadsides, dusting away fresh snow and even lifting damp leaves and moss from grass. Handleless besom brooms are even used in industry to sweep away scale from red-hot steel and were actually believed to improve the steel's quality.[80]

Clog Making

Desirable wood properties/qualities—Even grain (resistant to splitting), easy to work

Most promising species—Alder, willow, birch, beech, sycamore, maple

Special tools or machinery required—Block knife, a specialized carving tool resembling a meat cleaver, fastened to a sturdy wooden block

Materials volume required—One bolt 6 to 8 inches (15 to 20 cm) diameter by 16 inches (40 cm) long per pair. British alder rotations for clog soles spanned as long as 35 years.[81]

Skill level—Moderate to high

Price range—Difficult to find a craftsperson-made equivalent product, but all-wood clogs range from $25 to $100 online. Could easily sell for far more in the right market

Wooden-soled clogs date back to the Middle Ages. This durable, affordable, convenient footwear was commonly used on factory floors, in mines, and on farmland by both rich and poor. Clog makers fashioned each pair individually for a custom, comfortable fit.[82]

Clog makers preferred wood harvested in spring and summer. They rough-shaped the soles before leaving them to season for a minimum of 9 months, piling them up in conical stacks to maximize air circulation and accelerate drying. Because of the lag time, clog makers often travelled between woodlands converting raw materials into roughed-out soles. Once seasoned, the roughed-out blocks were sent to factories where they were shaped to fit.[83]

Hay Rakes/Forks

Desirable wood properties/qualities—Strong, straight grain

Most promising species—Ash, birch, alder, oak

Special tools or machinery required—Froe, shaving horse, drawknife, bending forms

Materials volume required—Minimal. One 4-to-6-foot (1.2 to 1.8 m) handle and a few additional short lengths

Skill level—Moderate

Time to produce—1 to 3 hours

Price range—$20 to $100+

For much of human history, haymaking has been a vast communal affair. After hand-mowing fields with a scythe, farmers use broad, lightweight wooden hay rakes to arrange the forage into windrows, later turning it to even out and accelerate drying. Willow, ash, birch, or alder handles were often steam bent to ensure straightness.[84]

Once hay was sufficiently dry, farmers used a wooden hay fork to pitch it into a barrow or wagon. While wooden hay forks will not tolerate heavy tasks, their lightness and flexibility make them ideal for this application. Wooden hay forks are often made from a single

Figure 5.21: Steam-bent wooden hay forks made by the author out of red oak.

section of riven oak or ash, steam-bent to form an ergonomic curve and to create the splayed tine angles. In France's Cevennes, craftsmen even grow forks by pruning young trees to three or four properly spaced branches and straightening the handles by steam-bending if necessary. Like some of the other crafts we've discussed, the practical needs for these tools may be somewhat limited, but they make beautiful, functional pieces of art, providing a window into an old-world way of doing things.

Tool Handles and Restoration

Desirable wood properties/qualities— Toughness, flexibility, riveability

Most promising species—Hickory, ash, fruit woods, hard maple, oak

Special tools or machinery required— Workbench, vice/shave horse

Materials volume required—2.5-to-4-foot (75 to 120 cm) lengths, 3+ inches (7.5 cm+) in diameter if riven

Skill level—Moderate

Time to produce—30 to 90 minutes

Price range—$5 to $30

Who doesn't have a modest collection of handleless tools lying around in need of some attention? Making tool handles requires only moderate woodworking proficiency and a few basic hand tools. And it's a valuable skill not only for homestead self-sufficiency but also potentially a small cottage industry.

Perhaps the most important step in crafting a functional, long-lasting tool handle is materials selection. Wood species plays a key role in a handle's longevity. Many tool handles require both flexibility and shock resistance. Most handles made in the United States today are either ash or hickory. Ash is strong, lightweight, and flexible. It has been the standard for baseball bats in American Major Leagues for good reason. Hickory is remarkably dense and known for its shock absorption properties. Axes, mattocks, picks—just about anything that will experience some type of regular impact—are all great candidates for hickory handles.

Avoid the temptation to make a tool handle using a sapling in the round. Unless it's carefully dried, roundwood moves and shrinks irregularly, loosening up over time. To ensure a durable handle, rive out the stock from a larger pole or log.

Riving preserves wood grain's continuity. The wood fibers should run clear along the length of the piece. Slightly oversize a riven handle made from greenwood and leave it to air dry for several weeks before final shaping. That way it won't shrink and loosen up later on.

Pay careful attention to the way the handle's grain orientation relates to the tool's plane of use. For maximum strength, orient the growth rings parallel to this plane. In the case of an ax, the growth rings should run parallel to the ax's sharpened edge. See Figure 5.22. To illustrate, imagine the handle's growth rings as layers of cardboard. When applying a striking force in the same plane as these layers, the stresses can eventually cause them to separate. In a tool handle, the growth rings would begin to delaminate from one another. If we instead imagine these forces applied across the edges of the cardboard, the depth of all of the layers together resists failure, distributing stresses much more evenly. Even if you don't make your own tool handles, remember this the next time you're choosing a new tool at the hardware store.

Making a greenwood handle involves two steps: rough shaping and finish fitting. If designing a replacement, take a close look at the original dimensions. When cutting a blank to length, leave a couple of inches of extra material at each end to accommodate any checking or

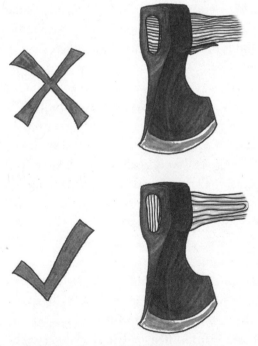

FIGURE 5.22: Growth rings should run parallel to the plane of use in a tool (bottom image). This helps prevent against the growth ring delamination that can occur in use as seen in the upper image.

other defects, and only cut it to length at the last possible moment. It's really hard to put it back on once it's gone.

Once sufficiently air-dried, fit the handle to the tool head. For round sockets (the hollow metal length that receives the tool handle), a lathe is highly accurate, especially if the socket is tapered. You can also use a shaving horse (or vise) and drawknife/spokeshave to shave the tenon to size. You'll need to do this for tools with oval eyeholes like axes, mattocks, picks, etc.

This step requires patience and a keen eye. Place the tool head over the mating end of the handle and tap it on firmly. If it fits perfectly to start, consider a career as a professional tool-handler! Most likely, it'll be too large. Remove the tool head and inspect the tenon. Areas with excess wood will show discoloration or abrasion, clear indicators of the high points. If it's difficult to see, run a pencil along the inside of the socket, leaving a layer of graphite on the metal. Next time you fit the head on the handle, the graphite will clearly mark the high points. Carefully continue until the handle fits snugly in the socket, taking care not to remove wood from areas that already fit. Round tool handles are often fastened with either a heavy-duty adhesive or fasteners. Use wedges, hammered into a saw kerf in the end of the handle on tools with wide, thin sockets.

As a diverse small enterprise, you could produce roughed-out riven tool handles sold for at-home fitting, provide tool repair and handle replacement services, or even collect, restore, and re-handle old tool heads.

Utensils

Desirable wood properties/qualities— Strong, dense, diffuse-porous, doesn't impart a flavor

Most promising species—Fruitwoods (apple, pear, plum), birch, maple, cherry, hawthorn, sycamore

Special tools or machinery required— Hewing hatchet, carving knife/knives, hook knife/gouge, adze (for bowls)

Materials volume required—2-to-4-inch (5 to 10 cm) diameter bolts, 6 to 15 inches (15 to 35 cm) long

Skill level—Depends on the quality of the end product

Time to produce—From 15 minutes to several days

Price range—$10 to $500+

Wooden spoons, knives, bowls, spatulas, and more are creative, functional, relatively inexpensive products with broad appeal. Finished items can vary considerably in quality, craftsmanship, and ultimately, cost. For food-related wares, dense, diffuse-porous, non-aromatic hardwoods are best since they're strong, less absorptive than ring-porous species, and don't impart a flavor. Late fall is said to be an ideal time for carving blanks since trees contain low levels of bound water, but many carvers just harvest materials as they need them.[85]

These products require just a few basic tools. With a saw, hatchet, and knife, you can fashion raw roundwood into beautiful, functional art. And with adzes, gouges, and hooked knives, you can sculpt bowl and spoon hollows with relative ease. Though not necessary, the drawknife and shaving horse simplify quick shaping and material removal. For $100 to $150, one can acquire a basic tool kit to get started making all sorts of wooden cutlery.

Wooden spoons vary widely in price from $5 to $100. High-quality artisanal spoons often

Figure 5.23: An assortment of tools and masterfully crafted wooden utensils by Mark Angelini, of Sedalia Sloyd, and Mountain Run Permaculture. Photo credit: Mark Angelini.

run between $40 and $60. While this may seem high to some, labor costs add up quickly, even at just $15 to $30 an hour. While many products require just a single small block of roundwood, carvers who have refined their craft are usually rather particular when it comes to choosing quality materials.

One of the best attributes of utensil carving lies in its simplicity. You can bring your work just about anywhere, enjoying wood's incredible workability. Well-crafted utensils should serve their users for decades if well cared for. They make excellent compact gifts, and they're a great way to build a relationship with hand tools and greenwood crafts.

Walking Sticks

Desirable wood properties/qualities—Strength, interesting forms/texture

Most promising species—Species is less important though they generally shouldn't be woods that are super soft; if riven from a larger piece and/or steam bent, oak or ash make good choices

Materials volume required—A single 4-foot (1.2 m) length 1 to 2 inches (2.5–5 cm) in diameter. Typically 3 years' growth for ash or chestnut[86]

Skill level—Low to moderate

Time to produce—15 minutes to a few hours

Price range—$5 to $100

Walking sticks are another ubiquitous coppice craft made from straight, small-diameter polewood and unique saplings with interesting shapes. These short sprout lengths should be slowly seasoned for 9 to 12 months before doing any finish work.[87]

Walking sticks were once so commonplace that there are a number of different categories based on use. A "walker" is tall enough to reach one's hip joint, while the taller "market stick" allows the user to rest and relax on it with

Table 5.9: Suitability for wooden utensils based on freedom from odor and taste once dry

Excellent	Acceptable
Apple	Cottonwood
Ash	Bald Cypress
Beech	Black Locust
Birch	Honey Locust
Catalpa	Oak species
Cherry	
Chestnut	
Dogwood	
Elm	
Hickory	
Holly	
Linden	
Maple	
Pear	
Tulip Poplar	
Sycamore	
Walnut	
Willow	

Adapted from Langsner, 1995, p. 36.

their hands or lower arm. The even longer "hill stick" aids hikers on trail walks and mountain ascents. The "crook" is a working stick used by shepherds with a hook at the tip to grab hold of sheep by the leg or neck. And there are even specialized crooks for work with geese, chickens, and swans.[88]

Baskets

Desirable wood properties/qualities—Straight growth, flexibility, colorful stems

Most promising species—Willow, dogwood, brown ash, white oak

Special tools or machinery required—Not particularly; occasionally forms

Materials volume required—Depends on basket

Skill level to produce—Moderate to high

Price range for completed product—$25 to $1,000+

In a world full of plastic containers, it's easy to take for granted the challenge of producing durable functional storage vessels. Baskets have played a critical role in cultures around the globe for millennia—to gather food, fuel, seed, and medicines; to store and transport goods; to collect and store water; and even transport children. In *Traditional Country Craftsmen*, J. Geraint Jenkins describes five different categories of European baskets: agricultural baskets used to harvest produce, market baskets for transporting products, industrial baskets used in factory applications, trade baskets for deliveries, and domestic and fancy baskets.[89] Coppicing yields straight, flexible, easily harvested stems, free of blemishes and pest damage, making it an ideal propagation method for quality basket materials.

We can classify baskets in two primary groups based on the materials they're made from.[90] In the first group, raw materials require minimal processing and are often of the highest quality when produced by annual coppicing. Willow, dogwood, western redbud (*Cercis occidentalis*), and buckbrush (*Ceanothus cuneatus*) are a few commonly used species.

The second group includes materials that require significant processing before use including white oak, hickory, and brown/black ash. Ash "splint" baskets were once common throughout the northeastern United States and

FIGURE 5.24: A beautiful collection of baskets made by Margaret "Pegg" Mathewson using coppiced willow.

Canada but have virtually disappeared today. Similar traditions in southeastern states use thin splints from white oak trees, a skill that continues today to some degree.[91]

For these types of baskets, weavers select high-quality poles 6 to 10 inches (15 to 25 cm) in diameter, riving them into splints the thickness of a single growth ring—either with a froe (white oak) or by pounding with a heavy mallet (ash). Ideal growth rings are about $\frac{1}{16}$ inch (1.6 mm) thick. Thinner rings produce weak splints while thicker rings resist bending. Depending on species, weavers must store and season materials for 1 to 4 years before use, but once finished, they make lightweight, seemingly indestructible baskets.[92]

The "trug" is another unique basket type often used in agriculture. These strong, light, boat-shaped baskets are fashioned using flat, slightly overlapping slats sprung into place between an oval steam-bent frame. Weavers split these thin slats from high-quality poles (often oak), later steam-bending or simply flexing the still-green wood into place. Because the frame holds the slats under tension, the trug is quite durable.[93]

And the outer bark from species like tulip poplar, basswood, elm, willow, and magnolia may also be used to fashion baskets by cleverly cutting, folding, and lashing the raw material. A 4-inch (10 cm) diameter tree yields enough bark to make several 6-inch (15 cm) diameter containers.[94] Drew Langsner's *Green Woodwork* provides detailed instructions on this unique craft.

Baskets come in a wide range of shapes, sizes, and purposes, and their quality can vary widely. All of these factors shape their market value. Visit craft shops and shows to see who's already selling baskets. What materials do they use? Where do they source them? Perhaps you can work with a weaver to choose which species/varieties you plant and harvest.

Baskets may range from a few dollars to several thousand depending on the degree of refinement, craftsmanship, and the relative abundance of the economy where they're sold. Several dozen $25 baskets or just a single $1,000 product could each garner the same return.

Rustic Furniture

Desirable wood properties/qualities—Strong, dense, attractive bark/grain, flexible

Most promising species—Hickory, hazel, birch, cherry, maple, willow

Special tools or machinery required—Tenon cutter, drill bits, shaping tools (chisel, gouge, knife, drawknife, spokeshave)

Materials volume required—A chair may require as few as two 10-foot (3.3 m) poles, 2 inches and 3 inches (5 and 7.5 cm) in diameter

Skill level—Moderate to high

Time to produce—A few hours to days or more

Price range—$15 to $1,000+

Coppice poles' round natural forms are perfect raw materials for rustic furniture—part art, part functional craft. Rustic furniture's unique aesthetic highlights the raw nature of the materials, often left unshaven and in the round. Dating back to the emergence of the Adirondack style in remote New York state and other parts of rural North America, demand for rustic furniture once reached a point where midwestern farms actually produced straight young hickory saplings for raw materials.[95] Rustic furniture makers express their design aesthetic through careful selection of small-diameter polewood. With access to the right sales outlets, rustic furniture offers considerable value-adding opportunities to materials otherwise only useful for fuelwood.

Tables, chairs, dressers, pergolas, arbors, picture frames, fences, etc. made in a rustic style are uniquely free-form. Innovation and creativity are two of the most important traits of a successful rustic furniture maker. Rustic furniture makers often "see" the finished piece in the standing tree, and because of this, selecting the right materials is one of the most important steps in construction. Author and rustic furniture-maker Daniel Mack writes, "A good rustic chair is like the transcript of a conversation between a person and a tree. Both the chair maker and the tree are recognizable in the final result… The tree and the maker have collaborated."[96] Stylistically, rustic furniture includes:

- **Stick/sapling furniture** using small-diameter polewood
- **Tree or log work** uses larger wood in its round or semi-processed state to create heavy, substantial furniture. This is the "official architectural style" of the US National Parks.

FIGURE 5.25: Various types of roundwood rustic furniture made from hickory polewood and also featuring seats made using hickory inner bark, harvested and woven by the author.

- **Bentwood furniture** uses long, young, straight saplings to fashion bent forms fastened to a roundwood frame. Flexible, fast-grown woods like willow, cottonwood, alder, and hazel often work best.
- **Split work, mosaic, and swiss work** uses full or half-round saplings and branches to form frames or decorative panels.
- **Bark work** usually features birch, cherry, or cedar bark fastened to a wooden frame or panel. Often used on cupboards, tables, picture frames, wall panels, desks, chests, etc.[97]

Most types of rustic furniture require similar raw materials. Furniture framing and structural stock usually measure 1.25 to 2.5 inches (3.8 to 6.25 cm) in diameter, while bentwood furniture also incorporates thinner flexible pieces between 0.75 and 1.25 inches (1.9 to 3.8 cm) in diameter for arms, backs, and seats.[98] The real challenge lies in the joinery.

Mortise and tenon joints form the strongest construction. Because roundwood dries differentially and shrinks accordingly, all raw materials should be thoroughly dry before shaping and assembly. Generally assume that wood air-dries at the rate of 1 inch per year.[99] So ideally allow a 2-inch (5 cm) diameter sapling 2 years to fully dry. For most post-and-rung-style construction (tables/chairs with horizontal members mortised into legs), it's most important that the rungs are fully dry so the tenons won't shrink, loosen, and weaken the joint. As with all furniture construction, joinery details determine the product's longevity. A well-built piece should feel strong, durable, and functional. If it's flimsy and wobbles, add diagonal braces to stabilize the piece and eliminate racking.

A rustic furniture tool kit requires little more than a saw, chisel, drill, drill bits, knife, hand pruners, tape measure, hammer, fasteners, and/or glue. Most rustic furniture makers today also use some type of tenon-forming tool—essentially a drill-mounted pencil sharpener that reduces roundwood diameter to a consistent dimension so it fits snugly into the corresponding mortise.

Rustic furniture is really an open-ended stylistic approach to functional wood art. As Daniel Mack writes, "There really aren't many rules, mostly ideas and suggestions."[100] For some inspiration, check out any of Mack's several books on the subject.

Green Woodworking

Desirable wood properties/qualities—Riven crafts: wood with coarse, long fibers, easily split; Carving/hewing: short, even-grained wood

Most promising species—Most hardwood species, especially oak, ash, hickory, maple, birch, cherry, beech

Special tools or machinery required—Froe, wedges/sledgehammer, drawknife/shaving horse, pole lathe/lathe

Materials volume required—One 3-foot (90 cm) pole 6 inches (15 cm) in diameter should contain enough wood to build an entire ladderback chair

Table 5.10: Common western wood species for green woodworking

Red Alder	Mulberry
Oregon Ash	Olive
Aspen	Pear
Black Cottonwood	Pecan
Pacific Dogwood	Persimmon
Madrone	English Walnut
Big Leaf Maple	Redwood
Mesquite	Pacific Yew
California Laurel (Myrtle)	Fruit Woods
Oregon White Oak	Eucalyptus
Black Walnut	Holly
Willow	

Adapted from Langsner, 1995, p. 41.

Harvesting Tree Bark

Bark is a fantastically functional material for all sorts of purposes and is often completely overlooked as a resource. Useful for weaving, leather tanning, waterproofing, food, fiber, medicine, and more, it's best to harvest tree bark during spring and early summer, when flowing sap helps it cleanly slip from the underlying wood. To separate bark from a tree, first slice through the outer bark using a knife, saw, or ax. Start by working around its circumference at the top and bottom of the length you'd like to harvest. Next, follow with a vertical incision connecting these two circumferential cuts. Using a dull ax head or a wide, wedge-shaped piece of wood, carefully pry underneath it, gradually separating the bark from the tree.[101]

If harvesting bark for decorative purposes, spread it out flat while still pliable, sandwiching 10 or more pieces together, and clamp them between lath strips so they dry flat. Once they are dry, you can fasten them to a surface using small-diameter nails.[102]

Scandinavian builders traditionally used birch bark to create a waterproof membrane beneath sod roofs.[103] As long as bark is kept moist, its oils minimize cracking, preserving its water-shedding qualities. Shingle lapping bark sheets on top of 2-inch (5 cm) thick sheathing boards fastened to the rafters, builders covered the bark with about 4 inches (10 cm) of soil, seeding it to grass and other herbaceous vegetation.[104]

In many parts of the world, humans have harvested fiber from the inner bark from *Tilia* species (linden/lime, basswood) to make rope.

Oak bark harvested for hide tanning was once an immensely valuable woodland product in Europe. Young oak actually contains the highest levels of tannin concentrations, and oak coppice was often cut on 20-to-25-year cycles. They stored harvested bark under cover to prevent rain from leaching out the valuable tannins.[105]

And some woven crafts, including baskets and chair seats, use inner bark from hickory trees. After harvesting and drying, bark coils can be stored indoors indefinitely. When ready to weave, soak it in warm water for 30 to 60 minutes to make it pliable. Hickory is so durable that 100-year-old ladderback chairs still have their original woven seat intact. Although hickory is strongest, tulip poplar, sycamore, pecan, magnolia, elm, basswood, and smooth willow species can also be used.[106]

Skill level—Moderate to high

Price range—$25-$1,000s; Varies widely depending on product and locale

"Green woodworking," a term coined by Baltimore chairmaker and joiner Jennie Alexander, describes the tools and techniques used in these traditional woodcrafts. It's "green" because most craftspeople start with fresh, unseasoned (or green) wood. Green wood is much easier to work with hand tools, and it's also considerably more flexible than seasoned wood, which lends itself well to many types of woven and steam-bent crafts.[107]

Green woodworking can make very efficient use of high-quality logs and polewood. A 6-foot-long (1.8 m), 6-inch-diameter (15 cm) white oak sapling may yield 6 to 8 steam-bent firewood carriers, a dozen medium-sized baskets,[108] or 3 or 4 ladderback footstools. Most traditional woodworking tools are either sharp blades used to shave through wood fibers or splitting tools. They perform better and remain sharper longer in wet wood. Shaping greenwood can sometimes feel almost like peeling a woody carrot. In most cases, green wood projects begin by splitting parts out from a larger section of a log or pole.

Riving

Green woodworkers rive materials from round-wood to minimize checking and relieve tension. Riving, or "cleaving" as it's called in the UK, is a controlled splitting process. Instead of swinging an ax at a target, you place a wedge, ax, or froe along the plane you intend to split, and strike it with a hammer or wooden mallet. Riving can be a remarkably fast way to deconstruct polewood and logs into useful parts while relieving the tension that would otherwise cause checking during drying.

Riven wood is actually stronger and more durable than sawn wood of the same species. Splitting pries apart wood along its fibers, producing stock with no "run out." While the lumber industry considers boards with a grain run out of 1:15 (along a 15-inch board, the grain shifts 1 inch) as prime material, riven stock has virtually no run out.[109] Because of this, riven material is more flexible and resistant to shear forces. Prying wood apart along its fibers also exposes minimal end and surface grain, making it more durable and decay-resistant than sawn stock.

Some woods rive better than others. Ring-porous hardwoods and woods with

Riven vs. Sawn Wood

FIGURE 5.26: There are several key differences between wood that's been processed by riving vs. sawing, perhaps the biggest being that riven wood has little if any fiber/grain run out. That is, the fibers run clear through the piece from end to end. Compare the dotted line that follows a growth ring on the sawn stock to the left with the consistent grain pattern of the riven wood on the right. Riven wood's fiber/grain continuity makes it very strong compared to sawn stock of a similar dimension.

conspicuous, well-developed rays (oak, beech, etc.) have natural planes of weakness, causing them to split cleanly and reliably. Some of the best-riving woods include ash, hazel, oak, and chestnut.

Like most crafts, riving is as much an art as a science. Understanding a few basic principles will dramatically improve results.

Principle 1: Read the pole. How straight is it? Will it make a difference if you split it along one plane or another? Are there branch scars indicating buried knots? Does the bark run straight, or is it twisted? These observations help inform how best to approach a split and optimize raw material use.

Principle 2: Start at the center of the log or pole. This means, orient your split through the wood's pith (the biological center) so that it will divide the piece into two equal halves. The pith is particularly weak. It's also the point where most checking originates. Eliminating the pith relieves accumulated tension in the roundwood.

Principle 3: If possible, always aim to split stock in half. If there's equal mass on both sides of the split, it's much more prone to carry through straight and true. Sometimes it's worth foregoing this rule to try to split a piece into three parts, but halving a log or pole is incredibly reliable in producing consistent, quality splits.

Occasionally splits begin to run out towards one side of the piece. Don't fret. It's often possible to control and redirect the split and bring it back to the center. To do this, build a riving brake, a stationary tool used to hold a billet firmly in place while the woodsman

Table 5.11: Wood species' relative riveability

Good to Excellent	Fair	Poor
Ash	Alder	Elm
Chestnut	Apple	Eucalyptus
Butternut	Beech	Hornbeam
Bald Cypress	Birch	Persimmon
Hickory	Catalpa	
Black Locust	Cherry	
Honey Locust	Cottonwood	
Oak, Red and White	Dogwood	
Osage Orange	Hawthorn	
Pecan	Linden	
Redwood	Maple	
Walnut	Mulberry	
Willow	Pear	
	Sycamore	

Adapted from Langsner, 1995, p. 36.

guides the split along. When a split begins to run towards one side of a piece, orient the thicker half on the bottom, and use one hand to apply downward pressure towards that half, levering the split open with the froe. This will carry the split towards the side under greater tension—the thicker side—steadily recentering it.

Shaving

Shaving is the other fundamental process involved in green woodworking. Greenwood shaving tools include (from coarse to fine) the hewing hatchet, drawknife, gouge, chisel, plane,

Figure 5.27: A selection of green woodworking tools arranged from coarse to fine from left to right. From left: a steel wedge, then froe and wooden mallet, hewing hatchet or side ax, two different drawknife patterns, and last, the spokeshave.

Figure 5.28: Orientating an edge tool by skewing it across the workpiece helps ensure it slices cleanly, leaving a smooth, even finish with less effort. Note the long, tightly spiraled shaving. Photo credit: Ammy Martinez.

spokeshave, and scraper. The shaving horse, workbench/vise, and spring pole lathe are the basic implements used to hold wood as it's shaped.

When shaping wood with edge tools remember two basic principles: always work with the grain and use a "skewing" and slicing motion. When working with hand tools, you must follow wood fibers' orientation. Perhaps clearer than working with the grain, you want to "carve down grain." Saws, rasps, sanders, etc. enable woodworkers to shape materials with no consideration to fiber direction. By pushing a hand tool "uphill" into wood fibers, the tool will bite in the wood, splitting it ahead of the edge. By reversing the tool, starting at the uppermost fibers and working downhill, it will cleanly slice without biting or binding. To become proficient in the use of hand tools, you must develop the ability to read wood grain orientation and respond accordingly.

The "skew and slice" principle helps create a clean, smooth finish with minimal effort. Skewing a tool means to orient the blade at an oblique angle to the workpiece. It's similar to a switchback trail up a hillside. A skewed blade exposes a longer edge surface to the workpiece. See Figure 5.28. The leading-edge tip engages in the wood, shearing across the piece. By sliding the tool's edge across the piece from one side to the other, you achieve a "slicing" effect, leaving a smooth, glassy finish (assuming you're working with a sharp tool). These two basic principles are invaluable to novice green woodworkers.

With a modest investment in a few basic tools, it's possible to set up a fully functional greenwood workshop in a space as small as 10 feet by 10 feet (3 m by 3 m). Completely human powered, these tools require no electricity. You can even set them up in the woods in the open air, just like the bodgers. Unlike conventional woodworking, green woodworking produces very little dust, instead yielding a pile of hardwood shavings useful for mulch, bedding, compost, and some of the best kindling imaginable. But be careful, it's addictive!

Building and Construction Materials
Thatching Spars

> **Desirable wood properties/qualities—** Straightness, flexibility, easy-riving
>
> **Most promising species—**Hazel, willow
>
> **Special tools or machinery required—** Billhook, froe
>
> **Materials volume required—**6, 000 to 12,000 spars for an average roof
>
> **Skill level—**Moderate
>
> **Time to produce—**1,800 to 2,000 spars per day from good-quality stock (1,200 to 1,600/day with poor/average materials)[110]
>
> **Price range—**In the UK, £80 to £100 per 1,000 spars (varies depending on location/market)[111]; little existing precedent in US/Canada

While the European thatch roofing industry has declined considerably in recent decades and is essentially nonexistent in North America, there was once an enormous demand for thatching materials generated by coppicing. Thatching requires three primary materials: **sways**, **spars**, and **liggers** (and the thatch of course). Sways span thatch bundles horizontally and are secured in place by spars. Liggers secure thatch to the roof along the ridge and eaves.[112]

Liggers are made from bolts 3 to 5 inches (7.5 to 12.5 cm) long and 1.5 inches (3.8 cm) in diameter, split into quarters. Thatchers shave a flat on each quarter's inner face and fashion long pointed ends. Spar blanks run 21 to 26 inches (53 to 66 cm) long and 0.5 to 2 inches (1.3 to 5 cm) in diameter and are riven into 0.5-inch (1.3 cm) diameter sections with pointed ends. A master thatcher can rive 20 or more spars from a single 2.5-inch (6.3 cm) diameter rod![113]

A single roof requires between 6,000 to 12,000 spars (400 per 100 ft^2 and 50 per lineal foot of ridge), usually made from hazel (although sometimes willow). A skilled spar maker can make up to 1,600 per day. Assuming they could produce about 200,000 spars annually, they'd earn about £16,000 in 2013 ($24,500 USD). One to 1.25 acres of hazel coppice managed on a 6-to-8-year rotation could meet a thatcher's annual material needs.[114]

Thatchers traditionally bought a year's worth of unprocessed coppice rods in early spring, transforming them into roofing materials when poor weather precluded working on the roof. Despite the substantial demand, thatch material production often complemented other coppice crafts, like wattle hurdle making.[115]

Garden Structures

Desirable wood properties/qualities— Durability, straightness, riveability

Most promising species— Chestnut, black locust, white oak, osage-orange, mesquite, redwood

Special tools or machinery required— Framing chisel/gouge, drill, drawknife

Materials volume required— Depends on product

Skill level— Moderate

Time to produce— A few hours to several days or more

Price range— $75 to $3,000+

Unique, functional, artistic garden structures like arbors, pergolas, and trellis panels offer excellent opportunities to add substantial value to polewood. Adding diversity and character to landscapes, these built elements provide inviting points of convergence and contemplation.

Because they're sited outdoors, 4-to-8-inch (10 to 20 cm) diameter polewood from rot-resistant species are usually best. Larger diameter poles can be riven in half or quarters. The joinery that connects framing members can be rough, using long screws or heavy nails, or refined, using mortise and tenon joinery held in place by wooden pegs. The level of detail depends on your skill, interest, aesthetic, and budget. For an overview of roundwood joinery techniques, see Ben Law's *Roundwood Timber Framing*.

Most garden structures require longer coppicing cycles than many of the products we've discussed. Depending on species, anticipate rotations of anywhere from 8 to 25 years.

FIGURE **5.29:** A creative, artistic installation built using coppice materials along a footbridge in Devon, UK.

FIGURE B1: (Top left) Sweet chestnut coppice cant and product piles (Prickly Nut Wood).

FIGURE B2: (Top right) Woodland ride recently cut back to allow more light to penetrate the understory (Prickly Nut Wood).

FIGURE B3: (Center left) Marcus Smart riving chestnut fence rails (Prickly Nut Wood)

FIGURE B4: (Center right) High tensile fence using coppiced sweet chestnut posts.

FIGURE B5: (Bottom left) Wood stockpile for charcoal production (kiln in background) (Prickly Nut Wood).

FIGURE B6: (Right) Fresh batch of completed hardwood charcoal.

FIGURE B7: (Center) Continuous wattle fence, Weald and Downland Museum, Sussex, UK.

FIGURE B8: (Left) Wattle wall panels in process for mushroom shade house, Vermont.

Figure B9: (Top left) Hardwood logs for shiitake production, Vermont.

Figure B10: (Top right) Shade house for shiitake mushroom cultivation.

Figure B11: (Center left) Stripping hickory (inner) bark in spring for weaving material.

Figure B12: (Center right) Coils of fresh-peeled hickory bark and drawknife.

Figure B13: (Left) Willow coppice and low pollards, Marget "Pegg" Mathewson's Withyhenge, Oregon.

FIGURE **B14:** Bundles of fresh peeled willow rods, Coates Wetlands and Willows, near Taunton, UK.

FIGURE **B15:** (Center left) Sweet Chestnut fence panels, Chevrier homestead, near Limoges, France.

FIGURE **B16:** (Center right) Exposed wattle and daub wall, Bulgaria.

FIGURE **B17:** (Right) Managing willow coppice for propagation stock (hardwood cuttings).

Figure B18: (Top) Michael Elstermann with a bundle of saplings used for making holiday wreaths and other decorations.

Figure B19: (Center left) Newly laid pleacher in a living hedge.

Figure B20: (Center right) Finishing up a layed hedge, Sussex, UK.

Figure B21: (Right) Catalpa speciosa pollard, Virginia, USA. Photo credit: Dave Jacke.

FIGURE **B22:** (Top left) Willow pollards in bottomland forest, Bulgaria.

FIGURE **B23:** (Top right) Candelabra-shaped pollards. England.

FIGURE **B24:** (Center) Pollarded sycamore street trees, Madrid, Spain.

FIGURE **B25:** (Right) Young black locust coppice stand, Vermont.

Figure B26: Black locust coppice stand as canopy begins to close, Vermont.

Figure B27: Black locust silvopasture, Vermont.

FIGURE B28: (Top left) Young developing shelterbelt with diverse plantings, Vermont.

FIGURE B29: (Top right) Two months regrowth on pollarded hybrid poplar, Vermont.

FIGURE B30: Mixed species coppiced hedgerow, Vermont.

Integrating form with function, garden structure construction couples especially well with landscape design and installation work.

When I apprenticed with Ben Law, garden structures were a particularly high-value product with consistent customer demand. As we harvested a coppice stand, Ben instructed us to select and separate materials based on form and size. For garden and polewood structures, we sorted poles into two diameter classes—2 to 3 inches and 4 to 5 inches (5 to 7.5 cm and 10 to 12.5 cm)—cut to 8-foot, 10-foot, and 12-foot (2.4, 3, and 3.6 m) lengths (or longer). By sorting materials during harvest, it was easy to maintain an organized inventory available as needed during the building season.

And even if you don't build these structures yourself, local landscapers and builders may have a need for these materials. So, if you already grow fairly straight, high-quality, rot-resistant polewood, simply sorting the nicest stems and marketing them as a building material should generate a much higher return than fuelwood, while requiring even less processing.

Buildings

Yurts

Desirable wood properties/qualities—Straight growth, strength

Most promising species—Maple, alder, ash, birch, willow, chestnut

Special tools or machinery required—Bending jig and steam box for roof ribs and compression ring

Materials volume required—Depends on size: for a medium-sized yurt, 72 poles for the trellis, 36 for the roof ribs, and two planks for the compression ring[116]

Skill level—Moderate to high

Time to produce—about 2 weeks for a 16-foot yurt[117]

Price range—$1,000 to $20,000, depending on size, level of craftsmanship, and design details

The yurt has quickly become an exceedingly popular lightweight shelter, and young coppice growth offers a superb construction material. Originally developed by nomadic tribes in the broad Mongolian highlands, yurts offer portable, adaptable housing capable of accommodating seasonal climatic variation.

Figure 5.30: A section of a yurt lattice wall under construction using short-rotation sweet chestnut coppice at Prickly Nut Wood, West Sussex, UK.

Traditional yurts feature a cylindrical lattice wall framework and a conical self-supporting roof structure, bound at the center with a circular compression ring. Most modern yurts are covered with waterproof canvas or vinyl. Traditional Mongolian yurts use layers of felt, adding or removing layers as temperatures demand. A wood stove customarily adorns the center of the yurt, with the flue exiting through the central peak.

Yurt poles and yurt kits have increasingly become a lucrative product for coppice workers. Two-to-6-year-old regrowth provides nearly all the materials necessary to build these ingenious structures.

A roundwood yurt's lattice walls generally call for a framework of 6.5-foot-long (2 m) by 0.75-to-1-inch-diameter (1.9 to 2.5 cm) coppice rods. Laying them out in an overlapping diagonal pattern, builders fasten poles to one another by drilling holes at each intersection and stringing the joint together with a length of twine knotted at both ends. A 16-foot (4.8 m) diameter yurt requires three of these panels to span its circumference.

Yurt rafters or "roof ribs" should be as straight as possible, peeled of their bark, and 1 to 2 inches (2.5 to 5 cm) in diameter, with the tip at least 1 inch. Length depends on the structure's diameter. Nova Scotia-based Little Foot Yurts uses coppiced maple, ash, and birch for the roof poles. The roof ribs' tips fit into mortises drilled into the central compression ring. Some builders create a more secure joint between the roof ribs and the compression ring by shaving the tip of the poles into a 1-inch-square cross-section that fits snugly into squared holes in the compression ring, locking them firmly in place.

The roof ribs can be used straight, or boiled or steam-bent to create a roof with a conical shape and more headroom around the structure's perimeter. After an hour or so of steaming or boiling, the poles are limber enough to place in a bending jig where they are left for 3 to 4 weeks while the bends set.

Cordwood Construction

While cordwood construction originated in eastern Europe, its relatively recent revival in the US by Rob Roy has helped inspire its spread around the world. Cordwood building utilizes air-dried lengths of cordwood (firewood) stacked like stones and held in place with mortar to fashion a building's wall system. Cordwood buildings can have a charming aesthetic quality, combining the soft textures of wood with the stable mass of mortared stone.

FIGURE 5.31A: Cordwood can be used to create a beautiful, inexpensive, durable wall system. Photo courtesy of Rob Roy.

While somewhat labor-intensive, cordwood can be a very inexpensive building technique. Unfortunately, standard cordwood construction offers only modest insulative performance with R-values ranging from 0.75 to 1.5 per inch wall thickness. To improve this, builders can install an insulative infill within the cavity between mortar beads. To learn more about the details of cordwood construction, check out any number of Rob Roy's several books on the subject.

Roundwood Timber Frames

Roundwood timber frame construction uses pole- to sawlog-sized trees (6 to 12 inches+, 15 to 30 cm+) fastened with mortise and tenon joinery and wooden pegs. This ancient craft makes much more economical use of wood than log cabin construction, creating an open-walled structure awaiting infill by any number of materials. Historically, most timber frame buildings were built using logs that were either hand hewn or sawn square. While there are many variations on framed polewood structures in the annals of vernacular architecture from around the globe, roundwood timber framing is a craft that continues to evolve today.

In a timber frame building, the joined timbers bear the weight of the roof, allowing builders to choose from an array of wall infill materials. Traditional European timber frames featured wattle and daub or light straw-clay infill (loose straw coated in a liquid clay slip solution, tamped between form boards). Precisely joining round, irregular polewood requires considerable skill. Ben Law describes his approach to this technique in his *Roundwood Timber Framing* book and video.

Modern Product Potential

As we explore opportunities to develop novel coppice wood products, we're faced with both challenge and opportunity. Often, consumers are unfamiliar with many of the artisanal products that coppicing can generate. Despite these products' age-old origins, we need to educate our clients and develop new outlets for them. For many of us already working in the relatively uncommon realms of non-timber forest products, ecological landscaping, organic agriculture, and natural building, we once again find ourselves at a very familiar crux. We need to *create* and *educate* our market.

Part of the allure of rethinking silviculture lies in a movement towards meeting more of our material needs locally. In places where coppicing traditions still persist, products like local artisanal barbecue charcoal already occupy

Table 5.12: British survey results of coppice workers' time allocation for various business-related tasks

Task	Percentage of Total Hours
Coppice Wood Harvest	30.4
Value-Adding	41.3
Marketing and Distribution	8.5
Office Work	10.4
Education and Demonstrations	9.4

Adapted from Collins, 2004, p. 94.

a familiar place in many people's lives, creating opportunities for viable land-based livelihoods. Here in North America, where there's little cultural memory of many of these traditional crafts, we need to do our own R and D to develop practical, functional, beautiful products that meet today's needs. And we don't yet know what forms many of these products will take. This is the exciting (and challenging) prospect coppice provides for us. Perhaps the most financially viable and ecologically regenerative product for your bioregion is one that hasn't been created yet. To help stimulate your creativity, here are a few more issues to keep in mind when designing your coppice-based business.

Add Value!

While we've discussed this theme throughout the chapter, it's hard to overstate its importance. It's usually only at a very large scale that raw material sales will support a healthy livelihood. Wherever possible, find opportunities to transform raw materials into something useful. It boosts the feasibility of your enterprise and also can add diversity to your operation.

Imagination and Innovation

Originality, vision, artistry, resourcefulness… These are the keys to a successful coppice enterprise. How can you develop something that's functional, beautiful, and durable that meets the needs of people and businesses in your community? What can you make from coppice materials that hasn't been produced before? How can you modify or improve on existing products to reflect a new aesthetic, an

evolving context, or a complete reinterpretation? Uniqueness will set you apart from what's already out there.

Look at What's Out There

What types of products/crafts already exist? What needs do they meet? Examine your own personal possessions. What would they look like if made from coppiced wood? Visit craft shows, surf the web, peruse catalogs. We're surrounded by ideas and inspiration. Keying into them is the first and most important step.

How Can You Make It?

Once you find inspiration in a form or recognize a clear need, design it using coppiced wood. What tree species or wood properties will function best for each application? What species do you have access to? Can you improve on the design to make it more elegant, functional, or better-suited to the materials you have at hand?

Develop a Range of Products with Varying Costs

This is critical to a successful craft enterprise based on both observation and experience. Having built greenwood ladder-back chairs for over 15 years, I've learned that developing a wide product base helps augment your business's appeal and accessibility. Many potential customers are attracted to the aesthetic and practice of greenwood craft but will find higher-end products too costly.

Develop products that cost $10, $20, $50, $100, $350, $1,000… In this way, you can connect with a much broader customer base, buffer yourself against instability in the market, provide diversity in the work you do, and be better positioned to offer products in a wide range of settings.

Network

This almost goes without saying, but ally yourself with other craftspeople who make complementary products or who may have a need for products you make. While producers often lack advertising and promotional skills or motivation, investing time and energy into building connections with craft guilds, galleries, farm stands, markets, and colleagues will help expand your networks, promote your work, more clearly identify unfilled niches and saturated markets, and most importantly, build cooperative community. The more we work together, the better off we all end up.

Craft fairs, agricultural shows, and open days at museums offer opportunities to demonstrate craftwork and connect with potential customers, while at the same time educating them about the value of your skills and trade. Often this face time serves more to promote and connect coppice workers with potential clients and allies than making direct sales. While many passersby are not prepared to purchase products at a public market, they very well may remember their experience, tell others about it, and reach out to place an order as the need arises.

In Britain, several organizations maintain trade directories that list coppice workers along with the products they make. Linda Glynn of the Wessex Coppice Group developed a national web database of over 500 coppice workers and several marketing and training programs. These directories identify potential markets, connect supply with demand, and assist individuals to make direct sales while increasing the visibility and appreciation of locally made crafts and woodland products.[118]

Branding

Branding products helps producers differentiate their work from others, creating an identity that customers learn to trust and associate with quality. It identifies your unique individuality in terms of materials, sourcing, design, and aesthetics. Branding products helps craftspeople further add perceived value to their work, helping increase sales and build customer loyalty.[119]

Focus on Function

By producing goods that meet a direct and immediate need, we can largely rest assured that they'll find a ready audience. This isn't to say that there's no room for creative expression in a coppice worker's livelihood—in fact, it's hard to imagine craft work devoid of artistry. Rather, when we focus on products that meet specific needs, our work becomes practical and useful to a clearly identifiable clientele.

In this era of innovation, some of the best opportunities for new product and market development emerge from the efforts of "secondary producers"—land managers with diverse livelihoods that usually rely on at least one off-farm income source. This financial stability frees them up to explore their personal interests in products and management that need not be economically viable at the outset. They have more room to innovate and experiment. Likewise, hobby producers with a relatively small land base similarly demonstrate a tendency towards innovation and risk-taking that's often less feasible for full-time land managers.[120] So let's explore some emerging market opportunities for sprout growers to explore as we envision a renewed, 21st-century land-based economy. And remember that perhaps the best products for your unique context may have yet to be discovered.

Arborsculpture

Desirable wood properties/qualities— Vigorous growth, roots easily from cuttings, grafts easily

Most promising species—Willow, poplar, boxelder (*Acer negundo*)

Special tools or machinery required—None

Materials volume required—Modest but depends on size; can be just a few to a couple of dozen stout hardwood cuttings

Skill level—Moderate

Price range—Individual rods (*Salix viminalis*): 6 feet (50p) to 9 feet (£1.60); living willow fence: from £2.50 per square foot

Arborsculpture is a horticultural technique where growers train living trees and shrubs into various configurations, grafting them together into a sculptural form. Arborsculpture is often used to fashion attractive, functional garden structures like arbors, trellises, domes, chairs, benches, and shelters. Over 100 years ago, John Krubsack of Embarrass, Wisconsin, created beautifully crafted furniture using planted and trained boxelder seedlings.[121] In some ways, European laid hedges are also a form of arborsculpture. Fast-growing hardy species like willow and poplar respond well to this type of management, producing vigorous growth, rapidly filling out the living sculpture's form. They can also be reliably propagated by hardwood cuttings.

The living bridges of Cherrapunji, India, are some of the most awe-inspiring examples of functional arborsculpture. Spanning rivers at lengths of up to 100 feet (30.5 m), they can support as many as 50 people at once, and some bridges are 500 years old! These bridge builders use the hollowed-out half-round lengths of betel nut poles to guide and support the growth of roots of the *Ficus elastica* tree which grow to form the bridge.[122]

Though not necessarily coppice per se, arborsculpture relies on many of the same sprouting properties, and coppice stools can produce propagation stock for arborsculpture installations. These installations are a great fit for community spaces, parks, schools, and other public places where a broad spectrum of people will regularly experience and engage with it.

It's difficult to project arborsculpture's financial viability. In addition to the initial design and installation work, maintenance and management may offer an even more stable, long-term income stream. Each year these installations require weaving, pruning, and/ or training to either form the desired shape or control excess growth. And creative entrepreneurs may also find product opportunities by creating arborsculpture kits that include a bundle of dormant hardwood cuttings and instructions for installation and maintenance.

To establish an arborsculpture installation, first lay out the shape on the ground. For species like poplar and willow, simply insert one-year-old hardwood cuttings in the soil at a depth of 4 to 8 inches (10 to 20 cm), being sure to keep the soil around them moist. In dry climates or places with irregular rainfall, consider drip irrigation for efficient, low-cost water security that helps ensure their survival. Once new shoots are large enough to train to shape, bend them to the desired form, binding overlapping stems together so that they develop a grafted union and begin to grow as one. As the sculpture matures, prune it regularly to maintain its form. Leftover prunings are useful for a variety of purposes: livestock fodder, mulch, garden stakes, and more propagation stock.

FIGURE 5.32: A living willow fence crafted by Margaret "Pegg" Mathewson at her farm, Withyhenge, in western Oregon. She owns and operates the Ancient Arts Center.

Case Study: Bringing Resprout Silviculture to the Landscaping World

Name: Tom Girolamo, Eco-Building & Forestry, LLC

Location: Mosinee, Wisconsin

Ecoregion: 51b Central Wisconsin Undulating Till Plain

Koppen-Geiger Climate Zone: Dfb—Humid continental mild summer, wet all year

Precipitation: Roughly 33 inches (84 cm) per year, well distributed;

USDA Hardiness Zone: 4a

With a career that has spanned several interconnected professions, Tom Girolamo's training has led him to degrees in urban forestry, forest management, and permaculture design. Following his undergraduate years, Tom started his career as a municipal arborist. He later began a forestry consulting business, eventually transitioning to the design, installation, and maintenance of ecological landscapes. He approaches his work less as a traditional landscaper but rather an advocate of holistic landscapes that benefit the homeowner, community, and the environment.

Although Tom learned about coppicing and pollarding in forestry school, like many American students, he was told never to use them in urban forestry. But today, he sees coppicing and related techniques as valuable tools to rejuvenate landscapes using already-existing plants and patterns. He does this in a number of ways.

First and foremost, Tom points out that many customers are drawn to particular plants because of their unique forms and habit, although they often don't necessarily realize this. Conventional landscapers tend to maintain shrubs' size by "rounding them over." But this practice completely obscures the unique look that often initially drew the client to that species. For example, take ninebark (*Physocarpus* spp.), a ☞

FIGURE 5.33: Tom Girolamo using coppicing as a tool to manage a customer's landscape. Photo credit: Tom Girolamo.

FIGURE 5.34: Coppice-managed industrial parking lot screening. Photo credit: Tom Girolamo.

plant known for its flowing vase shape, becoming more and more umbrella-like as the plant matures and the ends tip over. Instead of giving it the classic round-over shearing, Tom coppices one out of every three shrubs each year on a rotating basis. Sometimes this results in a 3-year coppice rotation for all land-scaped shrubs, but if growth is slow, he may skip a year or 2, stretching the interval to 5 years.

Tom also encourages clients to appreciate the value of coarse woody debris as a nutrient-rich mulch that can be chopped and dropped in place. For clients interested in a natural-looking landscape, he simply chops up prunings, and leaves them to decay right there. He's found it's often best to do this in spring, so the plant continues to colonize the space throughout the winter months. He likes pairing serviceberry with dogwood, two common, multifunctional landscape species. In this case, dogwood serves as the mulch plant, so he really doesn't need to worry about diminishing its vigor. It's a companion to the serviceberry, and within 2 years after cutting for woody mulch, it easily regrows 6 feet tall.

Tom also often finds ways to use coppicing to balance out landscaping installations planted too densely from the outset. Instead of shaping plants by top pruning to create more balanced use of space, he encourages clients to allow him to cut them back to their base and regrow. This can turn a wild overgrown landscape into something that appears organized and well-kept.

Several years ago, Tom and his crew designed and installed over ½ mile of low-maintenance species to help screen the parking lot adjacent to a massive industrial complex. Their design included a mixture of ornamentals and fruiting trees and shrubs. The random staggered pattern included plants spaced 4 to 10 feet apart, with a broadcast understory of flowering native prairie plants. They use coppicing as their main tool to prune and reshape the shrub layer, and they continue to maintain the installation as an aesthetically pleasing natural habitat, providing a dynamic living screen surrounding the complex.

And as a professional landscaper, Tom often finds himself with massive quantities of woody debris he needs to find a home for. As a remedy for this abundant resource, he began to construct his first hugelkultur bed in his plant nursery about 20 years ago—originally 5 to 6 inches high, 30 feet wide and 60 feet long. He essentially created a massive pile of organic matter including all sizes of material, branches, chips, excess soil/sod. While it wasn't exactly pretty initially, the biomass has transformed the site's 6 to 8 inches of sandy loam into a highly productive swath that grows some of the greenest and most productive grass on the entire property.

Tom's business shows how coppicing can be as simple as the informed use of woody plants' resprouting capabilities to create aesthetically pleasing forms and landscape management techniques producing woody materials for many different uses. While in this case, the wood products are really more of a by-product of the management, his integration of this approach to a major US industry shows how much potential there is to transform this market and achieve far more productive ends.

Ecological Restoration

Resprout silviculture techniques can aid ecological restoration efforts in the form of prescribed burns to rejuvenate forest ecosystems and manage forest fuel loads, stream bank stabilization to minimize soil erosion on impaired slopes, improved habitat and enhanced ecological functions along riparian buffers, and phytoremediation to break down and/or uptake contaminants in water and soils.

Prescribed Burns

Many landscapes actually require some form of disturbance to remain a balanced, healthy ecosystem. As we discussed in chapter 1, across much of this continent, Indigenous land management practices created regular disturbance events that stimulated the co-evolution of diverse, novel, dynamically stable ecosystems. But during the past century, federal management policies focused on fire suppression put a stop to these intermittent low-intensity fires, inadvertently creating the conditions for many of the catastrophic wildfires plaguing much of the West today. But a shift in perspective is increasingly leading towards the readoption of this traditional ecological knowledge through the renewed use of prescribed burns as a tool to restore and rejuvenate ecosystems.

Ecological Engineering

Short-rotation coppice systems can produce materials for watershed restoration projects and erosion control measures. Once again, willow

FIGURE 5.35: In the small city of Hellevoetsluis, along the Haringvliet inlet of the North Sea, these Dutch workers are using an interwoven network of willow fascines to help stabilize the lower submerged portions of dykes. (1956) Source: Pot, Harry / Anefo photographer, Dutch National Archives, http://proxy.handle.net/10648/a951d996-d0b4-102d-bcf8-003048976d84

is the most common species, with cultivars like *Salix purpurea* "Streamco" chosen especially for stream bank stabilization.

Often, these types of short-rotation stands yield both live stakes and brushwood bundles called fascines. These tightly bound bundles of branchwood and brush are similar to the faggot bundles we discussed earlier. Once installed, fascines help prevent erosion, diffuse runoff's erosive energy, and act as a silt trap. While their size may vary, fascines are often 7 feet (2.1 m) long and 12 to 16 inches (30 to 40 cm) in diameter and bound tightly along their length.[123] They're usually either staked in place or buried in an on-contour trench along vulnerable stream banks and erosion-prone areas. The live stakes are either soaked in water until they root or installed directly in the soil and allowed to root in place.

Habitat Restoration

Resprout silviculture practices can help revitalize wildlife habitat while yielding useful wood and non-timber forest products. As we discussed in chapter 3, conservation organizations manage a considerable portion of contemporary British coppice to preserve habitat for rare and endangered species. Managed hazel coppice supports bird populations at least four times greater than neglected coppice and five times higher than that of high forest.[124]

Similarly, some foresters employ strategic patch cuts to initiate coppice regeneration and create prime habitat for bird and small mammal species. In Massachusetts, Division of Fisheries & Wildlife forester Brian Hawthorne prescribed coppicing as a tool to regenerate a stand of big-tooth aspen (*Populus grandidentata*), also calling for the retention of black cherry, white ash, and red oak for their structural diversity and mast production.[125] On our farm woodlot in Vermont, we received a five-year NRCS Conservation Stewardship Program grant to carry out a patch cut in a low-grade stand of white pine and quaking aspen to create between-habitat diversity for birds and small mammals that thrive in early-successional habitats. Not only did we receive funding to do the logging work, but we also were able to harvest more than 14,000 board feet of sawlogs that we milled into lumber.

Floodplain forest habitats, with their long history of deforestation and human management, can also benefit from resprout silviculture. In Europe, many of these essential wildlife migration corridors were extensively cleared and settled from prehistory through the Middle Ages. Intact floodplain forests help connect fragmented forest blocks while providing ecosystem services like runoff infiltration, stream bank stabilization, riparian shade, and buffers against seasonal flooding.

Many European floodplain forests were once managed as coppice with standards with exceptionally high levels of biodiversity. In the Czech city of Olomouc, records dating back to the 16th century show floodplain forests sometimes generated as much as twice the revenue from the sale of acorns than wood sales! These multipurpose woodlands were also grazed extensively until the practice was abolished around 1850. By preserving and/or reestablishing these managed ecosystems, we might create economic incentives to maintain these vital habitats, along with their ecosystem services while also generating diverse yields.[126]

Phytoremediation

Sadly, some of today's most important ecological restoration work lies in remediating contaminated landscapes, and coppicing can play a valuable role in **phytoremediation** (the use of living vegetation to clean up contaminants on-site).

Phytoremediation occurs through a number of mechanisms—facilitating contaminant degradation, improving soil oxygen levels, absorption and/or translocation of contaminants, and immobilization of contaminants. Certain plants are actually capable of accumulating heavy metal contamination from water and soils.

Willow species' easy propagation, fast growth, dense root systems, coppiceability, and ability to accumulate pollutants make them especially well-suited to phytoremediation.[127] Often tolerant of nutrient-poor soils, willows can establish on highly degraded sites with scarce topsoil like mines, gravel pits, quarries, waste sites, road sides, and industrial spoils. At the same time, willow species support active soil biological populations. They appear to tolerate at least four of the seven most important heavy metal contaminants in soil: cadmium, copper, lead, and zinc, as well as cesium, one of four radionuclides.[128]

Additionally, willows act as living filters, maintaining high evapotranspiration rates, rapidly uptaking nutrients, producing voluminous biomass, and tolerate flooded conditions. Engineers have used willow plantings in phytostabilization efforts to help minimize contaminants entering aquifers adjacent to landfills, sewage treatment plants, waste dumps, and steelworks. In these scenarios, designers either locate plantings on top of contaminated zones to reduce rain and runoff from leaching contaminants into the soil or they encircle contaminated areas with willow plantings to impede the flow of contaminated runoff.[129]

Biomass production on marginal and/or contaminated land offers the potential to generate revenue while concurrently restoring degraded ecosystems, a unique property as compared to other restoration strategies. Researchers in Sweden and Belgium irrigated short-rotation coppice with wastewater, increasing biomass yield while offering tertiary wastewater treatment.[130]

Studies also suggest that SRC may offer a clean-up strategy on radionuclide-contaminated sites. One experiment monitored goat willow (*Salix caprea*) and aspen's (*Populus tremula*) ability to remove cesium (137Cs) and strontium (90Sr), two contaminants with relatively long half-lives that risk introduction into the food web. Both species accumulated these radionuclides from two different contaminated soil samples. Concentrations of 137Cs peaked in tree roots, while researchers found the highest concentrations of 90Sr in tree leaves.[131] Ultimately, they concluded that analyzing short-rotation coppice's viability as a phytoremediation tool would require long-term, large-scale field investigation under natural conditions.

Propagation Stock

Desirable wood properties/qualities— Species that have useful properties, are readily propagated from cuttings, fruit/ nut tree cultivar scionwood or rootstock

Most promising species—Fruit/nut producing, ornamentals

Materials volume required—From a few stems to hundreds+

Skill level—Moderate to high

Time to produce—Minimal

Price range for completed product— Scionwood/cuttings: $1 to $5 per 8-to-12-inch stem; grafted trees: $15 to $50 each

Coppice stools or managed pollards can generate rootstock and scionwood for grafted trees as well as cuttings for vegetative plant propagation. These propagation techniques rely on a supply of healthy, pest- and disease-free first-year shoots—materials coppicing readily produces. Some nursery growers propagate apple and pear rootstocks by layering or stooling, and several *Corylus* species can also be propagated by layering.[132]

Many woody plants including elderberry, currants, gooseberry, mulberry, dogwood, willow, and poplar can be easily propagated from hardwood cuttings grown on managed stools or pollards. British forest gardener, author, and useful plant expert Martin

Crawford manages a nursery stacked with low pollards that produce scionwood for his grafted fruit tree nursery. Either sold by the stick or as a bare-root or potted plant, these are all very high-value uses for one-year-old wood.

Woody Cut Stems

Desirable wood properties/qualities— Attractive flowers, interesting stem shapes, vibrant stem color

Most promising species—Forsythia, flowering quince, lilac, willow, dogwood

Special tools or machinery required—None

Materials volume required—As little as just a few stems

Skill level—Moderate

Price range—$1 to $10+ for single shoots

Some species produce foliage, flowering stems, seeds, fruits, shoots, and pods with aesthetic qualities that make them especially appealing to the floral industry. These products are commonly referred to as woody cuts. What's more, many of these species will readily resprout when cut, forming upright, vigorous, insect- and disease-free new growth perfect for this specialty market. These products can extend the growing season and also add unique diversity to floral arrangements.[133]

Some of the best species cut for forced blossoms include pussy willow, flowering quince, forsythia, fruit trees, filberts, beech, birch, witch hazel, redbud, lilac, magnolia,

Table 5.13: A selection of species grown for woody cuts

Genus	Species	Common Name
Buxus	sempervirens	common box
Chaenomeles	speciosa	flowering quince
Cornus	sericea	red osier dogwood
Corylus	americana	American hazel
Forsythia	× intermedia	hybrid forsythia
Hydrangea	arborescens	smooth hydrangea
Physocarpus	opulifolius	ninebark
Salix	gracilistyla	Japanese pussy willow
Salix	matsudana	curly willow
Symphoricarpos	× chenaultii	Chenault coral berry
Philadelphus	microphyllus	littleleaf mock orange

Visit appendix 2 for a far more comprehensive list. Note that for several of these, numerous cultivars and related species also exist.

rhododendron, and red maple.[134] The Pacific Northwest-based Kirk company began marketing cut stems from evergreen huckleberry, salal, and sword fern to local retail florists in 1939. The region's industry has continued to grow ever since to include evergreen species like Oregon grape, holly, and red-tipped huckleberry. Because these species retain their leaves after harvest, they're sold and shipped across North America and as far away as Europe.[135]

A number of willow species work well due to their adaptability to a wide range of soils, interesting growth forms, bud color, catkin longevity, extended forcing period, and low production costs. If grown from cuttings, blooms begin after just their second growing season, though growers should wait 2 to 3 years before their first harvest. In subsequent years, expect each plant to yield 10 to 30 stems annually. Once established, willow beds actually require annual harvest and maintenance, or they begin to develop side spurs, reducing their marketability.[136]

The optimal harvest window depends on species and the desired plant part. Stems are cut to lengths of at least 18 inches (45 cm). Harvest stems for their color when they're dormant. Stored dry, they'll last for months. Harvest species that produce ornamental fruits once the fruits reach maturity.[137] Species cut for forced blooms, like quince, fruit trees, and forsythia, should be harvested when the buds are tight, whereas rhododendron, camellia, witch hazel, hibiscus, Mahonia, viburnum, and lilac should be harvested as the buds begin to open. Growers generally harvest willow species for catkin blooms between January and April and stems for their shape, bud, or bark color between November and March.[138]

Many ornamentals set next spring's flower buds during the previous summer prior to dormancy. If cutting these plants for forced blooms, help ease their transition by harvesting during late winter, ideally during above-freezing temperatures. It's possible to extend the window for marketing forced blooms by more than a month by either forcing branches early or cutting them early and placing them in cold storage before forcing.[139]

Begin forcing stems for blooms at temperatures around 50°F (10°C). After cutting, place them in warm water (100° to 110°F, 38° to 43°C). After a few weeks, either speed up flowering by bringing them into a warmer space, slow the process by maintaining 50°F (10°C), or halt it altogether in cold storage at 35°F (1.7°C).[140]

Many growers space plants 2 to 6 feet (0.6 to 1.8 m) apart within rows with adequate between-row spacing to allow for field operations like mowing, cover cropping, grazing, etc. Some growers use staggered double rows in 5-to-6-foot (1.2–1.8 m) wide beds with 6-to-8-foot (1.8 to 2.4 m) sod paths between.[141]

Considering woody cuts' position along the value-added continuum, we see that it's perhaps one of the most value-dense materials we could possibly grow, fetching $1 to $8 or more for just a single dormant stem. Of course, not every community has access to these high-end markets, but woody cuts clearly demonstrate the substantial financial benefits of creativity and added-value potential.

Case Study: Woody Cuts for the DC Metro Area

This summary is based on notes and audio from a 2011 site visit and interview conducted by Dave Jacke.

Name: Leon and Carol Carrier, Plantmasters

Location: Gaithersburg, Maryland, 30 miles from Washington, DC.

Land Base: At the time of the original interview, their enterprise consisted of 2 acres, along with 5 neighbors that allow them to grow on their land as well, increasing the total production area to probably 3 to 3.5 acres. Although they're largely surrounded by farmland, they live in a more residential zone, so they manage patches of neighboring lots, minimizing neighbor's mowing needs, and allowing them to cut their own stems as needed. In recent years, the Carriers purchased another 5-acre farm and now have 5 large hoop houses along with many rows of annuals and woody shrubs.

Ecoregion: 64c - Piedmont Uplands

Koppen-Geiger Climate Zone: Dfa—Humid Continental Hot Summers

Elevation: ~ 300 to 400 feet (91–122 m)

Average Annual Precipitation: 43.22 inches (110 cm), relatively evenly distributed with 3 to 4 inches (7.5–10 cm) per month

USDA Hardiness Zone: 7a

Soil Types: Clay-rich; they've improved it over the years with lots of organic matter

Site History: Cow pasture back in the 60s and 70s

Special Tools/Equipment/Infrastructure Used: 20-by-20-foot glass and aluminum greenhouse, chosen largely due to aesthetic concerns in the neighborhood. Walk-in cooler. Gas-powered hedge trimmer for smaller-diameter sprouts. Electric chain saw for larger stems.

Training/Background of Manager: Leon earned a degree in ornamental horticulture from the University of Maryland. The industry was very new when he first began, so there was little precedent and limited resources. ☞

FIGURE 5.36: A row of flowering hydrangeas at the Carrier farm. Used with permission courtesy of Leon Carrier.

Leon has been in the cut flower business since graduating in 1980 when he started off doing farmer's markets, primarily selling tropical plants that he could easily propagate. In addition to his livelihood as a woody cuts grower, he also has an interior landscaping company. There he saw farmers bringing in buckets of lilac and peony stems and thought one day he'd

like to do just that. Today, he's got tens of thousands of peonies and so many lilacs that he'll always have enough. He and his wife, Carol, began growing flowers over 30 years ago on their property.

Leon and Carol grow many species, including lots of different hydrangea varieties, lilacs, pussy willows, winterberries (10 varieties), dogwoods, viburnums, rose hips, *Symphoricarpos*, snowberries, plum, cherry, crabapples, flowering quince, magnolias, and forsythia.

All their woody cuts are destined for the floral products market. They harvest some for their flowers, others for ornamental berries, and some as backing greens for arrangements. As far as sales channels go, they primarily sell their products retail at eight different farmer's markets each week. They do some limited wholesaling to local florists, but they don't look for it, and do not deliver.

They have seasonal employees that begin work in mid-February and have a total of roughly 10 employees. It's very difficult to determine how much they earn from the actual woody cuts portion of the business since they also buy flowers from wholesalers. But Leon did share a few details on local product values. Keep in mind that these numbers date back to 2011 and are also in a very well-to-do part of the country.

Lilac stems sell for $10 to $12 a bunch. They sell 5 *Ilex* stems for $15 each. Quince stems are sold in bunches of 3 for $10. On a single 20-year-old quince tree, Leon estimates he could harvest 60 to ☞

80 bunches, meaning that that single plant could generate $600 to $800, assuming they could sell it all.

⚒ All the plants are just a little different. Some get cut back to the ground, while others are cut at 3 feet. Leon's not particularly careful with his cuts. He said that most stems seem to sprout.

🕐 While a few species may be cut annually, his rotations usually span 2 to 5 years. It all really depends on the growth rate and vigor of individual plants and his ultimate product goals.

📜 Their plantings are pretty much all grouped together, so the management isn't all that difficult.

Leon and Carol have clearly created a comfortable life for themselves, doing what they love. While it isn't making them rich, Leon said, they've never had to worry about money, even in lean times. They've learned to roll with the vicissitudes of life and, in so doing, bring lush vegetative beauty into the lives of their customers, producing perhaps one of the most value-intensive woody crops per unit area.

Weed/Pest Challenges: Leon says that woodies are a lot easier to manage than herbaceous perennial flowers—no digging with less weeding and fertilizing. They're low input: just cut them back hard every year or every 3 to 5 years.

Leon is a big mulch fan, mainly using it for weed suppression, moisture retention, and soil enrichment. He applies a thick layer—up to 5 inches (12.5 cm)—to plants either every year or every other year. Pokeweed, bindweed, and wild grape have been the most persistent weeds, but they don't tend to be a major problem.

They've been lucky and have had very few issues with deer browse. When deer have visited the farm, they've only been interested in red osier dogwood and pussy willows.

Understory Production

Desirable wood properties/qualities— This product is less about wood than the understory species growing beneath established tree cover

Most promising species—Goldenseal, ginseng, and black cohosh (*Cimicifuga racemosa*)

Special tools or machinery required— None

Materials volume required—Depends on scale

Skill level—Low to moderate; note that it can be difficult to establish some high-value understory species

Price range—Varies widely depending on species and quality, from $5 to $500 per pound

Developing diverse, multistory coppice woodlands may perhaps represent the holy grail of resprout silviculture system design. When we examine traditional coppice systems, there are few examples of profitable systems that make use of the understory as much as the canopy. Some of these non-timber forest products include edible and medicinal mushrooms, medicinal and pharmaceutical products, berries and fruit, floral products, herbs and vegetables, and craft and landscaping products.[142]

Often non-timber forest products (NTFPs) reach the marketplace via skilled wildcrafters who identify and harvest wild populations of useful plants left to reproduce naturally. When done appropriately, this hands-off production system can be elegantly sustainable, but increasing demands for many of these specialty products has put greater pressure on wild stands.[143] Because human-assisted regeneration of wild stands is not always successful nor will it always adequately account for the age of the plants harvested, if not careful, wild plant populations will suffer from growing pressures.

Table 5.14: Promising/valuable understory species

Latin	Common Name
Cimicifuga racemosa	Black cohosh
Caulophyllum thalictroides	Blue cohosh
Panax quinquefolius	Ginseng
Hydrastis canadensis	Goldenseal
Ulmus fulva	Slippery elm (bark)
Valeriana officinalis	Valerian
Aristolochia serpentaria	Virginia snakeroot

Adapted from Teel and Buck, 2004, p. 14.

We can help minimize these pressures by harvesting only renewable aerial plant parts, though this is only an option if they possess the same active constituents as the roots. It appears as if this may be the case with goldenseal (*Hydrastis canadensis*). An unpublished chromatographic comparison of goldenseal root and rhizome to the leaf and stem by Herb Pharm found that the plant's two primary alkaloids, berberine and hydrastine, exist in significant proportions in both the aerial and subterranean parts.[144]

Goldenseal, ginseng, and black cohosh are some of the most valuable understory herbs in North American woodlands. Both goldenseal and American ginseng have suffered from overharvest and are regulated in appendix II of the Convention for International Trade in Endangered Species of Wild Fauna and Flora (CITIES). But although growers widely cultivate ginseng, goldenseal is still fairly uncommon. Though not without a suite of challenges, high-value understory herb production does offer growers a significant opportunity.

Healthy goldenseal plants require particular soil and shade conditions, and as with many forms of understory forest farming, problems tend to arise when growers attempt to cultivate plants in conditions that deviate from their native ecology. To try to avoid some pest and disease problems, some growers interplant companions like bloodroot (*Sanguinaria canadensis*), wild yam (*Dioscorea villosa*), jewel weed (*Impatiens* spp.), black cohosh, and spicebush (*Lindera benzoin*).[145]

American ginseng (*Panax quinquefolius*) occupies a native range that covers more than 20 northeastern states. This slow-growing plant requires part to full shade, preferring acid soils with a high calcium content. Long-valued by the Chinese for its energy-boosting ability, by the mid-19th century, extensive forest clearance and ginseng wildcrafting for export had put a significant dent in wild populations. Today wild ginseng root may sell from $200 to $500 per dried pound. While it can be successfully cultivated, if grown under artificial shade, its market value plummets. Because of this, growers strive to simulate natural growth conditions, utilizing sites matching ginseng's natural habitat. Ginseng grown this way is said to have comparable qualities to uncultivated wild roots.[146]

Black cohosh is used to help treat menopause, pre-menstrual syndrome, estrogen deficiency, dull pains, and some forms of depression. It prefers good, deep, loamy soil and will tolerate considerably more sun than goldenseal. In fact, it's possible to grow black cohosh under open field conditions with little to no shade.[147]

The United Plant Savers is one of the best resources for growers interested in understory cultivation. This decentralized national organization helps promote and research the production of non-timber forest products and other understory species with partner members throughout the country.[148] Also check out Ken Mudge and Steve Gabriel's book *Farming the Woods* for a much more in-depth look at the potential for several forest farming practices.

Education and Professional Training

Despite the decades-long decline and virtual disappearance of numerous crafts and land-based skills, the recent resurgence in "re-skilling" means that perhaps one of the most potent products to emerge from the managed woodlands of the future is information.

Education provides skilled craftspeople with another potential income stream, often at least as lucrative as the tangible goods they produce. It also helps build a network of talented individuals capable of contributing to the growth and spread of ecological forestry and handicrafts. Given that there's little coppicing and pollarding to teach from here in North America, we again can start by looking overseas for some inspiration.

In Britain, organizations like the British Trust for Conservation Volunteers (BTCV), the Green Wood Trust, the Devon Rural Skills Trust, Clissett Wood Trust, Wessex Coppice Group, Coppice Association North West, and the Weald and Downland Museum offer courses in coppice crafts and woodland management. Additionally, numerous individual craftspeople and coppice workers also offer hands-on training opportunities.[149]

The Bill Hogarth Apprenticeship Trust Scheme offers an especially in-depth three-year apprenticeship in coppice skills, connecting students with educational opportunities through three different outlets: existing courses hosted by other organizations, customized short courses and instructor days, and field work with mentoring woodsmen and/or craftspeople. As apprentices advance, they choose an area of focus and progressively spend more time with individual mentors.[150]

From daylong workshops in specialty crafts, to season-spanning apprenticeships in ecological forestry, there are countless opportunities to diversify your income stream while passing on skills to future generations. These types of trainings and workshops are generally hosted in one of several ways: through existing continuing-ed type institutions that provide a working campus where students can put the ideas into practice, at more conventional schools of all levels, on-farm or woodland events hosted by the instructor or interested community members, or long-form internships and apprenticeships. Educational events like these are fulfilling and enriching for all parties and are an essential step towards creating a talented and informed network of woodland stewards.

MATCHING PRODUCTION SYSTEMS TO REGIONAL CLIMATE AND PRODUCTION NEEDS

So, after exploring close to forty different products that make use of woody resprouts, we see how much opportunity these management techniques provide us. And this brings us back full circle to the beginning of our discussion on the economics of resprout silviculture. How does our management generate materials that meets our needs or the needs of our community, while at the same time directly responding to the ecological realities of our bioregion? Each of this continent's primary physiographic regions have unique characteristics, opportunities, and constraints that help direct us to

harmonious design solutions. For our resprout silviculture systems to be economically viable, they ultimately must respond to these realities.

Here in the cold temperate Northeast, we're blessed with abundant evenly distributed rainfall, four well-defined seasons, high population densities surrounding major metropolitan areas, and a wide diversity of woody species. The systems we create here will be very different than those well-suited to the deep fertile soils of the plains states or the abundant sun and often scant precipitation of the arid Southwest. Taking the time to identify your production goals is very similar to the whole site design process we'll discuss in our next chapter, and it's a process that we ignore at our own peril. So, take some time, jot down a list that includes your production goals, your site's resources and constraints as well as the market opportunities you see most promising in your locale. The time you take to do this will go a long way towards developing systems that generate healthy products, people, profits, and landscapes.

FINANCIAL ASSISTANCE TO SUPPORT COPPICE MANAGEMENT

While I'm currently unaware of any conservation initiatives that specifically support coppice system development, many of the practices we've discussed may meet the broad objectives of existing conservation and land stewardship-based grant and cost-share programs. In

TABLE 5.15: (See page 283) Adapted from Krome, Maurer, and Wied, 2009.

Table 5.15: Potential funding sources and support organizations for resprout-based agroforestry systems in the US

ATTRA	Managed by the nonprofit National Center for Appropriate Technology (NCAT), ATTRA's 30+ specialists research questions they receive and provide a written summary of their results along with any supporting literature.
Biomass Research and Development (BR&DI) Program	Financial assistance, grants, and contracts to entities to carry out research, development, and demonstration of cost-effective methods, practices and technologies used to produce biofuels.
Community Wood Energy and Wood Innovation Grant Program	Supports local, state, and tribal governments to install community wood energy systems or to build wood product manufacturing facilities. Competitive grants for community wood energy system development, technical assistance, woody biomass resource assessments, and monitoring plans.
Farm Service Agency (FSA)	Conservation loans to qualifying borrowers and projects including forest cover establishment and conservation practices including shelterbelts, windbreaks, riparian buffers and filter strips, and living snow fences.
Conservation Reserve Program (CRP)	Voluntary programs offer economic incentives to land managers to improve the soil, wildlife, and hydrological resources of farmland by temporarily removing it from production. Includes annual rental payments and cost-share assistance (up to 50% of project costs) to establish perennial cover on land previously in row crop production.
Conservation Reserve Enhancement Program (CREP)	USDA agreements with individual states offering farmers additional funding to develop practices addressing locally identified conservation needs. Typically requires that growers agree to a 10–15 year period during which they take their land out of agricultural production in exchange for an annual rental rate, an FSA-determined maintenance incentive payment, up to 50% cost-share funding for the installation of the conservation practice. Frequently applies to the installation of forested buffers and filter strips along waterways.
Conservation Innovation Grant Program (CIG)	Aims to stimulate agricultural producers' application of innovative conservation strategies.
Environmental Quality Incentives Program (EQIP)	Provides financial and technical aid to implement conservation practices (up to 75% cost-share) over 1–10 years.
US Forest Service's Forest Products Laboratory (FPL)	A wood research laboratory and information clearinghouse on wood use. FPL staff pass along relevant information to customers upon request, will generate new information, or point customers to other relevant sources of information and expertise.
Value-Added Producer Grants (VAPG)	Facilitates the development of value-added producer-owned businesses. Funds may be used to develop feasibility studies and business plans or to acquire working capital to operate a value-added business or alliance. The program requires cash and/or in-kind matching fund.
Wildlife Habitat Incentives Program (WHIP)	Technical assistance and cost-share payments of up to 75% of the total costs of establishing wildlife habitat improvement practices including tree plantings to improve aquatic and riparian wildlife habitat, windbreak establishment, and wildlife corridors. Agreements usually last 5–10 years.

my own experience, personal relationships are essential in connecting with these opportunities. Many of the management tools we've discussed here are quite different from what more conventionally minded service providers are familiar with. It's up to us to educate them, share our excitement, outline how and why it's relevant to regenerative land management, and push to see them recognized and valued.

Numerous funding opportunities already exist on national, regional, and local levels and may simply require some up-front legwork to build connections and identify potential opportunities for support. These programs can play a significant role in helping kickstart projects that enhance habitat, research new products, generate renewable energy supplies, enliven regional economies, etc. It'll be up to you to seek out your local and regional funding opportunities, but here are a few guidelines, recommendations, and programs that already exist nationally via the USDA to help support agroforestry projects.

Sven Pihl, Agroforestry Technical Service Provider with the Savannah Institute, has put considerable energy into decoding the many USDA programs that already exist and how they can be leveraged to help support the spread of agroforestry. Table 5.15 provides a list of some of these programs, along with brief descriptions of their scope and the type of projects they support. If you plan to pursue any of these opportunities, start by familiarizing yourself with the specific details of each.

As Sven points out, if you're going to receive funding, you need to follow the rules. And in so doing, you'll need to learn to look at these practices through each institution's lens. Each program has different payment structures and different motivating factors in different states. It can be a difficult process to navigate from the outside, and you're going to have to do a lot of your own homework. But by taking time to review these practices, learning their terminology, and identifying a few programs that seem to align well with your goals, you'll be primed for an in-person meeting with an organizational representative.

Once you've done this homework, contact representatives from your local NRCS, university extension, or eco-minded nonprofits. Tell them about your vision and needs, the existing programs that you feel are a good fit, and ask them what programs or funding they know of that might also help you get your project off the ground. These relationships can often take some time to develop, but will likely pay dividends many times over once well-established. By articulating a clear vision, with focused questions and grounded information, you'll set that relationship off on the right foot.

To help with this process, Sven has been crafting an interactive pdf document called *Using NRCS Practices to Design and Fund Agroforestry Systems*. This complex resource links relevant funding streams to different agroforestry practices. It will be available for download at coppiceagroforestry.com. It's an excellent tool to help you begin to navigate what can often be a confusing landscape of fiscal support. But fortunately, we don't have to go it alone.

FORESTRY COOPERATIVES

Forestry cooperatives facilitate silvicultural resource and skill sharing. They offer member-owners numerous benefits like access to woodland via collective woodland management and aggregated marketing and product distribution, helping individuals better position their products. Similarly, trade associations and cooperatives can provide growers and craftspeople with collective marketing and outreach benefits, helping build customer awareness and demand.[151]

It appears that these cooperative models are far more developed in Europe than here in North America. Village residents in Borgotaro, Italy, communally own their local forested land. These associations of forest owners and local authorities sell daily, weekly, monthly, and seasonal permits for wild mushroom collection. While local members usually receive permits free of charge, tourists and commercial pickers must purchase them. Permit sales help support improved forest management, forest road and trail maintenance, tourist facilities, and control of illegal harvesters.[152]

Also in Italy, researchers compared two different organizational models for marketing mushrooms: traditional and net-system approaches. In a traditional approach, a single enterprise produces a large volume of product for the mass market. The net-system approach integrates small- and medium-scale operations that supply relatively small quantities of high-quality niche market products. These small and medium-sized businesses often do not produce the same product, but instead complement one another as part of a larger strategic network, hence the name "net-system." The collective benefits from members' skills, product diversity, flexibility, and net production allows them to compete in regional markets and beyond.[153] We see these types of networks emerging in the local foods industry via initiatives like food hubs.

If these networks do exist in your region, they probably aren't built around the products generated by resprout silviculture, but that doesn't mean that the fruits of your labors wouldn't complement an already diverse product line of an existing cooperative. And if something like this doesn't already exist but the community of producers does, exploring these organizational structures can build stronger community relationships while helping support a larger network of land-based livelihoods.

CONCLUSION

Robust, resilient silvicultural systems must address social, ecological, and economic needs. Coppicing survived many waves of European expansion and development because it directly supported the local community. These practices will only find relevance in our culture today if we develop systems that reward our investments. They need to pay their way.

We inherit a remarkable collection of useful products to convert our coppice-grown materials into. Whatever products you generate and whatever forms your systems take, remember it's all about balancing what you have (species, landscape, ecology) with what you and your community need (income, fulfillment, livelihood, products). *This is the true art of coppice agroforestry design.*

So, at this point, we've laid the groundwork for a thorough understanding of the cultural, biological, ecological, systemic, and economic aspects of resprout silviculture. In part II, we explore the design, establishment, and management of resprout silviculture systems. We transition from a deep dive into what's been done through time to how these practices fit your own unique situation. It's been quite a journey to get here, and yet our process is only just beginning...

Part II:

21st Century Coppice: A Modern Integration

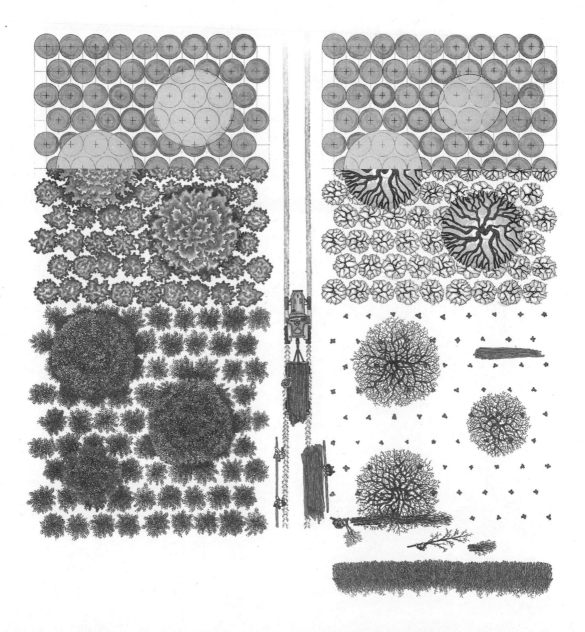

Chapter 6:

Listening to the Land:
The Art of Ecological Design

S o, we've arrived. Up till now, we've examined all the fundamental characteristics of resprout silviculture: the 5 W's of coppicing (who, what, when, where, why). And now we get to the fun part! And the scary part … because we've reached the most important part of the book: the how. And this is something that's unique to each and every one of us and our site-specific context.

In this chapter, we discuss how to put these concepts and practices into action and how to make them your own. You'll learn a strategic and flexible process to guide you that nests our systems framework within well-worn ecological design processes.

The design process can be a confusing and intimidating realm. It's easy to feel stuck at times. We've all been there. But design is as

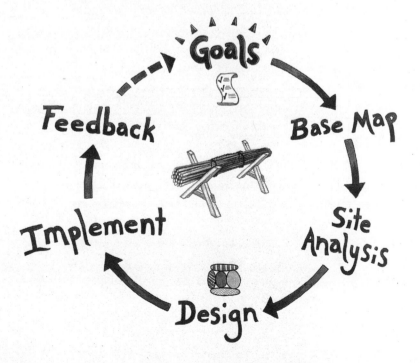

FIGURE 6.1: While the specifics may vary, most design processes involve these fundamental stages. The key to our process is that our design emerges as a result of clear, well-articulated goals and deep and well-informed site analysis. And the process never ends because feedback and changing conditions constantly inform how we nimbly navigate a fluid future.

much about process as it is about product. We can't foresee all of the potential implications of our design decisions, but we can do our best to plan for them. We make mistakes—and we learn from them. Design is iterative. And we're constantly receiving feedback.

Woodlands don't need management to survive. But we may choose to humbly and conscientiously steer them towards our management goals. This chapter will help you navigate your way along this path. It can be confusing, intimidating, and overwhelming, but remember—trust the process. It will guide you.

A SYNTHESIS OF STRATEGIES

Most design processes follow the same basic steps: goals, base mapping, site analysis, design, implement, feedback.

In this chapter, we explore a systematic approach to design that integrates forest management planning, our resprout silviculture systems framework, and a holistic approach to ecological design. Let's start with a look at the foundation of modern forestry planning.

THE SILVICULTURAL STANDARD: FOREST MANAGEMENT PLANS

Forest management plans organize, describe, and direct management decisions. Most forest management plans include several key pieces:

- Management objectives for the property
- Location, boundary, and stand maps
- Stand management goals
- Stand descriptions
- Site descriptions by stand
- Recommended treatments by stand
- Plan update procedures[1]

Parts 1 and 2 cover the entire property, while parts 3 to 6 explore each individual forest stand. A **forest stand** is a contiguous area that includes trees of similar species, age, size, and condition. After a quick overview of parts 2 and 4 to 7, we'll circle back and discuss goals.

Location maps place the parcel within the context of the broader region. Boundary and stand maps (usually represented on a single map) indicate the property boundaries along with the location of individual forest stands throughout the woodland.

Stand descriptions provide a summary of tree species present (with particular emphasis on the canopy dominants), their diameter, age, and the quality, density, volume, and projected growth rate of the stand.[2] For a primer on forest communities, review Natural Communities and Forest Cover Types in chapter 3.

Site descriptions describe each stand's aspect, slope, and site index along with unique features like rock outcroppings, hydrological features, and any other qualities that may influence management. As we discussed in chapter 4, the site index is an estimate of a stand's productivity. More productive sites grow taller trees faster. It's expressed as a tree's average height at a predetermined age (generally 50 years). In New England, an index of 45 or less is poor, 55 to 65 is average, and 80 or beyond is excellent. This means an area with a site index of 70 could produce 70-foot trees in 50 years.[3]

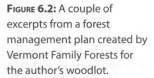

FIGURE 6.2: A couple of excerpts from a forest management plan created by Vermont Family Forests for the author's woodlot.

Recommended treatments are the action part of the plan. It outlines any suggested tree felling, interventions, infrastructure improvements, their purpose, and their relative priority.

Plan update procedures describes the frequency of plan updates (usually 5 to 10 years) as well as the type of data and observations that should be recorded upon revision. This may include changes in the condition of the stand, the outcomes of previous management activities, and any revisions in the timing or nature of future management.

Working with a Professional: The Role of Foresters

While few foresters are familiar with the nuances of resprout silviculture, their extensive experience with forest management and intimate knowledge of local ecosystems make them a valuable resource worth connecting with. They may have important insights, suggestions, ideas, or reflections that help ground and improve your design. Several types of foresters serve woodland owners.

Extension foresters work for state agencies and provide landowners with information on local resources and government programs and help connect property owners with consulting foresters. State-employed *county foresters* provide consulting services for private landowners and may help administer federal cost-sharing programs. Each state's forestry department usually employs one forester per county. As a result, their services are often in high demand, but their services also come free to you. Last, *consulting* or *private foresters* offer contract services including site assessments, management plans, and forest road design and will also often oversee a logging operation to ensure the logger follows the plan properly.[4]

For readers planning to integrate working trees into fields and grassland, the resprout silviculture systems you design are perhaps more akin to landscape design than forestry. While agricultural extension and the NRCS Technical Service Advisor services may be available free of charge, few of these advisors have much knowledge or experience with resprout silviculture. Towards the end of chapter 5, we discussed how these entities may have technical and monetary resources available to assist with agroforestry installation and management, but it would generally be uncommon to find personnel with enough knowledge of resprout silviculture to assist you with design work.

In that case, you may contact experienced farm and agroforestry design professionals who do have the technical know-how to aid you in your process. If you're a novice, the input they provide will likely save money, time, and energy. It's always a good idea to get a fresh set of experienced eyes on your project, so at the very least, consider consulting with an experienced practitioner if you're new to this realm.

GOALS AND MANAGEMENT OBJECTIVES

Goals drive design. They describe your motivations and needs and demand clear, refined vision. They identify values and provide context and constraints. They inform an action plan and offer valuable guidelines for monitoring and reflection. You simply can't do design well without well-articulated goals.

Clear written goals require us to identify the materials, techniques, timelines, and outcomes we desire and require. And they enable us to communicate them to others. So how do you start? Well, presumably, many of you have already begun.

What do you need from your landscape, and what are you passionate about producing? Perhaps you make a particular craft. Or you'd like to grow all your own heating fuel. Are you supplementing livestock feed with woody browse? Or growing propagation stock for a nursery? Is this a hobby, a side gig, or are you planning to build a full-on business?

As you begin to formulate a list of your primary goals, try to keep your options open by focusing on the function rather than specific forms. For example, instead of "growing willow withies for basketry," you simply "manage short-rotation coppice for small-diameter rods." This broader framing opens up the possibility of producing live stakes for arborsculpture and bank stabilization, stock for artist's charcoal, materials for woven fencing and walls, and stakes for all sorts of gardening needs, and leaves you much more open to research whether there are other species better suited to your needs. So, while you should certainly begin by identifying the specific forms you're most drawn to, keep your options open so that you may perhaps discover new and unknown uses for your raw materials.

And don't forget to consider the needs of people and businesses in your community and bioregion. Is there an existing demand for materials or products you can supply using coppiced wood? Will you need to educate customers about the strength and utility of these types of materials? Answering questions like these helps focus and hone our options as we proceed towards design.

You'll probably find that you'll need to develop two "tiers" of goals: broad general goals that articulate your long-term vision for the landscape and your own life as the land manager and stand-specific goals that identify the qualities of each uniquely managed patch in your landscape. This division helps identify specific near-term task work within the long-term context of a well-planned timeline.

As you hone in on more specific goals, you'll presumably begin to identify the actual quantities of materials you'll need. How many cords of wood do you need each winter? How many chairs, baskets, fences, etc. do you aim to produce each year? This level of detail begins to ground truth your vision, helping you determine the area and rotation length you'll need to meet your production goals.

MAPPING

A good design begins with a detailed base map that denotes the key features on a site and provides the foundation for thorough site analysis and design work. A base map drawn to scale allows you to accurately calculate the size of a prospective coppice cant or the length of a hedgerow, project anticipated yields, determine the number of trees needed for planting or the length of road required to reach a management area. Without a base map, you're flying blind.

Begin by locating a boundary survey that denotes shape and extent of your property. If you don't already have one, check with the local town hall or municipal offices where a survey should be available upon request. Unfortunately, surveys usually provide minimal information: access roads, boundaries, notable landscape features (fences, stone walls, etc,), existing buildings. It's also worth reaching out to local colleges and universities, agricultural extension agencies, your county forester, or the NRCS as they should have access to software that enables them to generate an accurate base map with high-quality aerial imagery, parcel boundaries, topography, and more using GIS mapping technology.

GIS Mapping

Various types of GIS mapping tools are becoming increasingly available to the public. GIS, Geographic Information Systems, is a method of cataloging and organizing geographic and spatial data so it can be represented and analyzed in map form. GIS maps allow users to overlay individual data layers and examine patterns between them. Many US states have their own web-based portal that allows users free access to various clickable data layers (property boundaries, topography, hydrology, geology, soils, historical imagery, etc.), measure area and length, and create a simple base map. This is a great place to start your design process if your locale provides this service. Try a web search for your state/province/county name and GIS. Additionally, Google Earth Pro offers the general public a powerful mapping and design tool that integrates well with available GIS datasets. And for the computer savvy, QGIS is a popular open-source GIS program that offers powerful functionality without the significant purchase and licensing costs of professional software like ArcGIS.

One of the benefits of GIS mapping is the ability to quickly and accurately measure length and area. It's easy to discover the length of a proposed access road or the area of an existing field or forest stand. This allows you to play with different design solutions and analyze their suitability in meeting your goals.

A GPS-enabled (Global Positioning System) device allows you to identify waypoints (marker trees, property boundaries, wellheads, exposed ledge, etc.) or catalog tracks or pathways you can later upload to a GIS-generated map. If you want to delineate boundaries between various forest cover types in an existing woodlot, use your GPS tool to track your path as you follow the boundary between the two stands. You could also use a track to mark out the location of field/cant boundaries, hedgerows, access routes, etc.

Handheld GPS devices are relatively inexpensive, although the cheaper the unit, the lower the accuracy. For rough GPS data collection, a number of different smartphone apps provide roughly two-meter accuracy, assuming you've got adequate cell or Wi-Fi service. While the costs of high-resolution GPS devices (with up to sub-centimeter accuracy) have rapidly fallen with improvements in computing technology, they're still too expensive for most landowners to justify. If this is something you believe your design requires, contact a local surveyor, civil engineer, landscape design professional, or excavation company as they may be able to provide you with these services. But

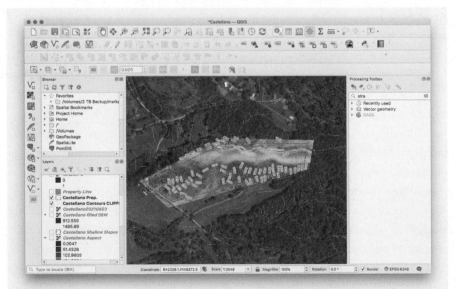

FIGURE 6.3: Screenshot of a base map created by Mark Angelini using the open-source GIS program QGIS. Photo credit: Mark Angelini, Sedalia Sloyd and Mountain Run Permaculture, used by permission.

in most cases, we don't need nearly that level of accuracy for this type of planning work.

Mapping Forest Stands

On forested properties, you may want to map the extent of forest cover types or ecological communities. For most readers, a consumer grade GPS unit or cellphone app will often provide decent enough accuracy, but you can also map the forested stands on your landscape the old-fashioned way. You'll need a magnetic compass and a knowledge of the length of your pace.

For scale's sake, remember that an acre is 208.7 feet by 208.7 feet. That's roughly 4,840 square yards, 70 yards by 70 yards. Assuming an average pace of 1 yard, one side of an acre is about 70 good paces. Knowing your pace allows you to quickly measure stand sizes, rough out coppice cants, or estimate the number of trees or shrubs required for a planting.

Begin at a known point that you can confidently locate on the map. Take a compass bearing that directs you through the area you'd like to map. If property boundaries are linear, consider choosing a bearing that parallels the property line.

Starting from your known point with your compass adjusted, sight along the compass to a clear landmark and walk towards it. Either run a tape measure along this line or track your paces as you go, taking note of changes in species, age distribution, slope, or other significant landscape features. When you reach your landmark, use the same compass bearing to locate your next destination. Continue this same process from point to point, leapfrogging your way along the line. Don't forget to record your observations along the way. By taking multiple transects that run parallel to one another at predetermined distances, you'll develop a data set that illustrates the distribution of ecological communities and features across the landscape.

However you choose to generate a base map, the most important thing is that you have one to guide your design. It should include all of the most important features on the landscape, including primary access points, existing structures, property boundaries, water features, and field and woods boundaries. With base map in hand, we next analyze the resources, challenges, and opportunities on our sites before starting to determine how to best make use of them.

SITE ANALYSIS

Ecological design is built on in-depth understanding of the landscape. While our goals guide, inform, inspire, and constrain the design from one direction, the site pushes and pulls in both similar and different ways. A successful design must respond to the site's ecological realities and emerge from the synthesis of this analysis.

While there's no substitute for on-the-ground observation, don't forget to utilize the myriad printed, electronic, and human resources available. The type of GIS mapping tools we discussed earlier including topographic and soils maps, weather and climate data, property boundaries, potential hazardous sites, previous ownership, and land use data can often be found in online databases or at local municipal

UAV/Drone-based Mapping, by Mark Angelini, Sedalia Sloyd and Mountain Run Permaculture

As with GIS, consumer-grade drones are increasingly prevalent and relatively affordable. An aerial photograph of your property—what used to take either a pilot's license and airplane or a high-dollar service—can now be taken as-needed with a piece of equipment costing anywhere from several hundred dollars to just over $1,000. And with a few extra pieces of software, the drone can be turned into a mapping powerhouse with uses far beyond a simple aerial photograph.

Drones are unmanned aerial vehicles (UAVs). Most consumer and commercial-grade drones are propelled by four horizontal propellers, equipped with a high-resolution photo and video capable camera, GPS, and a suite of intelligent features such as obstacle avoidance. They're operated via a radio controller (RC) linked to a smartphone or tablet. The phone or tablet operates the flight software while the controller communicates with the drone, similar to any other RC vehicle.

There are many drones on the market today, from cheap toys to wildly expensive commercial drones used in high-precision surveying and broadacre farm monitoring. For the average grower or consultant, the best models from a cost/function standpoint are made by DJI. They have several folding, highly portable models which I recommend. On the lower end is the Mavic Mini for around $400, or the Mavic Pro 2 for closer to $1,800. Both are well made and packed with features and can be found on the secondhand market in great condition at a very reasonable price.

So, what is drone mapping? It's the process of combining the autonomous functions of the drone, its camera, and GPS capabilities with software to make high-resolution and accurate aerial maps. You simply load flight planning software onto the smartphone or tablet—Drone Deploy or Pix4dcapture are some of the easiest to use—create a flight plan outlining the area you wish to map, then send the drone on its way. The software will autonomously guide the drone along a predetermined flight path while capturing images at a set interval to the drone's memory card. You'll be left with a series of overlapping aerial photographs.

To process these photos into a coherent map, they need to be run through an aerial map processing software. There are many options out there, but one of the best and most affordable is a web-based service called Maps Made Easy. The service is free for small sites, and around ten cents per acre for larger sites. Simply upload the photos and follow a few prompts to get them processed, which usually takes half an hour for small sites and a few hours for larger sites. The processed image, called an orthophoto—meaning a geometrically corrected image with uniform scale—is a high-resolution scaled aerial image of the land. This can now be printed and used as a base map, or for further analysis and design processes, loaded into a GIS or CAD software.

While drone mapping is a somewhat technical process, it will yield a very high-quality result at a considerable discount to a survey or comparable ☞

service. And beyond the map making functionality of a drone, there are several other ways a drone can be a valuable tool for design and assessment work.

- The drone can be used seasonally to monitor growth and changes across a site, both in photo and video form. There's something powerful and gratifying about seeing an aerial photo of a site before development and after 1, 5, or 10 years.
- The high-quality camera can be used to make photographs and video for education or marketing purposes. Just like with monitoring a site's progression over time, documenting seasonal tasks

from the bird's-eye view helps communicate a lot about managed landscapes.

- Through a suite of more advanced software processing tools called photogrammetry, the drone's images can be used to produce elevation maps, plant health maps, and several other analytical datasets.

Lastly, be aware that there are flight regulations that apply to the legal and safe operation of a drone, especially where there is an intent to engage in commerce. Familiarize yourself with these regulations before purchasing or operating a drone.

offices. But don't forget about the neighbors! Tapping into the deep experience and knowledge of longtime residents of an area can teach you about localized weather patterns, land-use history, and the challenges and potential of local land management.

Landscapes are immensely complex. We can learn a lot about a place by even just casually ambling off and taking in information as we observe it. But this approach leaves much to chance and little likelihood of a comprehensive survey. Without a strategy to organize and contextualize our observations, they can get lost in a bottomless sea of information. My own design practice grew leaps and bounds after learning a strategy to filter, focus, and guide my observations. By breaking the landscape into parts, our design process becomes more systematic and fluid.

Yeomans' Scale of Permanence

Australian mining engineer and farmer P.A. Yeomans (whose work we'll explore in greater depth in later chapters) developed comprehensive design methodology called Keyline Design in the 1950s and 60s. He sought to improve land-use planning by developing a strategic approach to the design of landscapes based on their topography. Placing particular emphasis on regenerative soil and water management, Yeomans organized his system using his ingenious "Scale of Permanence."

The Scale of Permanence provides designers with a categorical continuum of the features and influences that shape a landscape. It arranges these characteristics based on the time and energy required to modify them. The Scale of Permanence has been adapted by a number of practitioners, but the overall concept and the key pieces remain the same. We won't get into the details here; they're well fleshed out in Darren Doherty's *Regrarians Handbook* (in press) and *Edible Forest Gardens*, Volume 2.[5] Table 6.1 provides a hybridized version of the Scale of Permanence, modified to fit our context designing resprout silviculture systems.

You can use the Scale of Permanence to guide your site analysis by applying a technique called data overlay. Chart your observations for each of these characteristics on individual sheets of tracing paper. You'll often group similar characteristics together. For the GIS capable, create separate data layers for each element of the scale.

With your in-the-field analysis complete, toggle between different combinations of these datasets and look for patterns. Where do access and microclimate and soils all preclude good planting sites for intensively managed sprout stands? And where does the existing vegetation, its topography, and proximity all point towards the ideal site for your installation?

Distill the essence of each of these layers of data onto a single map. This is not an easy process. Give yourself a few drafts before you try to perfect it. Where might you make the least change for greatest effect, and which challenges require the most immediate attention? Use bright, bold colors that make your observations stand out. You're telling the story of the site. After a couple of drafts, you'll have distilled a dazzling diversity of complex observations onto a single sheet of paper (okay, or maybe a few). This deep and thorough site analysis process reveals clear landscape and land-use patterns that point us towards the overall layout of our design.

Because this book focuses on the design and management of resprout silviculture systems, we'll specifically examine some of the most important factors for you to consider.

Table 6.1: Scale of Permanence

Most Permanent/ Greatest Amount of Energy and Effort to Change	Climate
	Landform/Geography
	Water
	Access and Circulation
	Vegetation and Wildlife/Forestry
	Microclimate
	Buildings and Infrastructure
	Zones of Use/Fencing
	Soil Fertility and Management
	Economy
	Energy
Least Permanent/Least Amount of Energy and Effort to Change	Aesthetics and Experience of Place

Based on P.A. Yeomans' original Scale of Permanence; also incorporating Darren Doherty and Dave Jacke's revisions/additions.

Stand-specific Assessments

For people interested in woodland conversion to coppice management, understanding a site's soils; aspect; accessibility; tree age, height, and diameter; and stand stocking tell us a lot about its ability to yield wood products. Sites with good access and favorable conditions for tree growth may produce as much as two or three times more wood annually than poor sites.[6]

Soils

Soils affect productivity, access, water retention, erodibility, likelihood of seedling survival, and windthrow probability.[7] Texture, structure, horizons, organic matter, pH, fertility, biology, and compaction are some of the most important soil properties to consider. In the United States, virtually the entire country's soils have been identified and mapped. You can access this data using the USDA's Web Soil Survey. There you can identify your soil type and uncover loads of information on its suitability for all sorts of applications. Once you've identified and researched your soil type, it's a good idea to build on this information by doing soil testing for any areas you're considering new plantings. Private labs and many agricultural universities provide soil testing services for an average of $15 to $25 per sample.

We don't have space for a detailed overview of soils and their management, but it is important to understand at least a few fundamental characteristics and the ways they influence productivity, species composition, access, and more. We first introduced the concept of the site index in chapter 2 and revisited it in greater detail in chapter 4. But essentially, the character and quality of a site's soils is one of the key factors that determines productivity. In other words, better soils grow healthier trees faster.

In general, the deeper the soils, the stronger and more robust the vegetation. First and foremost, this relates to the depth of the topsoil or the "A horizon"—the shallow mantle of biologically active soil that supports upwards of 80% of all plant roots in temperate ecosystems. Unfortunately, several centuries of clear-cutting, overgrazing, tillage, and general mismanagement have caused severe topsoil erosion around the world, and many of our landscapes are just a shell of what they were pre colonization. So be sure to treasure whatever precious topsoil remains on landscapes you steward, keep it covered with living and decaying organic matter as much as possible, and continue to encourage biomass cycling by allowing leaf litter and coarse woody debris to accumulate and decay.

A site's overall soil depth is another key limiting factor and is due to the presence of restrictive layers like a compacted hardpan, or shallow bedrock. These features tend to restrict drainage, leading to a high water table, limiting the success of many species. Shallow soils also restrict rooting depth, limiting water and nutrient uptake and rendering plants more susceptible to windthrow. Sites with an undulating "pit and mound" topography often indicate poor drainage or shallow bedrock.[8] This type of landform develops when overturned root systems of wind-topped trees leave hollow pits adjacent to mounds of soil that develop as their root ball decays.

Soil texture also has a major influence on the species that colonize a site and should strongly influence the species you select for your design. The bulk of soil's volume of solid particles (remember that up to 50% of a healthy soil is actually the pore spaces between soil particles) are comprised of clay, silt, and sand—soil's literal building blocks. Most soils will at least have some proportion of all three of these particle sizes present, and they each contribute different characteristics to the overall matrix. But not all soil types are good for all species. Sand provides excellent drainage but is prone to drought and nutrient leaching, making it difficult to sustain health and fertility. Clay is far better at retaining nutrients, but its dense, plate-like microscopic particles restrict drainage and limit available oxygen. An "ideal" soil contains a mixture of all three particles, providing good drainage and water-holding capacity as well as an effective reservoir for mineral nutrients. But few of us are blessed with these optimal loamy soil types, so we should adapt our designs accordingly.

Start by identifying which native species already colonize similar soils in your area. This is your best guide. Many species can deal with excessive drainage but dense, heavy soils can be far more limiting. When selecting species for these clay-rich soil types, choose carefully. The roots of many species adapted to upland sites can't tolerate periodic inundation and anaerobic conditions, so consider choosing species that tolerate these conditions like alders, willow, swamp oaks, silver and red maple, bald cypress, dogwood, sycamore, and poplar, to name a few. Also, on these heavy soil types, be sure to provide thorough ground preparation to break up compaction and improve drainage.

The existing vegetation on the site also reveals many clues about the nature of the soil that lies underneath. **Indicator plants** like sedges, alder, plantain, and Queen Anne's lace provide numerous clues about the soil's texture, hydrology, fertility, and pH. By taking the time to do some careful species analysis and becoming familiar with plant preferences, we can begin to understand the types of soils we're dealing with before we've even dug our first hole.

And despite all this, remember not to let the perfect be the enemy of the good. A 1980 Kansas study exploring the economics of establishing a biomass energy plantation found that stands planted on the best soils required the most intensive weed management, and over the course of 5 years produced lower growth totals than stands planted on upland sites.[9] Everything works both ways—sometimes even soil texture and fertility.

Aspect and Elevation

Aspect—a site's slope orientation in relation to the sun—has a marked effect on productivity. In the northern hemisphere, sites with a southerly aspect more directly intercept solar radiation, compared to the shady faces of north-facing slopes. This means longer, hotter days—often perfect for maximizing production in agricultural applications like market gardens, small fruit production, and many types of orchard culture. But this can also lead to excessive evapotranspiration and water stress, causing droughtiness (especially in moisture-limited regions) and compromising a stand's growth potential.

Similarly, a site's elevation may affect both the texture and depth of the soils, along with climatic factors like temperature and wind patterns. Lower bottomland elevations tend to have much deeper, well-drained soils than the upper slopes of hill and mountain country, where shallow bedrock and thin soil can slow and impede growth. Upland sites experience cooler daytime temperatures, but bottomland zones can become frost pockets, collecting cold nighttime air as it settles overnight. Once again, everything works both ways.

Aspect's Effects on Productivity

FIGURE 6.4: Hillsides with opposing aspect in eastern Bulgaria, near the Macedonia border. The mountainside in the background is north-facing, making it cooler but also helping conserve moisture. It's primarily sweet chestnut. In the foreground lies an oak-dominated woodland. Because of its generally south-facing exposure, it's much hotter, drier, and less productive despite being just a few miles away.

FIGURE 6.5: A screenshot of the Sun Seeker app's 3D view that allows you to project the sun's path at different times of year wherever you're standing in the field (of course you need a cell or Wi-Fi signal). The bottom left line (which is blue in the app) indicates the sun's path on the winter solstice, the middle line (green in the app) the equinoxes, and the topmost lines (yellow and red, respectively in the app) represent the sun's path on the current day and the summer solstice. This screenshot was taken just a few days after the summer solstice, so these two topmost lines are almost perfectly aligned. This tool helps you understand site-specific day's length details so that you can plan your designs accordingly.

In many areas, even the temperate northeast, lower northeast-facing slopes provide some of the most productive sites for wood production because they tend to have deep soil and remain moist.[10]

Of course, we don't get to choose our site's aspect or elevation. It's something we must respond to. So, taking time to consider the benefits and drawbacks of different orientations can help you steer your plans to the best-adapted species and land-use practices. I'm a huge fan of the Sun Seeker smartphone app. In addition to a number of other solar analysis functions, it allows you to use your phone's camera to illustrate the arc the sun follows throughout the year. It takes much of the guesswork out of aspect interpretation.

Tree Height, Diameter, and Age

These next few observational criteria are especially useful to readers interested in transitioning woodlands into resprout silviculture managed systems. The stature of trees in a forest patch helps us understand productivity, successional stage, soil characteristics and their suitability for coppicing. Tree height, diameter, and age are some of the key species-level characteristics to focus on.

Perhaps more so than diameter, tree height is an important indicator of a forest stand's productivity. When forest stands grow densely, we see slow increases in diameter due to the competition for resources. Tree height is largely unaffected by stand density, so it serves as a more reliable indicator of plant vigor and site quality.[11] This is why the site index is actually a function of height.

Diameter

A tree's diameter helps indicate its maturity and provides an estimate of the volume of wood it contains. Because a tree's diameter varies with height, and many species' stems tend to flare towards the base, we measure it at breast height to maintain consistency. The easiest way to measure diameter at breast height (dbh) is by either holding a log rule horizontally against the stem and carefully sighting it across the face of the trunk, or using calipers or a diameter tape measure. If you'll be doing any significant tree felling, a diameter tape measure is really handy. Usually, one side of the tape is used to measure length, while the other is scaled to measure diameter. They can be easily clipped to your belt loop so it's at the ready when you need it.

Open-grown trees put on far more overall growth than trees growing in a dense forest stand, especially in terms of diameter. In the UK, most open-grown healthy broadleaf trees increase 1 inch (2.5 cm) in girth (circumference) per year. At this growth rate, an open-grown tree with a 6-foot (2.4 m) circumference (~30 inches or 75 cm in diameter) would average about 100 years in age. In forested conditions, competition for light, water, and mineral resources can cause a 50% reduction in growth rate to 0.5 inches (1.5 cm) per year or less.[12]

Age

Tree age can be difficult to estimate accurately. Soil fertility and site quality, precipitation patterns, length of growing season, microclimate, and species characteristics all affect a plant's

Table 6.2: Relationship between age, diameter, and girth for open-grown mixed broadleaf tree species in the UK (assuming 1 inch or 2.5 cm/yr increase)

Dbh	4″ (10 cm)	8″ (20 cm)	12″ (30 cm)	16″ (40 cm)	20″ (50 cm)	24″ (60 cm)
Girth/Circumference	12.2″ (31 cm)	24.4″ (62 cm)	37″ (94 cm)	49.6″ (126 cm)	61.8″ (157 cm)	74″ (188 cm)
Estimated Age	12	25	37	50	63	75

Adapted from Agate, 1980, p. 43.

growth rate. Tree growth rates also vary with age. Many species put on considerable annual growth during the early phases of development, steadily slowing as they increase in age and diameter.

The only way to definitively determine a tree's age is in cross-section—either by felling the tree or removing a core from the tree's stem using a tool called an increment borer and counting the number of growth rings. Although we know that most woody plants resprout most vigorously when they're in their juvenile stages, we don't need to know their precise age to make good decisions. For most species, the prime window to begin coppicing is when they're 8-to-10 inches (20–25 cm) dbh or less.

Stand Stocking, Structure, and Canopy Cover

These qualities are all important patch-scale woodland characteristics. "Stocking" is a measure of a forest stand's density, while stand "structure" refers to the distribution of age or size classes of trees in a stand. We measure stocking in a few ways: the total number of stems per acre, the basal area per acre, and the canopy cover.

Table 6.3: Estimated age at which forest-grown trees reach 4-inch dbh increments (in the UK)

	4″ (10 cm)	8″ (20 cm)	12″ (30 cm)	16″ (40 cm)	20″ (50 cm)
Beech	35	55	75	95	115
Oak	30	50	70	90	120
Sycamore, Ash, Cherry	20	30	40	60	
Birch, Alder	15	25	35		

Adapted from Agate, 1980, p. 44.

When measuring the number of stems per acre, we break them up into the same age class groupings we discussed in chapter 4 (seedling; sapling; polewood; small, medium, large sawlog). Although this grouping helps us express the distribution of tree age and size in a patch, it doesn't give us an easy way to compare density between two different stands.

Basal area is the sum of the cross-sectional area at breast height of all of the trees in a stand of a known area, extrapolated to the total number of square feet of trunk per acre (or square meters per hectare). The higher a stand's basal area per acre, the greater the total amount of wood contained in the stand. See Figure 6.6.[13]

To calculate stand basal area, measure a tree's dbh and convert it to area using the equation ($\prod r^2$). Do this for all of the trees in a known area and convert it to square feet of basal area per acre to determine the total basal area. As you might imagine, this can be a painstaking process.

Fortunately, a prism (and several other forestry tools) can provide a quick and relatively accurate measure of stand basal area.

To use a prism, choose a fixed point in a forest patch. Holding the prism at eye level and arm's length, align it with a nearby tree. The prism refracts the light passing through it, causing the stem section within the prism to appear offset. If the section of the stem visible within the prism still lines up with the stem above and below it, the tree counts as a "hit." See Figure 6.7. If the stem section within the prism is so offset that it lies outside the stem's plane, it's a "miss." Rotate 360 degrees around the fixed point you've chosen and tally the total number of hits. Multiply this number by the basal area factor (BAF) of your prism (usually either 5 or 10) to get a rough measure of the stand's total square feet of basal area per acre.

FIGURE 6.6: Basal area per acre gives us a valuable metric to measure density in a forest stand. Essentially, it is the sum of the cross-sectional area at breast height of all the trees in a one-acre stand (depicted by the white circle cross-section of each tree). The figure is expressed in square feet basal area per acre (or m^2 basal area per hectare).

Together, trees per acre and basal area per acre characterize a forest stand's state at a point in time. It's quite possible that two very different stands could each have the same basal area. One might be populated by a great number of small-diameter trees, while the other has just a handful of large mature trees. For example, an even-aged stand with 325 12-inch (30 cm) dbh stems has the same basal area per acre as 1,300 stems with 6-inch (15 cm) dbh (100 ft²/acre or 23 m²/ha basal area).

Theoretically, each forested acre has a fixed potential productivity. This means that each unique site has a limited carrying capacity.

Figure 6.7: Two views through a forestry prism. This ingenious tool allows you to gain a quick and generalized figure for current stand basal area. The image on the left shows a hophornbeam tree (*Osytra virginiana*) that would not count as part of the tally since the stem as viewed through the prism doesn't intersect with the trunk above and below. The white oak (*Quercus alba*) on the right, on the other hand, would count because the refracted view of the trunk still intersects with the actual stem. Tally all of the hits (or trees that count) around a 360-degree radius, and multiply by 10 (this prism has a Basal Area Factor [multiplier] of 10) to get a general idea of the relative stocking density of the patch in square feet of basal area per acre.

Table 6.4: Target basal area per acre ranges for two different fully stocked stands

Average dbh (Inches)	Northern Hardwoods Sawlogs	Spruce/Fir Sawlogs
5–9	55–75 ft²/basal area/acre	70–120 ft²/basal area/acre
10–13	75–90 ft²/basal area/acre	120–170 ft²/basal area/acre
14+	90–95 ft²/basal area/acre	170–200 ft²/basal area/acre

Adapted from Beattie, 1993, p. 109.

This growth potential can be distributed any number of ways among the trees present. Crowded sites with dense tree growth contain many small trees because of the intense competition between individuals for light, water, and nutrients.[14]

In even-aged stands, tree growth progresses steadily until the expanding crowns reach canopy closure and competition for light causes a decline in the growth rate. Slower-growing trees begin to die out. While the basal area of the stand steadily increases, the density of tree stems determines the rate of trees' diameter growth. As the stand begins to thin due to tree death or removal, the total annual accrual of basal area is concentrated between a smaller number of remaining trees. This thinning leads to a faster increase in stem diameter.

Stocking density is a comparison of current stand density in basal area to an ideal that depends on the management objective. Stands are classed as either over-, under-, or well-stocked. The optimum stocking for a stand varies with stem diameter. Stands stocked with

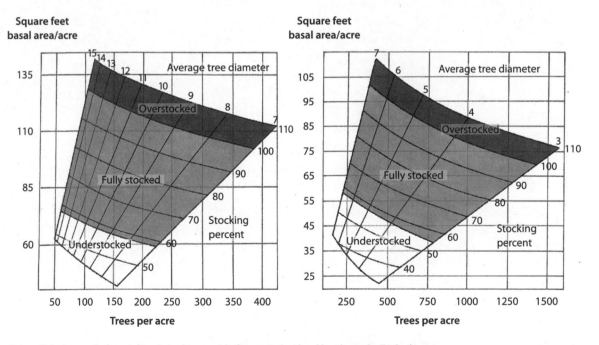

FIGURE 6.8: A sample forest stand stocking guide for central upland hardwoods. Each chart shows the relationships between average tree diameter, the number of trees per acre, and the stand basal area. With two of these bits of data, it's possible to determine the relative stocking density of a forest stand and whether it's over-, under-, or fully stocked. The chart on the left is used for stands where the average tree diameter is 7 inches or larger, while the chart on the right is used for stands where average tree diameter ranges from 3 to 7 inches dbh. Adapted from Strobl, 2000, p. 108.

larger trees will have fewer trees per acre though the optimal basal area increases.[15]

Given the predictability of this pattern, research foresters have developed stocking guides that describe the optimal stocking levels for different tree species. These guides include three lines that depict benchmarks for foresters to compare a stand's current stocking against a projected ideal. The "A-line" indicates a concentration of too many trees for desirable basal area growth; these stands are overstocked. The "C-line" describes stands with good diameter growth per tree but too few trees to maximize the stand's potential, an understocked stand. The "B-line" depicts the optimal stocking level, maximizing individual tree growth rate and per acre basal area growth. Stands stocked satisfactorily lie between the A and B lines while those between lines B and C should reach adequate stocking within 10 years. To use the stocking guides, you need estimates of two of three stand qualities: stand basal area per acre, number of trees per acre, or mean stand diameter. See Figure 6.8.

Canopy cover is another measure of tree density. It describes the percentage of an area covered by a canopy. In a stand with 60% canopy cover, only 40% of total available sunlight penetrates the forest canopy. This available light is a critical factor for seed germination and seedling growth, and it's also an essential guide of appropriate tree density to ensure silvopasture systems receive adequate sunlight for pasture growth.

Landscape Scale Pruning: Stand Structure

More often than not, traditional coppice systems were comprised of two or three layers of vegetation: an even-aged canopy (or standard tree-based canopy and even-aged coppice mid-layer) and a ground cover layer. By utilizing seed tree and shelterwood cutting strategies, you may steadily shape stand structure by opening sizable gaps and leaving either individual trees or pockets of woodland growth. This "lumpier" landscape texture will support more overall species diversity and a more complex ecosystem overall.

Despite the ease of management of even-aged monoculture coppice stands and coppice-with-standards, a more varied system will support more species and structural diversity and be suited for a range of different productive landscape uses. Depending on your goals, your target canopy cover may vary widely. Polewood producing silvopastures usually call for wide tree spacing with canopy cover as low as 15% to 30% and rarely more than 50%. For intensive coppiced polewood production, you'll be creating shifting patches with little to no canopy cover and perhaps with just a few high-value standard and/or seed trees to help repopulate the understory. It's a continuum. At one end, you have an open field and, at the other, a closed canopy forest. Your goals determine which pattern best suits your needs. Think of it as pruning at the stand or even landscape scale.

Wood Volume

Volume measurements provide useful quantitative data to compare stand productivity between species, sites, and climate zones; guide management decisions; and calculate the economic value of the wood contained within a stand. Most of the world expresses wood volume in cubic meters and the volume of a stand in cubic meters per hectare. In the US, we use board feet to express the volume of wood as if it were converted into lumber. One board foot is equivalent to a length of wood 1 inch thick, 12 inches wide, and 12 inches long. Conversely, one board foot could also be a board 2 inches thick, 6 inches wide, and 12 inches long. We can also express wood volume in cords (128 ft^3) or by weight. We'll generally use volume measurements throughout this book as a means of comparison between coppice stands.

Calculating wood volume in a stand can be relatively complex. If you only need a rough estimate, use Table 6.5.

In existing coppice stands or in woodlands already showing some form of stump sprouting, estimating the total number of rods or poles can provide another indicator of site quality and offer a projected yield (poles/rods per acre) that is perhaps more useful to the coppice craftsperson than a board foot/volume measurement. Start by determining the spacing between stools both within and between rows to figure out the number of stools per acre. Then find the average number of rods or poles per stool. Multiply these two figures to get the number of poles per acre. Table 6.6 indicates the product yield potential of various types of coppice (in the UK). This type of data can be very helpful in projecting a stand's economic prospects.[16]

SYSTEM DESIGN

So, everything we've discussed to this point has involved a massive information intake. We've identified and honed our production goals.

Table 6.5: Rough wood volume totals per log length and dbh in cubic feet (cubic meters in parenthesis)

Log Height	4" (10 cm)	8" (20 cm)	10" (25 cm)	12" (30 cm)
9.8' (3 m)	.85 (0.024)	3.3 (0.094)	5.2 (0.147)	7.5 (0.212)
13.1' (4 m)	1.1 (0.031)	4.4 (0.126)	6.9 (0.196)	10 (0.283)
16.4' (5 m)	1.4 (0.039)	5.5 (0.157)	8.7 (0.245)	12.5 (0.353)
19.7' (6 m)	1.7 (0.047)	6.6 (0.188)	10.4 (0.295)	15 (0.424)
23' (7 m)	1.9 (0.055)	7.8 (0.220)	12.1 (0.344)	17.5 (0.495)
32.8' (10 m)	2.8 (0.079)	11.1 (0.314)	17.3 (0.491)	25 (0.707)

Adapted from Agate, 1980, p. 45.

Table 6.6: Projected yields of different coppice systems in the UK

System	Age	Stools/ Acre	Average Stool Spacing	Projected Rods/Acre	Total Product per Acre
Neglected Hazel	19 years	400	10' x 10'	4,500	2 cords, 1000 stakes, 60 bundles pea sticks, 80 bundles thatching rods
Managed Hazel	9 years	600	8.5' x 8.5'	11,000	300 wattles, 250 bundles pea sticks, 10 bundles bean rods or 500 bundles of thatching rods
Managed Chestnut	16 years	590	~ 8.5' x 8.5'	2,350	6.5 cords (roughly 8 tons) fuelwood, 1500 stakes, 603 bundles pales, 30 bundles pea sticks
Mixed Species Coppice	17 years	440	~ 10' x 10'	3,600	4 cords (roughly 6 tons) fuelwood, 800 stakes, 60 bundles thatching rods, 30 bundles pea sticks
Mixed Species Coppice	50 years	320	~ 11.5' x 11.5'	2,620	24 cords (roughly 30 tons) fuelwood, 150 stakes, 15 bundles thatching rods

Adapted from Tabor, 1994, p. 155.

We've cataloged the resources, challenges, and characteristics of our site as a whole, deepening our understanding of the site and the specific characteristics of unique patches in the landscape. This is where it all comes together and we begin to make decisions both general and specific. And here's where the systems framework we discussed in chapter 4 becomes an essential guide.

As you may recall, the five main characteristics outlined in the systems framework include:

- Species and their characteristics
- Products and purposes
- Disturbance severity/practices
- Disturbance chronology/management timing/age structure
- Landscape pattern

How have your production and management goals and your site analysis helped point you towards more concrete decisions for each of these characteristics? You'll likely need to do a bit more research at this point, but you also probably have a pretty clear idea of your preferences and needs in some of these arenas. It's not a linear process. Begin with what you know and hone in on what you don't.

Consider making a photocopy of the systems framework and writing in or circling those details you already know, while making notes about the information you're lacking for each characteristic. For example, if you know you'd like to grow your own firewood (products and purposes), you'll probably want to choose to plant either species known for rapid growth and dense wood or sprouting species already present and thriving on your site, even if they're not the most perfectly suited for your needs. In this case, your end product (fuelwood) and your starting point (renovating existing woodland or starting from scratch) really help you narrow your species selection options.

And because you're planning to grow fuelwood, you're probably going to want to harvest polewood rather than short-rotation resprouts. For most of us, this means that we'd be managing the system by coppicing entire stools (Disturbance severity/practices) using patch-scale clear-cuts (Disturbance chronology/management timing/age structure and landscape pattern) to harvest the material on an 8-to-20-year rotation.

If you've been able to get specific about your production goals, you can proceed to the next level of detail and begin to map out how this disturbance chronology interacts with the landscape pattern. If you anticipate a mean annual increment of ½ cord of wood per acre per year, you can work in several directions to determine if you have adequate land to meet your goal, how big your patch cuts should be, how long your rotation needs to be, and how frequently harvests need to occur. With this information, you can begin to compartmentalize your landscape into stands harvested on a consistent rotation.

You may well find, after navigating the systems framework, that you ended up where you already knew you were headed—in this case, a fuelwood producing coppice system with … sized cants on a … year rotation. But hopefully the process helped clearly guide you through your decision-making, considering

each characteristic of the system one at a time. And perhaps it also helped you to contemplate different strategies to reach your goals or recognize new opportunities for complementary products that you weren't previously aware of.

In earlier chapters, we've discussed many of the considerations for each characteristic of the systems framework. However, we've yet to examine the spatial and temporal qualities of resprout silviculture from a design perspective.

Cant Size, Rotation Length, and Layout

If you're designing a more traditional coppice system, you'll need to determine the size of each cant, your rotation length, and the physical layout of your system. Additionally, your production goals, land base, slope, aspect, access, and species composition will all help guide your design. There are no hard and fast rules, just common sense guidelines. If you've taken the time to clearly articulate your goals, analyze your site, and consider how all of these pieces fit within the systems framework, cant design becomes infinitely easier.

Let's start off with scale. A coppice cant could be as small as an individual plant (that receives close to full sun) and as expansive as several acres. Cants generally range from 0.25 acres (0.1 ha) to as large as 3 acres (1.2 ha). The smaller the patch, the more considerable the effects of shading from neighboring trees and the more vulnerable to wildlife browse.[17] In fact, in many cases, cants of less than 0.5 acres (0.2 ha) are generally inadvisable because they're unlikely to receive adequate sunlight for good regeneration.[18]

Your chosen species, production goals, and land base provide the context to determine how many cants are best. If you have a projected annual material need, you can extrapolate from the available yield data to determine the optimal annual cant size. You could also base cant size on the projected timeframe for sprouts to reach market size.[19] If you want 10-year-old materials and have 6 acres (2.4 ha) available, 10 0.6-acre (0.24 ha) cants could provide your annual harvest. This may be too small an area to meet your needs in any given year, so you may instead need to adjust your rotation to every other year (or every third year).

🕒 Or perhaps you manage just a single cant for fence post production on a 15-year rotation. Or a nursery and landscaping business that requires stout 5-year-old tree stakes, managing two 0.5 acres (0.2 ha) cants on a 5-year rotation. A 6-foot-by-6-foot stool spacing amounts to 1,210 stools per acre. Assuming 3 usable poles per stool, this stand produces roughly 3,600 poles per acre. If each pole produces a single stake per 5-year rotation, they could harvest 1,800 tree stakes every 2 to 3 years.

In the UK, good-quality hazel cut for wattle hurdles is usually managed on a 6-year rotation. In southeast England, hornbeam is cut for firewood on a roughly 20-year rotation, and oak for charcoal and bark (for tanning) every 20 to 25 years.[20] Basket willow rotations seldom run more than 1 year, whereas chestnut with its many uses, may be as short as 5 or 6 years and as long as 20 to 25, and in some cases even longer.

For woodlands 15 acres or smaller, 3 acres generally ought to be the largest cant size so you can maintain an annual or semi-annual rotation. While smaller cants will yield fewer poles per harvest, they provide more balanced annual production. At an urban or suburban scale, cants must be very small, or you'll need to wait several years in between harvests. For example, one or two 200-square-foot cants of basket willow cut annually or every other year could conceivably produce enough materials to support a modest livelihood.

Coppice cants are usually cut in sequential order, moving between adjacent stands until the rotation is complete, returning back to the original patch. This patchy landscape mosaic makes wildlife migration between these zones of shifting canopy cover far easier, especially for invertebrates and other slow-moving species.

Layout

📃 Once you've chosen a target cant size and rotation length, you'll need to determine the layout of these patches in the landscape. Topography, hydrology, orientation, aspect, access, and soils should help direct you towards a well-informed plan. Essentially, the primary principles of cant layout and patterning are:

- Provide optimal solar exposure and minimal shade by creating cants more square than rectangular in profile (with some exceptions)
- Adjust your layout to respond to the topography to minimize erosion and facilitate access

- Work from south to north to reduce the effects of shading on stands along northern edges (in the northern hemisphere)
- Minimize edge to protect against browse

Optimize solar exposure and minimize shading by carefully considering the shape and orientation of your cants. Square-shaped cants minimize edges and can help reduce the effects of consistent, daylong shading by trees south of the stand. If you opt for rectangular-shaped cants, keep in mind that the narrower they are, the more pronounced diurnal and seasonal shade patterns will be. Rectangular cants with their long axis running east-west will receive even sunlight throughout the day, although the southernmost stools will largely be shaded by stands to their south, whereas north-south rectangular cants will receive a more even balance of shade distribution throughout the day. See Figure 6.9.

When it comes to new plantings, a perennial question in agroforestry layout and design lies in the orientation of tree rows. In the most over-simplified scenario, the two main options are north-south vs. east-west. Each have pros and cons, and the best option ultimately depends on several site- and goal-specific factors. As just mentioned, east-west row orientation creates a long but relatively narrow belt of shade north of the plantings. In some cases, this can be useful, perhaps providing shade for livestock, a cooler microclimate for more cool-season vegetables or other crops, etc. It also creates a hot sun-trap-like zone on the south side of the planting rows, with the alley ways between the plantings receiving considerable sunlight throughout the day, even more so as the spacing between rows increases.

Conversely, north-south row orientation helps to balance sun and shade patterns more evenly across the entire day, with the east side

Shading Effects of Different Cant Proportions

FIGURE 6.9: A relative comparison of three different cant shapes all roughly the same overall area, examining the varied effects on shading. In this case, shadows were drawn assuming surrounding cover is roughly 30 feet tall, and its solar noon on the equinox (roughly a 45-degree solar altitude). We see that the longer and narrower the cant, the more significant the daylong effects of shading to the south.

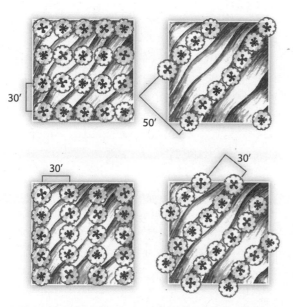

30'

50'

30'

30'

Sample Layouts for Agroforestry Plantings

Figure 6.10: Four different configurations of pollarded trees with an estimated 20-foot canopy on a 100-by-100-foot sample plot. The site slopes from the upper left to the bottom right, and contour intervals are 1 foot. Top left: Trees are planted in an east-west grid pattern with no attention paid to landform, 30 feet between-row spacing. Bottom left: Same between-row spacing but pollards oriented north-south, again no attention paid to topography. Bottom right: pollards oriented in more or less parallel rows along contour with 30-foot row spacing. Upper right: same contour-type pattern but 50-foot spacing between rows.

of trees and the rows between them receiving sun during morning hours, and the balance later shifting to the west side of plantings after noon. Understanding these differences is the first step towards choosing your preferred orientation. And ultimately, this decision should also respond to careful consideration of the site's topography.

Orienting cants along contour as appropriate helps minimize erosion, creating stable, efficient access and extraction routes. For example, on an idealized south-facing site with a gradual slope, a primary access road with switchbacks as needed ascends the hillside at a 5% to 10% grade. Woodland rides following contour separate relatively shallow cants (75-to-200 feet deep). The harvest rotation begins at the base of the hill, climbing upslope to a new cant each year. This minimizes shading since the stand immediately south of each cant is just 1 year old when the next successive cant

is felled. Of course, in reality, cant layout will be far more complex, but thorough site analysis and strategic design planning will guide your own unique layout and sequence.

Finally, choose cant shapes that minimize perimeter edge to help protect regrowth from browse. The longer and more sinuous the edge, the better the wildlife habitat, exposing more of the cant to these edge effects and becoming more challenging and costly to fence.

Integrating Standard Trees

Historic standard tree densities ranged between 5 and 40 trees per acre (12 to 100/ha) with 16 per acre (40/ha) as the most common according to Rackham.[21] Modern sources generally recommend 10 standards per acre (25/ha) and advise against more than 12 per acre (30/ha).[22] This is roughly equal to about 65-to-70-foot (20–21 m) spacing on center.

FIGURE 6.11: Newly coppiced sweet chestnut stand with oak and beech standards. Prickly Nut Wood, West Sussex, UK. Note the epicormic sprouts along the oak in the foreground, probably the result of increased light entering the stand following previous coppice cycles.

Table 6.7: Optimal age/size distribution of oak or ash standards in hazel coppice

Standard Age/Size	Number per Acre	Number per Hectare
0–25 year old saplings	7–8	20
25–50 year old young trees	4–5	12
50–80 year old semi-mature trees	3	8
80–125 year old mature trees	2–3	6
110+ year old, ready for felling	1–2	4
Total	17–21	50

Adapted from Hampshire County Council, 1995.

In reality, standards' canopy cover is probably a more important influence on coppice growth than the actual number of standards per acre. Coppice-with-standards systems suffer when standard trees occupy too much canopy cover. To balance polewood yields with standard production, aim for standard canopy cover of between 10% and 30%.[23]

If you'd like to create a sustained coppice-with-standards system, you'll also need to retain replacement standards to fill the gaps created by harvest. To do this, plan to reserve roughly twice as many young stems as mature standards. Five to ten young standards are said to cast equivalent shade to one mature tree. As these young standards mature and cast more shade, harvest the lowest-quality stems and only retain the highest-quality trees.

Hazel coppice-with-standards systems are said to be most productive with ⅓ canopy cover at the start of the cutting cycle, and as much as ⅔ by the cycle's end. This is roughly equivalent to 11 large mature standards per acre (30/ha) or as many as 58 young standards per acre (150/ha).[24]

Standard species that leaf out late and cast a dappled shade help create an early spring niche for many ephemeral flowering woodland plants. Because of this, Europeans preferred oak and ash.[25] Linden, chestnut, rowan (*Sorbus acuparia*), wild cherry, and crab apple are also preferred standard trees in the UK.[26] Because beech and hornbeam cast particularly dense shade, they are much less common.[27]

Case Study: Black Locust Coppice with Standards

My wife and I steward 52 acres (21 ha) of field and forest in Vermont's Champlain Valley. Our south-facing site has a gentle to moderate rolling topography, heavy clay soils, roughly 35 inches (89 cm) of annual precipitation, and although we like to think we're in USDA hardiness zone 5, we've seen -28°F (-33°C) almost every year we've been here, firmly planting us in zone 4.

Our small-scale farm enterprises are varied and diverse, and I've had a particular vision of establishing a stand of black locust for rot-resistant polewood, fence posts, and dense high-value fuelwood for years. Something of a mimic of the long-rotation chestnut coppice with standards systems I helped manage during an apprenticeship with Ben Law back in 2001. So, from a design point of view, I've been very clear on my species and products goals since long before I'd arrived here back in 2012. (Note: I wouldn't necessarily recommend this because designs should, and generally, must take into account local site conditions. In our case, our heavy clay soils really don't pair well with the soil types preferred by black locust.)

In terms of site selection, I identified a somewhat geographically isolated patch about 0.6 acres (0.25 ha) in size with convenient access, otherwise removed from the heart of farm activities as a prime location. Since this type of system requires minimal management once established, it made perfect sense to site it slightly off the beaten path, across an ephemeral tributary bisecting the property.

Because 5-to-8-inch diameter polewood is this planting's primary yield, I intend to manage the trees by coppicing on a to-be-determined rotation. Although I'm hoping trees will reach their target size after 10 years, I'm prepared for it to take as many as 12 to 15, so the rotation will be determined not so much by a prescribed interval but instead emerge based on site productivity. Because it's a relatively small patch, it'll either be treated as a single cant, harvested all at once, or split into halves and harvested on a staggered rotation.

The patch layout began with access. I left a 20-foot-wide swath along the eastern boundary ☞

FIGURE 6.12: A view of a 4-year-old black locust coppice with standards system in early spring in Vermont. Tree rows are laid out according to Keyline patterning. Note that the canopy is only just beginning to close and the understory is still vigorous and productive.

with our neighbors (the highest and driest section in the patch) as a primary access way. Along this edge, we planted fruiting species like elderberry, plum, apple, mulberry, and serviceberry. As far as interior patterning, I laid things out along a slightly off-contour pattern informed by P.A. Yeomans' Keyline system. Upon selecting a starting point, I used a laser level to lay out the first row, dropping slightly off-contour at a 1% grade from the valley to the adjacent ridge. With the pattern established, all other rows are equally offset, making it easy to manage herbaceous growth in between with a tractor if I so desire.

In terms of spacing, I chose to split the difference between typical long-rotation coppice, roughly 6.5 feet (2 m) within the row and 10 feet (3 m) between rows. This gives me ample space between rows, but still keeps the stand dense, optimizing wood production while encouraging straight, upright growth. Within the plot, I've added 15 to 20 standard trees of several different species (hardy pecan, *Carya illinoinensis*; butternut, *Juglans cinerea*; and bur oak, *Quercus macrocarpa*) at roughly 35-to-45-feet (10.7 to 13.7 m) spacing, knowing full well that they'll be quickly overtopped by the locust and some will struggle to survive. I intend to thin them, retaining only 4 to 7 standards following the first polewood harvest.

$ In the meantime, the stand continues to provide valuable shaded pasture for our small flock of sheep during the heat of July or August and the locust blossoms produce abundant nectar for honeybees and other pollinators. I've also been experimenting with currant and gooseberry plantings between the locust trees in every other row, to see if they can sustain fruit yields under a dense canopy.

In this part of the world, black locust fence posts are relatively uncommon and may fetch $2.50 to $3 per foot. There are roughly 200 locust trees in the stand. Assuming each one produces a 10-foot pole over 10 years' time ($25 per pole x 200), this 0.6-acre parcel should generate roughly $5,000 worth of product following the first harvest cycle. And I'm anticipating that number should increase by at least 50% following the first true coppice cycle when each stool should produce 2 or 3 stems. And last, the standard trees provide a long-term mast yield and valuable timber for the next generation of land managers here.

The Circulatory System: Access and Extraction Routes

It's nearly impossible to overstate the importance of safe, reliable access. For forestry endeavors, a complete access design should include roads, drainage structures (ditches, culverts, waterbars, etc), any necessary water crossings, and material landings. In most cases, your system's physical layout and pattern should begin with access. Everything else can respond accordingly.

As a general rule, in terms of sheer distance, the shorter the better. Strive to create a network that causes the least amount of disturbance, while creating convenient access to as extensive an area as possible, especially stands uphill of

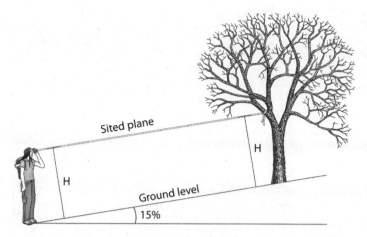

Sited plane

H

H

Ground level

15%

Understanding and Measuring Slope

FIGURE 6.13: A clinometer is a simple tool that allows you to measure the slope of a landscape. To use it, you approximate a plane parallel to the ground level by sighting to an object on the horizon at the same level as your eye (H). In this case, she has hung flagging on a tree branch at a height equal to her eye. By sighting through the clinometer and aligning the crosshairs with that point on the object, she can measure the slope of the ground by simply reading the number indicated in the object's lens. These tools usually indicate slope in both percentage and degrees.

the road. You want just enough of your land-scape devoted to access so you can reach as large an area as possible.

Road construction and maintenance is often the most destructive aspect of logging opera-tions, causing erosion, fragmenting habitats, and interrupting water flows. But this essential infrastructure also provides valuable benefits, so design it well.

Slope and distance determine an access route's viability with slope the most limiting factor. Aim to maintain a 5% to 7% grade, exceeding 10% only under rare conditions. A slope as shallow as 3% will minimize standing water collection, while slopes greater than 6% will require proper drainage structures. Although generally unadvisable, skid roads may accommodate grades up to 20% for distances of 100 feet (30.5 m).[28] See Figure 6.13.

Wherever access ways converge, cutting back the woody growth along the corners creates open glades that diversify habitat and support the wider turning radii of truck and

Table 6.8: Recommended distances between logging roads and bodies of water

Slope Between Waterway and Road Location	Minimum Distance Between Them
0%	25' (7.6 m)
10%	45' (13.7 m)
20%	65' (19.8 m)
30%	85' (25.9 m)
40%	105' (32 m)
50%	125' (38.1 m)
Avoid steeper slopes	

Adapted from Beattie et al., 1993, p. 193.

tractor traffic. These larger patches of more consistent full-sun conditions often support herbaceous vegetation-rich grassland that has been key to the creation of varied habitat types in European woodlands.[29]

Your access infrastructure should also include some sort of landing to sort, store, and transport harvested wood. Locate landings on

Table 6.9: Suggested minimum distances between drainage structures on logging roads

Grade of Roadway	Suggested Distance Between Water Bars	Suggested Distance Between Culverts	Suggested Distance Between Broad Based Dips
1%	400' (121.9 m)	450' (137 m)	500' (152.4 m)
2%	250' (76.2 m)	300' (91 m)	300' (91.4 m)
5%	135' (41.1 m)	200' (61 m)	180' (54.9 m)
10%	80' (24.4 m)	140' (43 m)	140' (46.7 m)
15%	60' (18.3 m)	130' (40 m)	Not recommended
20%	45' (13.7 m)	120' (36 m)	
25%	40' (12.2 m)	65' (20 m)	
30%	35' (10.7 m)	60' (18 m)	
40%	30' (9.1 m)	50' (15 m)	

Adapted from Beattie et al., 1993, p. 195.

level sites with well-drained soil. The landing may also provide a working area to process raw materials (rive poles, point posts, chop mortises, etc.).[30]

Because logging roads are costly, difficult to change, and significantly impact your landscape, consider consulting with a professional to lay out or review your design. This could be a great opportunity to tap into the expertise of your county forester. Even though they probably know little about coppicing, they should be able to offer some useful feedback on your access design.

Water Sources and Supply

During establishment, adequate water can make the difference between success and failure. Non-brittle humid climates often receive well-distributed rainfall ensuring the survival of a young system with minimal irrigation needs, while dryland and drought-prone climate zones will likely need to consider investing in this infrastructure. Water system design and layout are beyond the scope of this book. Check appendix 1 for recommended resources.

More than anything, the season chosen for planting can have a huge effect on success. Early spring or late fall ensures a much higher survival rate than summer months when temperatures are highest and evaporative losses greatest. Lower temperatures and moist winter weather mean that autumn plantings often have no need for supplemental watering, even in dry zones and especially Mediterranean climates with dry summers and wet winters. In central and eastern North America, spring is often

Key Access Design Principles

Consider several primary principles when laying out forest roads:

- Begin by identifying the most valuable forest stands and chart the shortest, safest, and most effective route possible.
- Minimize area devoted to roads, and don't exceed more than 10% of the total area.
- Avoid wet sites, ledges, or slopes of more than 50%.
- Keep roads set back from streams, increasing the distance as the slope increases-refer to Table 6.8. (A minimum setback of 25 feet (7.6 m) in small watersheds and 50 feet (15.2 m) within larger areas).[31]
- Minimize stream crossings. Site all crossings at right angles to stream-beds and at points with firm, low, gentle banks. Properly size and install any necessary culverts or construct bridges to minimize streamflow interruption.
- Aim to maintain grades between 3% and 6% and closer to level on sharp curves.
- Minimize long, straight, sloping road sections and provide adequate drainage via ditches and culverts as necessary. See Table 6.9.
- Prevent water collection on the road using water bars (diversions that direct water off the road surface) or broad-based dips (a more elongated variation on a water bar).
- Ditches and diversions should direct runoff flows into the woods or a siltation basin where sediment loads can settle out. They should not feed runoff directly into streams.[32]

the preferred planting season due to the cool temperatures, reliable rainfall, and an entire growing season ahead for young plants to get established.

Fodder and Fence System Design

$ **≋** Fodder-based silvopasture systems can provide a reliable, high-quality feed supplement to grazing livestock and often allow for continued understory grazing. These systems take three primary forms: pollarded trees distributed within existing pasture; a dense dedicated patch of managed coppice sprouts, shredded trees, or pollards; or hedgerow belts surrounding or traversing pastures and farm fields.

Silvopasture

Plant spacing within a silvopasture depends on the species, the type of livestock, annual rainfall and seasonal distribution, and of course, your production goals.

In some European wood pastures, the topography dictated the placement of pollarded trees in a pasture. On flat or gently sloping sites, pollards were often grown singly or in small stands, forming pollard meadows. On steeply sloping land, farmers often oriented pollards along narrow terraced banks.[33]

Silvopasture planting densities range from 100 to 400 trees per acre. Depending on the understory, canopy trees will need to

Table 6.10: Number of trees per acre for different silvopasture spacings

Between-Row Spacing	6' In-Row Spacing	8' In-Row Spacing	10' In-Row Spacing	15' In-Row Spacing
15' (4.6 m)	484	363	290	194
20' (6.1 m)	363	272	218	145
30' (9.1 m)	242	182	145	97
40' (12.2 m)	182	136	109	73
50' (15.2 m)	145	109	87	58
100' (30.5 m)	73	54	44	29

be thinned to keep canopy cover below the understory's maximum shade-tolerance level. Some cool-season grass species still produce good forage yields with up to as much as 60% shade. In general, warm season grasses perform best with 25% to 45% cover, while cool-season grasses prefer a range of 40% to 60%.[34]

If you're thinning an existing woodlot, aiming for roughly 60 square feet of basal area per acre should open enough light to support healthy understory forage development. Keep in mind, unthinned forest stands often contain 100 or more square feet of basal area per acre. It will often require several thinnings to reach this density without risking damage to remnant trees including windthrow, epicormic branch formation, or the proliferation of unwanted understory vegetation.[35]

Fodder Woodlands/Banks

💲 A fodder bank is a block of vegetation managed to yield high-quality livestock feed. It could be as simple as a brushy mid-succession field edge or as complex as a high-density plantation. Fodder banks serve as a nutritional supplement for livestock and are especially valuable in helping tropical graziers bridge the scant herbaceous forage available during annual dry seasons. Fodder banks are most common today in the tropics where rapid woody regrowth cycles allow for harvests as frequent as every 1 to 2 months.

🪓 Farmers generally manage these systems by either allowing animals direct grazing access or in a cut and carry system. Each has pros and cons. Direct grazing minimizes labor but also leaves the block far more susceptible to damage by livestock. Whereas cows primarily graze the leaves, sheep and goats will also strip tree bark, potentially causing significant damage to the plants. When animals are allowed direct access to graze, they should be managed as one would in a typical rotational grazing system, cycling them through paddocks to limit their impact and ensure that the stools or pollards have time to fully recover before regrazing. Cut and carry systems prevent unchecked livestock damage to

Case Studies: Traditional European Wood Pastures

Ribaritsa, Bulgaria

The Bulgarian village of Ribaritsa spans level lowlands and steep slopes from 1,970 to 2,300 feet (600 to 700 m) above sea level. Frost-free from mid- to late April through late October, winter temperatures dip to about 10°F (-11°C). Averaging 38.7 inches (982 mm) precipitation annually, the wettest months fall between April and early August.

Pollarded trees border many of village's meadows. Farmers pollard trees during dry seasons when the hay harvest falls short of their needs. They harvest leaf fodder between mid-July and October from hedge maple (*Acer campestre*), Norway maple (*A. Platinoides*), hornbeam (*Carpins betulus* and *orientalis*), ash (*Fraxinus excelsior*), black mulberry (*Morus nigra*), oak (*Quercus* spp., except for *Q. cerris* which the animals do not eat), white willow (*Salix alba*), large-leaved linden (*Tilia platyphyllos*), and elm (*Ulmus* spp.). In years of severe feed shortage, they also use European beech (*Fagus sylvatica*) and, less commonly, black alder (*Alnus glutinosa*).[36]

Every 3 to 4 years or so, farmers harvest 3.3-to-5-foot (1 to 1.5 m) shoots with an ax. They bundle the leafy branches and store them in a barn or on an elevated platform. Two workers can harvest 141 cubic feet (4 m^3 or 5.2 yd^3) of fodder a day. Unless annual hay yields fall well short of needs, it's primarily fed to goats. In the region, an average sheep or goat requires about 220 pounds (100 kg) dry hay or 35 cubic feet (1 m^3) leaf fodder per winter.[37]

Valdagno, Vicenza, Italy

Elena Bargioni and Alessandra Zanzi Sulli's article "The Production of Fodder Trees in Valdagno, Vicenza, Italy" offers a detailed look at the management of a small livestock farm in the Agno Valley of northeastern Italy. Located along the Agno River at an altitude of 2,635 feet (800 m) in the village of Castelvecchio, the area receives 58.6 inches (1,489 mm) annual precipitation. Southeast-facing slopes in the area were tilled; meadows were relegated to areas with less fertility, poor drainage, and colder microclimates; and wood pastures and woodlands occupied northwestern slopes.[38]

From the 1920s until 1992, the farm they profiled produced ⅓ of their annual fodder needs from managed pollards. Primarily ash (*Fraxinus excelsior*) but also including black alder, black poplar, (*Populus nigra*) cherry, maple, and elm, managed pollards were distributed throughout the pasture either singly, in groups, or in rows.[39]

Rows of ash pollards cut at 15 to 16.5 feet (4.5 to 5 m) height formed property boundaries or divided tracts of land, while hazel, hornbeam, and *Laburnum anagyroides* were coppiced for fuelwood and browsed in the field during dry summers. Farmers harvested branches every 3 years, pruning the trees' lateral branches to form 6-to-8-inch (15 to 20 cm) stubs called *gropi*, starting 6.6 feet (2 m) from ground level and ideally spaced every 20 inches (50 cm).[40]

☞

$ They raised 4 or 5 Burlina dairy cows, a local breed that produced 2.5 gallons (10 L) of milk per day each, 25 to 30 chickens, a pig, and 2 or 3 sheep for wool. From November to May, the farmers fed meadow hay, leaf fodder (or *frascari*), and grass from the woods that they collected as needed.[41]

Farmers harvested two types of tree fodder: *broco*, fresh leaves stripped from shredded/pollarded trees (often from the thin branches of the crown), and *frascari*, branches and leaves that they harvested, dried, and stored for use in winter. Ash was the most common species, though they also used alder, poplar, and hazel fresh.[42]

🕐 Using a billhook, or *cortelon*, they harvested frascari every 3 years during the second half of August.

They left the stems on the ground to dry for a day, later creating 10-to-12-inch (25 to 30 cm) diameter bundles that they stored under cover. It took 3 to 4 hours to collect fodder from each of the farm's 200 to 300 trees. Each tree produced approximately 17.5 cubic feet (0.5 m³) fresh leaves and 8 to 10 bundles; 85 to 100 trees produced 700 to 1,000 bundles annually.[43]

Before feeding it to livestock, farmers left the fodder out during the morning hours to absorb moisture. They then stripped the leaves from the branches to feed to the cows, saving the twigs for fuel. A cow normally consumes one bundle per day. A productivity analysis of conventional pasture versus wood pasture found that wood pasture supports one more cow per hectare (0.4 more cows per acre) than a landscape only producing hay.[44]

the stools, but they require a lot more labor in harvest and fodder transport.[45]

🐍 Some especially valuable tropical woody fodder bank species include *Acacia* spp. *Leucaena leucocephala, Mangifera indica, Morus alba, Musa* spp, *Cajanus cajan, Tamarindus indica, Stylosanthes guianensis, Centrosema pubescens,* and *Desmodium* spp.[46] Note that white mulberry (*Morus alba*) is particularly unique in that its range extends well into cold temperate climate zones.

🌿 Tropical systems are often planted very densely, with spacings ranging from 2 inches by 2 inches (5 cm by 5 cm) to 3.3 feet by 3.3 feet (1 m by 1 m). Yes, you read that right. It's basically the same type of spacing you'd use for many herbaceous crops. Planting double-row stands can help improve access. In this case, within-row spacing can range from 2 to 20 inches (5 to 50 cm), with 20 inches (50 cm) between rows, and 3.3 to 5 feet (1 to 1.5 m) between double rows.[47]

Plants are usually established by seed or cuttings and allowed adequate time to establish before harvest, usually between 9 and 21 months. Once established, rotations range from 6 to 18 weeks. Longer rotations give stools lengthier recovery periods and produce more total biomass, but cause an increase in the proportion of woody material while digestibility tends to decline. Shorter rotation lengths yield more palatable, nutrient-dense forage, but less overall volume, and can be a bit more taxing on the managed plants.[48]

There are few examples I'm aware of that mimic these intensive short-rotation fodder banks in the temperate world, but increasing interest in the value of silvopasture over the past decade has inspired considerable experimentation. Nevertheless, in colder climates, plant spacing would tend to be wider and recovery periods would need to be longer. But the practice offers fantastic inspiration for a retooled approach to dormant-season livestock forage.

Case Study: Mulberry for Fresh Fruit, Fodder, and a Woody Alfalfa Alternative in Southern California

Name: Loren Luyendyk, BA Biology UCSB 1996; Certified in Permaculture Design and Education, Biodynamic Agriculture and International Society of Arboriculture Certification; Licensed Nursery California Department of Food and Agriculture; www.sborganics.com

Location: Southern California, Santa Barbara County

Ecoregion: 85a, Santa Barbara Coastal Plain and Terraces

Elevation: Sea level to 8,000 feet

Average annual precipitation/Distribution and relative brittleness: 12 to 18 inches (30–45 cm), November to April, moderately brittle, Coastal Mediterranean climate with cool season precipitation

Hardiness zone: 9a

Project Description

🧬 💲 This project, still largely in a conceptual design phase, will explore various planting patterns and pruning training for mulberries to provide fresh fruit for market, fallen and culled fruit for hogs and poultry, and pruning wood or tree hay for ruminants. Mulberry, a common super food in health food stores, can fetch up to $15 a pound fresh and dried, especially species like *Morus macroura* 'Pakistani'. The fruit contains high levels of resveratrol, which has been shown to decrease incidence of some cancers. It is also high in protein (10% to 14%), making it a good animal fodder, especially for fattening hogs. Chickens could follow the hogs, eating the seeds (high protein) and providing vector control for fly larvae.

In addition to fruit yields, tree-based livestock fodder systems are more appropriate than annual hay systems in Mediterranean climates where pasture is often lean in the mid- to late summer. Mulberry leaf has long been used as hay for livestock. In recent lab tests from the UCANR, *Morus macroura* 'Pakistani' had 23.6% crude protein, making it a forage on par with alfalfa. A 25% ration has been shown to increase cow's milk production and gestation rate. Pruning stimulates mulberry, and trees can grow 10 to 20 feet a year after pollarding with little supplemental irrigation.

As a hay crop, mulberry may be able to replace alfalfa, California's largest agricultural consumer of water. Its yields rival alfalfa, with high-density ☞ plantings producing up to 10 tons/acre (22.4 t/ha), on just a fraction of the water. Replacing alfalfa with

FIGURE 6.14: Mulberry Fresh Fruit and Tree Hay Silvopasture Test Plot, Orella Ranch, Goleta, CA. Photo courtesy of Loren Luyendyk, used by permission.

mulberry tree hay could potentially cut statewide water use by 25% or more. Note that baling techniques need to be developed for this transformation to become feasible though.

Select varieties of forage mulberry could be planted at very to ultra-high density—up to 1,500 stems per acre (3,700/ha). The orchard design could be adapted from Central California's ultra-high-density olive orchards allowing for mechanical harvest and pruning. Olive pruning equipment could keep mulberry trees low so they can be browsed by

livestock and/or cut for baled leaf hay. In the dormant season, trees would be pollarded or cut back to the same spot once per year.

What's more, mulberry trees can be planted directly into existing alfalfa fields, potentially further increasing yields. The alfalfa will increase mulberry yields, providing income until the mulberry can be pruned. This would require existing irrigation systems to be modified.

Mulberry is very tolerant of a wide range of soils and climates, from heavy clay to fine sand, and from tropics to subarctic, and is very easy to propagate from seed, cuttings, and grafting. Vegetative propagation preserves precociousness and trees fruit early and heavy. Long-lived orchards can be established rapidly from on-farm stock and quickly begin producing fruit and tree hay.

System Details

Species - *Morus alba, M. macroura, M. serrata, M. nigra, M. rubra*

Purposes/Products/Projected Yields:

- Fruit: 10,000 lb/ac (11,208 kg/ha); leaf fodder: 3,000 lb/ac (3,363 kg/ha)
- Value in 2021 USD: fresh fruit up to $15/lb retail, dried $13/lb
- Other yields: 3 to 6 hogs/ac, 100 to 200 chickens/ac (broilers) or 100 to 200 layers, 2 to 4 cows per acre

Type of Management

💲🪓 Fruit is harvested in spring/early summer for fresh, frozen, and dried products. Culled/dropped fruit is gleaned by hogs post-harvest. Hogs are then ☞

followed by poultry to finish cleaning orchard floors, and for fly reduction, either broilers in pens following hogs or layers paired/grazed simultaneously with hogs. After orchard cleanup, trees are pruned (midsummer) using a front-mounted hydraulic boom sickle bar mower, and branches are laid in the row to be grazed in situ by cattle/sheep/goats or stored for winter hay. In fall, trees are pollarded back to the main stem and trained with side branches, like cordons of grapes/vineyards. Fall pruning can be fed fresh or dried for winter hay. Baling summer prunings for winter hay would require a forestry baler and some specialized/custom hay rakes.

Existing implements and machinery for baleage of other hay crops may work well to bale tree hay. Conventional square balers may be able to handle new growth of mulberry prunings provided it's not too woody or long. More research needs to be done.

Stand Structure and Disturbance Frequency/Timing

🕐 🍂 Blocks should be planted in single varieties of 1 to 10 acres, with multiple blocks of different varieties to extend fruit harvest. Fresh fruit for market and culls for livestock feed could be available for 3 to 6 months, with sequentially ripening varieties beginning in late spring (as early as March in mild winter regions) into mid- to late summer. Of course, season length depends on hardiness zone and growing season.

The hogs can be let into the orchard after fruit drop and be followed by chickens (layers or broilers) that will clean up the remaining fruit and disperse the hog manure. Next, the trees can be summer pruned,

FIGURE 6.15: Integrating pigs into mulberry silvopasture test plot, Orella Ranch, Goleta, CA. Photo courtesy of Loren Luyendyk, used by permission.

keeping them to a manageable height and providing fresh leaves for fodder. This tree hay can be fed to livestock directly in the pasture following pruning or left to dry in the field before baling. The trees can also be dormant-pruned yielding ramial whips with high concentration of sugars.

Pattern

🍂 Trees are trained like hedges or vineyard rows. No trellis needed although some stakes/supports to train cordons may be useful. Density would vary according to site, species, and pruning regimen, varying from a recommended 220 to 1,200 trees/acre (543–2,964 trees/ha): 3 to 10 feet (0.9–3 m) in row, rows 12 to 20 feet (3.7–6.1 m) apart (varies by species and context).

Hedgerow Design

Before we begin, a special thanks to Jim Jones of the Hedgerow Co. and the University of Waterloo for his review and recommendations for this passage. To design a hedgerow, you'll need to begin by choosing your species, spacing, hedge width, and landscape pattern. Of course, hedgerows can mean different things to different people, and the ultimate objectives your hedgerow is intended to serve will help guide and inform these characteristics. When it comes to plant selection, choose multipurpose species that yield fruit or nuts, respond well to coppicing, and are well-suited to your site's sun, moisture, soil, and wind conditions.

Adaptable, hardy, relatively compact, well-armed with sizable thorns, berry-producing and palatable to livestock, hawthorn has been the most common species in managed hedgerows throughout the past several centuries.

Table 6.11: Suggested species mixes for a conservation hedge in the UK

Species	Composition
Hawthorn (*Crataegus monogyna*)	60%
Hazel (*Corylus avellana*)	10%
Blackthorn (*Prunus spinosa*)	10%
Field Maple (*Acer campestre*)	5%
Common dogwood (*Cornus sanguinea*)	5%
Purging buckthorn (*Rhamnus cathartica*)	5%
Dog rose (*Rosa canina*)	5%

Adapted from Maclean, 2006, p. 73.

Hawthorn comprises 80% or more of many English hedges. During the 18th and 19th century, the majority of newly installed enclosure hedges consisted of densely spaced single rows of hawthorn and blackthorn. But in areas with apple and pear orchards, be aware that hawthorn acts as an alternate host for fire blight, a bacterial disease that also affects these species.[49] On this continent, osage orange and black locust provide a similar analogue (be forewarned of black locust's suckering habit though), and in dryland ecosystems, mesquite fixes nitrogen while also producing edible protein-rich pods.

Most European hedgerows are single or double rows of trees and shrubs, rarely exceeding 6 feet 6 inches in width (2 m). Close plant spacing helps ensure that any gaps that form will be quickly filled in by neighboring plants. If adequately dense, even single-row plantings can serve as effective living livestock enclosures.

Single-row hawthorn hedges average 9 inches (22.5 cm) within-row spacing, while many double-row mixed-species plantings have 18-inch (45 cm) in-row spacing (though they may be as close as 9 inches) and 15 to 18 inches (35 to 45 cm) and up to 24 inches (60 cm) between rows. That's roughly 20 plants per 30-feet (9.1 m) length for a single-row hedge and 40 plants per 30-feet length for double rows.[50] Perhaps more simply stated, single-row laid hedges often include 4 plants per meter with 5 plants per meter in a double-staggered row configuration.[51] See Figure 6.16.

Hedgerows designed for fruit and nut production and wildlife corridors can be spaced more widely (4 to 8 feet, or 1.2 to 2.4 m) to encourage lateral branching and a broader architecture. To optimize the use of space, choose shrubs of varying form and size (elderberry, currants, hazel, gooseberry, etc.) to fill in the gaps between medium- and large-sized trees.

If you plan to manage your hedgerow by laying, a double row provides more **pleachers** (a managed living tree stem and stump in a laid hedge) to choose between. Double-row hedges also provide better protection from wind, sun, and escaping livestock. A single row of trees and shrubs will require at least 6 feet (1.8 m) width while a double line can occupy up to 10 feet (3 m) or more.[52]

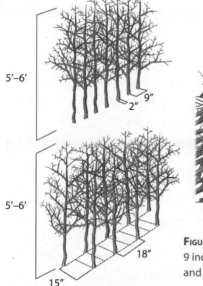

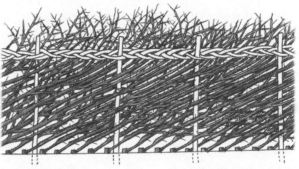

Hedgerow Spacing and Patterning

FIGURE 6.16: Top left: Single-row hawthorn hedge with trees spaced roughly 9 inches on center. Bottom left: Double-row hedge with 18 inch in-row spacing and 15 inch between-rows. Densely stocked, newly laid hedge on right.

Table 6.12: Number of plants required per single-row hedge length at various spacings

Within-Row Plant Spacing	16' (5 m)	33' (10 m)	66' (20 m)	98' (30 m)	197' (60 m)	328' (100 m)	656' (200 m)	984' (300 m)
6" (15 cm)	33	66	133	200	400	666	1333	2000
12" (30 cm)	16	33	66	100	200	333	666	1000
18" (45 cm)	11	22	44	66	133	222	444	666
24" (60 cm)	8	16	33	50	100	166	333	500
30" (75 cm)	6	13	26	40	80	133	266	400
36" (90 cm)	5	11	22	33	66	111	222	333

Adapted from Maclean, 2006, p. 60.

While a number of factors will inform your hedgerow's orientation, keep in mind that the shade cast by hedgerow plants affects neighboring vegetation located within one to two times the hedge height.[53] While east-west oriented hedgerows shade less of the overall landscape than those running north-south, their shade is concentrated along the hedge's northern margin. This means that it also creates a hot, exposed microclimate on its southern side. With this in mind, consider the value that these respective microclimates offer, potentially using the shaded space for access lanes or shade-tolerant crops or allowing livestock to gather along this protected edge. Hedgerows also help reduce evaporation and increase soil temperatures, moisture, and localized relative humidity.[54]

Although shade effects to adjacent plantings will reduce yields by 10% to 15%, hedges that also serve as a windbreak may result in yield and quality gains that stretch 2 to 12 times the height of the windbreak. Hedgerows designed as windbreaks are ideally about 40% permeable,

slowing prevailing winds, filtering the air, and avoiding the creation of turbulent airflow on the leeward side. Well-designed hedgerows reduce wind speed by over 20% over a distance 8 to 12 times the hedge height and provide modest benefits up to 30 times hedge height. To function properly, a windbreak's length should be at least 20 times longer than its height. Otherwise, they can accelerate wind on the leeward side.[55]

Two staggered rows of willow cuttings spaced 1 foot in-row and 3 feet (0.3 and 0.9 m, respectively) between-rows make a quick and inexpensive shelterbelt. This affords ample space for the developing plants and also allows the grower to harvest one row at a time on a regular rotation, while still retaining some wind protection and visual screening.[56]

If your design requires year-round privacy and wind protection, consider including species like beech, hornbeam, and oak (as well as coniferous species which cannot be laid) that retain much dead foliage through the winter months, a characteristic that's unique among

Table 6.13: Wind speed reductions at varying distances by different types of barriers (expressed as a function of barrier height)

A comparison of different types of wind reduction structures' effectiveness in slowing wind over varying distances. H equals the barrier's height. For example, a single-row deciduous windbreak, 25 feet tall, will reduce wind speeds by 35% 250 feet beyond the windbreak. Adapted from Gabriel, 2018, p. 147.

	5x H	10x H	15x H	20x H	25x H	30x H
Single Row Deciduous	50%	35%	20%	15%	5%	0%
Single Row Conifer	70%	50%	40%	25%	15%	5%
Multi-Row Conifer	75%	65%	35%	15%	10%	5%
Solid Barrier	75%	30%	10%	5%	0%	0%

broadleaved species. Also locate hedges a safe distance from roads and field access ways so they don't impede passage or winter snow removal and restrict driver's viewsheds.[57]

Stockproof Hedges

Stockproof hedges must be capable of resisting animals' efforts to escape to greener pastures. Hawthorn, blackthorn, holly, and other species armed with aggressive thorns are some of the best-adapted to browse and damage from rubbing. Stockproof hedge design varies depending on the type of livestock you're trying to enclose. In the UK, the two main divisions are single- or double-brushed hedges, sometimes referred to as sheep fences or bullock fences. It's difficult to design one that contains both. Single-brush styles like the Midland bullock hedge were designed to withstand cattle leaning and rubbing against the hedge, with pleachers and brash facing into the field. The dense, thick structure of double-brush hedgerows helps deter inquisitive lambs and sheep from finding their way beneath and between gaps.[58]

In some systems, farmers allow animals to browse on a hedge's young tips and lush growth. Though this approach requires little effort on behalf of the grazier, it's important to prevent over browsing to the point where it compromises plant health.

The alternative is the cut and carry system where growers lop off branches and/or coppice or pollard shoots, delivering this material directly to their livestock. This requires far more work and reliable fencing, but it does help limit overbrowsing.

Wildlife Hedges

Hedgerows designed for wildlife habitat support a wider diversity of species when left to grow tall. British studies have found twice as many bird species in a 13-foot-high (4 m) hedge than one less than 6 feet (1.8 m). Many native British tree species provide food sources and habitat for dozens if not hundreds of insect species, supplying a protein-rich food supply for birds and other wildlife.[59]

In England, mixed hedges well apportioned with hawthorn provide some of the best bird

Table 6.14: Number of insect species that feed on different British tree species

Tree Species	Number of Insect Species Hosted
Oak	284
Blackthorn	(153)
Hawthorn	149 (209)
Apple (Wild Crab)	93 (118)
Wild Rose	(107)
Elm	82
Hazel	73 (106)
Beech	64 (98)
Field Maple	(51)
Ash	41
Hornbeam	28 (51)
Holly	7 (10)

Note figures in parenthesis denote number of insect species that feed and live upon the trees. Adapted from Agate and Brooks, 1988, p. 18; figures in parenthesis from Maclean, 2006, p. 57.

cover. Hawthorn leafs out early, provides good protection, supports diverse insect populations, and yields edible berries. If the hedge won't be managed by laying, within- and between-row spacing can be as wide as 3 feet (1.8 m).[60]

Most of the existing hedgerows in the United States and Canada are rows of trees that sprung up along fencelines, where they were protected from browse and mowing. Though unmanaged, they provide shade and wind protection for livestock and crops and corridors for wildlife. They're generally ineffective as livestock barriers because they tend to be gappy along their bases.

To transform a mature hedgerow like this into a dense stockproof fence, you'd likely need to begin by coppicing most woody plants larger than a sapling to rejuvenate them and generate dense, lush, low growth. (Remember though that many woody species will sprout weakly once they are fully mature.)

Basket Willow Systems

Varieties of basket willow have been selected for cultivation for centuries. Britain's National Willows Collection contains nearly 300 species and hybrids! Today, the three most prominent basket willow species grown in the UK are *Salix triandra, Salix viminalis,* and *Salix purpurea.*

Amazingly, willows may grow up to 1 inch (25 mm) per day under optimal conditions. Though tolerant of a wide range of soils, basket willows grow poorly in saturated acidic soils and prefer a pH near neutral. Commercial-scale willow production often occupies lowland sites with fertile, deep, well-drained clay-silt soils. While total yields tend to decline on poor soils, they may actually produce better-quality rods.[61]

The almond-leaved willow (*Salix triandra*) produces pliable rods with minimal pith and a gentle taper that grow to 5 to 8.2 feet annually (1.5 to 2.5 m). It yields 6.75 tons/acre (15 tonnes per hectare) and is the industry's most important commercial species. The variety "Black Maul" comprises 80% of domestic production in Somerset, UK.[62]

Less common commercially, the purple willow (*Salix purpurea*) produces slender, tough, hard rods, 3.3 to 5 feet (1–1.5 m) in height. Great for small high-quality basketry, *Salix purpurea* produces up to 4.5 tons/acre (10 tonnes per hectare).[63]

Salix viminalis has a large proportion of pith and a steep taper towards the tip. It's used for larger projects like woven hurdles and garden structures. Stems of this hardy prolific producer reach 13 feet (4 m) and yield up to 11.25 tons/acre (25 tonnes per hectare).[64]

Willow growers also cultivate a wide variety of hybrids due to their disease and pest resistance, though most of them are well-suited to little other than coarse basketry and larger woven projects. In Germany, Poland, and Spain, growers cultivate the North American *Salix eriocephala* extensively. Originally chosen for its disease resistance, over time it has grown more susceptible.[65]

Plant spacing depends on species, soils, and management strategies. The denser the spacing, the higher the per acre yield, the straighter the stems, and the faster the willow shades out herbaceous competition (but the more difficult weed control will be). When cultivation work is done by hand, growers space plants as close as 1 foot by 1 foot (30 cm by 30 cm)—over 40,000 stools/acre (100,000/ha). For mechanical cultivation, Somerset (UK) growers space plants 1 to 2 feet (30 to 60 cm) within row and 2 feet (60 cm) between rows (10,000 to 20,000/acre or 25,000 to 50,000/ ha). The French commonly use 8-inch-by-32-inch (20 by 80 cm) spacing. Other common spacings include 10 inches by 18 inches (25 by 45 cm) requiring 40,000/ac (100,000/ha) or 8 inches by 16 inches (20 by 40 cm), equivalent to 54,000/ac (135,000/ha).[66]

🪓 While most basket willow coppice is cut clear to the ground, some growers manage willows as mini-pollards elevated 1 to 3.3 feet (0.3 to 1 m). They require less bending to harvest by hand, make weed control easier (though impossible by machine), and afford some protection from rabbit browse and disease. Some claim mini-pollards produce rods prone to undesirable basal curves at their base, while others say it's not a significant problem.[67]

🕐 Management cycles range from once a year after leaf fall to every 2 or even 3 years for heavier-duty crafts like hurdles and woven structures. Cutting plants following their first year in the ground helps encourage vigorous new growth in year 2. Most stands usually don't produce a salable crop until year 3, reaching peak yield between years 5 and 7.[68]

FIGURE 6.17: A low willow pollard at Margaret "Pegg" Mathewson's Withyhenge farm in Oregon. Note the straightness of the rods and the prolific sprout production.

Managing for Wildlife

The resprout silviculture systems you design will inevitably create wildlife habitat. And if enhancing wildlife habitat is a particular goal of yours, you may approach it in one of two key ways: either intensively by working to support certain select species or extensively by maximizing diversity among habitat types, species, age structure, and edge. If you have species-specific habitat goals, learn their needs and preferences. How does your landscape currently support them? How might you make changes to better meet these needs?[69]

Different tree species support different wildlife species. For example, aspen or poplar trees offer excellent food and cover for grouse, woodcock, hare, rabbit, deer, and turkey. They grow quickly, tolerate high-density spacings, and respond well to coppice management. In fact, some foresters prescribe coppicing *Populus* to improve habitat quality for wildlife.[70] Nut-producing species like oak, beech, chestnut, hickory, pecan, etc. produce abundant food for diverse wildlife populations. Incorporating these trees as standards allowed to grow to maturity diversifies habitat structure while creating abundance throughout the woods.

Many wildlife populations thrive along edge-type ecosystems (field and forest,

Table 6.15: Relative values and parts of various woody plants consumed as food by birds and mammals

Relative Value	Tree Species	Bark	Twigs	Buds	Foliage	Mast
Excellent	Oak species	x	x	x	x	x
	Black Cherry		x	x	x	x
	Apple	x	x	x		x
	Dogwood		x	x	x	x
Good	Maple species	x	x	x		x
	Aspen species	x	x	x	x	x
	Serviceberry	x	x	x	x	x
	Beech		x	x	x	x
Moderate	Birch species	x	x	x	x	x
	Hickory species	x	x		x	x
	Alder species		x	x	x	x
	Elm species		x	x	x	x
	Willow species	x	x	x	x	x
	Tulip Poplar		x		x	x
	Hawthorn		x	x	x	x
	Ash species		x		X	x

Adapted from Beattie et al., 1993, p. 50. Original source Gutierrez et al., 1979, *Managing Small Woodlands for Wildlife*, Information Bulletin no. 157, Cornell University. Note that these are relative, generalized values that vary throughout the season and depending on other available options.

water and uplands, early- to mid-succession ecosystems), so look for opportunities to appropriately optimize edge between stands.[71] The patchy landscape mosaic coppice management creates does exactly this. And if possible, leave some unmanaged, mature uneven-aged woodland (at least 5% to 10% of the total forest area) to create a more balanced habitat mosaic.

Keep in mind that different species require unique base range areas to find food, water, and cover. It will not be possible to meet the needs of some species within your individual woodland. In this case, you'd instead need to work to create more contiguous habitat between adjacent properties.

Encouraging structural, species, size, and age diversity increases the quality of the habitat. For instance, to support a diverse songbird population, maintain three distinct types of vegetation: a ground layer 2 feet (60 cm) tall or less, a mid layer from 3 to 25 feet (0.9–7.6 m) in height, and a canopy layer reaching 25 feet + (7.6 m +). Each of these layers offers habitat for different woodland bird species. Light thinnings that remove less than 20% of the basal area during a stand's early life help promote the emergence of an herbaceous layer.[72]

As we discussed in chapter 3, deadwood also provides a crucial resource for many species, creating the foundation for a healthy forest life cycle. Many insect species depend on deadwood in some form, either feeding directly on the deadwood or on saprophytic (decay-causing) fungi, parasitizing or predating invertebrates that feed on deadwood and decomposer fungi,

Table 6.16: Select wildlife species' estimated base ranges

Species	Range Estimate with Good Habitat
Cottontail Rabbit	½ acre
Squirrel	1–2 acres
Snowshoe Hare	10 acres
Ruffled Grouse	40 acres
Woodcock	50 acres
White Tailed Deer	640 acres
Wild Turkey	1000 acres
Bobcat	2 mile radius
Moose	2 mile radius

Adapted from Beattie et al., 1993, p. 47.

or by building nests in empty burrows. In the UK, up to 17% of all organic matter in virgin forest is deadwood that covers between 6% and 25% of the forest floor.[73]

To improve deadwood resources, leave wood standing dead, downed, and decaying, especially large pieces that decay slowly as well as brush and slash piles. Larger-diameter deadwood (16 inches+; 40 cm+) offers food, den, and nesting opportunities for insects, fungi, mammals, reptiles, and birds. Brush piles offer nesting sites for wrens, robins, and other small birds; food for fungi; and protected hunting sites for small mammals. Avoid sprawling brush piles in cut coppice stands since they may cover large areas of the forest floor and choke out herbaceous ground flora. And last, whenever possible, retain snags (ideally 5 per acre), the larger the better. Their limited shade won't compromise timber production, and they support many wildlife species.[74]

A COPPICE PATTERN LANGUAGE: DESIGN DETAILS AND CRITERIA

In many disciplines, a shared vocabulary emerges to describe the products, processes, and practices unique to the field. This common language embodies the nuances of the problem set they address and the solutions at their disposal. As this pattern language develops, the craft reaches a new level of complexity and organization with concise phrases that describe varied options to different scenarios. While the resprout silviculture pattern language is far from developed, here's a list of some promising coppice-based patterns.

Solo Stool

In particularly small-scale intensive applications, a *solo stool* produces edible leaves, medicinal products, and small-diameter polewood for garden stakes, weavers, kindling, and other needs. Because the relative risk is quite low, this may be the best way to experiment with coppicing and pollarding. It's also a great way to learn about the habits and uses of specific species.

The linden/basswood (*Tilia* spp.) is perhaps the best-known temperate salad-producing coppice species. At the Agroforestry Research Trust in Dartington, Devon, UK, agroforestry expert Martin Crawford maintains several low pollarded lindens cut at 2 to 3 feet (60 to 90 cm) high as a perennial salad green. While the tender leaves produced in early spring are typically best, trees produce fresh foliage throughout the year,

---

Case Study: Growing Fertility in an Agroforestry Setting

Twenty years ago, I spent several formative months apprenticing with Jerome Osentowski of the Central Rocky Mountain Permaculture Institute in Basalt, CO. At 7,000 feet (2,134 m) elevation on the dry, dense red clay soils of the pinyon-juniper complex, Jerome forever inspired me with a lifetime's worth of investment and transformation, creating a wildly productive forest garden ecosystem. One of the key principles I learned from him was his use of living plants to help cycle nutrients and build soil on-site. In recent years, I've come to understand that the practice of "syntropic agriculture" is largely based on this very same premise—using natural ecosystems as a model to guide and inform the creation of agriculturally productive, closed loop ecosystems.

$ 🌿 When it came time to plant out our first small-scale commercial berry production system, we integrated much of this knowledge into the design. This small 0.25-acre planting consists of six 120-foot rows of berries and trees following a symmetrical pattern based on Keyline Design principles. Using a rented excavator, I carved out ditches on the uphill side of each row, using the soil to build a planting mound below that gives crop plant roots double the topsoil, an elevated berm above the seasonal ☞

high-water table, and a water harvesting earthwork all in one.

🧬 The spacing of the berries and trees depends on individual species preferences. What makes this planting relevant to coppice management is what occurs on the uphill (in this case north) side of each berm. Here we planted four different nitrogen-fixing woody species in an irregular offset pattern flanking the trees and shrubs. This includes Siberian pea shrub (*Caragana arborescens*), sea berry (*Hippophae rhamnoides*), autumn olive (*Elaeagunus umbellata*), and buffalo berry (*Shepherdia argentea*). Each of these species produce fruit or seeds that we could harvest (and we do with the autumn olive and seaberry), and they're also nitrogen fixers, which means they host symbiotic bacteria on their roots capable of fixing atmospheric nitrogen in a form that's available to plants in the soil.

💲 🕐 🪓 As these plants begin to shade out and crowd the crop plants, we selectively chop and drop them, creating our own woody mulch in place and working to enhance soil fertility and plant fruiting and growth. These are sacrificial plants, located so that they don't shade the crop but instead can provide

added fruit yield, along with additional biomass to help enhance stand fertility. It's a very simple concept, and something that can pair well with just about any woody installation that seeks to optimize food production using on-site resources as much as possible.

FIGURE 6.18: Nitrogen-fixing woody chop and drop species. Sea berry (left: *Hippophae rhamnoides*) and Siberian pea shrub (right: *Caragana arborescens*) are both excellent woody nurse crops. Tolerant of many soil types, nitrogen-fixing, relatively compact in stature, both these species can tolerate regular coppicing and/or lopping while also producing pollinator fodder, berries, pods, etc.

requiring little more than harvest to maintain steady production. It's a wonderful system for small-scale coppice management in tight places.

Successional Shelterbelt

Shelterbelts absorb and deflect prevailing winds from crops, livestock, and buildings. They usually consist of at least two layers of woody plants. These plantings could be managed by regularly planned thinning of the entire band of growth, harvesting of individual rows on predetermined cycles, or cant-style clear-cuts that remove the vast majority of the woody growth along an individual section of the belt each year.

Chop and Drop

Fast-growing biomass producing plants and/or nitrogen-fixing species are coppiced, pollarded, or regularly pruned, and the woody debris is used as ramial mulch for more high-value crops like fruit- and nut-producing species. Coppicing these nurse plants can improve soil conditions for neighboring plants, both from the small-diameter prunings laid down as mulch and the dieback of the coppiced plant's roots.

Sucessional Suntrap

A suntrap is an arc of vegetation that opens to the sun and creates a buffered microclimate on the sunward side. By sequentially coppicing patches of woody vegetation along a field edge, you can capitalize on this valuable microclimate while avoiding any major interruptions of the adjacent field.

Graywater Garden

Household graywater is an excellent nutrient-rich resource. While there are several great ways to utilize graywater in your landscape, fast-growing, wet-adapted tree and shrub species will also thrive when used to convert this fertility into biomass. Willows, alder, dogwood, and poplar species all generally perform exceptionally well in wet conditions. By planting them in a location where they can safely make use of household graywater, you can benefit from the enhanced fertility, without needing to feel concerned about ingesting their plant parts.

Decorative Deer Fence

Use arborsculpture, the horticultural practice of creating living sculptures as a visual screen, a decorative partition, or even a living deer fence. Willow species' rapid growth and ease of propagation make them especially well-suited to this type of management. Once established, their stems can be interwoven so they graft to one another and form a dense, continuous barrier.

Pollarded Posts

This pattern uses managed live trees as fence posts. Though it will take several years before they are strong enough to serve their purpose, upon maturity, they can be managed as pollards, providing biomass yields for craft, fuel, and fodder. They could also be planted along an existing fence whose posts have begun to decay or along a future field partition.

DESIGNING CULTIVATED ECOLOGIES: MULTISTORIED, MULTIFUNCTIONAL GUILD DESIGN

When we examine the vast majority of traditional coppice systems we've inherited, we see a few primary patterns: "design by default," "adaptive modification," and "extreme simplification" (aka monoculture). Each of these systems have emerged for good reason.

Design by default is the unplanned outcome of ecological succession following forest clearance. Some stumps sprout, some don't. If you value the sprouts, you protect them and allow them to thrive. This is presumably the way many ancient coppice systems initially developed.

Adaptive modification occurs as humans choose to favor certain species, enhance plant spacing and density, and carefully shape their growth forms. The tools of adaptive modification include pruning, singling, layering, planting, and removal. They transform stand structure to meet production goals.

Extreme simplification involves the design and implementation of entire ecosystems, usually via planting. A shift towards industrial processes, specialization in the workplace (and the marketplace), and a drive towards growing productivity set the context for managed woodland plantations composed of the most valuable species at that point in time. Usually these systems contain one or just a few key species.

How about a new approach to coppice system design: the creation of "cultivated ecologies" or "planned interconnection?" This process should involve aspects of each of the previous strategies influencing resprout silviculture system design while also looking to natural ecosystems as models to guide our plans.

In reality, our process ideally integrates aspects of all three of these strategies. Because the species already present on a site indicate their tolerance for the current soil, light, and moisture regimes, design by default can help inform preliminary species selection. If existing vegetation is of particular use, we can choose to highlight the species mixes already present. And sometimes existing site conditions leave something to be desired. Whether it's the dominant species, system diversity, spacing, or any number of other factors, we may need to apply adaptive modification to optimize the site's potential and meet our goals. Wherever possible, aim to make the least change for the greatest effect.

TEMPORAL DESIGN: CANT ESTABLISHMENT AND HARVEST SCHEDULES

The resprout silviculture systems you design are perhaps some of the most lasting investments you can make in the future. It will take years for them to reach fruition—and they will hopefully continue to yield for generations to come. But even with a detailed design in hand, it's a big leap to begin implementing that plan on the ground. Often, the best place to begin is with an area that you already frequent regularly. This helps ensure that you observe it often, monitor regrowth, and watch for signs of browse and other damage. You can build on your successes and lessons learned as you expand your management in the future.

Especially in the early stages of implementation, strive to keep things small and manageable. It's better to take twice as many seasons to do the job right than to be overwhelmed by unreasonable goals. How much time and energy do you realistically have to devote to management? The length of your rotation will largely determine the time required to fell, sort, and extract polewood. Shorter rotations of just a few years may take as little as 3 to 5 days per acre, whereas rotations of 10 to 20 years can easily require 1 to 2 months or more per acre.

We'll discuss strategies for establishment and implementation in chapter 7, but at this point, you should at least begin to flesh out a rough timeline that projects the phased implementation of your plan. While your completed design should probably, at a minimum, indicate access ways, the extent and distribution of management patches, details on harvest cycles/ rotation length, the species present or planned and the ways they'll be managed, how you get from here to there is perhaps the most important part of your design.

If you're starting with open field and will be planting a new system, which zones will you start with this year, what type of site prep will you do, and when do you estimate it'll be ready for harvest? How will you break your plan into phases so that you leave room to receive and incorporate feedback and don't end up overwhelmed? What types of maintenance work will be required during the establishment phases, and will you have adequate time to keep up with it during this extra-intensive phase of site development? And if you're transforming existing woodland, which patches are priority zones? How do these priorities interface with your design's layout? Make sure you're not painting yourself into corners.

Taking the time to strategize and plan out your implementation over the next 1 to 3, 3 to 5, 5 to 10, 10 to 20 years is ultimately the most important stage in the design process. It's where you chart your path from concept to reality. Although you'll inevitably deviate from this road map as you progress, your process, along with your ability to adapt, is far more important than your ability to implement it to a T.

PUTTING DOWN THE PENCIL AND PICKING UP THE SHOVEL

Discovering best practices for converting high forest into well-stocked productive coppice or creating cycling patches of working trees are probably some of the biggest mysteries we strive to unlock in this endeavor. Because most traditional coppice systems emerged as relics of native woodland reshaped by adaptive modification, many of us will need to develop a modern analogue to these traditional, multigenerational processes.

One particular challenge we face given our cultural impatience and desire to see rapid results lies in the lengthy implementation timeline and the delayed return on investment of tree-based systems—especially in temperate climates. Although we will enjoy some of the fruits of our labors, our investments will continue to appreciate in value for many years to come.

So be sure to keep sight of this time frame when selecting species, layouts, harvest cycles, and patterning. And remember that this process is full of feedback loops and opportunities for improvement. Let this realization be both relieving and humbling. Our great-grandchildren will thank us.

We have much to learn. We have little experience with coppice system establishment. Experimentation will help us determine today's best practices. And we experiment by putting these ideas into practice. So how do we start? In chapter 7, we'll examine practical resprout silviculture establishment strategies and look at what it takes to bring our vision to fruition and transform fresh-cut stumps into vigorous green sprouts.

Chapter 7:
Getting Started:
Establishing Coppice on Your Land

So, how's that design coming along? Because this is where you're going to put it into action. After taking the time to analyze your site and clarify your management goals, you've created a plan that points you towards the landscape you aim to create.

Generally, this leaves us working from one of two starting points. We have either standing woodland/forest, succeeding oldfields, or hedgerows/treelines where the species, spacing, and structure are largely predetermined, or open field, leaving us to choose the details, prepare the site, and install our plantings. With the former, we usually need to intervene and redirect some of the key variables to better meet our goals. With the latter, we have the burden of choice. We'll call these two strategies *woodland conversion* and *field transformation*. Each has its

Table 7.1: Pros/cons of woodland conversion vs. planting

	Woodland Conversion	Field Transformation
Pros	Can result in a coppice response the season after cutting	Assures the desired species are present
	Inexpensive since largely relies on existing vegetation	Creates ideal stand spacing and patterning
	Generates a wood yield even upon stand establishment	All plants begin from the same point in time
Cons	Stands aren't necessarily comprised of desired species	Costs money—for site prep, planting stock, installation labor
	Spacing is probably inconsistent and gappy	Seedlings take several years to mature and get established
	If in existing woods, patch must be large enough to provide good light	May require watering and weed control
	May result in very uneven growth since tree age may vary widely	

unique challenges and advantages, and regardless of which scenario we start from, most of us will probably find that each intervention offers some very useful tools and perspective. Let's begin with a look at some of the considerations involved with transforming high forest to resprout silviculture management.

WOODLAND CONVERSION

As best we can tell, most traditional resprout silviculture systems evolved as a result of our ancient ancestors' conversion of high forest into systems focused on producing woody resprouts. For many of us, we begin our journey from a similar point in forest succession. Like other issues, as a management strategy, woodland conversion has pros and cons.

Buoyed by the vigor of established trees' root systems, woodland conversion can provide a much more rapid return on investment than planting a new stand because the newly coppiced stools respond with rapid regrowth in their first season after cutting. There's no need to nurture seedlings through their early years. Management efforts can instead focus more on maintenance, protection, and enhancement.

Woodland conversion also has its share of shortcomings. It's pretty unlikely that the patch you've selected has your desired density and species mix without a fair bit of modification. Stool spacing in typical coppice stands is often as tight as 6 to 10 feet (1.8 to 3 m). Many forests just simply aren't this dense. To optimize the use of space and encourage upright growth with minimal branching, you'll need to do some restocking and fill in gaps.

Also, there's a good chance that some of the species present aren't particularly useful to you. In this case, something will need to change. Either broaden your goals to make better use of these species or work to remove or outcompete them, planting out the gaps with species you do want. So, despite the quick road to resprouting, it'll probably take some time to reshape existing high forest into the coppice stand you desire. And in some cases, you also may need to recalibrate your goals to better accommodate these on-site realities.

When assessing suitable sites for woodland conversion, pay close attention to the species composition, spacing between trees, tree age, stand density and structure, relative presence of young seedlings (especially undesirable opportunistic/invasive species), existing and potential access routes, solar orientation, and microclimates. By starting with patches that have relatively convenient access and already include the species and density you desire, you'll better conserve your time and energy and enjoy the benefits of this positive feedback.

By the same token, compare the relative productivity and quality of different sites to be sure to identify a stand or stands that are well-suited to your production goals. A healthy, diverse uneven-aged forest stand on good soils should be left that way and managed using the most appropriate silvicultural tool. In many cases, this will be selection management. And at the same time, a dense, overgrown, early successional oldfield on productive soils may well be a high-priority site for conversion. Needless to say, it can take a lot of energy, time,

A Woodland Conversion Strategy

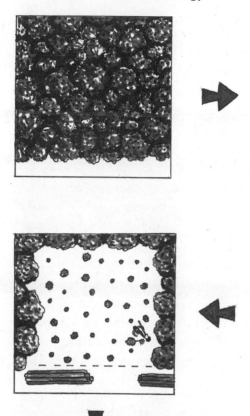

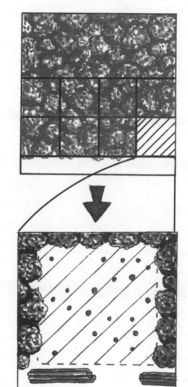

FIGURE 7.1: This illustration sequence demonstrates one strategy for transforming existing high forest or mid-succession woodland into a coppice stand. Top left: A larger-scale look at the forest stand prior to division into coppice cants. Top right: A simplistic layout for 8 square-shaped coppice cants. Middle right: The distribution of stumps (soon to be stools) that remain following the initial clearance cut of just a single cant. Middle left: The same cant following replanting and/or layering into gaps to improve stand density. Bottom left: The first coppice growth rotation in the cant. Bottom right: The cant following coppice pole harvest. Note the increasing size/number of of the established stools.

and resources to improve a site, so consider the site-quality values of each stand as best you can before choosing where your coppice systems fit.

Similarly, keep in mind that not all trees resist disease and heal wounds equally. When considering prospective stands for coppice conversion, know that healthy and unhealthy trees

Table 7.2: Things to consider— woodland conversion

Species Composition
Tree Age and Health
Stand Density/Spacing
Accessibility
Site Quality/Index
Aspect
Relationship to Future Managed Stands

often occur in clusters. Try to key into these distribution patterns to help choose which stands show the greatest promise for healthy regrowth. Trees with cankers (a localized lesion or dead spot caused by bacteria, fungi, physical damage, or other damage) are often weak and susceptible to infection. Cankers are often a symptom of other problems.[1] In general, don't invest time managing trees that already exhibit some type of significant defect. If possible, replace them with individuals with superior genetics.

In an ideal scenario, your woodland already contains a stand of healthy trees of the species you prefer, at an ideal spacing and an age where they'll respond with a vigorous coppicing response. In this case, woodland conversion amounts simply to a patch-scale clear-cut, carefully cutting all trees down to within a few inches of the soil surface. We'll discuss some recommended best practice techniques as to how to do this in

Case Study: Making Lemonade from Our Do-nothing Poplar Patch

Life doesn't always deliver what you thought you needed when you need it. And sometimes it takes some time to realize that what you already have could be just what you need, if you can only see it that way. On our small farm in Vermont, a small colony of quaking aspen (*Populus tremuloides*) continues to teach us this very lesson.

Quaking aspen is renowned for its ability to form an expansive colony via the formation of root suckers, new sprouts that form from adventitious buds on the plant's roots. I'd always learned to look at aspen (also commonly known as poplar, or here in Vermont as "popple") as something of a trash tree with soft, rapidly decaying wood with little commercial value. And that may very well be the case—until you can see it in a new light.

When we first moved onto our property, this colony of quaking aspen owned the northeast corner of our lot, clearly the offspring of one grandmother tree ☞

that towered over everyone else. For a number of years, I simply watched the colony slowly expand and decided to adjust my seasonal brush hogging to let it gradually colonize an increasing arc along our land's edge.

$ It wasn't until we started a small log-grown mushroom operation and began grazing sheep that I realized some of aspen's hidden values. While we've focused our mushroom production on shiitake that prefer dense, often slow-growing hardwoods, it turns out that oyster mushrooms actually prefer to grow on soft hardwoods like aspen, cottonwood, and boxelder. This simple recognition transformed my relationship to this clonal colony of trees already well-established.

In contrast to the thousands of trees we've planted and worked hard to ensure they establish themselves in the face of herbaceous competition and wildlife pressures, these poplar stems just keep on growing, the colony expanding and more wood accruing each year. And in so doing, they create a shaded boundary, intercept and slow winds, and provide free visual screening. As new sprouts emerge and expand the colony's spead, I selectively lop leafy branches to toss to our sheep as their grazing rotation takes them through this zone. The shade they cast also gives them some respite from the heat of the summer sun. And all the while, I continue to watch as these fast-growing sprouts add more and more diameter growth each season, knowing in another few years it'll be time to start thinning the patch and turning their soft, fast-grown wood into delicious oyster mushrooms.

While I wouldn't have planted these trees in this part of the property, they've freely offered themselves up to us. This do-nothing poplar patch is perhaps one of the most exciting and rewarding outcomes of our steady and open observation-based approach to ecological design. And in many ways, this to me feels like an ideal strategy to developing regionally appropriate and site-specific agricultural ecosystems. We work with what the landscapes give us, participating in succession and gently steering it towards useful outcomes for us and the larger ecological community.

FIGURE 7.2: This colony of quaking aspen continues to expand as far as we allow it. As the stems grow a bit larger, we plan to selectively harvest poles for oyster mushroom substrate.

chapter 8, but just remember that the stool that remains after cutting a tree is at least as valuable as the wood you just harvested. So, you'll want to protect them from damage as much as possible, both during felling operations and materials extraction. This is a major distinction between more conventional forestry operations where there's little concern for the health and integrity of the remaining stumps during and after logging.

If you'd like to create more of a coppice-with-standards type structure, then you'll need to identify and retain probably 10 to 25 trees per acre, and allow them to grow on as a long-term timber yield.

Unfortunately, this "ideal species, right age, perfect density" scenario is probably quite rare, and most stands will require varying levels of intervention to transform them into their true potential.

Existing Stand Dynamics and Their Effects on Cant Development

How do these details inform the specifics of our woodland conversion plan? Let's look at each individually.

Species Composition

It can take a huge amount of work to transform the species composition of an existing forested stand. Whenever possible, it's far easier to work with what we already have. So before choosing to liquidate a less than optimal standing forest, spend some time learning about the utility of these undesirable species, or at least begin to consider an alternative path. You may stumble onto some exciting surprises.

Stem and Stand Density

In coppice stands, stem density determines total overall production and also contributes to stem quality. The optimal stocking and stem density should vary with local climate. In humid zones, dense spacings (6 to 10 feet or 2 to 3 m) encourage straight growth, restrict competition, and should produce more poles. In drylands and moisture-limited ecosystems, optimal density is more a function of the available water resources.

A general recommendation for mixed species coppice in the UK is about 400 stools per acre (1,000/ha) on a 15-year rotation. This is equivalent to spacings of roughly 10 feet (3 m) between stools.[2] In many cases where precipitation isn't a limiting factor, increasing the stocking density can help to maximize short- and medium-term yields.

More often than not, newly established cants that result from woodland conversion will require a fair degree of restocking. The extra effort this requires also offers numerous opportunities to diversify the stand, improve stand structure, and create additional yields. If you aren't relying on natural (seed-based) regeneration, you'll either need to plant out gaps or layer in young coppice shoots.

Natural Regeneration: Restocking Forest Stands

Some stands will need significant modification before showing promise for efficient coppice management. In these cases, you might consider working to renovate the stand using some of the conventional silvicultural

Table 7.3: Short- and long-rotation coppice system characteristics

Species	Products	Disturbance Chronology	Spacing	Stools/Acre (ha)
Short-rotation Systems				
Willow	Baskets, sculptures	1–3 years	1–6′ (0.3–1.9 m)	1210–43,560 (3000–109,000)
Chestnut	Walking sticks, etc	3 years	6′6″ (2 m)	900 (2250)
Hazel	Hurdles, thatching materials, bean poles	7–10 years	8.2′ (2.5 m)	600–800 (1500–2000)
Long-rotation Coppice				
Alder	Turnery	10–20 years	10′ (3 m)	450 (1100)
Sycamore	Turnery	10–20 years	10′ (3 m)	450 (1100)
Ash	Tool handles, turnery	10–25 years	10′ (3 m)	450 (1100)
Chestnut	Fencing	15–20 years	11.5′ (3.5 m)	320–400 (800–1000)
Birch	Turnery	15–25 years	11.5′ (3.5 m)	320–400 (800–1000)
Hornbeam	Firewood	15–35 years	11.5′ (3.5 m)	320–400 (800–1000)
Linden	Turnery	20–25 years	11.5′ (3.5 m)	320–400 (800–1000)
Oak	Fencing	18–35 years	15′ (4.5 m)	80–200 (200–500)
Mixed Species	Firewood and fencing	15–20 years	11.5′ (3.5 m)	320–400 (800–1000)

techniques like seed tree and shelterwood cuts that we discussed in chapter 4.

While there doesn't seem to be much of a precedent for this in the literature, that's probably because there have been very few efforts to transform high forest into various forms of coppice management here in temperate North America. These techniques have a proven silvicultural track record. Using them could help pave the way to restock and redirect a high forest stand into a managed copse. Know that these interventions are complex with many variables at play, and they can be risky. They do appear to be the best tools for the job, but they may not always be the best practice for your specific patch of woods. If you're in position where you have existing forest that you'd like to convert to coppice management and anticipate it's going to require a lot of work to renovate, take some time, revisit your goals to make sure this is the best strategy for you and your woodland.

Seed Tree and Shelterwood Systems

To apply either seed tree or shelterwood cuts, we begin by identifying healthy, mature canopy trees of desirable species that we want to retain to disperse seed into understory gaps we create while clearing the surrounding woodland patch. These trees will help repopulate the stand, hopefully increasing the stem density of the species we want to encourage.

Seed tree cuts can either be uniform (retaining individual trees scattered throughout the stand) or grouped (preserving patches of seed trees). Whether uniform or grouped, make sure seed trees are spread throughout the stand to encourage even regeneration. Once again, be sure to keep the highest-quality individuals of the most desirable species so they're the ones that get to pass on their genetics. Optimal seed tree density depends on many factors, including the quality of the stand, the maturity of the existing trees, and the desirability of the species already present. General recommendations range from 1 to 15 trees per acre (2 to 25/ha).

Table 7.4: Common seed-bearing age of selected forest-grown trees

10 Years or Less	10–19 Years	20–29 Years
Red Maple	Paper Birch	Hickory
Aspen	Ash	Hophornbeam
Cottonwood	Black Walnut	Oak species
Black Locust	Tulip Poplar	
	Black Cherry	
	Basswood	
	Elm	

Adapted from Beattie et al., 1993, p. 106.

While the seedlings of many species will germinate in shaded conditions, heavy shade will significantly slow their early growth. Seed trees are usually removed as early as possible (generally within 3 to 5 years) to help kickstart the new stand's growth rate, leaving a young even-aged stand in their wake.

Shelterwood cuts take a more incremental approach to stand conversion than seed tree cuts, removing mature trees using a series of two or even three cuts to moderate the climatic effects on the understory while helping repopulate it. The main difference between these two systems is that shelterwood cuts involve a series of several harvests and can transpire over the course of 15 to 20 years or more. In most cases, shelterwood cuts probably will not be as appealing an option as seed tree cuts since the progressively thinned patch will tend to be shadier longer, slowing resprout and seedling regrowth and probably also encouraging the strongest growth among mostly shade-tolerant species. But keep it in mind as yet another tool in your tool kit.

The Kentucky Department of Forestry compared two shelterwood treatments to find the optimal number of seed trees to retain. They kept either 20 or 34 canopy trees per acre (49 and 84 trees per hectare respectively) on each of twelve 2-acre (0.8 hectare) sites of 60-to-90-year-old white oak (*Quercus alba*). They found that stands with 20 reserve trees per acre had twice as many regenerating seedlings than stands with 34. They concluded that the difference was the likely result of more available light for seedling germination and growth.[3]

If you're considering using either of these natural regeneration techniques to enhance the structure of an existing woodland for coppice management, keep in mind a few key principles.

- First check for signs of active seedling regeneration already taking place. On some sites, natural regeneration may be very difficult for a number of reasons. High wildlife pressure, poor seed production, erratic precipitation, etc. may make it difficult to get a new stand going naturally.
- Try to time felling operations with your desired species' seed production cycle. Many species have an irregular annual seed production pattern. Producing huge quantities of seed or "mast" every 3 to 5 years, followed by several years of little or no seed production, they help keep seed predator populations in check.[4] In many cases, you can observe seed development from the ground with a set of binoculars by late June.
- Many species will not tolerate competition with herbaceous vegetation. That said, many early successional species, along with oak and black locust, are exceptions and can tolerate herbaceous competition and even mature in grassland. You may need to manage weed growth for the first 2 to 3 years following cutting to ensure the development of strong, healthy seedlings and even keep a protective eye tuned in for up to the first 10 years, watching for excessive competition from shrubby and climbing weed growth.
- After felling, aim to complete forestry work before the spring to avoid damaging young developing seedlings.

Some Principles of Ecoforestry

The practices and philosophy of ecoforestry offers some great guidelines to help guide your management.

- *Retention:* Before choosing which trees to remove, first identify which trees you want to keep. Perhaps these are seed trees that you retain to increase stand density and also become long-term standard trees in your system.
- Minimize impact to riparian zones and other sensitive areas (provide a 50-foot+ (15 m) buffer minimum) and preserve existing drainage patterns as much as possible.
- Preserve structural diversity in forest stands and retain snags, mature trees, and fallen trees.
- Use low-impact removal methods whenever possible. Minimize road construction and ensure roads are small in scale, shallow in grade; pay close attention to the gentle redirection of water off the road bed into adjacent spongy forest soils.
- Leave coarse woody debris to decay, store water, and create habitat.
- Engage in a full cost accounting of management activities.
- When in doubt, don't![5]

Likelihood of Stump Sprouting

Woody plants' maturity has a major influence on their coppicing response. As a general rule, adolescent trees (at ages between 5 and 30 years and diameters of 2 to 10 inches (5 to 25 cm) respond with vigor. The disparity in resprouting vigor between young and mature plants presumably has to do with mature trees' reduced populations of and investment in viable dormant (preventitious) bud development.

While some species coppice readily at just about any age (California bay/myrtle, *Umbellularia californica*; bigleaf maple, *Acer macrophyllum*; coast redwood, *Sequoia sempervirens*), others including birch, some willows, and red alder may produce few if any sprouts when cut once mature. Unfortunately, the data on this characteristic among individual species is scant. It's generally safe to assume though that the more mature the tree, the less vigorously they'll coppice. For a much more in-depth overview of this topic, revisit Sprouting Ability and Time and Factors Affecting Sprouting Vigor in the second half of chapter 2.

In some parts of the US, oak species' regeneration has largely come to rely on stump sprouting due to increasing forest density, lower light levels, interspecies competition, deer browse, acorn predation, and fire suppression, among other factors.

Sands and Abrams monitored the number and height of stump sprouts from three oak species (black, chestnut, and white oak, *Quercus velutina*, *Q. montana*, and *Q. alba* respectively) on a 741-acre (300 ha) clear-cut in east-central Pennsylvania.[6] Overall, increasing diameter led to a decrease in stump sprouting in all three species. In other words, smaller stumps generated more sprouts, a pattern that we find with many sprouting species. Black and chestnut oak produced the most sprouts per stool. White oak stumps in the 4-to-8-inch (10 to 20 cm) diameter class produced the most sprouts, while black oak sprout numbers were highest among stumps 8-to-20-inch (20 to 50 cm) in diameter. Stumps with diameters of 27.5 to 31.5 inches (70 to 80 cm) produced the fewest sprouts. Research in eastern, mid-western, and southeastern regions found similar patterns.[7]

In a 1978 study in a Tennessee white oak forest, researchers found that trees over 8-inches (20 cm) dbh and 60 years of age produced few or no sprouts. They also found a wide variation in sprout formation for trees smaller than 8 inches (20 cm). The most vigorous sprouting occurred among roughly 40-year-old stools followed by steady declines until to the age of 60, where only 10% of stools resprouted. In their first year of regrowth, stools averaged about 13 sprouts each, ranging in height from just a few inches to over 4 feet (120 cm), averaging 2.5 feet (75 cm).[8]

Tools and Techniques for Filling In Gaps

Some woodlands only need modest improvements to reach their full potential as coppice systems. In these cases, the work really revolves more around filling in gaps than restructuring or repopulating the stand as a whole. In these scenarios, our main options are to plant out gaps using purchased stock or propagate new plants right there in the cant using vegetative propogation techniques.

Planting

To make a stand more dense and diverse, you'll need to either purchase and transplant seedlings or propagate them yourself. Many wholesale nurseries offer volume pricing on a large selection of tree and shrub species. Though economical, it's often far more preferable to acquire locally adapted stock when possible. Developing these connections and networks can take time to cultivate. Reach out to local nurseries and conservation associations, plant networks, or perhaps better yet, identify and collect your own seed from healthy local stands.

If you have the time, search your community to locate specimen trees with the qualities most valuable to you and collect seed. By propagating them yourself, you'll save on the cost of plants and also know their genetics are adapted to your locale. To maintain genetic diversity, be sure to collect seed from at least three individuals. The more the better. (We'll discuss how to start your own woody plants by seed later in the chapter.)

In some cases, it'll be faster and more efficient to propagate new plants vegetatively right there in the stand (or in a nearby patch). You won't be adding to the genetic diversity of the stand, but you don't need to collect or purchase anything from off-site, and you can make several new plants from a single stool. Two of these techniques are called layering and stooling.

Layering

You can easily create a clone of a plant from a branch or sprout using a technique called layering. For many species, layering happens naturally as their lateral branches bow down under their own weight (or that of heavy snow or fallen branches), bringing them in contact with the earth and stimulating the development of adventitious roots.

We can mimic these same conditions to inexpensively increase stand density using clones from a mother plant. The process is simple, and many first- or second-year coppice stools produce excellent sprouts for layering. You

Propagation by Layering

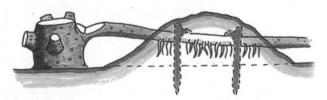

FIGURE 7.3: Top: Newly coppiced stool with one sprout left for layering. Bottom: The sprout has been bent so as to come in contact with the soil (while often not necessary, note the relief cut made on the top side of the shoot to ease bending). The stem is pinned to the ground, covered with soil, and left to form adventitious roots before cutting free from the stool and left to grow in place or transplanted to a new location.

just need to wait for the shoots to grow large enough to pin down.

Select a young, supple stump sprout located on the side of a stool near a gap in the cant. Gently bend it down towards the ground and scrape away a thin section of bark about 3 to 4 inches long (7.5 to 10 cm), exposing the cambium at the point where it meets the ground. If possible, to help prevent drying, excavate a short trench 1 or 2 inches (2.5 to 5 cm) deep and bend the stem down. Pin it to the ground with a forked peg, stone, or other weight. Mound soil up over the stem. Cut the tip of the stem, leaving only about 6 inches (15 cm) protruding (or 5 or 6 buds) beyond the peg to help stimulate root production. It can be a good idea to flag these layered stems so they can be easily found next season. See Figure 7.3.

The following autumn, check the shoot for root development. If it has taken, use a set of pruners or loppers to cut it free from the stool. You now have a clone of the original coppice stool. If it's too close to the parent plant, dig it up and move it to a better location. Because layering isn't always successful, it's a good idea to start up to 50% more than you actually need.

You could also use layering to provide a steady supply of transplants by managing a row or patch of stools in a more convenient location closer to home. This can help eliminate the lag that occurs when you layer first-year sprouts in the same cant where they develop. Instead, you'll already have rooted plants ready to install, helping speed up stand restructuring.

Stooling

Stooling or stool layering is another relatively straightforward vegetative propagation technique. Mound soil over and around the sprouts on a first-year coppice stool and allow them time to develop adventitious roots. The following spring, pull back the soil and check for root development. Cut each newly rooted stem free, and plant them as you would a bare-root tree. Again, be sure to mark these transplants clearly so you can care for them while they're still small. (For an illustration and more in-depth description, revisit chapter 4 and Figure 4.8).

These two low-cost vegetative propagation techniques have one main drawback. They create genetic clones of the parent plant, so you lose the variability that comes with plants of seed origin. In some cases, this can be a benefit because, ideally, you're propagating new plants from an individual that you know displays some desirable characteristics. But keep in mind that, if you're using the same few individuals to populate a large area, you're creating a stand with limited genetic diversity, which can leave it more vulnerable to pests and disease.

Of course, since these transplants are just one year old, they won't be as vigorous as existing coppice stools. The mature stools will tend to outcompete transplants, causing them to struggle from the competition. To accommodate this, you may choose to use a follow-up "renovation" clear-cut of the cant after 3 to 5 years to help transplants catch up with the existing coppice stools. This can lead to fuller, more consistent growth across the stand.

In brittle ecosystems where moisture is a key limiting factor, stand structure should probably resemble more of a savannah-type ecosystem with 20 to 60 feet (6–18 m) or more spacing between trees and shrubs. While this dramatically reduces per acre polewood yield, it opens up far greater opportunities for understory production, including berries, herbs, flowers, and pasture.

Prescribed Burns

Controlled or prescribed burns may be the most low-input, cost-effective woodland conversion strategy. They help reduce and manage fuel loads, cycle nutrients, control persistent pests and disease, alter stand composition, and stimulate vigorous regrowth.

Fire's role in ecosystem processes has largely been misunderstood for the past few centuries. As we discussed in chapter 1, Indigenous people around the globe have used fire for thousands of years to mimic the effect of brushfires and reset ecological succession.

Modern management efforts of targeted fire suppression have led to massive fuel load accumulations, creating the hazardous conditions that often lead to catastrophic fires. Prescribed burns during seasons when the potential for fire spread is low, on the other hand, keep fuel loads under control, and help maintain native forest cover and mid-successional plant communities.

Today there is a resurgence in exploring fire as a tool to rejuvenate ecosystem health, revitalize forest structure, support diverse wildlife populations, and reduce fuel loads, while also producing quality craft materials.

All that to say that this book will not explore best practices when it comes to safe management of controlled burns. It's a huge topic in and of itself, and not one that I'm particularly well versed in, living in the humid Northeast. For readers in western states and provinces, and even the prairies and humid eastern regions, fire is a powerful tool for ecosystem management.

Of course, never underestimate fire's destructive potential if not managed properly. If you're unfamiliar and inexperienced with safe techniques for prescribed burn management, seek the assistance of a skilled professional to plan safe fire containment strategies and control measures. We're all better off safe than sorry.

CONCLUSION

Converting standing high forest to coppice management requires considerable forethought, planning, and management intervention, but it can be one of the quickest and most cost-effective establishment strategies and is your most efficient option if you're working with a landscape with existing forest cover.

It certainly has limitations and may be a poor management decision if you have mature, high-quality high forest or exceedingly sparse, low-grade scrub. But no management strategy meets all needs in all scenarios.

For some of us, either we lack high forest to convert or the appeal of resprout silviculture lies in creating a completely novel ecosystem from scratch on open land. In these cases, we have the chance to design our systems from the ground up. And we also take on the considerable work and responsibility required to transform field into woodland. Let's take a look at what's involved with this approach to system establishment.

FIELD TRANSFORMATION

For some of us, our journey begins with open field where we assume the responsibility to design a productive ecosystem from the ground up. Despite higher up-front costs (seedlings, ground prep, stakes, tree guards, planting, watering, etc.) and considerable maintenance and monitoring during the first several years of establishment, field transformation allows you to choose the species, spacing, and layout. So, while it may take longer for your system to reach full production, you won't have to deal with many of the challenges associated with

Table 7.5: Field transformation— things to consider

Species
Uses/Purposes
Type of Planting Stock
Sources
Size/Maturity/Provenance/Genetics
Planting Site Quality/Fertility
Site Prep
Spacing
Landscape Pattern
Planting Tools/Techniques
Weed Control - Mulch/Mowing
Pest Control—fencing, shelters, etc.
Irrigation?

adapting and restocking an already established forest stand.

Today, we enjoy relatively abundant access to inexpensive nursery-grown seedlings, quality tree-planting tools, and water systems that help significantly reduce the likelihood of seedling mortality. It's actually pretty easy and inexpensive to plant a tree, not to mention an orchard or a woodland block. That doesn't mean it's easy to nurture and protect those plants so that they thrive and achieve their true potential. Reforestation or afforestation can be a major investment with high stakes and a long time frame. So being sure to take time to select the proper species, with the best genetics, suited to local conditions, can make the difference between great success and painful failure. Because at the end of the day, planted trees have but three options: grow, adapt, or die.[9]

Species

There's no single most important quality to inform species selection—it's up to you to prioritize your needs and the needs/qualities of your site using all the information available to you to make a well-informed decision. Because you've already taken some time to flesh out the details of the resprout silviculture system/s you envision, you've likely already answered a number of key questions. What do you want to produce? What qualities do you require in a woody plant to meet those production goals? What species already grow well on your site/ in your soils? Which species do you know will not tolerate particular characteristics of your landscape?

A Field Transformation Strategy

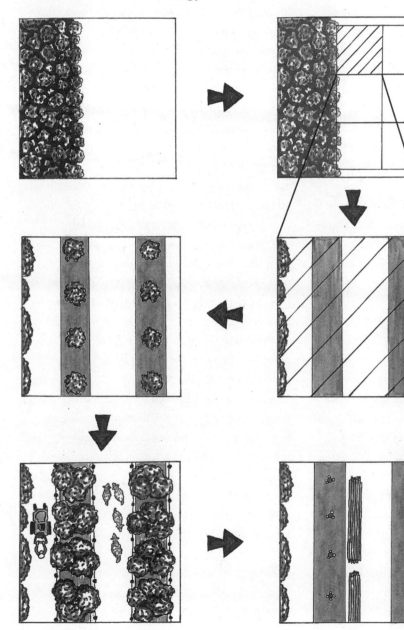

FIGURE 7.4: An idealized sequence depicting one approach to establishing a new coppice or silvopasture system in an existing field. Top left: The area of interest at a larger scale. Top right: A conceptual representation of the area broken up into individual cants for management. Middle right: A single sample cant with planting beds prepped using tillage and a bed shaper. Middle left: The same cant, following tree planting (note that the scale has been shrunk down far smaller than a typical minimum cant to show detail). Bottom left: The cant as the planted trees steadily mature, with herbaceous vegetation in alleys managed by periodic mowing or grazing. Bottom right: The cant following coppicing.

Choosing the species that populate your resprout silviculture system is perhaps one of the most important steps in your system design. This is certainly something you can't easily change after establishment. Depending on your familiarity with woody plant species and the specifics of your bioregion, you may already have answers to many of these questions. But if you don't, cast a wide net, get to know each option as best you can. Are there people nearby already growing them? What challenges have they faced? Many of us will find that we'll need to compromise at least somewhat. This is where we'll need to do some deep and honest analysis and clarify our priorities while identifying those areas where we're not willing to compromise.

We often begin by clearly describing the specific functions and/or products we plan to produce. Which of these are most important: rot-resistance, heat value (BTUs per cord/ton), density, growth rate, multifunctionality, fruit/nut production, wildlife value, nitrogen fixation, etc.? The species-by-use tables in appendix 2 offer a great starting point for many of these production-related decisions.

At the same time, pay careful attention to the natural communities in your locale. While some of the species you discover may not be particularly useful, perhaps it leads you to learn about a related plant in the same family or genus that has more valuable properties or qualities and is well-adapted to your specific site conditions. This strategy of identifying "ecological analogues" can be especially useful in developing species polycultures.

Because the soil, moisture, and microclimatic characteristics unique to your soil will take quite a bit of time and energy to transform, they should provide you with several key selection criteria. After learning more about these fundamental aspects of your site, research which species prefer those conditions and also learn about the ones that simply will not readily grow there. As the process leads you towards a more and more consolidated species list, it'll become much easier to narrow down the ones best-suited to your goals and your site.

If you have the time and space, consider planting several potential species, monitoring their performance over a few seasons. Though it may seem like a long time to wait, the feedback this experimentation provides will often prove a wise investment.

As a general target, narrow your options down to the most promising three to five species. If plant sources and financial resources allow, consider sourcing plants with varied origins. This may mean purchasing seedlings from several nurseries or collecting or purchasing seed gathered from a number of different established stands.

Keep in mind that vigorous sprouting is probably a characteristic few nurseries select for, so gathering different genetics and comparing their responses to each other may also help shed light on which coppice with the greatest vigor. Little if any work has been done to develop quality cultivars of coppicing species, so recognize that this is an additional benefit of the work you do while establishing novel systems.

Planting Stock

In most cases, young (1 to 3 year old), small-diameter bare-root trees will be the most economical stock for a new installation. Besides the price, transplanting young bare-root trees has numerous advantages over more mature potted and/or balled and burlapped trees, including the following:

- Transplanting is stressful. Young trees have a better ratio of roots to aerial parts and are more capable of putting on new root growth. The smaller the tree, the less dramatic the stressful effects of transplanting.
- Young trees are much easier to transport and care for before planting. They can be kept in cold storage until you're ready to plant in spring and they take up very little space. A bundle of 100 plants might be just 2 to 4 feet (0.6 to 1.2 m) long and 3 to 5 inches in diameter (7.5 to 12.5 cm).
- For larger installations, young trees can be notch planted or mechanically planted (as opposed to pit planted) saving considerable time.
- They are less vulnerable to wind damage and often do not require staking.[10]

Don't underestimate the value of quality stock with good genetics.[11] It's something you simply can't change later on. Remember that less than 1% of the trees that begin life in a forest ever achieve the opportunity to reproduce. And now consider that we aim for 80% or higher survival rate in our plantings, despite the fact that many nurseries do very little to weed out the weakest individuals. This means we're often planting trees that would have little likelihood of survival in a natural forest.[12]

Provenance describes the geographic origin of planting stock. A plant's provenance affects its hardiness, growth rate, form, tolerances, etc. Whenever possible, seek plant stock of local provenance with qualities well-suited to your site and your materials needs. If you're purchasing seed or seedlings from a commercial grower, ask where the original seedstock was collected. It's quite possible they won't be able to answer your question. If that's the case, you'll need to decide how important this is to you. If you're collecting seed yourself, choose healthy trees with good form and vigor, and avoid any trees that appear diseased.[13]

In most cases, we procure our planting stock in one of three forms: seed, bare-root, and cell-grown stock.

Seeds vs. Seedlings

Somewhat counterintuitively, field trials at the University of California at Davis in the late 1970s found that smaller plants often outgrow larger ones once in the ground. Several tree species started from seed, direct-planted in an unirrigated field outgrew well-watered trees started in flats, transplanted into liners, and potted up into one-, and later, three-gallon pots over the course of 3 years.[14] Often these growth patterns are especially pronounced with taprooted tree species, including most nuts and many oaks. Because taproots rarely survive transplanting, direct seeding is one of the only

FIGURE 7.5: A vibrant selection of bare-root trees, ready for planting. Note that the roots have been wrapped in wet newspaper and loosely enclosed in a plastic bag to prevent drying.

ways to grow plants with an uncompromised root system.

But direct seeding has its own share of challenges. Without some level of bed prep, direct-seeded woody plants can suffer significantly from the competition with herbaceous vegetation as they germinate and attempt to establish. They can also be very difficult to keep track of considering that herbaceous vegetation can easily reach 3 to 4 feet (90–120 cm) within 6 weeks of the start of the growing season. Without a good way to mark them and provide some weed control to help jump-start their development, direct seeding into an existing field is a very risky prospect. So, if you're interested in starting plants from seed, it's probably best to grow them out in nursery beds, transplanting them to their permanent location after one or two growing seasons.[15]

Table 7.6: Pros/cons—seed-, bare-root, cell-grown stock

	Starting from Seed	Bare-root Plants	Cell-grown Stock
Pros	Inexpensive	Relatively inexpensive	Often high-quality stock
	Requires very little space	Easily Shipped	Readily transplanted mechanically
	Relatively easy to trace provenance	Can be high-quality, adaptable young stock	Seedlings retain fine root hairs inside container
	Introduces genetic diversity to stock	Ready to plant upon arrival	Ready to plant upon arrival
Cons	Delays stand installation for a year	More expensive than seed	Often slightly more expensive than bare-root for plant stock and shipping
	Narrow window for spring planting	More difficult to trace provenance	Tend to be smaller plants
		Limited planting season	

Bare-root Stock

Next to purchasing or sourcing your own seed, bare-root trees are probably the most economical option for establishing a new system. Usually grown densely in in-ground beds, nurseries lift their seedlings in late autumn or early spring before sorting and bundling. Once dug, the plants are then either placed in cold storage or shipped direct to customers. Seedlings are usually sold either by height class in one-foot increments, or by their age, and in some cases, both of these characteristics are listed.

Nurseries often transplant or undercut young seedlings to encourage stronger, more fibrous root system development and sturdy, bushy growth. Many wholesale nurseries describe their stock using a two-number system that indicates the number of seasons they grew in their original bed along with the number of seasons they grew after transplanting. This means 1+1 trees have been grown for a year, transplanted and then grown for an additional year, whereas 1+0 plants just spent one season in their original planting bed before lifting. The 1 + 1 plants tend to have sturdier root systems, better able to survive transplanting than plants that have not been dug and replanted at the nursery before sale.[16]

While more expensive and time-consuming to plant, larger seedlings will usually outperform smaller seedlings, so if budget allows, it's worth buying the largest stock possible. Keep in mind that the quality of bare-root stock is more a function of the quantity of fibrous root material and root collar diameter than the actual height of the plant. Seedlings with stout root collars tend to be much more resilient.

Bare-root plants are generally only available during spring and fall months while plants are dormant and the ground is not frozen. If you dig your own or an order arrives from the nursery and you can't get to them for a few days, you'll want to keep them from breaking dormancy. Carefully protect their roots by heeling them in in a shallow trench. Choose a moist, protected, shady site with good soil, excavate a trench, and loosely lay the bundle of plants into the trench diagonally, taking care not to damage their roots. Lightly cover them with soil, loosely and carefully filling the spaces in between. Water them thoroughly and keep them moist until you're ready to plant. Otherwise, make sure their roots are well hydrated, wrap them in wet shredded newspaper, sawdust, or some other moist media, tightly wrap the root balls in a plastic bag, and store them in a cool, dark place like a basement or root cellar. Ideally, you'll plant your bare-root stock before the buds have swollen and they've begun to break dormancy.

Container-grown Stock

Container- or cell-grown seedlings are typically grown for one year in specially designed pots with a tall, narrow form. Often the cells feature vertical grooves that help promote root development without spiraling. The open bottom end causes the roots to "air prune" (see sidebar on page 360) once they reach the air, helping initiate lateral root growth. While cell-grown stock cost more to produce than bare-root seedlings, they can increase transplant success, especially for large taprooted species such as hickory, beech, walnut, and oak.[17]

Cell-grown stock are often planted in the autumn after their first season, though in some climate types, spring planting is preferential. Because their roots are protected by the soil within the cell, they can be planted quickly, suffering from less transplant shock than bare-root trees since their roots are more or less undisturbed. Because of the additional costs associated with producing them, cell-grown seedlings tend to be more expensive than bare-root trees and also take up more space in transport, leading to higher shipping costs. Also, they tend to be smaller than their bare-root equivalent, requiring thorough weed control since they can be easily overwhelmed by competing herbaceous vegetation. Because of their relatively small root system, cell-grown stock may require watering during dry periods.

Cell-grown stock is typically grown in hoop houses or greenhouses and hardened off by moving outside a few weeks before planting. If storing plants prior to planting, keep them well watered in a shady, sheltered location. If cell-grown plants must be stored in their containers over winter months, their root systems must be protected from frost in cold climates or they may suffer high mortality rates.[18]

Compared to bare-root plants, container-grown plants retain their fine root hairs inside the container. They often suffer less from transplant shock than field-grown stock, which have to regrow root segments that are lost when lifted from nursery beds. The root damage and loss that occurs when digging bare-root plants can be significant. Robert Woolley of Dave Wilson Nursery in Hickman, CA, estimates that digging machines may sever up to 50% of a pecan seedling's roots and 15% of the root systems of other species. By preserving seedlings' root tips, transplants can immediately absorb nutrients and water instead of investing energy in regenerating these damaged tissues.[19]

Setting Up Your Own Nursery

If you're looking to start your own nursery, you'll probably be propagating plants from seeds or cuttings. Both of these starting points are very space efficient and inexpensive. While you'll be one year behind where you'd otherwise be if you purchased seedlings, you could use that season to prep your planting sites for the following year and likely come out ahead of where you'd otherwise be with far less cost and greater confidence about the quality and origin of your planting stock.

Seedlings require very little space with densities of up to 10 to 12 plants per square foot (107 to 118 per m^2).[20] Compared to cuttings, seed propagation can be somewhat unreliable, with some species producing offspring with widely diverse genetics. But as we've discussed earlier, depending on your goals and context, this can be an asset as well as a drawback. It's not uncommon for plants propagated by seed to show a fair amount of variability, with some similar to the parent stock in growth rate, stem quality, and other characteristics and others lagging behind. For our purposes in resprout silviculture, this genetic variation may result in a diverse population that expresses different growing qualities, tolerances, and forms. If, on the other hand, you need known genetics from a mother plant with particular qualities, you'll generally want to propagate plants vegetatively.

Many good resources are available on best practices for starting your own nursery. Perhaps the best approach is to visit a nursery in your area that grows the types of plants you're interested in, using techniques that align with your goals and values. Over the past decade, small-scale nurseries have been growing in popularity. Akiva Silver has built a very successful nursery in New York State's Finger Lakes area on a half-acre and has been generously sharing his insights and experience on his website and blog. His model of high-density planting in fertile living soil in in-ground nursery beds aligns very well with the needs of many growers interested in coppice agroforestry. In a blog post titled "Starting a Nursery Business," he lays out a number of key principles to keep in mind.

- Just do it. Don't wait. You're destined to make some mistakes, and you'll also have much success. Start small with a few species you're interested in and build from there.
- It's all about the soil. Invest in compost and mulch and create a rich healthy matrix to give your seedlings a strong, healthy jump start. Create permanent beds full of organic matter and keep them permanently covered with woodchips or other organic matter.
- Grow your own or identify stock plants that express valuable characteristics you can continue to collect propagules from year in

and year out. Always keep your eyes out for quality specimens; ask the landowner for permission and build and nurture these relationships with people and plants.[21]

Be aware that some species have particular requirements that must be met before they germinate. You'll need to be sure you create these conditions before planting. Some seeds have dense hard coats that require some form of degradation or scarification before they can absorb water and begin to germinate. Some examples include black locust, honey locust, albizia, Kentucky coffee tree, and yellowwood. Weathering (physical, microbial, or chemical, sometimes achieved by fall planting, allowing natural conditions to do the work), mechanical abrasion (a file or abrasive wheel), and a hot water soak (steeping seeds for 6 to 12 hours in 190°F (88°C) water) can all be used to scarify seed.[22]

Many seeds also require a period of cold stratification, kind of like an after-ripening period for the embryo. It occurs at temperatures ranging from freezing to 55°F to 60°F (13°C to 15.5°C), with 41°F (5°C) an apparent ideal. Direct-sowing seeds in outdoor beds in the fall is the easiest stratification method, although it leaves them vulnerable to hungry wildlife. Some growers cover seeds with sand in flats, storing them outdoors or in cold storage underneath a layer of boards. Many growers simply store them in a moist medium (sand, topsoil, peat, etc.) and refrigerate them at 34°F to 41°F (1.1°C to 5°C) until planting. Plastic bags work well, retaining moisture and allowing gases to diffuse.[23]

Air Pruning Beds

Air pruning occurs when growing plant roots are exposed to air by growing through either the base of specially designed pots or the slatted bottom of dedicated planting beds. This causes the exposed root tips to die back, stimulating the development of a strong, healthy fibrous root system and also helps prevent roots from spiraling in the container, and constricting their development. Growing trees in air pruning beds or pots can help encourage a full, healthy root system, rapid growth, and even earlier yields.

Many woody plants spend much of their first season investing energy in a long taproot, only later producing the fibrous, nutrient-accumulating lateral roots. In the 1970s, Oklahoma State University horticulturalist Carl Whitcomb found that air pruning the taproot encouraged strong fibrous lateral root formation. Plants grown by air pruning have been found to reach twice the height, girth, and biomass above and belowground as bare-root transplants that weren't air pruned.[24]

To produce container-grown air pruned plants, select specially designed pots that help direct root growth to prevent circling and cause roots to air prune. These usually feature solid walls that offer easy seedling removal, hold enough soil to support strong growth, have ribbed sides to direct roots downwards, and have openings at the base that cause downward-growing roots to die back.

Over the past few years, I've seen several different variations on air pruning bed designs that allow growers to produce a large volume of vigorous strong plants in a very compact area. A friend and colleague of mine, Erik Schellenberg, owner of Black Creek Farm and Nursery in Ulster County, NY, shared a design that appears very simple, modular, effective, and easy to replicate with readily available materials.

He fashions the base of these beds using a commonly available shipping pallet. Make sure to select a pallet that's in good condition so it will last for years to come. Over the top of the pallet, he lays either ¼-inch or ½-inch hardware cloth mesh; this is the permeable bottom that causes downward-directed root growth to air prune. The sides of the planting bed are formed with lumber. At a minimum, build a box using 2 × 8 stock, and for deeper rooting plants, perhaps stack boards two layers high to make the bed as deep as 12 or 14 inches (30–35 cm). Fill the cavity with a mix of high-fertility substrate with good moisture retention. Erik prefers well-rotted woodchips blended with compost and mineral soil, covered with leaf mulch. To help ensure seedlings overwinter without damage from frost, he banks up bags of leaves that he collects from neighbors.

On top of the 2× material, he builds a frame covered with hardware cloth, using scrap oak stakes from a nearby vegetable farm to protect developing seedlings from rodents and other wildlife. It's easily removable, making weeding convenient. Plants are either fall seeded or, for extra-high-value varieties, spring planted after cold stratification over winter.

The bed in the photo contains roughly 300 seeds of a named chestnut cultivar. These seedlings should reach 3 feet high by the end of the season.

These high-value plants could fetch $15 each the following spring and digging them couldn't be ☞

easier. And because of the high moisture-holding capabilities of the substrate, they require minimal watering throughout the season. What a supremely compact, highly productive, value-dense production system!

FIGURE 7.6: Erik Schellenberg's modular air-pruned seedling bed design: 2x8 raised bed sides sit atop a hardware cloth-covered bottom elevated off the ground (not visible in this photo). This causes tree roots to naturally air prune as they stretch downwards beyond the bed's base. An easily removable top protects against girdling and browse but allows for convenient weeding. Photo courtesy of Erik Schellenberg, used by permission.

Cuttings

You can clone an existing plant and maintain and pass on characteristics of the parent plant using cuttings. Cuttings are taken from either roots or stems. Stem cuttings come from either hardwood or softwood shoots.

Research suggests younger growth will root more reliably than older growth. The theory is that plants increase production of rooting inhibitors as they age. Schrieber and Kawase tested this, monitoring the rooting success of cuttings taken from different parts of 12-year-old American elm (*Ulmus americana*). Collecting cuttings from the top of a tree, from sprouts off 6 to 7.5 foot (1.8 to 2.3 m) high stumps, and from 1 foot (30 cm) high stumps, they achieved rooting success rates of 38%, 64%, and 83% respectively. With this understanding, some growers who work with otherwise difficult to propagate plants prune them hard to keep them in a juvenile state.[25]

Softwood Cuttings

Softwood cuttings are taken from tender first-year growth and have a fairly narrow window for successful rooting. After collection, softwood cuttings must be kept cool and moist. Depending on local climate, the softwood condition usually lasts for 2 to 8 weeks between May through early July. It's generally advisable to avoid taking cuttings from stems during the flowering stage.[26] The ideal window and best practice for collecting and rooting softwood cuttings can vary by species, so it's difficult to make blanket recommendations. If you're interested in specific species, research their particular needs.

That said, in general, softwood cuttings should have several nodes and be 3 to 6 inches (7.5 to 15 cm) long. They will usually take anywhere from 1 to 5 weeks to root. Their bottom end is usually dipped in rooting hormone; gently set in a lightweight, moist, porous medium; and carefully kept moist until rooting, often by covering with a plastic dome or cellophane. To check if a cutting has rooted, give it a firm tug and feel if there's resistance. Rooted cuttings can be fall planted in a container, directly into the field, or overwintered in beds or hoop houses.[27]

Hardwood Cuttings

Collect hardwood cuttings while plants are dormant, beginning after they've lost their leaves in October-November and continuing on through late winter before new spring growth begins. Hardwood cuttings should be 6 to 20 inches long (15 to 50 cm) and ¼ to ⅜ inches in diameter (6.35-9.5 mm) to ensure they have adequate energy storage. Many cuttings root more reliably if treated with a rooting hormone. Cuttings that begin to leaf out before roots develop drain energy from the base of the cutting where rooting must occur. These often die because they can't yet uptake water to support developing leaves.[28]

Once again, the requirements and recommendations vary for different species. For many species, propagate cuttings using a porous, lightweight substrate, including bark, sand, pumice, soil, peat, or perlite. Rooting containers must provide adequate drainage and should be 3.5 to 4 inches (8.9 to 10 cm) deep.[29]

Willow, poplar, and elderberry are some of the easiest species to propagate from hardwood cuttings. In soils and climates where there's adequate spring moisture, you can often simply push them straight into the ground in their planting location, protect them from herbaceous competition, and they'll root and form a new plant. Select 6-to-15-inch (15 to 38 cm) lengths of healthy dormant stem. The larger the diameter the better, although they can be as small as 0.2 inches (5 mm). When starting hardwood cuttings, try to make the upper cut just above a bud. Square-cut tops make it easier (and less painful) to insert cuttings into the soil by hand.[30]

Start hardwood cuttings as early in the growing season as possible. Roots begin growing when soil temperatures reach 41°F (5°C), and it's important that they have a

chance to develop before the plant puts much energy into aerial shoots. This also avoids the potential for drought damage in summer if late-planted cuttings produce shoots that exceed the root's capacity. Late fall planting may increase cuttings' likelihood of rot over the winter or being heaved from the soil with spring

FIGURE 7.7: A vigorous, well-rooted one-year-old black currant hardwood cutting.

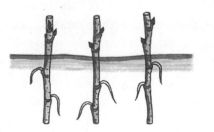

Willow Propogation using Hardwood Cuttings

FIGURE 7.8: A single shoot from a willow stool cut into 8-to-12-inch (20 to 30 cm) lengths, (top left) either left to soak in a bucket of water to form roots prior to transplanting, or inserted directly into the soil to root in place. Note that not all species will root this easily. Some have more species-specific requirements.

freeze/thaw cycles. Depending on the length of the cutting, leave anywhere from 1 to 4 inches (2.5 to 10 cm) protruding from the soil surface. The shorter the aboveground projection, the more drought-resistant the planting. If the soil is stony, you may want to use a tool to perforate a cavity for the cutting. This helps avoid damage to the cutting when planting. And of course—be sure the top of the cutting is facing upwards![31]

Root Cuttings

Root cuttings are also collected during dormancy, when their carbohydrate reserves peak. Plants that produce root suckers are particularly successful when propagated from root cuttings. And as we discussed earlier, root cuttings collected from younger plants tend to root most reliably. If possible, collect cuttings in the early morning when air temperatures are cool. Store them in moist plastic bags, wet burlap, or a cooler before planting out.[32]

The length of root cuttings depends on root diameter: for $\frac{1}{16}$-to-$\frac{1}{4}$-inch (1.6 to 6.35 mm) sections, take cuttings 3-to-4 inches (7.5 to 10 cm) long. For $\frac{3}{8}$-to-$\frac{1}{2}$-inch (9.5 to 12.7 mm) diameter roots, collect 1.5-to-3-inch (3.8 to 7.5 cm) cuttings. Thicker cuttings tend to produce shoots more reliably, while those from thinner roots often establish better. Plant them upright in a loose medium, either vertically with the proximal (the thicker end that came from a point closer to the plant's collar) end level or slightly proud of the soil surface or in flats horizontally, covered with $\frac{1}{2}$ inch (12.7 mm) of medium. Keep them well watered and maintain even moisture by covering flats with plastic.[33]

You can actually produce root cuttings in situ during late fall. Using a sharp spade, cut a circle around an existing tree and then cut a second concentric circle 6 to 10 inches (15 to 25 cm) from the first. Leave the cut roots in the soil and allow them to form shoot buds at their distal ends once spring arrives. That fall or the following spring, dig them up for planting.[34]

Of course, we've only begun to scratch the surface of the many best practices for plant propagation. As you get clear on the species you plan to build your systems around, do some much more in-depth research to learn about the most effective ways to create new plants, taking care to collect your planting stock from high-quality, locally adapted individuals. The efforts we invest in selection of quality genetics are a huge contribution to the world of resprout silviculture, both in our own bioregions and beyond.

Once you've procured your planting stock, whether they be cuttings, seed, bare-root, or cell-grown, it's time to put your plan into the landscape.

Layout and Spacing

Layout and spacing depends on species' growth characteristics, the fertility and moisture characteristics of your site, and your chosen system and landscape pattern. Dense spacing encourages straight, upright growth, with less lateral branching, rapidly shading out competing understory vegetation. That said, wider spacings will help accommodate the more extensive root systems woody plants need to develop to meet their water requirements when moisture is limited.

In humid environments, standard coppice spacing recommendations should be perfectly adequate. Three-to-5-year short-rotation cycles can range from 5-to-8-feet (1.5 to 2.4 m) spacing with longer rotations between 8 and 10 feet (2.4 to 3 m). British Forestry Commission grant schemes require standard plant spacings of 910 trees per acre (2,250/ha), or 6'10" by 6'10" spacing (2.1 m × 2.1 m). Adjust these numbers to your unique site conditions and context. Some growers even feel that 6.5-feet (2 m) spacing leaves too much space between stools and suggest as little as 4.9 feet (1.5 m) to give the new stand a jump start.[35]

Patterning in polyculture systems can be even more complex due to variations in growth rates between species. The British Forestry Commission recommends planting clumps of the same species within mixed-species cants. They suggest at least nine of the same species at 6.5 × 6.5 foot (2 m by 2 m) spacing, or, for 10 × 10 foot (3 m by 3 m) spacing, groups of 5 to 7 of the same species.[36] Honestly though, there really is little precedent to follow here, and this is something we'll need to learn and perfect by doing.

A quick refresher on plant spacing calculations: to determine the number of plants per acre for a particular spacing, divide 43,560 by the square of the distance (in feet) between individuals.[37] A spacing of ten feet (10 by 10 feet = 100 square feet) therefore requires 436 trees per acre. (In metric, divide 10,000 square meters by the product of the between- and within-row spacing in meters. This is the number of trees per hectare.) So a planting with 3 meters spacing both within and between rows

Table 7.7: Spacing for different types of systems/rotations

	In-Row Spacing	Between-Row Spacing
Short Rotation Coppice	1–3' (0.3–0.9 m)	1–10' (0.3–1 m)
Long Rotation Coppice	6–12' (1.8–3.7 m)	6–15' (1.8–4.6 m)
Silvopasture	3–15' (0.9–4.6 m)	10–100'+ (3–30 m)

would include (10,000 m² / [3 m × 3 m]) = 1,111 trees per hectare.

If your objective is to create a more natural, multistoried silvopasture system with an open understory, begin by estimating the mature size of each species. In most cases, you'll want to choose a spacing between plants that keeps their crowns from overlapping within the row and probably allows for even more unshaded space between rows. While pollarding will help limit their spread, open-grown trees tend to colonize far more area than they would in a dense woodland. An alternative would be to plant clumps of trees with gaps of at least 80 feet (24 m) between the clumps so that growing tree canopies don't shade out understory growth.[38]

With your spacing chosen, you'll need to lay the pattern out in the landscape. It's up to you to decide how accurate you need this layout to be. If you'll be doing any machine work or alley-crop type management in between the rows, you'll probably want to ensure more precision. But if it's just grazing or episodic polewood harvest, it could be a bit more seat-of-the-pants.

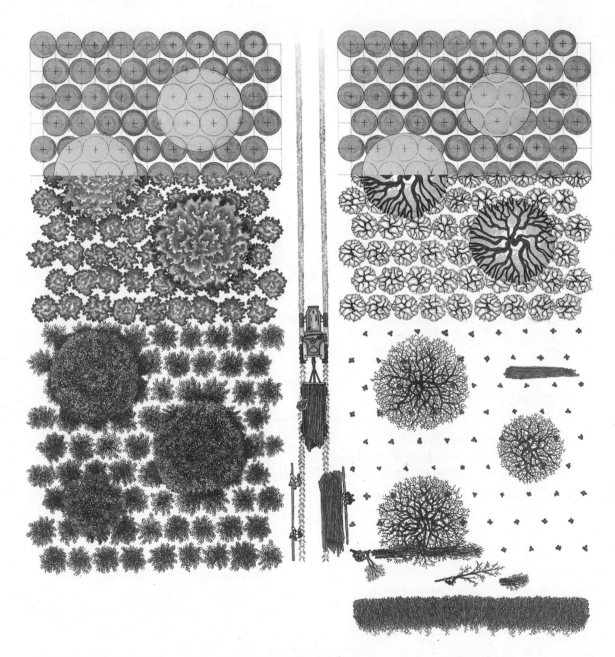

FIGURE 7.9: An example of a common coppice-with-standards layout on a patch that's roughly 1 acre. Coppice stools are on an offset 10-by-10-foot spacing, while standards are spaced roughly 65 feet apart for a total of 12 per acre. The central access lane is roughly 25 feet wide.

Start with a good supply of wooden stakes, bamboo rods, or landscape flags and a 200-to-300-foot reel-type tape measure. If you're especially lax with your layout, you could even rely on your pace, but it's very easy for inaccuracies to compound, so it's a good idea to measure to a reasonable level of precision.

Identify your starting point and run the tape measure along that row. Place your stakes or flags at each within-row planting location along the run. Depending on how large an installation you're planting, you may need to do it in segments, picking up the stakes as you plant and laying out the next section. In most cases, it makes sense to plant one row at a time, then stagger your next row offset to your chosen between-row spacing to optimize the use of space and solar capture between plants. In this way, you'll leapfrog from one row to the next, steadily filling in the patch.

While we won't discuss it here, know that technological improvements in GIS capabilities have made it increasingly possible to use high-resolution GPS equipment to translate a GIS-generated design directly into the landscape with sub-centimeter accuracy. This may be a wise strategy for large installations, layouts that aim to optimize water harvesting and topographic pairing, or those that involve complex and overlaid agroforestry land-use patterns. For many of us, however, this may be far more precise than we need.

Coppice-with-Standards

In a coppice-with-standards block comprised of just a single species (which I'd generally advise

against due to the myriad problems associated with monocultures), choosing standard trees within the stand has little consequence for the overall composition of the woodland. But if you're installing a coppice-with-standards system that includes standards of one species and coppice stools of another, you'll need to carefully design the layout to ensure you optimize the spacing and patterning throughout the whole system. This can become increasingly complex if you take into consideration the need to start with an overstocked pool of standards to accommodate for mortality, variation in growth rate, etc. Assuming the common recommendation of roughly 10 standard trees per acre, we can assume close to 65 to 70 feet (20 to 21 m) between mature standard trees. But they won't occupy that space until they're mature, so you may want to begin with spacings roughly half that dimension, anticipating the need to reshape the stand as it matures.

Bundle Planting

Although it's a relatively obscure practice, there may be scenarios where the practice of "bundle planting" may be useful. In parts of northern Germany during the Middle Ages, trees were occasionally bundle planted to produce multistemmed coppice-like clumps with wide canopies that produce mast (seed) and shade in wood pastures. They planted 2 to 7 seedlings of 1 to 3 or more species either in a single hole or individual holes spaced 6 inches (15 cm) apart. Whereas short-rotation pollards rarely produce seed during their management cycle and only create significant shade during the last year or

Idealized Site Preparation Strategy

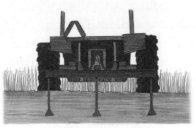

Subsoil plowing

Tillage

Bed forming/mounding

Cover crop

Mowing/crimping

Planting

so before harvest, bundle plantings provided more consistent, reliable mast for wildlife and shade for livestock.[39] Perhaps there are a few useful locations to employ this practice in your system design, or it simply gives you a useful way to make use of leftover bare-root stock following a large installation.

Ground Prep

Effective ground preparation can make the difference between a healthy, vigorous, thriving installation and slow, struggling, stunted growth. Good site prep improves plants' ability to survive during wet and dry times, outcompete herbaceous vegetation, and mature and bear fruit. This requires additional planning and up-front investment, but you only have one chance to make conditions as ideal as possible for your new planting. The most important issues to address when developing a comprehensive ground preparation strategy include fertility, compaction, weed pressure, and available moisture.

FIGURE 7.10: An ideal approach to site prep for tree planting using a compact or mid-size tractor and a few key implements. From top: Subsoiling plant rows (and the alleys in between if possible) to remediate any existing soil compaction; tilling the planting row with moldboard plow or rototiller to break up and loosen sod, incorporate existing vegetation, and create a clean bed for planting; (optional) follow up tilled bed with bed shaper, consolidating loosened soil from a 5-to-6-foot-wide tilled swath to a 3-to-4-foot raised bed; sowing annual cover crop to outcompete sod and existing soil seed bank; mow or crimp cover crop; plant seedlings into mulched bed in spring.

Fertility

In chapter 6, I described how US readers can uncover quite a bit of information about their soils using the USDA's Web Soil Survey (http://websoilsurvey.nrcs.usda.gov/). Enter your address, identify an "Area of Interest" and access an aerial map that indicates the prevailing soil types on the property. Keep in mind, though, that soils are highly variable even over very short distances, and this pattern-level soil-type identification is accurate only to a point. To learn about the actual physical and chemical properties of your soil, you'll need to do some on-the-ground testing.

Collect a soil sample for each ecologically unique zone you're planning to develop. Areas of varying slope, vegetation type, and aspect may differ in texture, structure, and fertility. At a minimum, select a lab that provides data on soil organic matter, cation exchange capacity, macro- and micronutrients, and pH. Be sure to clearly label each sample. Once you receive your results, compare your soils' properties with the fertility needs of the species you're considering planting.

When it comes to best practices, there's rarely a single correct approach. And soil improvement strategies are no different. In fact, it's a whole study in and of itself far beyond the scope of this book. Inevitably, we each end up deciding which approach best aligns with our philosophy, our site, our resources and capabilities. More than anything, remember that soil is far more than the sum of its parts (mineral matter, organic matter, pore space, air, water, and biology). It's an ecosystem. And it abounds with myriad complex processes and relationships.

Always strive to understand your soils from a whole systems perspective, considering not just the organic matter, relative proportions of macronutrients like nitrogen, phosphorus, and potassium, and pH. A few sources I've found particularly enlightening include John Kempf, Dan Kittredge and the Bionutrient Food Association, and Elaine Ingham, among others.

Taking time to properly address soil fertility and biology, the season before planting will pay long-term dividends. And remember that "optimal" is often a moving target—a destination, not an endpoint. We set off on a trajectory, monitor the results, accept feedback, and make adjustments.

Compaction

Cultivation, overgrazing, vehicle and foot traffic, as well as heavy machinery, all cause soil compaction. Compaction restricts water infiltration into the soil, reduces aerobic soil biological activity, restricts plant root development and overall productivity, and can last for decades or more. While over time, vegetation and soil biology may gradually undo these effects, breaking up hard packed soil, if left to natural processes, it can easily take many years to reverse.

Although primary tillage using a rototiller or moldboard plow can and, in many cases should be an important part of good site prep, aim to use them only during stand establishment. Subsoil cultivation is one mechanical strategy to relieve soil compaction and break up hardpan. Though it is a type of cultivation, subsoiling has a much lower impact on soil structure than tillage because it doesn't invert

FIGURE 7.11: Yeomans Keyline plow. This three-shank model (shanks aren't visible in this photo, they're buried in the soil) is decompacting a hay field.

soil layers. It instead shatters and loosens them. Repeated plowing and rototilling damages soil structure, can mix subsoil and topsoil layers, and destroys soil microbiological communities, while volatilizing soil organic matter.

Subsoil plows feature one or more long, heavy steel shanks fastened to a toolbar drawn by a tractor (note that there are also single-shank subsoiler designs that can be pulled by draft animals). The shanks usually have a narrow profile (between 1 to 2 inches thickness) and extend anywhere from 20 to 36 inches deep. There are a number of different subsoil plow designs on the market, and to varying degrees, they all do the same thing: soil decompaction through deep ripping. Well-known designs like the Yeomans Keyline Plow are recognized for their narrow shank profile and shallow ripping angle that helps minimize disturbance to soil horizons and reduces the

horsepower necessary to draw it along. Due to its rigid yet lightweight design, the Yeomans plow requires roughly 15 to 20 horsepower per plow shank and leaves an impressively clean finish in its wake. But if you can't locate a proper Keyline plow, just about any subsoiler is better than no subsoiler.

If you're planning an agroforestry system and suspect compaction is a limiting factor, consider subsoiling your tree rows as an important first step to your site prep. When you're planting long-lived woody plants, you just have one opportunity to decompact the planting row.

When subsoiling, remember that you're not only relieving soil compaction, you're also modifying the hydrological patterns of the landscape. Subsoil plow rips can function like small drainage swales, helping absorb and infiltrate surface runoff. So, think consciously about the pattern you choose to follow as you subsoil. Orienting plow rips down slopes may increase drainage on waterlogged sites, but it also increases runoff velocity, encouraging soil erosion and quickly ridding the landscape of this precious water resource. For water optimization and erosion prevention, contour patterning provides a far more ideal layout in both humid and dryland ecosystems. And while truly on-contour layouts slow and infiltrate water in place, Keyline patterning takes this concept one step further. Evening out drainage patterns through careful landscape analysis, Keyline patterning gently redirects water from concave valley landforms, downhill onto convex ridge landforms. For more details on Keyline patterning, check out any of P.A. Yeomans' titles on the subject, along

A Keyline Primer

Australian mining engineer, agricultural innovator, and land-use planner P.A. Yeomans developed a holistic land-use design system in the 1950s with the aim of developing on-site water security and actively building topsoil. He called his system Keyline. As a forward-thinking response to Australia's rapidly degrading agricultural landscapes, Keyline design responds to the topographic character of a landscape to inform the layout of farm, homestead, and community infrastructure. The result is a highly patterned landscape that slows, infiltrates, and evens out surface water flows, aiming to eliminate runoff and soil erosion while nurturing plants and reinvigorating soils.

The term "Keyline" refers to a physical plane in the landscape and derives from the location of the "keypoint of the valley." The keypoint is the inflection point in a valley where the character of the slope changes from convex to concave. One can often clearly identify a valley's keypoint because of a clear change in vegetation, plant vigor, or soil moisture.

In classic Keyline pattern layout, the contour line passing through the keypoint is called the Keyline. This contour line becomes the principal plane informing the orientation of subsoil plow passes and also often agroforestry rows, access roads, and trails, etc. While Keyline patterning follows a contour-like orientation, it typically only utilizes one true contour line, with other rows above and below the original Keyline, parallel, but slightly off-contour, often increasingly so with each respective pass. The ultimate goal is the evening out of water distribution across the landscape. Proper Keyline patterning should have the effect of gently redirecting water from points high up in valley landforms where it naturally concentrates, downslope across the adjoining ridges. This dries out valleys, which are often too wet for healthy crop production, and moistens ridges, which are more prone to droughtiness. It also has the added benefit of symmetry. Whereas most contour patterns are irregular, with variable spacing between contour rows, the Keyline pattern, once established, follows parallel both above and below the keypoint, making it far more well-suited to machine work.

In Oregon's Little Applegate Valley, permaculture and Keyline designer Tom Ward uses the pattern to inform the placement and grade of forestry trails. These paths fall 1:50 (a 2% grade) from the valley downhill out on the adjoining ridge. This gentle slope is suitable for both foot and bicycle traffic and feels almost level while providing a modest grade change across hillsides. This landscape pattern also helps break up otherwise complex landscapes into smaller compartments, perhaps helping distinguish boundaries between coppice cants or silvopasture paddocks with similar management regimes.

Yeomans' Keyline system also places particular emphasis on strategically sited irrigation ponds. While the details are beyond the scope of this book, multifunctional ponds with gravity-feed potential could be an important asset in holistic woodland management, especially for planted systems that may require irrigation during establishment. Often the keypoint of a valley provides one of the most efficient sites for gravity-fed farm ponds. Because the concave ☞

shape of the valley maximizes pond volume while minimizing the amount of earth moving required to build the dam wall on the downhill side, ponds sited at the keypoint can be particularly efficient surface

water storages. For more details on earthen pond construction, see K.D. Nelson's *Design and Construction of Small Earth Dams* and Darren Doherty's *Regrarians Handbook.*

with Darren Doherty's *Regrarians Handbook* (in press).

Managing Competition

If ever there were an appropriate time to use tillage, it's when establishing long-term no-till systems like agroforestry plantings. When young woody plants are installed into pasture, meadow, or grassland, they face an already established sward of vigorous herbaceous vegetation that can reach 4 feet (1.2 m) in height in the first 4 to 8 weeks of the growing season. Without doing something to break the toehold of herbaceous perennial roots, it's a tall order to plant young seedlings and expect vibrant, healthy growth.

We can choose numerous strategies to get a jump on this competition: targeted in-row tillage the season before, followed by sowing an annual cover crop; solarization or occultation using clear or black plastic; heavy mulch application; sod removal… you name it. The key is to do what's possible to give your new plantings a head start. Let's take a brief look at each of these.

Targeted tillage simply refers to rototilling or moldboard plowing the swath of land where seedlings will be planted. This should usually be done the season before planting for several reasons. In humid parts of North America,

bare-root tree planting season usually spans early spring months when soils are often wet and temperatures cool. This is often a challenging time of year to do any tillage since wet soils are highly prone to compaction. Also, site prep using tillage the season beforehand allows you to sow an annual cover crop to grow your own green manure mulch in situ and out-compete the advance of adjacent herbaceous competition. Tillage also helps loosen soils, making it easier to plant young seedlings. See Figure 7.10.

Solarization and occultation are techniques that smother and either "cook" or shade out existing herbaceous growth. Solarization uses a clear plastic tarp to heat up the soil below, kill existing vegetation, and even help reduce some soil pathogens. This technique initially stimulates weed growth, but ultimately cuts it off from precipitation and creates a hot microclimate at the soil's surface. During peak summer, the process can take as little as 2 to 3 weeks using 6-mm poly, but if you're starting with existing sod, you'll likely need quite a bit more time. In this case, occultation may actually be the better option.

Occultation uses opaque coverings to shade out weed growth. If it's done using a dark-colored material, it often reaches higher temperatures than clear-plastic covered soils.

The process does take longer than solarization, often requiring 4 to 6 weeks (or more if you're doing it over existing sod cover). You can use just about any type of tarp. Lumber wrappers are perhaps the cheapest option, often free for the taking at building suppliers, but they do tend to start to deteriorate after just a few seasons. Silage tarps are a relatively heavy, economical option and should provide several years of productive life. All of these materials will need to be well weighted down, using sandbags, stones, lumber, logs, whatever you have around. While they generally work best if they follow up primary tillage, they can create bare soil for planting if left in place long enough. When used on sod, be sure to mow the grass as short as possible before laying out the sheeting.

Similar in many ways to occultation, heavy mulch application can shade out and kill existing vegetation. However, it often takes a pretty heavy layer to do this effectively, and the sod often creeps back into the planting bed within a season or two. Large intact sheets of cardboard below the mulch can give an added layer of weed suppression. I often use whatever materials I can get free or cheap, including woodchips, shredded tree leaves, animal bedding (often free from horse farms), and mulch hay. You may have different biomass waste streams to tap into in your locale, but generally aim to use whatever you can get your hands on. When possible, apply it the season before planting as thickly as possible in as many alternating layers as reasonable.

Although relatively cumbersome, sod removal is a technique that actually creates less soil disturbance than tillage. I've rented a gas-powered sod cutter tool on several occasions to skim a strip of sod 18 inches (45 cm) or so wide by 1.5 to 2.5 inches (3.75 to 6.25 cm) deep to either roll up and allow to decompose, invert and use as mulch, or if oriented along contour or some other topographic pattern, flip downhill to form a berm to help capture and infiltrate runoff and passively irrigate young seedlings. Ideally, you'd still follow this technique with heavy mulch or a cover crop at the very least, but it does create short-term bare soil conditions relatively quickly

Whichever technique you choose to manage and set back existing herbaceous competition on your site, it's perhaps one of the most important steps you can take to prepare your site for planting.

Planting Techniques

There are many ways to plant a tree. They're each appropriate in different situations and at varying scales. As the old saying goes, the ideal is to "dig a $10 hole for a $1 tree." But this level of investment isn't always practical. Or necessary. Often, the scale of the planting will help point you towards which techniques are best. In this section, we'll discuss pit planting, notch planting, establishment using hardwood cuttings, mechanical augers, and tractor-mounted tree planters.

When it comes to planting, keep in mind a few basic principles.

- Try to time your plantings during the shoulder seasons, i.e., spring and fall. When temperatures are cooler, water is more abundant and seedlings experience less transplant shock.[40]

Case Study: Planting a Polewood-producing Edible Privacy Screen in Vermont's Champlain Valley

In addition to abundant sun, a rolling south-facing topography, heavy clay soils, a seasonally high water table, and some killer views, our home and farm lies starkly exposed along Vermont's largest north-south state highway. Of course, this is a blessing and a curse, and it's something I was very much aware of before deciding to purchase the property.

$ From my first visit here, I saw that the 1,100 feet (335 m) of road frontage screamed for a very

Figure 7.12: On-contour rows laid out and prepped using a sod cutter, removing an 18-inch-wide (45 cm) swath of sod, flipped downhill.

clear solution—a dense belt of multipurpose woody species to soften the road, act as a proving ground for diverse plant species, catch the eyes of passers-by, define our property boundary, and create a living classroom that provides habitat for birds and insects, while also generating food and wood products.

For this planting, I chose to follow an on-contour pattern, with in-row and between-row spacings of roughly 6.5 feet (2 m). I laid the rows out using a laser level and then rented a sod cutter to cut an 18-inch-wide (45 cm) swath of grass along that line, and fold it over downhill, creating a berm-and-basin-style infiltration swale.

I purchased close to 50 species of trees and shrubs from a wholesale nursery, planting almost 1,000 stems in a 30-foot-wide (9 m) band along the length of the hedgerow. My goal was to create a stepped windbreak belt that formed unique patches of diverse ecology for people to enjoy and us to learn from and utilize.

Following layout and sod cutting, we broadforked each planting site and then notch planted each seedling. The plants were tiny, 6 to 18 inches (15–45 cm) high. To ensure they were protected from girdling and browse, we made 18-to-24-inch-tall (45–60 cm) hardware cloth tree shelters for each plant. Due to the number of stems, we broadcast only a modest dose of compost around each plant, along with a cover ☞

crop cocktail to colonize the bare soil with desirable species.

It's now been 7 years since the system was installed, and we've gleaned many lessons in the interim. First, a clear set of winning species have emerged, including wild plums, elderberries, hybrid poplars, osage orange, bur and swamp white oaks, willow species, white mulberry, aronia berry, and several more useful species. And similarly, we discovered a number of, well, not winners, like American mountain ash and flowering dogwood.

Secondly, it's nearly impossible to overstate the importance of site prep. These trees went into unimproved, overtaxed, compacted clay hayfield. Without a whole lot of scything, mowing, weed whacking, you name it, many of these stems would have been swallowed up by herbaceous succession. Using the first season to try to break the perennial sod layer, improve soil structure, reduce compaction, and increase fertility likely would have made a big difference in the early establishment of many of these plants, in the long run saving on labor costs.

Choosing a spacing that works with your available equipment is very valuable. At that point, we didn't have any equipment, but once I got a tractor, the 6.5-foot (2 m) between-row spacing was virtually impossible to maintain with anything bigger than a walk-behind mower. All of our future planting installations have used between-row spacings of at least 10 feet (3 m).

And last, the value of organic fertilizer, mulch, and landscape fabric. The first few years we had limited resources, which largely resulted in a substitution of human labor for functions that could've been served by better planning, fertility, and weed control from the get-go. After a few years, we began to acquire much larger stockpiles of woodchips and mulch hay, each of which helped build soil in place and stem the vigor of herbaceous competition.

At the same time, we invested in 3-foot-wide rolls of landscape fabric, cutting squares to fit around each tree. These helped put a seasonal stop to some of the herbaceous competition that plagued seedlings from day one. I can only imagine what would've happened had we started the planting using these tools. ☞

FIGURE 7.13: Hedgerow at age 7.

【$】【⚒】 The stand has just begun to reach a point where we've begun to coppice individual plants as needed, sculpting this woody belt to serve numerous functions. The canopy has nearly closed, and the experience of the space has changed completely. From here on, we plan to continue to manage it as a thicket, letting some mature canopy tree species grow as standards and colonize the canopy like bur oaks and black walnuts; pollarding some trees like hybrid poplars, willow, and white mulberry for cut and carry livestock fodder; harvesting fruit off wild plums, elderberries, aronia, and sand cherry seasonally; top-work grafting crab apple, wild pears, and plums with cultivated varieties; and selectively harvesting chop and drop woody mulch and perhaps useful polewood as the stand continues to mature. The system is a bit of a hybrid. It's not intended to be an even-aged coppice rotation but rather a living sculpture that we can assess and engage with each year, continually learning from and improving it.

- Take time to evenly spread roots throughout the planting hole or notch. Rather than squeezing excessively long roots into a hole they're too big for, prune them off so they fit comfortably within the void you create for them.
- Prune off any damaged roots with a sharp pair of hand pruners before planting, leaving a clean cut to prevent infection and help stimulate fibrous root production.[41]
- Plant trees at the same depth as they originally grew (or just slightly deeper) and firm up the soil around the stem after planting. This means paying careful attention to the location of the root collar and setting them in the hole accordingly.
- Protect bare-root seedlings' roots from physical damage, sun, and wind. Keep them covered at all times while out in the field by wrapping them with wet newspaper, heeling them in to moist soil, or keeping them in a cool, shaded location with roots in a plastic bag so they remain moist and protected.
- Thoroughly water each tree after planting (2 to 4 gallons (7.6–15 L) each). One deep watering is better than several shallow ones.[42]
- Protect young seedlings from herbaceous competition.
- Insufficient planting holes, inadequate water, and excess fertilizer all compromise young trees' health. Remember, you're planting roots at least as much as the aboveground parts—handle them with care.

Keep in mind that the most active period of tree root growth only slightly overlaps with that of their aerial parts. Autumn-planted tree roots begin growing in short order, using nutrients stored in the stem. Their aerial parts are dormant and won't grow until spring. Root growth stops by late autumn/early winter. And as the soil warms in early spring, young feeder roots begin to grow although the aerial parts are still dormant. Soon thereafter, root growth slows as leaves and shoots begin to expand.

Table 7.8: Planting techniques pros/cons

	Notch Planting	Hardwood Cuttings	Pit Planting	Handheld Mechanical Transplanter	Auger	Tree Planting Machine (Tractor Mounted)
Pros	Very quick	Seconds to install	Potentially the best quality technique	Fast and easy way to plant cell-grown trees	Quick way to dig fairly large holes	1000s of trees in a day
	Minimal tools/equipment	Very inexpensive	Carefully orients roots in their new location	Easily plants hundreds of seedlings in a day	Loosens up soil in planting hole	
		Requires very little skill	Aids in amendment incorporation and decompaction		Allows you to carefully place soil around plant roots	
Cons	Notches are prone to opening up in expansive soils due to drying	Cuttings first need to develop a root system, so they can be slow to establish	The most time consuming and labor intensive	Sourcing the right planting stock for all species can be difficult	Equipment costs $500+ although can be rented	Requires a tractor and specialized equipment
	Fall planted trees prone to frost heaving	Require additional weed control early		Requires specialized equipment	Saves time digging but still requires careful planting	Can be difficult to follow complex patterning
	Can result in roots being forced to fit an inadequate hole	Can be difficult to keep track of			Should be avoided on wet heavy, clay soils	Stock can't be too big or too small
	Not well suited for larger bare-root plants					

This is why we tend to prefer to plant during the shoulder seasons. It helps give young seedlings a head start before they've begun to leaf out.[43]

To further kickstart bare-root trees, soak them in a mycorrhizal root dip solution prior to planting (note that you may even wish to supplement this with additional beneficial elements like molasses, effective microorganisms (EM), and more). Mycorrhizal fungi maintain symbiotic relationships with most plant species. This fibrous web-like fungal network vastly increases plant root's surface area and improves their ability to uptake water and nutrients (especially phosphorus). And it's a two-way street. In exchange, plants exude photosynthesized sugars through their roots, supporting the mycorrhizae's growth and development.

These fungal associations also can even connect individual plants living in close proximity to one another, enabling them to share nutrients and water resources, synchronize flowering times and other biological cycles, and alert neighboring plants to the presence of pathogens and pests. You can purchase powdered mycorrhizal root dips that include multiple strains of broad-spectrum endo- and ectomycorrhizal fungi. In healthy, fertile, biologically active soil, mycorrhizal inoculants may not be necessary because growing plants can acquire ample nutrients without assistance. But because these biological inoculants are a minor cost in long-term vitality of your planting, in most cases, it's likely a matter of being better safe than sorry.[44]

Pit Planting

Pit planting is the most careful and time-consuming planting technique. In general, holes should be deep enough to accommodate the plants' root system and 2 to 3 times as wide. Remember that $10:$1 hole-to-tree ratio? The care taken to custom dig each planting hole, loosen the soil, and carefully pack soil around the plant roots can take anywhere from a few minutes to 15 to 20 or more per plant depending on your soils and skill level.

My preferred tools for pit planting include a round-point shovel, a heavy-duty planting spade (a shovel-type tool with straight edges, a nearly flat blade, and often a short handle), and a mattock (sometimes called a pick-mattock). While any one of these tools can do the job well, I find that I can quickly dig a quality hole when using these three tools in tandem. If you'll be doing a lot of pit planting, it's well worth the investment to purchase a planting spade like the King of Spades. Typical round-point shovels are great for cutting and digging loosened material in a hole or a pile, but usually they aren't engineered for the heavy work that tree planting demands, especially in dense, rocky soils. Planting spades are solid steel with the handle welded to the tool head. They're designed for this type of work and can take a huge amount of levering and abuse. I find the mattock to be especially useful in skimming off any sod and also loosening up compacted soils, while the pick end comes in handy breaking in the bottom of the planting hole.

As you dig your hole, try to lay out the soil in the same order as you remove it. Be sure to keep track of the topsoil and save it for the last layers of backfill. If you're planting into existing sod cover, start by skimming it off and setting it aside. You'll use it at the end to form a berm on the downhill side of the tree to collect, direct, and infiltrate runoff right around the tree. Once you've dug your hole, make sure the depth is adequate for the tree you'll be planting. You don't want to have it ¾ planted only to find out that it's far too shallow and you need to start all over.

Carefully hold your bare-root seedling at the height you'd like to plant it, making sure that the base of the root collar is flush with the surrounding grade or slightly buried. Begin backfilling the hole using the soil in the reverse order it came out of the hole, gently but firmly packing the loosened soil around the roots as you go. Take time to train and direct the roots outwards in different directions, trying to mirror the pattern that each tree's roots would have followed if it were never dug up in the first place. Also make sure that you're planting the tree so that it's plumb in two perpendicular directions. You don't want to step away only to realize that your new tree has a major lean in one direction.

Once you've completely backfilled the hole, firm up the soil one last time. Unless you're expecting a storm, be sure to water it in well before calling it a day. Preventing water stress the first week following transplanting is very important to your plant's long-term health and well-being.

There are a number of different philosophies on whether to amend the planting hole or save fertilizer and/or compost for a final top dressing. I've come to believe in the latter strategy, and avoid modifying the soil in the pit, instead blanketing the surface surrounding your new transplant with nutrient-rich media. Roots tend to feed where conditions are best. Many tree planters maintain that the improved fertility and soil conditions in an amended planting hole can actually stunt root development because it discourages roots from exploring outwards into the unamended surrounding soil. And in dense, heavy soils, the loose absorbent texture of an amended hole can actually create saturated conditions that stunt and drown root hairs. Finally, the leaching effects of watering and rainfall will cause these nutrients to gradually leach deeper into the soil profile, so very little of this fertility is lost.[45]

Notch or Slot Planting

A planting (or dibble) bar is a heavy-duty, solid steel, narrow wedge-shaped spade, used to open up a notch in the soil for bare-root tree planting. While notch planting is perhaps the fastest and least expensive manual planting technique, it

FIGURE 7.14: A selection of useful tools for tree planting by hand. From left: Heavy-duty broadfork by Meadow Creature, pulaski used to skim off sod and loosen up soils, a mattock with long narrow pick for further decompacting planting holes, planting bar for quick notch-style tree planting, heavy-duty tree planting spade, round-nose shovel, and extra-wide fork for woodchips and other kinds of loose lightweight mulch materials.

Figure 7.15: A team of friends using planting bars to install bare-root seedlings into soil following sod removal using a sod cutter.

really only works well with seedlings and small transplants with compact root systems 6 inches (15 cm) wide and about 6 to 8 inches (15–20 cm) deep.

To use a planting bar, step on the spade's thickened edge, working it into the soil to create a cavity deep enough to accommodate the seedling's roots. Work the handle back and forth to open a wedge-shaped slit. Choose a bare-root tree whose roots will fit the notch, prune off any roots that are damaged or otherwise too large to fit the notch dimensions, and gently swipe them into the notch with a sideways wiping motion. Pull the plant gently upwards to encourage the roots to spread loosely inside the opening. Make sure the root collar is flush with

the surrounding soil (or up to 1 inch (2.5 cm) below the surface to accommodate for settling if the soil has been recently cultivated).

Reinsert the planting bar into the soil a few inches back from the original slit, and lever it towards the planting notch, closing up the opening. Often this is done twice in order to carefully firm up the soil around the root system and ensure no air gaps remain. And that's it. It can take just a minute or two per plant. For seedlings with larger, more developed root systems, use the planting bar to make two perpendicular notches in the shape of a "T" to better accommodate the roots.

Notch planting is best suited for light, friable, or recently cultivated soils and is more challenging in heavy clay-rich soils. Bear in mind that if heavy soils dry out excessively before young roots have had a chance to get established, the notch can open up and create air pockets that compromise root growth. Similarly, if notch planting into recently subsoiled ground, avoid planting directly into the rip lines. These zones are prone to shrinking and swelling which can also expose young tree roots.

While it certainly can be done, notch planting into established sod is risky because of the aggressive competition from herbaceous vegetation. To help give seedlings a head start, first remove the sod surrounding the plant and invert it to help smother the competition.

Planting from Cuttings

As we discussed earlier, some particularly vigorous species like willow, poplar, dogwood, and elderberry can be easily propagated from a 6-to-24-inch length (15 to 60 cm) of healthy young growth inserted directly into the soil, leaving just a few inches exposed. Cuttings should be stout (larger than ⅜ inch (9.5 mm) in diameter) and free of insect damage and disease. They can be inserted into the soil either vertically or at a 45-degree angle.

Cuttings planted vertically will tend to produce just a few shoots due to the central leader's apical dominance. When planted at a 45-degree angle, the cutting will often produce many more straight stems, all stretching vertically for light. Cuttings are a quick, inexpensive way to establish a coppice stand. Unfortunately,

Short-Rotation Coppice Establishment

Short-rotation willow coppice can tolerate a wide range of soils from mine spoils and gravel to heavy clays and sand, but grows best on well-aerated clay or sandy loam soils with sufficient moisture available within a 3-foot (0.9 m) depth.[46]

Establishing short-rotation coppice stands requires diligent weed control. Commercial growers either thoroughly till the planting site or they resort to one or two applications of a glyphosate-based herbicide (RoundUp), with the first application in mid-summer and a follow-up in autumn—a strategy that I avoid and do not personally endorse.[47]

Prior to planting, growers plow soils in autumn to a depth of 10 inches (25 cm), allowing the tilled soil to overwinter. (On lighter soils, it may be possible to hold off until spring). Cuttings for plantings stock comes from dormant one-year-old growth and is planted in spring. One to three shoots usually emerge from each cutting, reaching heights of up to 13 feet (4 m) during the first season. The following winter, growers coppice stems to ground level (within 4 inches (10 cm) or less of the soil surface), causing the stool to form 5 to 20 new sprouts the following year. These dense stands should reach canopy closure within the first three

months of their second season and are often harvested after 3 years.[48]

On a commercial scale, growers use either "step planters" or "lay-flat planters" to establish stands. The step planter cuts 4.9-to-8.2-foot (1.5 to 2.5 m) rods into 7-to-8-inch-long (18 to 20 cm) cuttings, inserting them into the ground and firming up the soil around each. Willows are usually planted in twin rows at 2.5 feet (0.75 m) spacing with 5 feet (1.5 m) between each set of rows. These systems average densities of 6,075 per acre (15,000 per hectare) or approximately 2'8" by 2'8" spacing. Lower-density plantings average 4,860 per acre (12,000 per hectare) or a roughly 3 feet by 3 feet spacing, and tend to result in thicker stems that yield larger chips. This becomes useful when chip quality is more important than overall stand yield.[49]

Short-rotation poplar plantations tolerate most soil conditions but grow best in deep fertile soils and do not tolerate shallow or waterlogged soils. Short-rotation poplar stands are often planted at densities of 4,050 to 4,860 cuttings per acre (10,000 to 12,000 per ha). Although they usually produce only 1 to 3 shoots per stool, poplars often outperform willow, in terms of total biomass produced, yielding fewer, heavier stems.[50]

only a few species are capable of establishing reliably this way, and most require ample water until they've had a chance to establish strong root systems, especially in more free-draining sandy soil types. Be sure to choose the source of your cuttings wisely, since you're essentially creating a colony of clones of the original parent plant.

Mechanical Planters

Handheld mechanical planters like the Pottiputki planting tube allow you to rapidly install cell-grown woody plant stock into prepared soil. There are a number of different mechanical planter designs on the market. These tools usually feature some type of hole-opening probe at the base, a chute to drop the seedling into, and a mechanical trigger that opens the gates at the base of the probe,

FIGURE 7.16: Action shot of two workers planting Christmas trees on a tractor-drawn mechanical transplanter. Photo courtesy of the Mechanical Transplanter Company in Michigan. This photo is of their Model CT-12.

releasing the seedling into the cavity. To use, you simply step onto the footrest at the base of the tool, drop the seedling into the chute, pull the trigger, remove the planter from the soil with a twisting motion, and firm up the soil around the seedling. These cleverly designed tools make hand planting quick and easy. They do have limitations in terms of the size of seedlings they can handle. Also they limit you to cell-grown stock, which is often less common in the conservation nursery trade, but for small and mid-sized plantings where you're able to source quality plants of the species you need, they're excellent tools.

Some growers like to use tractor-mounted or handheld gas-powered augers to rapidly excavate planting holes. Far faster than digging holes by hand, these tools usually have interchangeable augers ranging from 6 to 10 inches or more in diameter (15 to 25 cm). You'll still need to follow up and hand plant individual bare-root trees, but because the bulk of the work is often in the hole-digging and soil loosening, these tools significantly expedite the planting process. A note of caution though: take particular care in wet, clay-rich soils because their mechanical action can actually glaze the soil's surface and restrict water infiltration and root growth. But in the right conditions, these tools can work wonders.

And last, for large-scale plantings, by far the most efficient tool is tractor-mounted tree planting machines. These are especially common for Christmas tree farms and entities that do large conservation-type plantings. They're very similar to the waterwheel transplanters used by many commercial vegetable growers, and some growers modify this equipment to better-suit tree planting.

Featuring a coulter wheel that pre-perforates the soil, followed by a curved heavy-duty digging shoe that creates an 8-to-10-inch-deep (20–25 cm) slit, these machines accommodate 1 or 2 seated planters who insert plants, one at a time, into the opened seam, which is firmly closed up by a set of opposing press wheels. In good conditions, an experienced crew can plant more than 1,000 trees an hour.

A tree planting team ideally requires at least 3 people: the driver who carefully follows the planting line at a low, steady speed, 1 or 2 planters who ride on the machine and insert seedlings into the seam, and 1 helper who either supplies the planters with individual plants in the appropriate sequence (for multispecies plantings) or bundles of plants as they need them to keep the process moving along smoothly. If available, an additional worker can assist with maintaining the plant supply and follow up firming the soil around the plants.[51]

Purchased new, these machines are relatively expensive, ranging from $3,500 to $6,000 or more. But for projects where large-scale tree planting is an annual endeavor, they may easily pay for themselves many times over when compared to annual hand-labor planting costs. For smaller-scale growers, it's probably much more difficult to justify this expenditure. However, some local and regional conservation districts maintain inventories of specialized farm implements for rent and occasionally this includes mechanical tree planters. Before you

look into purchasing your own machine, do some research and contact your local ag extension, conservation organizations, or NRCS office to see if this tool is something that's available locally.

Planting a Hedgerow

Hedgerow establishment is much like any other form of agroforestry installation we've already discussed, but here are a few important concepts to consider. Because hedgerows are often relatively narrow tree belts planted in fields or along field edges, it's especially important to give plants a head start by knocking back herbaceous competition. Often hedgerow installations begin with some form of tillage (moldboard plowing or rototilling) to loosen the soil and give seedlings a boost. Again, it's okay to plant directly into established sod, but you'll need to be much more diligent with weed control during the first few years.

Because many hedgerows are often spaced very densely, it may be easier to rent a trenching machine or even hand dig a trench if necessary, instead of digging individual holes for each plant. For single-row hedges, excavate a trench to about the depth of a shovel blade (8 to 10 inches or 20 to 25 cm) and approximately 12 to 18 inches wide (30 to 45 cm). You could also use two appropriately spaced subsoil plow shanks to perforate and loosen the soil along the row, making hand excavation and planting much easier.[52]

On sites with shallow soil, poor drainage or other challenging conditions, you could try the "plowed ridge" or "crown" planting method.

FIGURE 7.17: A gappy neglected hedge recently laid with gaps planted out to increase density.

Using a moldboard plow, till a strip where the hedgerow will be planted and leave the inverted sod to settle and form a mound. This elevated ridge gives young seedlings deeper topsoil with improved drainage.[53]

Also, while we usually try to minimize the use of plastic mulch or landscape fabric, this could be an appropriate use to help kickstart establishment. Usually, plastic mulch is installed over a freshly tilled bed using a tractor-mounted mulch-layer. This could also be done by hand in a no-till situation, although it'll be far easier to bury the edge of the plastic if the soil is already tilled. Otherwise, it will require a lot of handwork to dig in and bury one edge of the plastic. Temporarily weight down the other edge using rocks, boards, sandbags, or whatever else you have at hand.

If using this technique, when it's time to plant, fold the sheet back to expose the bed and plant the seedlings. Once installed, cut them back to a height of 6 inches (15 cm), lay the sheet back across the bed, and carefully puncture it using the seedling tips. Finally, bury the loose edge of plastic in the soil. Once the plants are established, you can do your best to remove the plastic mulch. Unfortunately, plastic photodegrades over time and will be very difficult to remove completely without littering your landscape with plastic bits. For this reason, consider spending a bit more money and purchase a plant-based biodegradable plastic mulch instead.[54]

Of course, you can also plant your hedgerow using pit or notch planting techniques, but for densely spaced hedgerows, it will be a lot of labor. However you establish your hedgerow, many installers recommend cutting plants back as short as 6 inches (15 cm) after planting to encourage the development of strong, bushy vertical growth.[55]

MAINTENANCE AND PROTECTION

Adequate water, weed, and pest control are key to ensuring planting success, especially during the first few years. While thorough site prep (as we discussed in the Managing Competition section of chapter 7) will help get your plantings off on the right foot, it won't take long before herbaceous vegetation returns to rapidly colonize the disturbed soil. Because we avoid the use of chemical herbicides, this leaves us with a few key tools: annual cover crops, landscape fabric, and mulch.

Because you've created bare soil, this is your opportunity to colonize it with something that you'd like to get established. If you only select annual cover crops, knowing that they'll die back after a season (or in some cases winterkill) means that you won't be setting yourself up with more perennial herbaceous competition as young plants get established. Depending on the time of year and your local climate, buckwheat, field peas and oats, tillage radish, winter rye, and more can quickly establish a dense soil cover, create a biomass-rich green manure, outcompete wild vegetation, support pollinator populations, and buy some time for seedlings to colonize the site. You may need to mow or crimp the stand down as it matures since they may well overtop small bare-root seedlings within a couple of months, but cover crops offer a quick and relatively inexpensive weed-control tool.

For installations of more than 2 acres, mulch rapidly becomes cost, labor, and materials prohibitive. In my area, my go-to mulch options include mulch hay, horse (or other livestock) bedding, and woodchips. I like these materials because they're often abundantly available and cheap to free. If I can only choose one, I prefer woodchips since they're theoretically weed-seed free and dense enough to stay weighted down during windy days. They also provide great soil cover while helping nourish the soil fungal populations that are essential to growing healthy trees, and relatively long-lasting. Ultimately though, their weed suppression often only lasts a few months before grasses begin creeping in.

What's more, for any of these materials, you'll need a massive amount for most mid- to large-scale plantings. Assuming trees are spaced 6 feet (1.8 m) on center, a one-acre planting contains 1,210 plants. A 6-inch-thick (15 cm) layer of mulch laid in a 2-foot-by-2-foot (60 × 60 cm) patch around each plant will require 2 cubic feet (three 5-gallon buckets) each for a total of 90 cubic yards, about one tractor trailer load. That's a lot of woodchips!

Here's another way to look at this equation: Consider for a minute that a typical woodchip load from a local arborist is 8 to 10 cubic yards (6.1 to 7.6 m³). For simplicity's sake, we'll assume 10 yards (7.6 m³). That's 270 cubic feet. Now here's a very useful number to know. There are 7.5 gallons in 1 cubic foot. That means 1.5

Table 7.9: Volume of woodchips required per tree/patch

Mulch Ring Diameter	Mulch Thickness	Mulch Per Plant - ft³	Volume Per 100 Plants - ft³ (yd³)
2'	3"	0.25	25 (0.93)
	4"	0.33	33 (1.2)
	6"	0.5	50 (1.85)
3'	3"	0.56	56 (2.1)
	4"	0.75	75 (2.78)
	6"	1.125	113 (4.2)
4'	3"	1	100 (3.7)
	4"	1.33	133 (4.9)
	6"	2	200 (7.4)
5'	3"	1.56	156 (5.8)
	4"	2.1	210 (7.8)
	6"	3.1	310 (11.5)

5-gallon buckets is 1 cubic foot. I personally like to mulch young trees with at least three 5-gallon buckets of woodchips or roughly 15 gallons, 2 cubic feet. So that means that one arborist woodchip load is enough to mulch 135 trees. And for even better weed suppression, you may want to go a bit heavier. You also may need to renew it annually for the first few years.

Because of this, it may be worth considering landscape fabric to provide temporary protective cover and weed suppression. While this will keep herbaceous competition at bay for perhaps even the first full season, it doesn't take long before weed seeds begin to germinate on top and the root systems of running grass species and other spreading herbaceous plants begin to colonize the edges and work their way towards seedling stems. For this reason, I've found it's a good idea to remove the fabric after a season and re-lay it if it's still required.

One of the challenges with landscape fabric is that you need to pin it down somehow. Often this is done with landscape staples, long U-shaped metal pins that anchor the corners of the fabric into the soil. When possible, I've foregone the use of staples and instead laid woodchips on top of the fabric. This also helps protect the fabric from UV degradation and makes for a much more aesthetically pleasing look. But it also creates a more inviting vector for adjacent plant roots to creep in and colonize the mulch, so there's no perfect system.

Biodegradable options are available, including burlap, cardboard, coconut "coir" mats, and more. In the case of burlap and cardboard, I've found that they rarely last long enough

to provide sufficient protection, generally breaking down before the end of their first season. However, bare cardboard, not covered by woodchips, may last 2 or more years. I don't have personal experience with coir mats, but I've heard they can be effective both at slowing weed growth and competition and retaining moisture. Generally, ¼ inch or so thick (6 mm), these disc-shaped fibrous mats can easily be installed around seedlings. They come in a range of diameters, and presumably different qualities as well, and prices can vary widely, from 50 cents each to several dollars or more. Although they are made from imported materials, they likely have a much lower carbon footprint than any plastic-based mulch products and can simply be left in place to biodegrade once they've served their purposes. As always, there's no single ideal material. It's up to us to choose what works best for our unique application.

Tree Shelters

On many sites, tree shelters are essential to protect seedlings against browse and girdling by deer, elk, moose, rabbits, mice, voles, and squirrels. They are usually made from plastic or wire mesh (hardware cloth or window screen). Height depends on the wildlife threat. To protect against deer damage, you really need shelters that are 6 feet (1.9 m) tall.

There are quite a few different manufacturers and designs for plastic tree shelters. Many claim to have the added benefit of improved growth rate due to the warm greenhouse-like microclimate and increased humidity. There are even some reports of up to two to five times normal growth rates during the first few years.[56] In British Forestry Commission trials, unprotected oak transplants 8.5 inches (22 cm) tall reached 13 inches (34 cm) height in two seasons, whereas trees protected by a wire mesh guard reached 21 inches (53 cm), and those in ventilated plastic shelters reached 52 inches (132 cm)![57]

Plastic shelters do have some drawbacks. They each require a stake in order to stand upright, though this can often be as simple as an inexpensive bamboo rod. Regardless of material, most tree shelters also create a protected space for weeds, which can be difficult to remove without lifting the shelter first. Plastic tree shelters also force plant growth upwards with little opportunity for lateral branching, which, while great for polewood and timber trees, is less ideal for hedgerows and shrubs where more bushy growth forms are ideal. And in areas with hot, dry summers, tree shelters can actually cause heat stress, so it's important to choose shelters that are well vented. Depending on quality, manufacturer, height, and volume, plastic tree shelters can range from 75 cents each to several dollars or more. As with other plastic products, UV light will degrade them over time and lead to a shattered mess of plastic shards in the landscape.

Wire mesh (hardware cloth) shelters prevent against rabbit and vole damage, and if tall enough, though pretty uncommon, they could also protect against deer browse. One drawback is that side branches often find their way outwards through the mesh and begin to grow through the shelter, making it difficult to

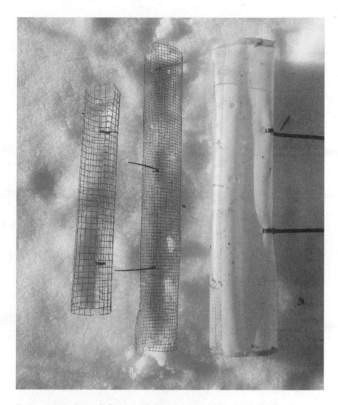

Figure 7.18: From left: ½-inch hardware cloth, 18 inches tall; ¼-inch hardware cloth, 24 inches tall; vented plastic tree shelters. Each material and design has pros and cons.

remove the guard as the tree develops. They allow for freer airflow and provide a bit of stability and wind screening.

Many growers make their own tree shelters relatively inexpensively out of rolls of ½-inch or ¼-inch hardware cloth. While ½-inch material is far more rigid, it's possible for rodents to squeeze their way through these larger openings. I've usually bought 3- or 4-foot-wide rolls, cut them in half lengthwise using an abrasive wheel mounted in a circular saw, and then used an angle grinder to cut sections to length, depending on the diameter opening I want to create. Zip ties are a quick and easy way to tie the edges together.

When installing tree shelters, be sure to work the base into the ground 1 to 2 inches (2.5 to 5 cm) deep to protect the stem from damage by burrowing rodents. Hardware cloth shelters that have been cut from half rolls of material have rough pointed ends that make embedding them into the soil quite easy. Be sure to check them a few times a year to make sure they're effective and haven't worked their way out of the ground.

Irrigation

Ideally, planting during the shoulder seasons minimizes the need for supplemental watering for new plantings. That said, rainfall patterns can be unpredictable, and water is essential to strong healthy seedling establishment. Most growers consider 1 inch (2.5 cm) of rainfall per week ideal for plant growth and development. If a dry spell lasts more than 2 weeks, water becomes a critical limiting resource for a new planting.[58]

Irrigation is an entire topic in and of itself. Usually after their first season, seedlings are well-established and don't require any water to survive, but irregular precipitation during the first season can lead to a lot of stress for everyone. Drip irrigation is just about the most efficient water delivery technique, but it requires a lot of plastic tubing and distribution infrastructure that's vulnerable to vehicle traffic and mowing. But it's a lot less time-consuming than hauling buckets to individual plants. There's no simple solution to this. Just do what you can to minimize water needs with well-timed planting, deliver additional water as required and you're able, retain soil moisture with mulch whenever possible, and hopefully you'll get your new planting off to a strong healthy start.

PROJECTING COSTS

Just like any other enterprise, if you're planning to develop a financially sustainable enterprise where you're not just coppicing for self-reliance or your own engagement with the landscape, you'll need to do some careful financial projections and accounting. For many agricultural crops, ready-made enterprise plans provide a template for growers to plot out their expenses and income. To my knowledge, there's little precedent for coppice product-focused enterprise plans, but we can look to similar enterprises for models that we can adapt to our unique context.

Given the perennial nature of our crop, if you're establishing your system through field transformation, orchard crops probably offer the best analogue. Do a search for one of the more common agricultural crops in your

TABLE 7.10: (See page 389) This draft enterprise budget begins to flesh out the most important income and expense line items related to the establishment, management, and harvest of black locust coppice during the first projected 10-year rotation. Keep in mind that these numbers will shift with successive rotations as all establishment costs will have to be realized following the initial harvest, and ideally established stools will produce 1.5 to 2 times as many poles as the first rotation. Again, these numbers are general figures that should be adjusted to local recent figures, but this can serve as something of a template to begin to plan a financial strategy for sustained harvest from various coppice-based wood production systems.

Table 7.10: Draft enterprise plan for black locust polewood coppice

Tree Spacing	At 6'6" by 10' = 670/acre			
INCOME		**Total Number**	**Price Each**	**Gross per Product**
Building Poles	Year 10+	200	$50	$10,000
Fence Posts		300	$30	$9,000
Firewood Cords		2	$200	$400
Hay/Pasture	Year 1-5	?	?	?
Suckers/Seedlings	On-going	?	?	?
			Gross Income	**$19,400**
EXPENSES				
Fixed Costs		**Year 1 Costs**	**Projected cost over 10-year rotation**	
Property Tax	On-going	$50	$500	
Insurance	On-going	$25	$250	
Lease Payments	On-going	$40	$400	
Total		$115	$1,150	
Variable Costs				
1. Establishment				
Site Preparation	Subsoiling - 1 hr/acre, tillage	$80	$80	
Fertilization		$150	$150	
2. Planting				
Plants	$1/ea	$700	$700	
Tools/Equipment		$200	$450	
Labor	58 hrs at $15/hr	$870	$870	
Irrigation		$500	$800	
3. Maintenance				
Fertilization		$150	$300	
Pest Control		$75	$750	

Table 7.10: *Continued*

Weed Management	1 hr/wk/ for 20 weeks	$200	$1,000	
Pruning	58 hrs (5 minutes per tree) @ $15		$2,500	
4. Harvesting	$250/day for 10 days		$2,500	
5. Marketing				
Promotion			$200	
Travel/Shipping			$300	
Total Expenses		$2925	$12,900	
		Net Profit after 10 years	$6,500	

bioregion that includes that crop's name and the term "enterprise plan." The templates you uncover should have many of the same line items. Of course, due to the high establishment costs and delayed return on investment, these calculations must be multiyear and complex. A comprehensive plan should include both the fixed and variable costs that we discussed back in chapter 5.

Fixed costs may include farm insurance; utility bills relevant to the enterprise; any rent, taxes, or mortgage payments directly related to the production area; depreciation on equipment that you use as part of normal business functioning, etc. These costs exist and remain consistent regardless of the number of products you generate.

Your variable costs are associated with batch size, so they can vary from year to year. This is most complex during establishment when you'll need to account for the cost of plants, ground prep, fuel use, planting labor,

any fertilizers, compost or mulch, and irrigation systems along with weed control, fencing/protection, and general maintenance labor. Assign an estimated value for each of these categories, some vary by the hour, others by volume. Often, it's a good idea to choose a higher value than you realistically anticipate since costs are often more than we'd like to think (hopefully).

Next, you'll need to project annual costs during the intervening years before harvest. This will be much simpler, yet there will still usually be some variable costs associated with maintenance and upkeep. And of course, those fixed costs stay the same.

And then there's the profit side of the equation. Here's where you project how many wooden widgets you produce, assign retail and wholesale costs for them, and determine what your total gross income is. Hopefully you factored in all of your labor into your variable costs along the way. The difference between your income and expenses reflects the net profit of your enterprise. Like the land-based design process we followed in chapter 6, it's this financial design process that allows you to measure your success in dollars and cents, compare one enterprise or product to another and provide you with some baseline data to ground truth your business performance in years to come.

One especially valuable aspect of coppice agroforestry compared to annual agricultural enterprises is that your establishment expenses are theoretically just a one-time cost. Once your system is up and running, you're able to sustain productivity for years and even decades to come. That's the benefit of a vigorous, deep, long-lasting root system.

And similarly, this is one of the benefits of woodland conversion establishment costs compared to new system planting. When you cut an existing woodland for the first time, you are already harvesting a valuable material. And in so doing, you stimulate the plants to regrow. So, you've essentially skipped the establishment phase—the landscape has already done it for you. Now, of course, that doesn't mean you've been dealt the perfect system from the get-go. There will be a new unique set of costs related to system enhancement that may be more complicated in some ways than simply installing whatever species you want onto a theoretical blank slate.

Even if you don't choose to build your own custom spreadsheet for your enterprise, I strongly encourage you to grab an old envelope, flip it over, and sketch out some rough line items and numbers. Personally, I'd rather get it roughly right than precisely wrong. The time you take to chart out your process will only make you that much more comfortable at understanding it as a whole system.

KEEPING RECORDS IS KEY

There's no better time to begin keeping good records than at establishment. Record keeping helps us measure the results and relative success of our management decisions. It ideally maintains a working log that chronicles each activity undertaken, the dates it was done, the condition of the landscape before and after, the weather and site conditions, any associated costs, the people involved; any wind, browse, girdling, or other types of stress or damage; and any other relevant details. Take notes on growth rates, lessons learned, challenges, weather patterns, etc. What would you repeat and/or do differently next time? Update these records annually, tracking the system's growth and development.

Fixed point photography using a digital camera, cell phone, or quality drone is a fantastic complement to written records. Be sure to gather some baseline perspective views, capturing several photos beforehand. Attempt to capture a viewshed that you can easily recreate in future seasons. To do this, identify a large stone, property corner post, easily identified tree, etc., and then use a magnetic compass to measure the bearing of your photograph. Record the starting point and compass bearing of each shot you take, along with the name of the photo file and save it as part of your management plan. Now you can recreate the photo in later years, creating a visual record of your work.

So, we've done it! Our design is installed, our systems established, and our trees growing! And… we've only just begun. Because this is when the real work begins, when we manage and maintain our investments in the landscape. In our final chapter, we'll explore best practices for making our first coppice cut, training and maintaining our pollards, laying hedges, and more. We'll examine the coppice worker's tool kit, the timing of the harvests, safe working techniques, material harvest, storage and extraction, management cycles, and techniques and strategies for protecting our investments. Trust me, the best is yet to come!

Table 7.11: Sample form for management record keeping

Stand Name/Number/Age		
Species		
Date		
Tasks completed		
Total labor		
Area of Stand		
Work completed by		
Materials Harvested		Volume Harvested + Notes
	Fuelwood	
	Species	
	Craft Materials	
	Species	
	Brushwood	
	Misc.	
Total Estimated Value		
Total Labor Costs		
Pest/Weed Problems		
Improvements Necessary		
Misc Notes		

Adapted from Beattie et al., 1983, p. 116.

Chapter 8:
Coppice Management and Harvest

The culmination of our establishment efforts arrives when we finally harvest the sprouts we've nurtured. And ensuring a successful harvest requires varying degrees of ongoing attention during the intervals in-between. Some of these tasks include selection, thinning, planting, propagation, pest and weed control, and access maintenance. And of course, there's everything involved with the harvest, collection, sorting, and transport of our raw materials. These practices all form the core of chapter 8. But first, we need to start with the very first step in the life of a coppice stool—the coppice cut itself.

FIRST HARVEST CYCLE: COPPICE ESTABLISHMENT

As we now well know, with a few notable exceptions, coppicing usually occurs when plants are dormant to minimize stress, ensure leaf litter return to the forest floor, minimize impacts to ground flora and forest soils (especially in cold climates where soils freeze), reduce the potential for disease and pest infestations, and leave regenerating plants the entire growing season to resprout and harden off before their next dormant cycle. (Tree hay harvest is the primary

exception, where you must cut sprouts while plants are leafed out.) Anecdotal evidence also suggests winter-cut wood is more durable than wood cut during the growing season. Because winter-cut wood contains less sugar, it may be less attractive to decay fungi.[1] It also has a lower moisture content, requiring less seasoning than wood harvested in summer.[2]

When to Cut

From the time when trees first lose their leaves until they break bud in spring, conditions are ideal for coppicing. Woody plants only really begin to add to their stored energy reserves in the current growth ring once their leaves have matured towards the end of their annual growth cycle. Just prior to dormancy, whole tree energy reserves peak. These reserves steadily decline throughout dormancy, bottoming out with the arrival of spring as new growth starts and leaves begin to form. This appears to be a particularly vulnerable stage in their annual life cycle. This is why it's important to avoid coppicing during this low-energy period between bud break and leaf emergence.[3]

In especially cold climates, it's also a good idea to avoid cutting during particularly cold weather—the idea being that it can cause damage to wood cells and compromise the cut face's ability to callus over.[4] Of course, our work schedule and the weather don't always align, but keeping these basic principles in mind may help you better strategize when you plan your harvest.

So, what can happen if you coppice too late? Early 20th-century research on chestnut stump sprouts in Pennsylvania compared sprout number, vigor, and quality among stools cut in winter and those cut in May. Stools cut in May produced more shoots, but they were low in quality and only reached about half the height and diameter of sprouts from winter-cut stools. The sprouts from May-cut stools developed soft shoot tips with wood cells imperfectly lignified in the upper 8 to 15 inches (20 to 37.5 cm), and 30% to 50% of their length suffered winterkill.[5]

A Harvard, MA, study on red maple stump sprouts found 100% of stumps cut between May 31 and June 30 produced sprouts, 4 out of 5 cut July 28 sprouted, while just 1 in 5 cut on August 24 produced sprouts. Sprouting stumps continued to produce new buds for 4 to 5 weeks after the emergence of the first bud, but beyond September 6, no new buds developed. Sprouts from winter-cut stools grew faster and taller than those cut later in the growing season.[6]

A French study of holm oak (*Quercus ilex*) found cutting during dormancy led to minimum stool mortality and maximized the number of new sprouts and sprout height and diameter, the only exception being a period of frost in January. Only 73% of these January-cut stools produced well-developed sprout clumps compared to 90% for other winter-cut stools. March, the month just prior to the start of their growing season, was the most favorable harvest month in terms of sprout development.[7]

In contrast, oak harvested for bark for leather tanning is usually cut during the early part of the growing season when the bark readily slips from the stem. This is the most vulnerable window for regenerating stools. Some theorize that low coppice yields towards the end of the 19th century were actually the result of these repeated early-season harvests.[8]

So, it appears that all signs suggest that, for optimal resprout growth and stool health, always aim to cut when plants are dormant. And despite this, it's probable that woodsmen historically cut woodlands throughout the course of the year as needed. During periods of high wood product demand, woodsmen sometimes couldn't afford to remain idle during a portion of the year.[9] But we know that this may unfortunately have had long-term costs for overall stand health.

Table 8.1: Effects of coppice timing on American chestnut height and sprout diameter

Timing of Coppicing	Average Sprouts per Stool	Average Sprout Height	Average Sprout Diameter
December-January	22	6.15' (1.9 m)	.42" (1.07 cm)
May	35	3.5' (1.1 m)	.23" (.58 cm)

From Mattoon, 1909, p. 40.

⚔ Where to Cut

Most coppice practitioners recommend cutting stools as close to the soil surface as possible.[10] In the case of hazel, new growth from low-cut stools often emerges from near or even below ground level, creating sturdy, well-rooted new shoots. This also helps minimize the "pistol-grip" sweep at the base of the stem that often renders the bottom 6 to 12 inches (15 to 30 cm) of the pole useless for most crafts.[11]

When I visited the sweet chestnut coppice managed by the Chevrier family in France's Limousin region, they showed me how their coppice resprouts emerged very near the soil surface, with little to no pistol-grip form (See Figure 8.1). They actually use a leaf blower to clear within and around each stool so they can cut it as low as possible.[12] Their technique produces top-quality chestnut poles, some of the highest-quality chestnut coppice I've observed.

Cutting as low on a stool as possible also helps minimize the likelihood of rot developing at the sprout's base. The development of **butt rot**—decay that originates at the base of a sprout—often corresponds with decay-causing organisms entering stump sprouts through two main points: the heartwood connection with the parent stump and dead sprouts.[13] By cutting sprouts as low to the ground as practical, new sprouts tend to originate low on the stool, reducing the chances of butt rot.[14]

And of course, there are exceptions to every rule. Oregonian master basket weaver Margaret "Pegg" Mathewson insists that cutting basket willow species low to the ground

FIGURE 8.1: Sweet chestnut sprouts emerging from a low-cut stool managed by the Chevrier family in Limousin, France.

encourages the exact same type of undesirable, asymmetrical growth (pistol-gripping) that longer-rotation coppice workers report occurs on high-cut stools. As a result, for a number of reasons, Margaret cuts her willow stools at approximately 24 to 30 inches (60 to 75 cm), something of a hybrid between a coppice stool and a pollard boll (See Figure 8.2). Given the intricate nature of her work, stem quality is of the utmost importance, and she confidently stated that she has found that stems emerging from an elevated stump produce materials better suited to her needs.

So which technique is right? Well, it appears that the jury is still out on this one. Based on what I've learned and experienced, I always try to make my final coppicing cuts as close to ground level as is reasonable. But if you're

FIGURE 8.2: Low willow pollards at Margaret "Pegg" Mathewson's. Despite these willow pollards having been cut roughly 3 feet above ground level, they produce very straight sprouts with little to no pistol-gripping. Also, it makes harvest much easier and more comfortable.

managing pollards, it's an altogether different prospect, and perhaps that's the reason for the difference in material quality. In the end, the only way we learn is by trying it and paying careful attention.

When cutting a previously coppiced stool, make fresh felling cuts in "new wood" between 0.5 to 3 inches (1.25 to 7.5 cm) above the site of the previous cut. Be sure to avoid cutting into older wood (the remnant of a previous coppicing), especially on old stools that have particularly thick bark or a limited number of viable dormant buds.[15] This helps ensure that a supply of new dormant buds remains ready to resprout and also avoids causing damage to any callus tissue that had compartmentalized the wood cells present at the time of the previous harvest.[16]

When harvesting coppice poles, always make clean cuts that slope away from the center of the stool. This helps direct moisture outwards, protecting the center of the stool from water accumulation and decay. Despite this widespread practice, I should note that research conducted by the British Forestry Commission found no evidence that the direction of the cut affects stump mortality, growth, or shoot numbers.[17] But ultimately, it just seems like a commonsense approach if you're concerned with stool longevity. In Mediterranean climates (in the northern hemisphere), orienting cut pollard faces to the north appears to

FIGURE 8.3: A neatly managed and well-organized coppice stand in Limousin, France. Poles have been cleaned up, sorted, stacked, and brush piled to the side out of the way.

help minimize the surfaces from drying, while in Scandinavia, growers often orient cut faces to the south to optimize exposure to sunlight.[18]

Thinning stools to just one or two sprouts early in the rotation appears to also help reduce the chances of decay.[19] According to Roth and Hepting, sprouts that develop high on a stool are more prone to developing decay. Any wounds caused by pruning these off young sprouts should be too small to create infection points.[20]

Some research also suggests that controlled burns after coppicing can kill buds that form high on stools, reducing the likelihood of rot. A Virginia study in a mixed oak stand found rot in just 0% to 17% of shoots after burning compared to 15% to 64% in unburnt stands.[21] And just 10% of black oak (*Quercus velutina*) sprouts originating below ground level had decay compared to 60% of those emerging 2 inches (5 cm) above the soil surface.[22]

And last, always avoid the temptation to store your ax or billhook off the ground by striking it into a living stool. This creates a wound exposed to pathogens and water. You want to leave a stool with as clean and intact a surface as possible.

When to Initiate Coppicing

Unfortunately, there's no definitive answer as to when it's best to make the first coppice cut on a tree or shrub. With some particularly vigorous species, coppicing after the first year's growth appears to possibly help multiply the bank of dormant buds, "programming" the plant to develop more sprouting capability in preparation for future cuts. However, it is important that young plants have enough time to establish a strong and well anchored root system before commencing coppicing. A number of sources suggest a minimum establishment period of 5 years and perhaps as long as 8 to 10.[23] When I asked Ben Law this question a long time ago, he suggested waiting until the stems have reached a size that's useful to you.[24] Seems like logical advice. For fuelwood this may be between 7 to 15 years or so.[25]

Most species sprout with greatest vigor during their juvenile/adolescent stages.

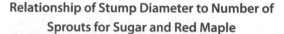

Relationship of Stump Diameter to Number of Sprouts for Sugar and Red Maple

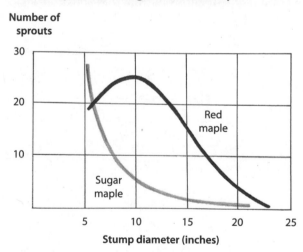

Figure 8.4: Here we see the general effects of stump (or stool) diameter on the number of sprouts formed on red and sugar maple. Broadly, the larger the stump diameter, the fewer the total number of sprouts. While more sprouts aren't necessarily better, this data does support the general concept that younger, smaller-diameter woody plants tend to respond to coppicing with greater vigor than older trees. Adapted from Solomon and Blum, 1967, p. 6.

Several studies suggest that sprout formation increases with increasing diameter up to 5 inches (12.5 cm) dbh, beyond which it decreases.[26] In a 10-year-long Missouri study comparing different forestry treatments on oak sprout regeneration, sprouts that originated from smaller-diameter stumps grew larger in diameter and taller than those from larger stools. The tallest sprouts emerged from stumps with a diameter of less than 6 inches (15 cm).[27] See Table 8.2.

And sometimes multiple coppice cuts in relatively short succession can be used to help stimulate a more robust sprouting response.

Table 8.2: Average 10-year height and diameter of the dominant stump sprout from three oak species by stool size class in the Missouri Ozarks

Species and Stool Diameter	Height	DBH
Scarlet Oak		
<6" (15 cm)	19.2' (5.6 m)	2.1" (5.3 cm)
6-11" (15–28 cm)	16.8' (5.1 m)	2.0" (5 cm)
>11" (28 cm)	8.6' (2.6 m)	1.1" (2.8 cm)
White Oak		
<6" (15 cm)	16.2' (4.9 m)	1.9" (4.8 cm)
6-11" (15–28 cm)	11.6' (3.5 m)	1.5" (3.8 cm)
>11" (28 cm)	6.4' (1.9 m)	0.7" (1.8 cm)
Black Oak		
<6" (15 cm)	13.9' (4.2 m)	1.6" (4.1 cm)
6-11" (15–28 cm)	15.6' (4.8 m)	1.7" (4.3 cm)
>11" (28 cm)	6.7' (2 m)	0.7" (1.8 cm)

Adapted from Dey et al., 2008, p. 31.

This process, known as "stumping back," is usually done on 1-to-3-year-old regrowth in newly established or neglected coppice stands as a way to rejuvenate stools and increase the total number of sprouts.[28]

Short-Rotation Basket Willow

In the case of newly established willow beds for basketry, it's common practice to make the preliminary cut the first year after establishment. Though they'll yield little usable material, it helps strengthen the plant's root system. Usually, it isn't until a willow stool's third year that it will produce a full crop of quality withies.[29]

A skilled willow cutter can harvest up to 30 bundles of withies per day, while a mechanical harvester averages about 2.5 acres (1 ha) per day. After harvesting, growers grade stems into three or four different quality and size classes. If rods are harvested during the warm window in early spring while the sap is rising and the buds are beginning to burst, it's possible to cleanly and easily peel the rods, producing bright white stock. During the rest of the year, they boil rods for several hours to soften the bark and make it easier to peel.[30]

POLLARDING

Sculpting a vigorous, well-balanced, productive pollarded tree has in many ways become a lost art. I'd like to extend a special thanks to goat/sow/cowherd and pollarder Shana Hanson of 3 Streams Farm in Belfast, Maine, for helping improve and inform this section. Shana has devoted considerable time to researching the history, art, and science of pollarding, integrating knowledge from authors Helen Read (England), Håkan Slotte (Sweden), Ingvild Austad (Norway), Michael Machatschek (Austria), and others.

While there are many parallels between coppicing and pollarding, there are also quite a few significant differences. Ultimately, pollarding is less forgiving than most coppice stool management, especially during the establishment phase. Coppice stools have a very short distance between the root system and the origin of new sprouts, making it relatively easy for the sprouts to draw sap upwards from the roots. A pollarded tree, on the other hand, needs to draw sap along a significantly taller and more extensive aboveground architecture, while at the same time requiring adequate sprouting in numerous locations along the stem in order to produce enough photosynthate to support the entire plant. The failure of multiple branches to sprout following pollarding, especially when adjacent to one another, can significantly impact the conductive capability and energy storage of the plant as a whole. Therefore, it's especially important to shape trees to an optimized and vigorous form before commencing pollarding.

Ultimately, to create a new pollard, you want to begin with a healthy well-rooted tree, with a robust canopy and a trunk stiff and tall enough to withstand livestock pressures. While trees that meet these criteria are well-primed for the severe cutting that begins the cycle of regular pollarding, the actual process of creating a productive pollard ideally begins much earlier in a plant's life cycle.

Preparing a Tree for Pollarding

Creating a new pollarded tree involves two main stages: the early pruning, shaping, and training to achieve an optimized architecture for your management goals; and the actual initiation of pollard cuts that establish the location and distribution of pollard "bolls, knobs, heads, or knuckles" (all of which are synonymous) that you'll repeatedly cut back to with each successive harvest. See Figure 1.4 in the introduction.

According to Michael Machatschek, the first 15 years of a young tree's life should be spent pruning and shaping it to develop a "rich crown." Often this involves pruning the central leader to initiate additional branching as needed, removing unnecessary side branches, and generally pruning the crown towards a balanced robust form. With this desired form achieved and the stem and branches adequately mature, it's time to make the initial pollard cuts that establish the pollard boll locations.[31]

Elizabeth Agate, author of *Woodlands: A Practical Guide*, recommends beginning new pollards by the time they've reached 4 to 6 inches (10–15 cm) in diameter, with some degree of success on trees up to 1 foot (30 cm) in diameter.[32] Austad and Hague write that Norwegian farmers begin pruning pollards to shape at an age of 10 to 15 years or roughly 4-to-6 inches dbh (10 to 15 cm).[33] In both cases, we'll presume diameter refers to diameter at breast height. Shana has heard anecdotally of the optimal diameter as that of a "wine bottle" or "a woman's lower arm" as the place to be cut.

Ultimately, best practices vary between the species, the local climate, and the quality of the site, among a host of other potential variables. And in reality, more important than the tree's age or dbh is the health, vigor, and size of the stem wood where you initiate pollard head cuts.

A number of observable characteristics indicate the relative vigor of a tree. Key among them are the length of the most recent year's shoot growth; the volume, color, and aroma of the foliage; and the foliage's relative palatability and desirability to livestock.

In general, as with coppice stools, sprout vigor and sprouting likelihood appears to decline as trees mature and lose vigor, which for some species, may be as young as 25 to 30 years. It's best to initiate pollarding before major branches have grown excessively large. This is important for three key reasons. Sunlight is crucial to sprout development and survival. Smaller-diameter young wood is usually located at the plant's periphery where there's better access to available sunlight, and hormonal activity tends to be strongest. Also, on many species, preventitious buds that lie beneath stem bark are more likely to break through the thinner, smooth young bark than thicker, furrowed older bark. And last, cuts into larger-diameter wood expose more end grain, increasing their susceptibility to fungal infection.

So as with coppice stools, we find that it's often best to initiate pollarding on adolescent trees. That said, there are a number of species can be pollarded at diameters up to 16 inches (40 cm). Trials at Essex, UK's Hatfield Forest found that open-grown ash, hornbeam, field maple, and pedunculate oak trees pollarded at 8.2 feet (2.5 m) successfully resprouted with no real correlation between stem diameter and likelihood of resprouting.[34] But more important than this initial sprouting is long-term sprout survival, and it's not uncommon to have low survival rates among young sprouts from

FIGURE 8.5: Young common buckthorn pollards, New Haven, VT.

large stems, perhaps due to poor or otherwise inadequate connection to the trunk's xylem and phloem networks. So, this again points us to the importance of shaping pollards early on in their life cycles.

🪓 How and Where to Cut

Often, people choose to pollard trees because it allows them to grow woody sprouts without a need to fence out livestock or wildlife. Because of this, pollard height can vary from 5 to 33 feet (1.5 to 10 m), but usually pollards begin at heights of roughly 6 feet (1.8 m) for deer, up to 7 feet (2.1 m) for cattle and goats (on their hind legs), and 9 to 10 feet (2.7 to 3 m) for horses. Keep in mind that winter snowpack will require cuts even higher into the tree canopy. Also know that regrowth will not always originate from the top of cut stems, so it's often best to initiate pollard knobs a bit higher than your ultimate goal in case sprouting occurs at a lower point.

When shaping trees towards an idealized pollard form, follow many of the same guidelines as you would if pruning a fruit tree or an ornamental. To encourage a full and balanced crown, inspect each branch carefully, and try to make your pruning cut above an outward-facing bud.[35] This encourages new growth out and away from the center of the tree, helping form an open, airy canopy while increasing the tree's spread. According to Shana, small young sprouts are often incapable of drawing sap long distances along the trunk or out along narrow branches, so aim for an economy of form that places pollard knobs closer to the main stem so as to help ensure sap reaches the plant's periphery.

Some of the key factors to consider when choosing the location of pollard knobs include the total available sunlight at each sprouting location, the estimated amount of available energy at each location and within each branch, and distance along the stem energy must be drawn through, since long small-diameter branches often do not sprout well. Available soil moisture and site fertility also affect pollard sprouting responses.

For some species, the more knobs you create, the less vigorous the sprouts. Shana generally recommends no more than 4 or 5 knobs on a young open-grown tree, and in some cases just 1, depending on species and goals. But remember, different species respond to management differently with some capable of supporting 12 to 20 or more.

Make pollard cuts at nodes, either just above a bud or ideally along a branch collar, whenever possible. Remember that cuts made at or between nodes leaves a decaying stub lacking any natural protection zones.[36] Often even well placed cuts made just above a bud may result in new sprouts forming below the cut tip, leaving a dead and decaying stub. If this occurs, it can be pruned off later as necessary. Helen Read's research of European pollarding traditions suggests foresters unanimously choose to start pollards with a cut above a fork or whorl in a tree to help stimulate a more even, spreading form.[37]

Often, cuts made towards the tree's top sprout quite vigorously, but lower, longer, and smaller-diameter branches may need some assistance to remain productive. Because of this, Shana advises against locating pollard knobs too far from the trunk, suggesting that for some less vigorous species they should be within arm's reach for surest sprouting results. In some cases, on large trees or trees in partial shade, it can even be best to leave the lowest branches uncut or lightly cut back. This maintains

FIGURE 8.6: A freshly pollarded sycamore in Switzerland shows the practice of retaining a sap riser sprout on each pollard knob as a means to help draw out sap reserves and stored energy from the tree to its periphery. Clearly, pollarded sycamores may support numerous pollard knobs. Photo courtesy of William Scott Bode, used by permission.

healthy foliar growth helping preserve a living ladder to access to upper sections of the tree. Here it can also be worth leaving single sprouts to serve as sap risers to aid in drawing sap up towards the branch's periphery. As the cut stem diameter increases, sap risers appear to become increasingly valuable, especially for species like beech and oak. See Figure 8.6. Also, Shana advises pruning or pinching off any sprouts that form on the pollard that are not in optimal locations for boll formation, just as you would water sprouts or epicormic sprouts on a fruit tree.

In the drier parts of southern Europe, and with early-successional species like birch, poplar, and many conifers, as in stump culture, it can be important to retain at least ¼ to ⅓ of a healthy canopy at any one time to limit stress and increase the likelihood of resprouting.[38]

Once you've established your pollard's architecture and the location of the pollard knobs, make future pollard cuts by cutting sprouts back near their collar along the pollard knob. Be sure to avoid cutting into old tissue below the sprout's point of origin.[39]

An idealized pollarded tree's form should balance crown geometry and mechanical loading. Pollarding redistributes mechanical loads to different parts of the tree that may be unprepared for these new weight distribution patterns. To minimize the risk of structural damage to the trunk, carefully choose the location of pollard heads, anticipate the weight under different weather conditions, and observe and respond to existing structural weaknesses that could lead to catastrophic damage.[40] But don't let the perfect be the enemy of the good. Even though a pollarded tree growing along a forested edge will tend to develop an imbalanced crown due to uneven solar exposure, these can be highly productive trees if managed well.

Last, when choosing a pollard's shape, consider the effects on sunlight and shading. Form is dictated by both the trees' need for plenty of sun at each sprouting location (especially a concern in a pollarded woodland), and the grower's understory management goals. Multiple pollarded branches will create a fuller, denser shadow that may be desirable for shading livestock in pasture, but less so for an alley crop-style vegetable production system. Pollards with a high single boll on a short-rotation cycle create small moving shadows, ideal for intercropping.

Accessing and Harvesting Pollards

Now, more practically, how do we physically access trees for pollarding? Pollarding introduces some management challenges since our materials are up off the ground in a low to mid-tree canopy. The strategies most suitable for your unique context will depend on the tools you have available and the height you choose to manage your pollards.

In many parts of Europe, farmers strategically retained branches on pollarded trees to create step-like access into the crown to facilitate sprout harvest. In Valdagno, Italy, farmers commonly left a 6-to-8-inch-long (15 to 20 cm) stub on each branch, called a *gropo*, both to produce fodder and to form this living ladder.[41] On particularly tall and/or columnar pollards,

retaining longer stubs that extend up to an arm's length from the trunk provide even more space for climbing and more comfortable perches from which to harvest.

Always strive to make the cleanest possible cut to help protect against fungal entry. At heights of less than 6.5 feet (2 m), a billhook, pruning saw, or a set of loppers may be all you need to cut pollard sprouts. Although difficult to do well, the smooth surface left by a sharp billhook leaves the best finish of the three. While easiest to use at ground level, loppers require two hands to use and depending on design and sharpness, may crush the bark on the underside of the cut, which isn't ideal to prevent infection.

To reach pollard heads beyond arm's reach, you'll need to elevate yourself to work safely and efficiently. Most simply this can be done using orchard-type ladders with a wide base and narrower open top, wide enough to nest around the pollard stem. This makes it easy to work within a tree canopy without needing to work above your head, which is often unsafe, physically demanding, and difficult to clearly see one's work.

At a larger scale, probably the two most efficient approaches I've seen are working from the bed of a pickup truck, loading lopped shoots right into the truck and/or trailer, and a tractor with front-end-loader/homemade cherry picker-type basket platform that allows the tractor operator to drive up directly to each tree and position a worker to harvest the sprouts from an enclosed platform. Both of these systems appear to be relatively efficient and safe without requiring any particularly expensive or specialized equipment. Of course, these vehicles do compact soils and also have limited accessibility in some applications, so they aren't appropriate in all circumstances.

And last, as we discussed in Loren Luyendyk's southern California case study in chapter 6, as the scale of the system increases, harvest volume and simplicity of system layout make other mechanized harvest tools suitable options. In this case, the idea would be to use a front-mounted hydraulic boom sickle bar mower to top the low pollards in mid-late June, leaving a remnant stub from that season's growth to generate a second flush of sprouts, harvested later that season by pollarding back to the boll. This system would generate two, and conditions permitting, perhaps as many as three cuts annually. Remember, though, folks, we're talking Southern California. Harvests this frequent aren't an option for most temperate climate-dwelling readers.

⏲ Timing

Depending on the product, pollarding usually occurs during mid- to late-summer or dormant months. While research conducted at Burnham Beeches in Slough, UK, found no significant difference in trees cut throughout the calendar year,[42] many experts suggest avoiding cutting during spring and early summer, especially during the month following the flush of spring growth. This is the point when the bulk of stored carbohydrates have been used up and will not be replenished until after the leaves have fully developed. Also cutting during

autumn when sapwood is driest may encourage fungal infection.[43]

In dry climates, high temperatures, drought, and extended periods of strong sunlight make summertime pollarding a riskier prospect. Some evidence also suggests delaying pollarding in the year after a pronounced dry spell, because persistent drought stress can jeopardize trees' starch reserves and affect sprout vigor.[44] Also expect some pollards not to sprout until the following spring, especially following dry seasons.

As with coppice stool management, when establishing new pollards, some growers consider it good practice to initially cut the tree twice, with just 2 years in between. This is said to help stimulate regrowth and promote new wood formation, enclosing the developing sprout and better anchoring it.[45]

And like with coppice stools, Scandinavians often angled cut faces south to receive more light; while growers in Mediterranean regions orient cut faces north to minimize surface drying.[46]

Disturbance Chronology

Pollards managed for tree hay are usually cut every 3 to 5 years, though it may be as frequent as 1 or 2 years and as long as 8.[47] While a few select species may tolerate annual or biennial harvests, frequent repeated harvests will stress most species on most soils in temperate climates. Recall our discussion in chapter 2 on the distribution and seasonality of woody plant's nonstructural carbohydrates (NSCs). Because the highest concentrations of these

NSCs lie in the branches and twigs, especially during summer months, excessive harvests may rob the plant of this important energy storage, steadily undermining their health and resilience. Repeated harvests also lead to reduced photosynthetic capability, compromising the tree's ability to feed itself, support new growth, and replenish carbohydrate stores.

According to Machatschek, in Austria it may be possible to cut pollards annually on a well-watered, nutrient-rich site, but it's usually best to leave 2 years between harvests. In Norway, 4 to 7 years tended to be the most common rotation, with 5 years considered a sustainable harvest cycle. This also sometimes included intermittent leaf stripping along with the collection of fallen raked leaves for fodder.

Once you've begun pollarding, it's best to maintain a relatively consistent rotation even if you don't intend to use the material. Pollarded trees develop an architecture built to support the weight of a regularly harvested crown. If unmanaged for an extended period, heavy outward growth may break off, often shattering attachment points and causing damage to the main trunk.[48]

In several parts of Europe, growers have been working to restore old pollards no longer in rotation. If pollards have gone unmanaged for some time, it is possible to revive them, though it must be done with considerable care, as some species, like beech respond quite poorly in the UK. Generally, trees that haven't been cut for long intervals require an incremental approach to canopy thinning.[49] In Essex, England's Epping Forest, woodsmen have repollarded over 1,900 trees since 1981, finding wide variability in success rates (91% with hornbeam, *Carpinus betulus*; to 53% for oak, *Quercus* spp.) and as low as 7% for beech (*Fagus sylvatica*). Some beech pollards didn't even produce buds until the second growing season, and in a few exceptional cases, regrowth didn't emerge until the fourth growing season after cutting.

Reports from Romania and Hungary describe woodsmen selectively harvesting individual pollard sprouts while retaining others. Helen Read reports this to be an important practice to follow with beech and oak trees, recommending the selective thinning of larger branches while leaving smaller ones.[50] So ultimately, it appears that best practices tend to be species-specific. Whereas the removal of all branches on a young beech pollard would likely kill the tree, and experience in Noway, Sweden, and Maine all suggest birches and poplars should only be thinned, vigorous species like willow appear to do just fine when all branches are harvested.

In some wood pastures, graziers give animals direct access to woody browse. While these fodder banks save energy and harvest time, it's important not to overbrowse and stress plants to the point that they cannot recover. In cool temperate climates, do not browse woody plants more than twice during the growing season to avoid unrecoverable strain. Also avoid browsing plants when they are especially vulnerable in spring as they are leafing out and as they enter dormancy and drop their leaves in the fall.[51]

Case Study: Sculpting and Tending an Air Meadow in Maine

Name: Shana Hanson and 3 Streams Farm

Location: Belfast, Maine

Ecoregion: 82f Midcoast

Koppen-Geiger Climate Zone: Dfb—Humid Continental Mild Summer, Wet All Year

Elevation: ~85 feet (26 m)

Average annual precipitation: 47.93 inches (122 cm) but increasingly unpredictable

Hardiness zone: 5

Site aspect: Rolling strips of well-drained ground alternating between streams and moist seams, running west-southwest to east-northeast, with a slight overall northward incline.

Soil types: Tunbridge-Lyman, B slopes

Special Tools Required: Hand pruning saws (Samurai and Wheeler), chain saw (using canola oil for bar oil), homemade lightweight wooden ladders, climbing ropes, and homemade rope harnesses. Airtight plastic barrels and buckets for small amounts of silage. Barn loft or lumber tarps for dried leaf fodder.

Training/Background of Manager: Shana noticed the toxic plumes of exhaust from passing cars as a teenage New Jersey bicyclist. Realizing the preponderance of fossil fuels in nearly every processed food, alongside the broken fertility cycles in agricultural production, she felt driven towards a self-sufficient lifestyle. In fall of 1982, Shana moved to Maine at age 19, using money she'd earned as a street musician to buy land and tend her own garden.

Shana has been pruning and grafting fruit trees since 1983; harvesting and later growing wild blueberries since 1989; living as a seed-saving, complete-diet subsistence farmer from 1983 to 1988; producing and saving seeds from 2000 to 2009; and producing milk since 2004. She's also taught her son and many interns and WWOOFers to know this way of existing. She broadly aligns her lifestyle and work with climate action, re-greening, enriching, and hydrating land surfaces[52] as a positivist reaction to mainstream practices that lead to the drying, killing, burying, covering, and exporting of soil and biomass.

Following challenging and unpredictable annual and biennial seed-saving conditions during an exceedingly wet 2008 and 2009, Shana transitioned towards tending wild perennial pastures and woodlands, increasing her herds of goats and geese. Around 2010, at her goats' request, she began pruning trees for fodder. She first learned that she was pollarding in 2011. She soon dove deeply into the practice, seeking resources from a number of experts in the field, including Helen Read, who sent along her study of pollards in eight countries, leading her to Håkan Slotte's doctoral study on leaf hay from 2000. When Dave Jacke and I referred her to Michael Machatschek's book *Laubgeschichten*, she sought the help of local translators who spent many hours helping decipher these foreign language resources.

In winter 2018, Shana attended and spoke at Colloque Trognes (Colloquium on Pollards) in Sare, France, later returning home to begin a SARE FNE18-897 farmer grant, which led her and interns Josh ☞

Kauppila and Emily MacGibeny to climb and pollard an acre of woodland into an "air meadow" for fodder, and measure the yields by monitoring the weight of her goats. The study also included chipped/ensiled and chipped/dried fodders, informed by Ingvild Austad's study of shredded tree fodders for sheep.[53]

Species: Mixed broadleaf and conifer species. In order of quantity: red maple, balsam fir, white pine, beech, white cedar, white birch, yellow birch, white ash, red oak, big-toothed aspen, quaking aspen, American elm, brown ash, hemlock, black cherry, American basswood, hophornbeam, fruit and nut ☞

Figure 8.7: Left: Goats visit the SARE FNE 18-897 Demo Plot at 3 Streams Farm on an attended walk, fall 2019, after 1 year of growth. Right: A different vantage from the same demo plot on August 22, 2020, depicting 2-year-old regrowth. Photos courtesy of Shana Hanson, used by permission.

trees. Also: Fruit and nut-producing shrubs, vines and canes; woodland and wetland shrubs, herbs, and ephemerals.

$ Products: Milk from fresh greenery of woodland and pasture (grain-free), occasional meat, high quality of life for her livestock, biodiversity via Foliage Height Diversity (FHD), and most importantly, knowledge.

Economics: In summer, 3 to 4 gallons of milk per day. She achieves this with about two 3-hour browse walks and/or 6 large armloads of cut tree fodder, or a combination, plus frequent pasture paddock moves. In winter, more tree fodder cutting, less milk, and

FIGURE 8.8: Goats at 3 Streams Farm enjoy the third harvest of a volunteer black locust on-site, July 19, 2021. Photo courtesy of Shana Hanson, used by permission.

2½ square bales per day purchased hay. See SARE FNE18-897 for tree fodder dry matter yield per acre.

Shana measures the value she gleans from re-sprouts in self-sufficiency terms rather than US dollars. Thanks mostly to tree-rich milk, she does not regularly purchase groceries or health care. Her direct farm sales of outdoor unlicensed milk unfortunately became illegal about 8 years ago, but this prohibition may not last. Thanks to her friend Susan Littlefield, she has also produced a stockpile of beeswax-coated aged cheese wheels in barrels in her root cellar.

⚒ Disturbance Severity: Pollarded woodland "air meadow"; pollarded edge trees along pasture and road; pollards in pasture. Browse layers managed by shepherding/goat-cow herding.

▨ Stand Structure: Multiaged naturally regenerating wild trees with a few trees (white oaks, apples, pears) planted in pasture; beds of primarily basswood natural regeneration protected from goats. The air meadow is largely comprised of 58-year-old growth with easily pollarded younger trees below (due to the previous owners' light selective cutting for firewood).

🕐 Disturbance Chronology: Shana gauges pollard harvest frequency based on tree energy and regrowth, and has been shifting from a 3-year towards a 5-year rotation.[54] She harvests areas sooner if growth looks headed for breakage and later if tree wounds are not closing or if trees appear weak. She also elects to retain more foliage when pollards are shaded, weak, or during early-season harvests.   ☞

Shana can access most trees in the pasture from the first few rungs of a short ladder, or by climbing into the tree. Small trees line many field edges with large trees behind them, requiring a longer ladder, and sometimes climbing rope and harness until new branches develop. Oak trees that receive sufficient sunlight generate new branches she can climb at second harvest.

Pattern: Shana pollards and harvests tree leaves from all layers in her woodland, small pasture, and road front, rejuvenating and improving fodder quality, while making subsequent harvests easier to reach. Some of this regrowth has failed due largely to competition from surrounding trees, so she continues to incrementally enlarge and complete these pollarded areas. In her experience, Håkan Slotte's recommended minimum of ⅓ day of sun[55] is a *bare* minimum and considerably less than ideal.

She works the pollarded patches in short daily rhythms, often with the goats present, instantly consuming the fodder she drops and ready to take her off wandering for their next menu-course.[56] Pollarding creates gaps that allow more sun to enter her woodland, increasing the lushness and density of understory layers. Goats, and sometimes her cow, enjoy this understory forage during twice-daily 2-to-4 hour wandering browse sessions. She typed her first draft of this case study while accompanied by browsing goats in a lush green paddock, wearing a leather visor she'd made on a previous goat walk. During blueberry harvest, she often ends up sleeping outside, relocating a couple of times during the night to ensure goats have full bellies!

Shana's future goals include creating winter tree hay storage, which she began during her SARE FNE18-897 grant project, as well as developing tools to mechanically separate leaf matter from brush during industrial tree harvests to help reduce the manual labor required to harvest tree fodder and make the practice more practical for modern farmers.

Weed/Pest/Varmint Issues: Shana honors the work of porcupines, who prefer hemlock, basswood, and red oak. She delays her harvest if they've recently been at work in an area. While they tend to thin inside the tree canopy, she tends to shorten branches to space regrowth closer to the main trunk; so long as they don't both overharvest, their strategies are complementary.

Browntail moth caterpillars arrived in 2019 and cause a poison ivy-like reaction. She says she cannot safely harvest oak, birches, cherries, and possibly poplars, nor other trees near or below them in spring until heavy rain, snow, and wind clean away the tiny caterpillar spines. Colin Yarnell, who lives at her farm, has trapped many egg-bearing browntail moths using a work light clamped to a canning pot full of soapy water at a key time when the moths were flying. She also removes their silky dangling winter nests off reachable trees in winter, squashing them or burning them in the wood stove.

MANAGING HEDGEROWS

🕐 🪓 Laid hedges are often managed with a flail mower to keep them to heights of 3 to 5 feet (1 to 1.5 m), but this practice is increasingly considered over-management, steadily undermining the health of many British hedgerows. To help inform a more sustainable approach to hedgerow management, the Hedgelink partnership encourages growers to follow a more flexible strategy, adapted to individual plant species, that allows them more opportunity to grow between management cycles. These recommendations encourage landowners to trim hedges every 2 or 3 years instead of annually, steadily increasing the height of cutting by a few inches with each successive cut, and trimming as needed but only laying a hedge once it is at least 10 to 13.1 feet (3 to 4 m) tall.[57]

Traditionally, landowners use several techniques to restore and maintain hedges including "brushing" or "siding" up, coppicing, trimming, and laying. Until the development of the mechanical hedge trimmer, different regions had uniquely patterned hedge trimming styles like the Warwickshire hedge and the Welsh double brush.[58]

Brushing or siding up involves the thinning and removal of spreading side branches from a bushy overgrown hedge. This creates a more compact growth pattern, allowing more light to reach the base of the hedge to encourage new growth. Where hedges become sparse or overgrown, coppicing them to ground level can initiate vigorous, new low growth and restore the hedge's density. While hand-operated or tractor-mounted mechanical trimmers are often used today to keep hedges low and dense, hedgelaying is a traditional technique that regularly renews the hedge, maintaining tight growth and reinvigorating the hedgerow plants.

As hedge plants develop, their lateral branches interlock and create a dense stock-proof barrier. But after about 20 years or so, plants begin to lose vigor and their bottom and inner branches die back. If left untended, this dieback can create large openings in the hedge and render it useless as a living fence.[59]

Though the styles, techniques, and intervals varied by region, hedges were traditionally managed by laying or coppicing every 5 to 15 years, providing farmers with fuelwood and craft materials. In British author William Marshall's *Planting and Ornamental Gardening* (1795), he explains that Norfolk hedges were cut at 12-to-15-year intervals, while farmers in Surrey and Kent laid hawthorn hedges every 7 to 8 years, and described rotations as short as 5 to 6 years in Yorkshire.[60] Rotations could even span longer intervals, although they'd eventually lose their effectiveness at enclosing livestock.[61]

The Art of Hedgelaying

Hedgelaying is best done while plants are dormant. In the UK, most hedgelaying occurs between late September and late March through early May. This timing varies considerably based on local climate. Try to avoid hedgelaying during hard frosts as it tends to dry out exposed wood, making it brittle and more prone to break when laid. Ideally all work is completed before bud break in spring.[62]

Many hedges require a fair amount of reshaping prior to laying. Hedges full of weedy and irregular growth may need a preliminary thinning to encourage more even growth to fill in gaps. This can take some time, especially for hedges that have gone unmanaged for long intervals. Trim back enough brushy growth so the stems can be conveniently laid, while also preserving enough brush to help deter browsing livestock.

Before actually beginning to lay the hedge, select the pleachers (the live managed tree stems and stumps in a laid hedge) you plan to retain, assuming you have enough to be selective. Try to favor individual plants instead of multiple stems from the same plant to maintain more even spacing between the pleachers. Begin by removing any plants located outside the hedgerow to help preserve a dense, tight belt. Also remove deadwood, old pleachers, and excess livewood that make it difficult to lay the remaining pleachers and might choke out new growth. Because derelict hedgerows often contain significant gaps after thinning and laying, save some of the leftover brush to fill them in and protect tender young regrowth.

The pleachers you retain can be interwoven to create a dense wall of vegetation. Stems longer than 8 to 12 feet (2.4 to 3.6 m) high are often too large in diameter for laying, while shorter stems result in a much lower barrier. About 2-to-4-inch (5 to 10 cm) diameter stems are ideal because of their ease of handling and high probability of regrowth. Trees larger than 8 inches (20 cm) in diameter should either

be coppiced or retained as hedgerow trees. Assuming you've got enough pleachers to work with, don't bother laying awkward stems with irregular forms. They're best coppiced to fill in the gap that's left behind. If you're going to require a considerable amount of deadwood to create a thick, dense hedge, consider coppicing the entire hedge and lay the resprouts once they've reached an appropriate size. Plant out any remaining gaps to thicken the line and encourage diversity. If there are any large open-grown trees, basal prune them by removing lower branches up to as high as 20 to 25 feet (6–7.5 m).[63]

When laying a pleacher, remember the old adage, "leave the bark and a bit more."[64] Cut about ⅔ of the way through the base of the pleacher somewhere between 1 to 4 inches off

FIGURE 8.9: Phil White laying a pleacher using a Yorkshire billhook. Note the heavy leather glove on his left hand to protect against the thorny spines on many hedgerow species.

the ground (2.5 to 10 cm), leaving a section of bark, bast, cambium, and some sapwood so the aerial parts continue to draw nourishment from the roots. The thicker the stem, the higher you'll need to start the downward cut. As a general rule, begin this cut at a height 3 to 4 times the stem's diameter. See Figure 8.10.

Leave the cut surface smooth so it will not collect water and cause decay. Clean up any

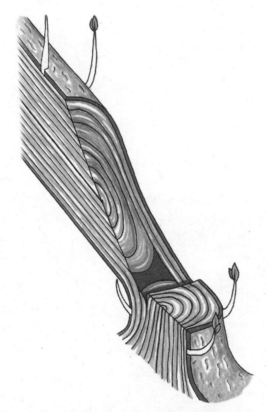

Figure 8.10: A newly laid pleacher in cross-section. The stem has been cut roughly ⅔ of the way through but still maintains a connection via the phloem, cambium, and a couple of growth rings. While often a split does continue down slightly into the stool base, it's best to avoid it to prevent water from collecting there. Sprouts soon form from the base and along the pleacher length.

ragged surfaces. Pleachers usually resprout from about 1 inch (25 mm) below the point where the stem is cut. Avoid cutting pleachers too high because it can leave a gap at the base of the hedge, forming high new sprouts that can be more difficult to lay in the future.

The Yorkshire billhook is a pattern designed especially for hedgelaying. This heavy double-sided tool has a long handle socketed into a steel collar with a long curved blade on one face and a shorter straight edge on the backside. The two complementary edges give the hedgelayer additional weight and versatility. See Figure 8.9.

To cut and lay a pleacher, use a billhook or ax for the downward cut and a saw to trim the remaining stub (or if you're particularly skilled, you can use a billhook for this cut as well). Start with an acute downward cut, slicing through the wood fibers. Aim to cut about ¾ of the way through the stem. Experienced hedgelayers can usually do this in about four cuts for a 3-inch (7.5 cm) stem. Cuts that are too short result in kickback or backsplit, an upwards split carrying through the pole. To remedy this, trim off the kickback and thin the remaining hinge to prevent further splitting. Excessively deep cuts risk severing too much cambium, killing the pleacher or leaving it vulnerable to damage by livestock, wind, or wildlife. You can always practice pleaching cuts on trees you don't plan to retain in the finished hedge.

With each progressive blow to the pleacher, use the other hand to gently pull it downwards. Some hedgelayers use their billhook to twist open the stem following each cut. Once you've laid the pleacher in its final location, trim off

FIGURE 8.11: Left we see a section of unlaid hedge. Note the visible knobs regrowth at about 3 feet high suggesting the hedge had been topped some time in the past 5 or so years. On the right, we see the newly laid hedge complete with stakes and binders. The sparse section in the center should get filled out as hedgelaying continues from the left.

the remaining stub. Start about 1 inch (25 mm) above the soil surface, cutting upwards towards the point where the pleacher bends over. Some hedgelayers even continue the main cut all the way to the ground level, causing the plant to sprout at the soil surface. This keeps growth dense and often makes it easier to lay in the future.

Carefully lower the stem into place. Try not to twist it as you can pinch off sap flow through the cambium. In a Midland-style hedge, lay pleachers about 9 inches (225 mm) off the main line that the hedge follows to create a wider tree belt and also open up space to install binding stakes in between. Try to lay pleachers at a 35-to-45-degree angle. While sap can flow downhill, the sharper the bend at the base of the pleacher, the greater the chance the cambium becomes crimped or damaged. If you have large gaps between pleachers, work in a "deadwood pleacher" to maintain its thickness. To do so, just insert the pointed end of a brushy length of deadwood into the soil at the proper angle, weaving it in as you would a live pleacher.

Once you've laid all stems in the hedge, work back and remove any awkward side branches and fill in large gaps between pleachers with lengths of deadwood braced between hedging stakes. You can also layer young shoots from coppice stools into any gaps.

While not absolutely necessary, hedging stakes and binders help support and stabilize a newly laid hedge, protecting the pleachers from damage caused by strong winds, wildlife, and livestock. Hedge stakes should be 5 to 6 feet (1.5–1.8 m) long, 1.5 to 2.5 inches (4 to 6 cm) thick with a pointed end pounded into the ground 1 to 1.5 feet (30 to 45 cm) deep. Just about any straight young poles will work for hedging stakes, though avoid willow as they will often root and become a living portion of the completed hedge (unless that's useful to you). Use about two stakes and two binders for every 39 inches (1 m) of hedge, although stakes can be spaced as wide as 32 inches (81 cm).[65] Drive the stakes in as you go, weaving the tops of the pleachers between them. Be sure to leave room for the binding along the top of the stakes.

Hedgerow binding helps hold the pleachers in place and keeps the stakes in line. Young coppiced hazel, ash, and willow make some of the best binders. Choose the longest stems you have available, at least 8 feet (2.4 m) long and ideally 12 to 15 feet (3.7 to 4.6 m) and no more than 1 inch (25 mm) in diameter at the butt end. It's best to use binders while they're still green, ideally within 6 weeks of cutting. The binding itself isn't so much a twist but a stiff roll, and the patterns vary depending on the materials and the regional tradition.

If you'd like to grow standard trees in a hedgerow, don't prune their leading shoots and allow them to stretch upwards. Encourage this vertical growth by pruning away lower side shoots early on in its life cycle.

A newly laid hedge is vulnerable to damage during the first year or 2 of regrowth. Farmers often use temporary electric fencing to protect it. Also, try to remove aggressive herbaceous competition as necessary. You can also help encourage dense, bushy woody growth by trimming the hedge back for the first 2 to 3 years.

Hedgelaying is labor-intensive and requires some skill to do well. Depending on the state of the hedge, a single skilled worker using only hand tools should be able to lay 33 feet (10 m) in 1 day. A professional hedgelayer using a chain saw can lay up to 132 feet (40 m) of hedge in good condition in a day.[66]

TOOLS AND FELLING TECHNIQUES

Harvesting poles and rods from coppice stools and pollarded trees involves four main steps: felling, limbing (snedding), cross-cutting (bucking), and transport. You really only need a few basic tools to carry out most of the work. Namely, the billhook, pruning saw/bowsaw/chain saw, tape measure, loppers, hewing ax, and sharpening stones. Let's begin with a discussion on several useful tools for tree felling.

We each develop our own particular preferences for tools and management strategies as our experience grows. While the specific tools may vary, we can generally organize felling tools into two main categories: hewing and sawing. Hewing tools (billhook, ax) are swung in an arc and used to sever wood fibers. In young coppice growth, a skilled hand can slice cleanly through a sprout in a single swing. Remember, the saw has only been in widespread use for the past few centuries. Throughout history, most coppice was traditionally harvested using hewing tools.

Heavier in weight, long-handled, and well-balanced for swinging blows, axes are a great complement to billhooks. Usually suited to larger-diameter stems, a healthy ax swing can make very short work of coppice sprouts. Requiring an ergonomic swing for effective use, axes can make it easier to "cut with the grain" and leave a clean cut. But skillfully wielding an ax requires experience, and it can take time to learn how to swing precisely and hit a target cleanly. A 4.5-to-6-pound (2 to 2.7 kg) axe with a 31-to-36-inch (80 to 90 cm) handle is said to be well-suited to coppice work. Less

Billhooks

Something of a cross between an ax and a machete, the billhook is said to be the precursor to hand pruners. And it also happens to be perfectly designed for resprout management. You can use a billhook to swiftly and cleanly fell, trim, rive, and sharpen rods.[67] Billhook blades are generally 8 to 10 inches (20 to 25 cm) long with a 4-to-6-inch-long (10-15 cm) handle. Incredibly, these versatile tools date back to the Bronze Age![68]

Local blacksmiths often forged custom tools based on the weight and profile woodsmen preferred for their work. Billhooks were widely manufactured throughout Europe for centuries, and as demand grew, unique tool patterns were refined for specific uses and regions. This led manufacturers to develop hundreds of specialized patterns that vary in shape and size.[69]

In 1930, the French tool company Talabot reported they had archived over 3,000 different billhook patterns.[70] The length and weight of the blade, curve of the hook, and overall balance of the tool all influence its use. At one time in the UK, every county (and some districts within counties) had a particular billhook pattern designed to match the needs of local coppice workers and the specific material they harvested. These patterns usually carried the name of that region including Yorkshire, Stafford, Newtown, Knighton, and LLandeilo, to name a few, all of which are still available today from the A. Morris and Sons of Dunsford company.

Until relatively recently, blacksmiths hammered billhooks from a solid steel billet, gradually thinning the blade from the back to the cutting edge and from the tang to the nose. This process produced a well-balanced tool with minimal shoulder (the bulk of the blade's body that supports the cutting edge). When sharp, these billhooks could cleanly sever wood with a single well-placed swing. In contrast, many modern mass-produced hooks are pressed from plate steel, resulting in a poorly balanced tool with a thick profile, often inferior to the hand-forged tools of the past.[71]

Billhooks get their name for the curved nose of the cutting tool. Raymond Tabor explains that billhook blades with concave edges are especially well-suited to trim twiggy material since the curve gathers ☞

FIGURE 8.12: There have been many hundreds, if not thousands, of billhook patterns developed throughout the ages, each designed to meet the unique needs of different tree species, products, and craftspeople. Here's a selection of four different styles.

the tips with each swing. He writes that billhooks with a short nose are best for felling rods formed on dense stools or in situations where one needs to cut downwards close to the soil surface. Billhooks with a long nose are especially useful for riving rods, a process that is fundamental to most wattle hurdle construction. If choosing a billhook pattern for felling work, consider a heavier tool to help carry it through the pole.[72]

Many hedgelayers preferred the double-bladed Yorkshire pattern. With a curved blade on one side and a flat blade on the other, the curved face is used for heavy cutting and the straight side for final trimming, topping stakes, and other light work. Alan Brooks and Elizabeth Agate, authors of *Hedging: A Practical Handbook,* also recommend the Staffordshire billhook for most hedging work. Lighter than the Yorkshire, it's less tiring to use but also has a double-edged blade that offers versatility for hedging work's various slicing tasks. Avoid using a well-sharpened billhook on deadwood because it will quickly dull the edge.

There are still a number of quality billhook manufacturers. A. Morris and Sons of Dunsford in Devon, UK, continues to produce hand-forged billhooks. They make several patterns that are useful for a range of work in the coppice and will reportedly make custom tool patterns if the customer supplies a cardboard template. Spear and Jackson also continue to produce billhooks and slashers (long-handled billhooks for hedging work). JAFCO manufactures billhooks and slashers with durable fiberglass handles, said to be 10 times stronger than hickory (but you don't get to enjoy the feel and grip of a wooden handle). And if you know the dimensions and proportions of a particular design, you could always reach out to a local blacksmith/toolmaker and ask them if they'd be willing to forge you a custom tool. A well-designed billhook will last generations, so the one-time investment in a quality tool will last many coppice rotations.

And just a brief a note on safety when using edge tools…. Always keep a safe distance from fellow workers, ideally at least three times the length of one's arm and tool at full extension. To reduce the likelihood of an edge tool slipping from your hand, try not to wear a glove on your tool hand or be sure to choose gloves with a tacky grip. Direct the arc of your swing away from your body so that an accidental bouncing blade doesn't pose a threat, and whenever possible, stand on the opposite side of a pole as you work, helping protect your legs and feet from a poorly placed swing.

experienced workers may prefer a lighter tool with a 2.5-to-3.5-pound (1.1 to 1.5 kg) head and a 30-inch (75 cm) long handle.[73] A hewing tool with a slender shoulder makes it easier to slice through the work and sever limbs cleanly from cut poles. See Figure 8.13. For your tool to work optimally, either select a tool that already has this profile or modify it using an angle or bench grinder.[74]

While a well-honed ax or billhook can make a clean cut on stems up to 3 inches (75 mm), it's a difficult skill to master, especially when trying to cut low on a coppice stool. As with all edge tools, cut with the grain, following the direction of fiber growth. Maintain a slicing action and cut across the thickness of the stem very gradually, as opposed to cutting perpendicular to the rod's thickness. See Figure 8.14.

A slicing cut with the grain helps minimize splitting, leaving a far cleaner surface finish, while also helping protect the stool against infection. Felling coppice rods with an edge tool is tricky business as you really need to use a wide arc to swing the tool upwards and sever the stem at the lowest point possible. This is very difficult to do, especially for larger-diameter stems. It is far easier to cut the stem from the top down, swinging downwards and across the fiber direction, but this leaves a ragged finish that will invite decay and reduce the stool's vitality. If you end up making a ragged cut, don't worry, just clean it up with a saw cutting below the extent of any splits traveling down through the remnant stub.

If you're using an ax or billhook on stems too large to sever with a single blow, start with

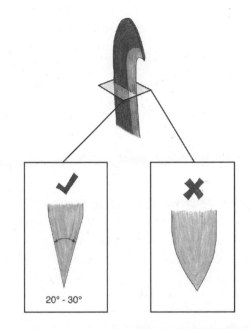

FIGURE 8.13: For most slicing tools, a 20-to-30-degree included angle provides the optimal balance of strength and thinness. Many tools often arrive factory ground with far too stout a shoulder and therefore require some degree of modification before use. Otherwise, they will not cleanly slice through wood.

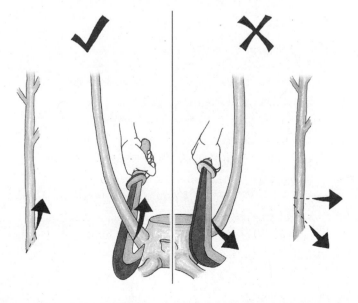

FIGURE 8.14: When using a billhook to cut coppice sprouts, you ought to always swing with the grain. In this case, you'll want to cut from the ground up to avoid tearing the wood fibers. This is a very difficult skill to master. You also want to follow a long slicing trajectory as opposed to an oblique angle close to 90 degrees. It definitely takes practice!

an upwards cut from below, then follow with a strike from above a few inches higher than the end of the first cut. Try to remove a wedge of wood with this second swing, reducing the thickness of the stem. Repeat this process until you completely sever the stem, ideally with a final upwards stroke. See figure 8.16.

Traditional knowledge suggests that hewing tools leave a far superior finish than saws or pruning tools like loppers, helping limit infection and decay.[75] French researchers compared the effects of different cutting tools and techniques on holm oak (*Quercus ilex*) stool mortality. The techniques included chain saw at ground level, chain saw at 6 inches (15 cm), ax at ground level, and a traditional technique used widely towards the end of the 19th century called *saut du piquet* ("stump breaking") where woodsmen cut the shoot at about 20 inches (50 cm) and then used a head of an ax or a sledgehammer to knock off the top of the shoot,

detaching it from the stump. They believed this process caused new shoots to form independent root systems. Their results showed little difference in stool mortality for ax and chain saw felling techniques (from 1.6% to 3.8%), but the saut du piquet caused high stool mortality (14.8%) and resulted in fewer, smaller sprouts.[76]

Specialized Pollarding Tools

Norwegian farmers used the sickle (common, lopping, and leaf), knives, axes, and even broken scythe blades to lop pollard shoots, while some shepherds carried a specialized tool called the **houlette** to harvest tree hay. Kind of like a sharpened spoon mounted on a long staff, the houlette extends your reach to harvest tree hay high on managed trees without the need for a ladder or tree climbing skills.[77]

Loppers and hand pruners are also versatile, efficient tools for coppice and pollard management—especially when selectively pruning

Tool Sharpening and Care

To safely and efficiently produce quality work, you need sharp tools. For a tool to be sharp, its two surfaces must meet at an infinitesimally small point. Your goal is to shape the steel on each face of the tool to such a fine point that you cannot physically see their intersection with the naked eye—to develop an arris (intersection between two faces) that is so fine, it will cleanly and easily slice through wood.

And at the same time, for a tool to function properly, the "enclosed angle" (the angle formed

between the two bevels of the tool's edge) must be appropriate for the intended use. Most woodworkers use two main types of edges: those used for shaving or slicing and those used for splitting. You can create a razor-sharp edge between two tool faces at a 90-degree angle. While sharp enough slice your finger, an enclosed angle this abrupt is useless for shaping wood. At the other end of the spectrum, a shallow angle of just a few degrees will easily slice through many materials, but there's little steel left to ☞

support this thin edge, and it's prone to dent and deform in use.

Most tools used for shaving wood (chisels, carving knives, drawknives, plane blades, spokeshaves, hewing axes, etc.) are sharpened with an included angle of between 20° and 30°. The steeper the angle, the more the tool resists cutting, but the stronger and stiffer the edge. Finer work tends to call for a shallower angle, while a steeper edge is better suited to work with denser woods. See Figure 8.13.

One of the most common questions novice woodworkers ask is "How often should I sharpen?" And of course, the answer depends. With use, friction will gradually blunt a sharp tool's edge, compromising performance. At this point (or rather, a little while before this), it's time to resharpen. The more you use a tool (and the harder the material you're working), the more frequently it requires sharpening. It takes experience to feel when a tool no longer performs properly, but in time, you'll know. While some folks prefer to frequently tune up the edge with a fine abrasive, constantly keeping it well honed, others use it until the edge has been worn to the point where you need to reshape the bevel with a coarse stone. What's your preference: frequent but quick maintenance or irregular but intensive reshaping?[78]

A note on tool care: A sharp, well-honed tool is a treasure to behold and should be treated as such. It takes time and energy to shape a steel edge to a fine point. Once it's there, do everything possible to avoid damaging that edge during use and in storage. When using a sharp tool, take care to never drive it into the soil or use it to cut materials tainted with sand or other abrasive material. Just one single careless stroke can cause a deep nick in an otherwise perfectly maintained edge. (Trust me, I've done it—several times.) Always clean away any soil surrounding a stem before you make a low cut and do everything possible not to drop a tool on its sharpened edge. (But never try to catch a sharp tool if you've dropped it! The cost or time investment in a new edge is far cheaper than the time lost recovering from a flesh wound.) If you know you'll be cutting dirty wood or in challenging conditions, consider using a beater tool that you're okay with damaging.

Ideally protect your sharpened tool edges with a leather or cardboard sheath. This can also protect you from cutting yourself accidentally. And lastly, many hand tools easily blend into the woodland landscape when set down even for just a few moments. To prevent tool loss and time wasted searching for a misplaced tool, mark tool handles with brightly colored tape or paint so they stand out against the myriad earth tones in the woods.

stems and stools or reaching into confined spaces in established hedgerows. While slower than hewing, these types of pruning tools create a clean, precise cut when sharp and well-tuned. Make sure you try to preserve the branch collar with each cut, fully and cleanly cutting through branches so that you avoid tearing off lengths of bark at the base of the cut. Be sure to use bypass pruners, as opposed to the anvil-type pruners that crush the bark at the base of the cut. They're designed for making cuts in dead wood. Often, simply taking the time to find the right position to make a cut can make all the difference in leaving a clean smooth surface.[79]

For all but the shortest-rotation regrowth, a handsaw is perfect for harvest and thinning tasks. There are quite a few different saw designs to choose from. In the UK, many coppice workers favor the bow saw, a thin, aggressive, inexpensive, replaceable blade mounted on a bent, lightweight steel handle. These versatile lightweight saws are capable of cutting stems up to 6 inches (15 cm) in diameter. For general use felling and cleaning up poles, a 21-inch (53 cm) bow saw is a great size.[80]

Bow saws' curved shape and pointed nose help fit between tightly spaced sprouts on a

FIGURE 8.15: A selection of tools for harvesting coppice poles, pruning pollards, and other woody management tasks.

stool. Their main drawback is their lightweight blade that can be easily pinched and bent if not used carefully. Keep several replacement blades on hand so it's easy to get back to work if you bend one. While bow saws make a thin kerf, straight cuts can be difficult, especially if the blade is bent or dull.

With a thicker blade and often more heavy-duty than a bow saw, pruning saws are great tools for felling 1-to-4-inch diameter (2.5 to 10 cm) coppice efficiently. With aggressive coarse teeth sharpened on three faces, a compact, lightweight, sharp pruning saw will make short work of a pole up to 6 inches (15 cm) in diameter. Because the blade is stiff, take care not to flex it in a cut as it's very difficult to re-straighten the blade once it's been bent. Even if it isn't your primary felling tool, pruning saws are super-useful for coppice work, especially when navigating narrow spaces between stems on a vigorous stool.

And then, of course, there's the chain saw. Because generally we're cutting material far smaller than most chains saws can handle, a small compact saw is usually plenty adequate and in most cases is probably a much better-suited tool than most professional-grade logging-type saws.

Advances in battery technology over the past decade have brought great improvements in cordless electric chain saws. With no need to carry mixed fuel into the woods, electric tools are quieter and can make for a calmer overall working environment since the chain stops running when you remove your finger from the throttle. I recently purchased a compact

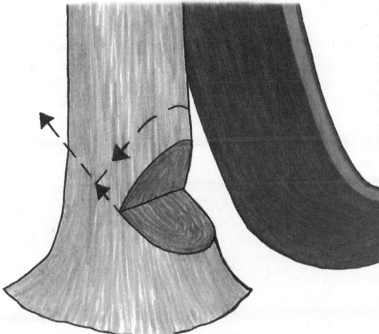

Figure 8.16: For larger-diameter material, you'll need to make multiple cuts to sever it cleanly. Start with an upwards swing with the grain, then follow with a cut from above, removing a notch of wood as you meet the end of the first cut. Repeat as needed, working incrementally through the piece, ultimately severing it with a final upward swing.

mini-chain saw made by the Milwaukee Tool company with a standard 6-inch (15 cm) bar and have found it to be very versatile, fun to use, and quite nimble. I could see it being an excellent tool for harvesting coppice rods up to a few inches in diameter. Of course, these tools still require bar oil to lubricate the chain, and they are just as dangerous as a gas-powered saw, so you should still wear appropriate protective gear. More on safety in a bit.

Last, the arborists at Colonial Williamsburg in Virginia actually prefer to use a sawzall to cut the sprouts from their pollards. They report that chain saws leave too rough a surface, creating a cut face more prone to infection. And because of their thin, narrow blade, they're easier to fit between tight-growing sprouts.

Felling

For poles up to about 3 inches (7.5 cm) in diameter, there's usually no need to make a bird's-mouth cut to direct the fall of the stem.[81] Just cut clear through it, ideally starting on the inside face of the pole so it falls out and away from the stool. Use light, relaxed strokes and the full length of the blade to make the best use of the tool. If the pole settles back onto the saw blade, push it away gently to keep it from binding. Take care to saw completely through the stem before it falls to keep it from splitting or tearing off a section of bark. For poles 3 to 5 inches (7.5 to 12.5 cm) in diameter, it's often best to make a small undercut across its face to relieve the tension that accumulates as it begins to fall.[82] For stems larger than 5 inches (12.5 cm) in diameter, you'll probably want to use a notched felling technique described later.

Because it's nearly impossible to cleanly fell poles close to the base of the stool with just a single cut, you'll probably need to do it using two separate cuts: the first one to fell the pole, the second to clean up the stool and leave a low-cut face. Always remember that resprout silviculture systems rely on maintaining the long-term health of the stool, so take time to carefully preserve each stool's integrity.

A Note on Safety

When it comes to time, it's hard to beat the speed and efficiency of the chain saw for harvest and processing work. But please be safe. Chain saw work can be very dangerous so learn how to be a careful and effective tool operator. Loggers have developed a number of safe, efficient tree-felling techniques, all of which can be appropriate in the right application.[83]

When using a chain saw, always wear proper protective gear: at minimum a helmet with faceguard, ear protection, and chain saw pants made from long polyester, Kevlar, or ballistic nylon fibers designed to bind the chain if it engages with the fabric. Before felling, be sure to thoroughly check for aerial hazards directly overhead, as well as the area where the tree will fall. Make sure there are no hung-up branches (widow-makers), dead stubs, forked trees, vines, or other hazards that might drop when the tree is felled. Choose and clear an escape route so you can safely move away from larger stems as they fall. And it's also a good idea to keep a well-equipped first aid kit nearby when working in the woods that includes bandages, dressings, adhesive tape, scissors, and clean water.

Here are a few other considerations to keep in mind while felling:

- Try to choose a felling direction that prevents the tree from getting hung up in a neighboring tree. Avoid obstacles within the felling path (downed trees, boulders, stumps) that could cause the tree to kick back upon impact, also possibly breaking the stem.
- Be particularly careful with trees that get hung up. They can be unpredictable and are unsafe until they're safely on the ground.
- If coppice poles up to about 4-inches (10 cm) dbh do get hung up, check to make sure there are no potential aerial hazards. If things appear safe, put your shoulder underneath the pole, lifting the cut end from the ground, and swiftly carry the stem away from the point where it's entangled. In most cases, the crown will quickly drop to the ground.[84]
- Inspect each pole for any embedded fencing wire, stones, glass, or other hazards that could damage the saw before felling.
- When felling a coppice stool, start with the most accessible poles around the perimeter of the stool, incrementally working inwards towards the center. See Figure 8.17.
- Take particular care on windy days. Strong gusts are unpredictable and especially dangerous. If you're at all uncertain, wait until conditions calm down.[85]

When it comes to felling larger trees, you'll want to use multiple steps to control the direction of fall and minimize damage to the stem. First, make a downward diagonal cut along the "front" or outer face of the tree. Carefully sight along this orientation and make sure you're pointing it in the direction you'd like the tree to fall. This dictates the placement of the hinge that controls the felling direction, so now is the time to make changes. The width of the bird's-mouth cut at its base determines the length of the hinge that will control the tree's fall. Generally, the hinge length should be about 80% of the tree's diameter at breast height, which means that an 8-inch (20 cm) diameter tree needs a 6.4-inch-long (16 cm) hinge.[86]

Next, make a level cut across the face of the tree perpendicular to the direction of fall. Stop once it reaches the base of your initial downward cut. These two cuts should converge perfectly with one another, removing a wedge of wood that forms an approximately 70-degree angle.

Finally, fell the stem with a back cut level with the bottom of the bird's-mouth. The depth of the back cut determines the thickness of the remaining hinge of wood. On average, the hinge should measure 10% of the tree's diameter at breast height.[87] Make the hinge even in thickness across its length, or the tree will not fall properly. If you've made these three cuts carefully, your tree should drop in the exact location you were planning. And always remember, if in doubt, don't.

There are some great chain saw and logging training courses available to help build your confidence with safe felling techniques, the

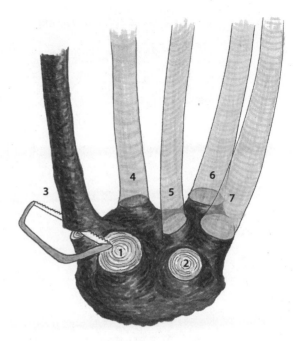

Figure 8.17: An example of a strategic sequence to harvest coppice poles form a stool. Proposed cuts are numbered 1 to 7. Begin by working around the perimeter of the stool, felling each pole so it drops away from the stool's center. If there are several poles in the center, it may not be possible to access them until the outermost poles have been harvested. While the 3-to-5-inch (7.5–12.5 cm) diameter stems depicted here may be large enough to require a bird's-mouth cut to help direct the felling, smaller poles can often be cut clear through in one pass.

Game of Logging being one of them. Never push yourself to do something you're not comfortable with. The associated dangers are rarely worth it.

SNEDDING OR LIMBING

With coppice poles on the ground, it's time for **snedding** (or limbing), the process of removing the side branches and brushy tops. This step also gives you a chance to inspect each pole for quality and utility. As a general rule, expect to spend twice as much time cleaning up a cant (snedding, cross-cutting, sorting, and stacking) as it took to fell it.[88] The billhook is hands down the best tool for snedding coppice up to 10 to 15 years old. For larger-diameter

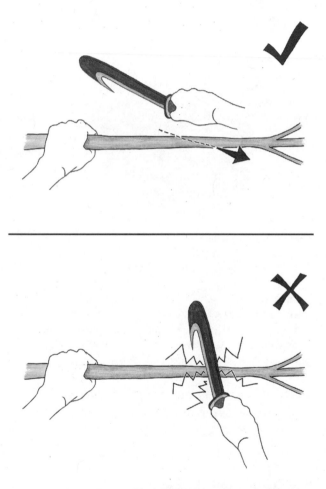

Figure 8.18: As we discussed earlier, always work to cut in the same direction as fiber growth, essentially from the base of a stem towards the tip. This helps minimize tearing as you sever through the stem. Also aim for a shallow angle that slices through the stem. Note that, when snedding, you generally want the billhook to ride cleanly along the stem (not indicated in this drawing).

polewood with branches bigger than 1 inch (2.5 cm) in diameter, a felling ax, hatchet, handsaw, or chain saw are probably more appropriate.

For many North American woodsmen, the ax has always been the tool of choice. Historically, this is likely due to the abundance of large timber trees here compared to Europe where the lighter, nimbler billhook better suited their needs. The ax's heavier head and longer handle allow the user to throw the tool with more force, making it a fantastic choice for limbing larger poles and small sawlogs.

When snedding, take care to remove side branches as close to the main stem as possible, leaving little to no protruding stubs. I learned to work from the butt end of the pole to the tip. This means you're always working with the wood's grain, allowing you to cleanly sever the branch with a single well-placed swing. Use a very shallow angle to slice through the wood. Hewing tools will not cut well at 90-degree angles. See Figure 8.18.

Of course, there are always exceptions. French coppice worker Chantal Chevrier told me her family actually prefers to work from the opposite direction—from the pole's tip to the butt.[89] With short-rotation coppice (up to about 5 years of age), this can work when the side branches are relatively small because they tend to snap off at the base, eliminating any trace on the pole's surface. This is why they prefer it. But, because this orientation actually runs against the grain, it's inadvisable to do this when snedding larger branches as it often causes the wood to split uncontrollably and can damage the pole.

To ensure smooth and safe snedding and cleanup operations, remember the following suggestions.

- Keep the site clear of brush so there's less to get caught on or trip over as you work.
- With larger poles, work on the uphill side to avoid being trapped if it rolls downhill as you remove branches that may be holding it up and supporting it.
- By removing any upwards-pointing branches, you can better balance the pole to prevent unpredictable shifting as you work.
- Wait to cut branches on the underside of the pole until the very end. They are under tension and will often spring back when cut. After you've removed all accessible branches on the pole's top and sides, roll it over to clean up the underside.
- Try to stand on the opposite side of the pole as you work so your leg is protected if the billhook bounces or slips from your hand. If you must stand on the same side of the pole that you're cleaning up, keep your body offset from your swing so you're safe if it does bounce unexpectedly.
- Try to cut cleanly against the stem without marking it up. Do this by swinging the tool at a shallow angle so that it rides flat along the pole. It's more efficient and minimizes nicks and cuts in the pole.[90]

When removing larger limbs, take them down in pieces, working from the outer edges inwards. These limbs may be under considerable stress and may need to be treated as you would a windblown tree, by making an undercut on the compression (bottom) side of the limb first to keep it from binding when you try to sever it from above.[91]

As you work, always try to think a few steps ahead. You'll be processing a large amount of material. The more organized, streamlined, and thoughtful you can be, the easier your job. The "brash" (brush or slash in the US) you remove can be useful for many things. Probably the least labor-intensive way of dealing with it is to leave it in place to decompose. Just be sure to think about how you pile it and how it might affect future access, complicate harvest, or otherwise be used to mitigate soil erosion or form a brushy field barrier. Brash can be used for making brush hedges around the stand, fascines, besom brooms (usually with fine branches like birch), fodder for livestock (when they still retain leaves), kindling, check dams, protecting regrowth on individual coppice stools, brushwood swales and other types of permeable erosion control structures, and of course it can also be chipped and used for all sorts of applications.

Traditionally, coppice workers burned this material in the cant as they worked. Maintaining brash fires takes a fair amount of attention. With most of the material green, it can take some effort to get the fire started, though once you've established a good bed of coals, you'll readily burn through freshly cut greenwood. This can require considerable focus, steadily feeding the fire and ensuring it stays under control. While the ash and charcoal can be broadcast to cycle some of the fertility, it really doesn't make the best use of this resource. Probably one of the best things about brash fires lies in the simple pleasure of burying a few foil-wrapped potatoes in the inferno an hour or so before lunch, so that you're greeted with a perfectly cooked steaming warm meal come lunch time.

Of course, use common sense when planning brash fires (or large burns of any sort for that matter). Check with your local fire warden to see if there's a burn ban in effect. Consider the moisture content of the soil and surrounding vegetation and avoid burning on days with strong winds. Rake away combustible material around the perimeter of the burn pile, exposing mineral soil to create a fire break and help prevent it from expanding outwards.

If you need to haul brash any distance from where it's cut, take time to arrange the stems so the tips all point in the same direction with butt ends lined up with one another. A consistent pile is so much easier to pick up and move comfortably and efficiently. Keeping the brash a fairly consistent size, say 4-to-6 feet (1.2 to 1.8 m) long, makes moving piles much more manageable.

CROSS-CUTTING AND PRODUCT SELECTION

With poles felled and snedded, it's time to select, cut, and sort materials for their intended use. Cross-cutting poles to length is the first step in converting raw materials into product. Develop a keen eye that's tuned in to the straightness and quality of each stem. For most value-added products, you really need straight lengths. But ideally, as we discussed in chapter 5, you'll have already thought up a wide range of uses for materials of different lengths, diameters, and quality.

For short-rotation coppice, a pair of loppers may be all you need. Loppers allow you to make a square cut quickly and accurately, far faster than you could with a saw. And you hardly have to bend down to reach the end of the rod. You could also use a billhook, but it's very difficult to cut to an accurate length, and it leaves a sharp pointed end, which often is not ideal. For larger-diameter poles, you'll probably want to use a pruning saw or chain saw. This is also another perfect application for a compact battery-powered chain saw.

When I worked for Ben Law felling 18-year-old chestnut coppice, we cross-cut pole lengths, selecting for ten or more different products, including:

- Round timber framing poles: Straight; 5+ inches (12.5 cm) in diameter; 10+ feet long (3 m+)
- Large building poles: Straight; 4 to 5 inches (10 to 12.5 cm) in diameter; 8, 10, 12 feet long (2.4, 3, 3.7 m)
- Small building poles: Straight; 2 to 3 inches (5 to 7.5 cm) in diameter; 8, 10, 12 feet long (2.4, 3, 3.7 m)
- Fence posts: Straight; 5 to 6 inches (12.5 to 15 cm) in diameter; 5.5 feet (split in half) (1.7 m) long
- Fence rails (for post and rail fencing): Straight; 6 inches (15+ cm) in diameter or larger; 10 feet long (3m) (split into quarters)

- Tree stakes: Straight; 2 inches (5 cm) in diameter; 6 feet long (1.8 m)
- Yurt poles, roof rafters: Straight; 1 inch (2.5 cm) in diameter at tip; 8, 10, 12, 14, 16 feet long (2.4, 3, 3.7, 4.3, 4.9 m)
- Yurt poles, wall lattice: Straight; 1 inch (2.5 cm) in diameter; 6.5 feet long (2 m)
- Wattle fence stakes: 1.5 to 2 inches (2.75 to 5 cm) in diameter; 3 feet long (0.9 m)
- Wattle material: 8+ feet (2.4m+) long, straight, flexible
- Unique pieces for rustic furniture, decorative structures
- Short useful lengths for rustic stools, chairs, tables, etc.
- Walking sticks
- Mushroom logs: 3 to 8 inches (7.5 to 20 cm) in diameter; wound-free, need not be straight
- Firewood: any length, need not be straight
- Charcoal stock
- Brash

Having such a wide array of products that span the value-added continuum creates lots of opportunity to make the most efficient use of materials. This is ideal from an economic standpoint—there is no waste. Everything has a potential use. You just need to optimize it.

You'll definitely want to carry a tape measure to select and mark quality sections from each pole. Because the end grain usually checks as it dries, consider cutting lengths 2 to 4 inches (5 to 10 cm) longer than needed. You can always cut it shorter later.

For larger poles, you'll need to cross-cut them in place, but short-rotation coppice poles

FIGURE 8.19: An assortment of sweet chestnut materials sorted for various uses. Prickly Nut Wood, UK.

can be brought to a landing where you can set up a cross-cutting station. When cross-cutting, keep these simple principles in mind:

- When possible, support the pole at both ends to avoid the tear out or splitting that occurs when an unsupported end falls as it's severed. If you can't support both ends of the pole, make a relief cut on the underside about ⅕ of the way through the stem. This prevents tear out and keeps the pole from binding on the saw blade.
- Make square cuts. This is a difficult skill to master, but the only way to fix a bad cut is with a better cut.
- Rather than trying to cut clear through the pole from top to bottom with a handsaw or chain saw, use a back and forth "rocking" motion, cutting towards the bottom edge of each side of the pole, before removing the remaining peak of wood in the center. This can help maintain a square even cut.

EXTRACTION

Before you start felling, create a clear convenient plan for material storage. Ideally, it's located below the cant so you always haul materials downhill. Because we want to minimize damage to coppice stools and woodland soils during felling and extraction, most coppice workers haul materials to the landing by hand.

Though green wood is heavy, even long poles can often be carried by hand with ease if balanced on your shoulder. See Figure 8.20. Stand a pole or poles upright and allow them to gently rest on your shoulder about ⅓ of the way from the butt end. Wrap your arm over the top, lift with the legs, and away you go. Using your hand, apply light pressure to the butt end as a counterweight to the mass of the pole behind you. This way one person can haul a pole 14 feet (4.3 m) or longer safely and comfortably.

Figure 8.20: Once you get the hang of it, it's pretty easy on the body and very convenient to carry a pole or several poles on your shoulder. It requires far less strain than holding it at waist level.

It takes a bit of practice, but once you get it, it's actually quite fun! Hopefully you've established good vehicle access so you never have to carry materials too far before reaching a place you can load a truck, tractor, or trailer.

Timber tongs are another excellent tool for moving pole-sized logs around the woods. This tool features two pivoting jaws connected by a handle. They allow you to pick up and balance or drag a log without even having to stoop down. With two pairs of tongs, you and a partner can conveniently pick up a pole at both ends.

For larger harvests, you'll probably need to use a tractor, pickup truck, ATV, horse-drawn wagon, or sled. A log trolley can serve as a human-powered version, enabling two woodsmen to move a large quantity of wood with relative ease.[92]

And as materials get larger and the scale grows, extraction becomes an even greater endeavor. At a certain scale, it may make sense to seek out a local logger with equipment suited for efficient extraction. Oftentimes, the bulk of the sorting work can be done by hand, hauling crosscut poles to strategically placed piles. This could also be done using a tractor with a logging winch that employs the tractor's PTO to operate a length of cable that pulls bundles of poles or individual logs up to 160 feet from the machine, so they can be skidded out of the woods.

At a certain scale, a forwarder or tractor and forwarding trailer may be the best tool for the job. These machines use a mechanical grapple to pick up loads and place them in a trailer bed. This helps minimize damage to the polewood, keeps it clean, and also prevents damage to forest soils and ground flora since the materials aren't dragged through the woods. Another benefit is that the grapple makes it easy to carefully off-load materials into organized piles at the log landing.

STORING PRODUCTS

Four-to-six-inch-plus (10 to 15 cm) diameter poles can be stacked neatly in piles. Try to elevate all wood piles on pallets or sacrificial poles to prevent waterlogging and decay. If you plan to use your materials for greenwood crafts, store them either in a sheltered location or even submerged in water until ready to use. For many products though, you ultimately want your materials to dry slowly and evenly.

For shorter lengths, or materials that you'd like to dry more quickly, you can stack them in alternating layers to improve airflow. With small-diameter materials, you can also bind them into bundles, using twine at both ends, that can be easily hauled by one or two people. This also makes it easy to keep track of inventory.

If you've harvested coppice material for cordwood, consider stacking it in the woods to season for a year, during the course of which it'll lose significant weight as it seasons. It can also keep your yard from filling up with greenwood. This allows you to collect air-dried firewood from your woodlot as needed.[93]

Fresh-cut green wood is softer than seasoned wood and is considerably easier to work. If you'll be peeling poles for a building or craft, doing it after felling but before storage is often far easier than once the wood has dried. The drawknife is a perfect tool for peeling saplings and small-diameter poles. It works even better with some type of holding device like the shaving horse. For larger poles, a sharpened spade or a log-debarking tool called a spud allows you work upright comfortably and efficiently.

To properly season wood, you need to slowly and evenly lower the material's moisture content so the internal moisture steadily migrates from the center to the end grain and outer surfaces, helping avoid checking and cracks.[94] To minimize checking, seal the end grain with a wax-based end-grain sealer like Anchorseal or a relatively impervious material such as paraffin, latex paint, or urethane varnish, helping slow moisture loss and protecting it from invasion by airborne fungi. This is probably only worthwhile for particularly valuable materials. If you do choose to do it, be sure to coat the ends as soon as possible after sawing. The weakness caused when a check forms in drying wood will be there forever.[95]

To encourage even drying, stack poles to ensure maximum air circulation on all sides. Sawmills usually use 1-by-2-inch (2.5 by 5 cm) stickers to separate courses and create space around and between boards. Stickers can be spaced every 16 to 24 inches (40 to 60 cm) along the length. Be sure to align them in vertical rows to evenly distribute the weight and keep them from warping.[96]

Carefully select your wood storage location to encourage gentle drying. Ideally this is an area sheltered from direct sunlight and protected from strong prevailing winds. Store wood up off the ground to avoid splash and absorption from the soil, and protect it from direct rain by covering with a tarp or metal roofing. If you like, you can get a rough idea of the wood's moisture content using a relatively inexpensive moisture meter, but for most greenwood crafts, it's really not necessary.[97]

In the eastern United States, the moisture content (MC) of air-dried wood should stabilize around 15% to 20%. In the drier western states, it's more like 10% or less. Kiln drying can lower the MC to about 5%.[98]

CANT MANAGEMENT

Efficient cant management requires a conscious, coordinated approach. Slope and access are the two details that should most directly inform your felling strategy. All cants should

Harvesting a Cant With 2 or More People Parallel to a Woodland Ride

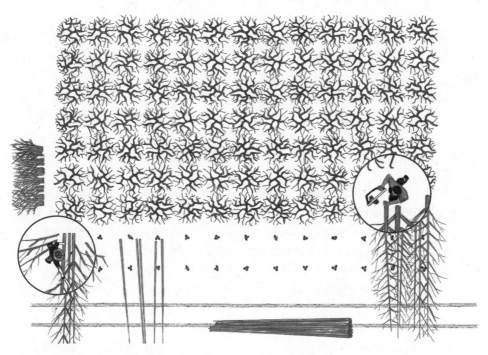

FIGURE 8.21: Here two people work together to cut and process coppice growth along a woodland ride, working in strips from right to left so as not to interfere with one another, steadily progressing upwards and away from the ride. By starting along the ride, they fell poles across the opening, making snedding and processing easier and opening up wider and wider access with each subsequent row. The worker on the right is felling coppice poles, followed up by a partner who cleans up the poles, sorts and hauls materials, and cuts products to length. This division of labor keeps the worker cleaning up and processing materials out of harm's way as poles are felled. With this approach, it's important to harvest the entire cant during the same season if possible so that the following season's regrowth doesn't box in any remaining stools from easy harvest and extraction. Adapted from Agate, 1980, p. 99.

be adjacent to some type of access, either a footpath or access road. In an ideal scenario, a wood's road frames each cant's downhill boundary. In this scenario, start by felling the stools along this edge, opening up better access to the stools uphill and making it easier to haul poles downhill towards the road. If stools vary in size, consider cutting the smallest first to open up space to fell larger stems later. Work along in strips, felling, snedding, cross-cutting, and hauling the materials in a section before advancing to the next strip.[99] Figure 8.21.

Woodsmen traditionally felled and processed coppice working their way through strips half a chain wide (11 yards or 10 m) that ran parallel to the access ride.[100] This

Harvesting a Cant Individually or in Pairs, Working Perpendicular to a Woodland Ride

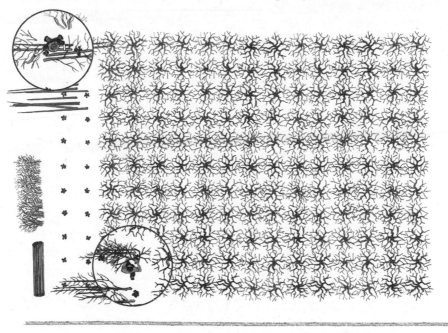

FIGURE 8.22: In this illustration, two people harvest coppice by working in wide strips perpendicular to the ride. This can be especially useful if there's any concern that it won't be possible to harvest the entire cant that season. Instead of blocking off access to any remaining coppice growth, you'll be able to preserve it along the cant edge. Here they've begun at the bottom left with one worker cutting coppice rods from the ride upwards, laying the cut material to the side. If working alone, they can then work back down the strip towards the ride, snedding the rods/poles, cutting them to length, and sorting materials into organized piles, before harvesting the next strip. If working in pairs, the second person follows along working up the felled coppice as the other continues to cut the rods/poles along the next strip. Adapted from Agate, 1980, p. 100.

coordinates the efforts of the working team, preserves access, and creates a safe, predictable working environment. These strips can even be as narrow as 15 feet (4.6 m). Hazel stands are often managed by cutting 33-foot (10 m) parallel strips.[101] More important than the specific strategy though is that you develop a clear, safe system.

In some cases, it might make more sense to cut wide strips that run perpendicular to the ride. This strategy preserves ride access to any sections you don't reach in a season instead of leaving isolated patches on the far end of a cant that lack convenient access to a ride.[102] Figure 8.22.

When harvesting young coppice growth, you can use a similar perpendicular strip method, but instead of starting at the access road, start at the far end. Fell the stools along a linear swath, and lay each pole on or next to the newly cut stool with the butt ends facing the ride for easy extraction. Be sure to lay the poles in the same direction with their butt ends lined up to make snedding and extraction easy. After felling, haul the stacked poles to the ride, starting with the closest stack and progressively working back to the point where you started felling. This maintains a clean jobsite and accessible pathways between stools.[103]

In most cases, standard trees should only be felled after harvesting coppice to avoid damaging any standing coppice poles.[104]

Although much of this work can be done by a single person, a team of three is often most efficient: one person felling with a chain saw, one with a billhook cleaning up the poles, and another hauling and sorting materials.

Generally, it's best to remove the lightest material first to quickly clear the area and ease the extraction of the larger poles.[105]

Working full-time, an experienced hazel cutter requires about 1 week to fell an acre, and depending on the product, up to an additional 5 weeks to add value to it.[106]

It's hard to overstate the value of keeping good records as you work. It's very easy to forget details after a few days or weeks pass, and knowing how many hours it took to complete a task, how many poles you cut, their average length, diameter, and quality, etc. will be super helpful to forecast labor and yields in future seasons.

MAINTENANCE INPUTS

After harvest, coppice stands are generally left unmanaged to regenerate until the next planned harvest. But time invested during the interim can pay off in improved productivity, reduced losses to pests, and less competition with herbaceous vegetation. We can break these maintenance activities into two main categories: stand improvement and pest/weed control.

Stand Improvement

Often, by increasing stocking density, we will improve total productivity and eliminate niches that could be colonized by unwanted species. Generally aim to leave no more than 11.5 feet (3.5 m) between stools.[107] Fill in any of these large gaps by planting or layering.

If a stand needs much restocking, cutting it again after 2 or 3 years will encourage more even, vigorous regrowth. Though it will delay yield by a few years, it can considerably improve sprout vitality and quality.

For standard trees and long-rotation coppice where knot-free timber is a goal, strategically basal-prune stems, removing branches before they become enclosed and form knots. Remove side branches before they reach 2 inches (5 cm). Basal pruning is often best done in two stages. When the tree is about 16 feet (5 m) tall, prune branches to 6.5 feet (2 m) height; at 26 to 33 feet (8 to 10 m), prune to 13 to 16 feet (4 to 5 m). Time spent basal pruning high-value long-term timber products can add considerable value to each stem.[108]

Short-rotation Willow

Short coppice rotations often exhaust plants' root systems, regardless of species, causing yields to gradually decline after 20 to 30 years. At this point, stands are often replanted to start anew. Some smaller-scale growers allow these senescing stands to succumb to herbaceous competition and use it for pasture for 4 or 5 years before later replanting.[109]

Pruning and Thinning

Pruning coppice stools, pollards, and hedgerow plants can improve yields, produce higher-quality products, stimulate above- and below-ground growth and stack yields in time. Though pruning and thinning (or draw-felling) is relatively common in the tropics, it appears to be pretty rare in temperate climates. When it did occur, stools were often pruned to a single stem selected for retention as a standard tree. But a well-timed thinning can help redistribute a plant's annual growth increment to the highest-quality stems, resulting in better poles and a shorter rotation.

Because stems on a coppice stool self-thin naturally as more vigorous shoots out-compete the stragglers (from as many as 20 to 150 shoots per stool in the first year of growth to just 5 to 15 after 15 years[110]), coppice workers rarely invested time and energy in this process. Also remember that most coppice workers were landless, so there was little motivation to intervene mid-rotation to improve the quality of a regenerating stand. If you do elect to do this to a stand, just be careful not to over-thin because large gaps may become colonized by fast-growing undesirable species.

F. Lambert, author of *Tools and Devices for Coppice Crafts*, suggests pruning coppice stools after 3 years' growth, removing any thin, dead, weak, or otherwise low-quality stems. This might include outer shoots with pistol-gripped bottoms as well as shoots emerging from clusters of dormant buds that can grow into one another and choke each other out. As a general principle, the longer the rotation (or the larger the desired pole), the fewer the number of stems should be retained.[111]

Oaks, red maple, sugar maple, tulip poplar, black cherry, basswood, sweetgum, and white ash have all been found to respond well to thinning. For these species, thinning often begins after at least 5 years' growth. The earlier the thinning, the larger the remaining stems become. In the case of red oak, delaying thinning from age 5 to 10 resulted in a 12% reduction in stem dbh by age 25 and a 23% to 30% reduction at age 15 or 20. Beyond age 25, it appears there are few benefits to thinning stools of most species, except basswood and red maple.[112]

In West Virginia, a 10-year-long study monitored stump sprouting of 50-to-60-year-old red oak, black cherry, tulip poplar, white oak, and chestnut oak. More than 80% of the stumps surveyed had sprouted by the end of the first season and still retained live sprouts after 10 years. Stools averaged 21 to 42 sprouts at the end of the first season but had self-thinned to 3 to 7 sprouts after 10 years. By the end of

Table 8.3: Average height and diameter of dominant stump sprout by species and silvicultural practice in the Missouri Ozarks after 10 years

Species and Silvicultural Practice	Height	DBH
Scarlet Oak		
Clearcut	25' (7.6 m)	3.1" (7.9 cm)
Group Selection	13.2' (4 m)	1.4" (3.6 cm)
Single Tree Selection	6.5' (2 m)	0.7" (1.8 cm)
Black Oak		
Clearcut	18.6' (5.7 m)	2.3" (5.8 cm)
Group Selection	12.5' (3.8 m)	1.3" (3.3 cm)
Single Tree Selection	5.1' (1.6 m)	0.4" (1 cm)
White Oak		
Clearcut	15.5' (4.7 m)	2.2" (5.6 cm)
Group Selection	13.6' (4.1 m)	1.5" (3.8 cm)
Single Tree Selection	5.2' (1.6 m)	0.4" (1 cm)

The size and extent of forest disturbance has a major impact on sprout development in these oak stands. Adapted from Dey et al., 2008, p. 30.

the first year, dominant red oak and black cherry sprouts cut during dormancy had grown twice as tall as those cut during the growing season, but there was no significant difference between them after 10 years. Because stools' dominant sprouts appeared to change during the first few years of regrowth, the researchers concluded it's best to wait to thin stools until a clear dominant stem emerges. They recommended waiting 10 years.[113]

In 4-to-22-year-old even-aged red oak stands in southwestern Wisconsin, researchers monitored the effects of thinning stools to just one or two dominant stems. These sprouts formed from clear-cut patches of trees between 80 and 110 years old. After 5 years, stools thinned to a single stem showed a 33% increase in dbh over sprouts on unthinned stools (2.03 vs. 1.53 inches), and stools thinned to two sprouts showed a 14% increase in dbh (1.74 inches). They also found that thinning stumps early should improve stem quality, helping reduce the sweep at their base as they grow away from the center of the stool.[114]

A study of red and white oak stands (averaging 14.3-inches [36.3 cm] and 13.3-inches [33.7 cm] dbh, respectively) in southern Ohio found the tallest white oak sprouts grew in stands that were thinned and intentionally burnt, reaching 78.3 inches (199 cm) in 2 years. They also showed much greater diameter growth.[115]

A 1953 study on Missouri's Ozark Plateau compared the effects of singling scarlet oak (*Quercus coccinea*), black oak (*Q. velutina*), southern red oak (*Q. falcata*), white oak (*Q. alba*), post oak (*Q. stellata*), and hickory (*Carya* spp.) stools. They monitored height, dbh, and mortality at 12, 16, 21, and 30 years after cutting. After 30 years, singling had increased average sprout diameter by 25% and stand basal area by 64% over the control treatment. Johnson and Rogers (1980) estimate that singled northern red oak stools in the Lake States could reach 12- inches dbh (30 cm) on good sites after 25 years and 8.3 inches (21 cm) on intermediate sites. The annual basal area increase peaked between the 12th and 16th year, with significant declines during the following years, suggesting that a crop tree thinning could have further increased diameter growth.[116]

If thinning coppice stands for timber, remove all but one or two of the best-formed, defect-free poles from a stool. Also, try to select stems on the same stool that are separated as widely as possible from one another and ideally attached to the stump at or below ground level. Favor sprouts of high quality over sprouts large in size. This helps minimize the potential for decay through the parent stump or dead companion stems.[117]

Pruning Hedgerows

Proper hedgerow pruning calls for a very different strategy. In most cases, your goal when pruning a hedgerow is to encourage full, bushy growth by stimulating lateral bud development. As with most pruning, it's best to start while plants are young to shape them from an early age and minimize large wounds and potential decay.

As hard as it may be to believe, many growers in the UK recommend pruning one year old autumn-planted hedge species like hawthorn, blackthorn, myrobalan, and privet as low as 6 inches (15 cm) above the ground to stimulate new shoot development in spring. The following winter, they cut back the previous year's growth by one-half, and in the third (or fourth) winter, they prune back all lateral and leading shoots to an even, balanced form. This helps develop a thick base at the bottom of the hedgerow.[118]

Pest and Weed Control

The two biggest challenges that woody plant growers face during establishment are competition with herbaceous vegetation and damage caused by wildlife. This is especially important for new plantings and freshly coppiced stools during the first 3 to 5 years.

Grass and forbs begin growing before trees break bud and continue well after leaf fall. They can aggressively shade out small young trees and compete for available soil moisture and nutrients, while creating great habitat for rodents. When engulfed with grass and other herbaceous growth, trees grow slowly and may even struggle to survive. Herbaceous competition is generally less of a concern in woodland conversion projects than new plantings because most stools in an existing forest plot are already well-established and the canopy cover had been shading out most of the understory competition.

Your best opportunity to get ahead of herbaceous competition is by taking the time to do good site prep. On sites where sod is well-established, it's a very risky prospect to plant trees without some effort to knock back the competition. We've already discussed this in chapter 7, but hopefully your preplanting site prep has helped give new plantings a head start.

While mowing does help knock back aboveground herbaceous competition, it also tends to stimulate more vigorous regrowth, so it's often best to limit mowing to one or just a few times per season.[119] Of course, with less frequent mowing, the taller, lignified growth often requires a larger, heavier-duty mower. For this, a scythe, string trimmer, brush mower, flail mower, sickle bar mower, or brush hog are probably the most appropriate tools. If you like, you can then rake up the organic matter and use it to mulch your plantings.

During the years following coppicing, dense bramble growth frequently creates a thorny barrier blanketing the understory. Despite this, British studies have found that by year 9 or 10, the dense shade cast by the closed canopy chokes them out, so there's usually little need for intervention.[120] We discussed the pros and cons of using heavy mulch to deter weed growth towards the end of chapter 7, so review that section to consider these soil-building and moisture-conserving alternatives to mowing.

Wildlife Protection

Unfortunately for us, many wildlife species recognize new plantings and young coppice sprouts as a flavorful, energy-dense meal. The damage they cause can be severe, and it varies between species, so you'll need to know who you're protecting your plants against to help stave off the damage. Depending on growth rates in your area, your plants may be vulnerable for the first 1 to 5 years. We can organize this damage into two main categories: browsing and girdling.

Let's start with the browsers. Deer, elk, rabbits, and moose can do major damage to supple young sprouts as they munch away on the high-energy, nutrient-rich leaves and tender softwood shoots. These young shoots are most vulnerable during the spring and early summer months. Fortunately, by the end of the first season, on good sites, coppice regrowth may have developed large enough to extend beyond browse height.[121] But when necessary, protection from browsers is paramount to ensure quality sprouts and a robust, healthy stool.

Rodents and rabbits pose a threat at ground level, girdling young trees by stripping away bark and eating the cambium that lies beneath. If the damage completely encircles a stem, it will cause the shoot to die back all the way to the base of the wound. As sprouts grow in diameter, they are less susceptible to girdling, though in areas with deep snow cover, many rodents will still gladly girdle stems 4 inches (10 cm+) or more in diameter.

While I've yet to personally experience their effect, in the UK, squirrels pose another unique girdling threat, chewing through tree bark and twigs early in the growing season as sap flow begins, sometimes continuing through late summer. Preferring smooth-barked species like maples, birch, beech, and chestnut, squirrels will also damage young oaks. Tree sap often concentrates below the wound and can lead to the formation of yeast colonies. These wounds create canker sites where young stems can break easily, causing additional damage to the underlying wood.[122]

Historically, some of this wildlife pressure would have been curtailed by larger numbers of people engaged in woodland management alongside far greater hunting pressure on wildlife populations. Also, wood banks, ditches, palisade fencing, and laid hedges helped protect managed coppice from larger wildlife and livestock.[123]

We can organize our potential responses to these wildlife challenges into three main categories: *exclusion/enclosure, culling, and deterrence.* While some managers report culling can positively impact pest populations and the resulting damage to coppice growth, according to Ratcliffe, it's unlikely that culling will prevent impacts on coppice regrowth or reduce deer populations in any significant way.[124] Here we'll limit our discussion to exclusion/enclosure and deterrence.

Exclusion/Enclosure

The most reliable browse protection strategy is also the most intrusive and costly. While the benefits of fencing are clear—the near complete exclusion of browsers—it has a number of drawbacks. Its mere presence transforms the character and experience of the wooded landscape, aesthetically and ecologically, fragmenting wildlife habitat, while also concentrating wildlife populations on adjacent unfenced land.[125]

Fencing's cost-benefit depends on the scale of your planting and the system you're establishing. The alternative is to protect plants individually using tree shelters, a strategy we discussed in chapter 7. Ultimately, the size and

Table 8.4: Potential solutions to wildlife challenges

Rodents, Rabbits	
Individual tree shelters	Plastic
	Hardware Cloth (¼" or ½")
	Window Screen
	Tree Wrap
Deterrence	Cats/Dogs
Deer, Elk, Moose	
Individual shelters	Plastic - 6′ minimum
Fencing - Physical	Woven Wire
	Plastic netting/'Tenax'
Fencing - Psychological	Electric netting
	Multiple stands tape/rope/polywire
	Multiple stands high tensile wire
	Woven Wire
Deterrence	Dogs
	Hunting

shape of the area you need to protect, along with the planting density, play a huge role in this decision.

We can break fences into two main categories: physical and psychological. Physical fences require less maintenance but must be able to withstand physical pressures of wildlife or livestock pushing against them or underneath them. They are more costly, but once installed properly, they should also be the most reliable. Psychological barriers use electricity as a deterrent. They are often lightweight, portable, and particularly useful for nonpermanent protection, but they also require routine inspection and maintenance to ensure they're working properly. The charge on electric fences must be consistent and very strong, as that is their only deterrent. The most effective fence for deer protection is an 8-foot (2.4 m) fixed-knot, high-tensile woven wire fence.

For a deer fence to be effective, it must achieve some ideal in height, width, or a combination of the two. Capable of jumping up to 8 feet (2.4 m), deer require a substantial fence. An alternative is two parallel 4- or 5-foot (1.2 or 1.5 m) fences offset by about 4 feet (1.2 m). Though deer could easily jump over a single 4-foot fence, they're unlikely to try to breach two fences in a single bound.

Electric fences provide a more lightweight option than a permanent woven-wire perimeter fence. Ben Law has developed a movable electric fence system to protect coppice sprouts during the first year or 2 after felling. It consists of permanent posts (chestnut harvested from the coppice stand that he's protecting) about 7 feet (2.1 m) high with post insulators that carry either electrified high-tensile wire or polywire spaced every foot, powered by a portable solar charger. To economize on wire, he occasionally uses bailing twine or another form of visible long-lasting line for the strands at 5 feet (1.2 m) and above. These act as more of a visual deterrent.

Wellscroft Fence Company in New Hampshire, USA, provides fantastic resources for growers. With decades of experience in the field, they recommend several key principles to ensure electric fences perform properly:

- Ensure that the fence is always energized and properly grounded.
- Maintain a perimeter of at least 6 feet to the outside of the area excluded to keep the fence visible and prevent vegetation from shorting out the fence.
- Bait the fence using either tinfoil-wrapped peanut butter or scent traps to train animals to respect it.
- Test fence voltage regularly using a fence tester and ensure it maintains at least 3,500 volts.

Black plastic netting is another movable temporary option. In the UK, it is available in 1.5- and 1.8-meter heights in 100-meter-long rolls (a 100-meter roll weighs 22 kg, 48.5 lb). It works well because it can be taken down once coppice regrowth is no longer vulnerable and re-erected to protect another cant. The netting is attached to 2 lines of high-tensile wire strained between posts. One person can erect a 164-foot (50 m) length in as little as half an hour, requiring minimal posts, tying it to existing trees with bailing twine where available.[126]

Another option is to create a dead hedge using the voluminous brush leftover following coppicing. While very time- and material-intensive, a well-built dead hedge is said to provide reliable deer protection. There are several ways to build dead hedges, each with pros and cons.

Table 8.5: Characteristics of different types of wildlife exclusion fences

Psychological Fence Options	Design/ Materials	Positives/ Effectiveness	Drawbacks	Longevity	2011 Prices
Three-dimensional Tape or Rope Fence	2 fences, 4' apart; One conductor at 30" on outside fence, 2–3 at 20" and 40–60" on inside	Highly; Difficult for deer to jump over/through	Uses more wire and posts, requires more land set aside for the fence and more maintenance, needs to be baited	5–10 years	50–90 cents/ foot
Electric Netting	Fiberglass posts with a woven net of polywire, comes in heights up to 68"	Easy to move, very effective	Not great for browsers due to the low maximum height. Expensive, not ideal in winter	10 years	$0.55–$2.60/ foot
8 Wire High Tensile - 6' high	Permanent fence posts and high tensile wire	Very permanent, good for low-medium deer pressure, may be set up bipolar for winter or dry summer conditions. Cost effective for large, straight areas with minimal access needs	Difficult to move, requires strong, well built corners. Must be baited regularly to train wildlife; requires a visibility lane outside so animals see and avoid it	20 years	$1–$1.55/foot
91" Tenax type plastic mesh with Offset Electrified Wires	Permanent posts, plastic anti-deer mesh, high tensile or polywire	Least obtrusive aesthetically. Easy to install	Must be baited, less effective in winter, vulnerable to ice, snow, and fallen limb damage.	10 years for mesh	$1.20–$1.55/ foot
4–6' Tall Woven Wire (2x4") with 1 top electric wire and 1 offset 12" up	Permanent posts, woven wire, high tensile or polywire	Very effective for all types of livestock, low maintenance, year-round protection	High materials costs	25 years	$1.70–$2.90/ foot

Table 8.5: Continued

Physical Fence Options	Design/ Materials	Positives/ Effectiveness	Drawbacks	Longevity	2011 Prices
Smart Net Anti-Deer Fence-8'x100' rolls	Posts, netting	Easy to install and move, has built-in polyester support line (only requires posts every 15–25')	Requires well braced corners. Moderately expensive	15 years	$1.60–$2.40/ foot
91" Tenax C-Flex anti-deer mesh	Posts, netting, 8 g polywire or 12 g steel wire	Low visibility, easy to install. Hangs on 8 g Polywire or 12 g. Steel wire only requires posts every 25'	If not electrified, may be penetrated, small wildlife may chew through it	10 years	$1.10–$1.35/ foot
6' Woven Wire (2x4")	Posts, woven wire	Appropriate for small areas with light deer pressure, posts every 15'	Requires well braced corners, and must be stretched	20 years	$1.30–$2.25/ foot
Fixed Knot Woven Wire (6', 8', 10')	Durable posts, woven wire	Most permanent, cost effective deer option. Resists ice and snow and deer penetration. Least maintenance	Must be well braced. Requires long-lasting posts due to longevity. Most expensive	25–30 year minimum	$2.25–$4.25/ foot

From Wellscroft Fence Systems, "Fencing Options for Wildlife Control," 2011.

The first involves simply laying brash around the perimeter of the cant with the tops facing outwards and the butt ends towards the cant interior. It's relatively quick to build and requires little skill, but it needs a massive amount of material. These hedges should be at least 6 feet (1.8 m) high and 8 feet (2.4 m) wide. This combination of height and width improve the hedge's effectiveness; however, it must be continuous. Unfortunately, it also creates optimal habitat for rabbits and other small mammals.[127]

To create a more solid brushwood alternative, surround the cant perimeter with two rows of 2-to-3-inch (5 to 7.5 cm) diameter offset fence posts spaced 4.5 feet (1.5 m) apart in rows and 1 to 3 feet (0.3–0.9 m) between rows. Ideally, the posts extend at least 6 feet (1.8 m) high. Densely stuff the cavity between the posts with brash at an angle so that the spiky tips fill out the top. Stomp down the material to create a thick, dense barrier. While more time-consuming to build, it's longer lasting, takes up less overall space, and creates less rabbit habitat.[128]

A narrower woven dead hedge uses a single row of stakes at least 6 feet (1.8 m) tall and spaced 3 feet (0.9 m) apart. Densely interweave brash and small-diameter stems between the stakes to form a solid narrow boundary that provides minimal rabbit cover.[129]

And last, though it will take a number of years before it's effective, you could also try to grow your deer protection using a dense row of hardwood willow cuttings, as we discussed in the arborsculpture section in chapter 5.

Deterrence

Your simple presence in a recently felled coppice woodland can help deter deer and other wildlife.[130] Coppice workers once spent the spring and summer months living and working in the stand they'd recently cut, converting harvested material into product. This served the coppice well, helping make certain the young stump sprouts were at least somewhat protected during this vulnerable time in their life cycle.

While that's pretty impractical for many of us today, you may find ways to adapt this pattern to suit your site. If you hire seasonal help or host interns, perhaps they could live in portable housing that can be moved between cants each season. Also, browsers tend to be most active during the early morning and evening hours, so even just making your presence felt during these windows could help somewhat. And dogs also have a similar effect.[131]

In Victorian times, woodsmen reportedly painted a mixture of cow manure, soot, and water onto stools to repel deer.[132] We know little about its effectiveness today.

Coppice cant design and layout can also help reduce deer damage. Deer are edge-dwelling species and tend to favor sites with some degree of protection from predation. Minimizing a cant's perimeter edge and optimizing the open exposed space within each cant's interior not only improves light levels and coppice regrowth but also creates less hospitable wildlife habitat.

Of course, depending on your production goals, pollard management helps solve these issues with browsers, since once established, managed sprouts lie beyond the reach of most wildlife. It's not nearly as convenient to harvest pollard sprouts up high than coppice sprouts at ground so this may not be a good substitute for designs focused on long-rotation polewood production, but for growers with high levels of wildlife pressure, product needs that align well with pollarding, and relatively modest material needs, this could provide a browse-proof alternative.

And amidst this discussion, remember that our goal is not to eliminate wildlife from working woodlands but rather prevent them from damaging our wood products. Many of us are managing ecosystems that are severely fragmented from neighboring habitat patches.

Often, top predator populations have been decimated or even completely eliminated. This, along with a general decline in hunting in many communities and the proliferation of edge habitats throughout urbanized landscapes, has led to an explosion in the populations of browsing animals like deer. There is a fine line between balance and destruction, with considerable grey in between. Please just consider how to best balance your own polewood needs with those of the diverse populations of wildlife with whom you share the forest.

Access

With time and use, access rides and footpaths will often suffer from erosion or rutting and require some repair to remain functional and safe. Before looking for a quick fix to patch a problem, take some time to analyze why it's occurring in the first place. Erosion often results from long stretches of steep grades or inadequate or improper drainage. Before repairing a problem that will only reemerge later on, can you refine the design and solve the problem altogether? If not, how can future use and access development minimize the likelihood of these problems in other parts of your woods? Maintaining good access is a constant work in progress, but it's essential for safe and efficient coppice management.

RESTORING OVERSTOOD COPPICE

For North American readers, this section is probably less relevant, but in parts of the world with longstanding coppice traditions, restoration of derelict (overstood/neglected)

coppice presents a pressing need, with vast acreages having gone unmanaged for decades or more. However, several challenges we face in restoring neglected coppice require many of the same strategies we've discussed to convert high forest into coppice.

Overstood coppice stands typically present the following key problems:

- Stools have been shaded out and died, creating gaps.
- This leads to irregular, low-quality growth, with large lateral branches reaching for light.
- Some stools are overmature, too large for conversion into their usual products, making harvest and extraction more costly, challenging and energy-intensive.
- For certain species, sprouting ability declines with age.
- Often, the initial restoration cut yields materials useful for little more than fuelwood due to low stem quality.
- Access routes are often overgrown and in a state of disrepair.

Overstood long-rotation coppice (normally managed on rotations of 25 years or less) may be stable and capable of rejuvenation for up to 40 years, but beyond this point will then begin to steadily deteriorate in quality for all but the most hardy sprouting species.[133] Overstood hazel coppice can be particularly difficult to bring back, and if over 40 years old, it may not regenerate at all. Hazel often suffers from the shade cast by excessive numbers of oak standards. To help reduce the effects of this competition, aggressively thin the standards to maximize available light.[134]

Restoration work is often heavier and slower going than in-cycle coppicing, so it's often recommended to keep restoration efforts small, one acre per year or less, until you've successfully reestablished the management rotation. The British Trust of Conservation Volunteers estimates it takes an average of 120 person days (for a skilled chain saw operator and several volunteers with hand tools) to clear and sort one acre (300 days per hectare) of derelict coppice. A professional team could probably complete the job in one third the time or less. So, restoring derelict coppice requires 2 to 3 times as many days as it takes to cut and process 15-to-20-year-old growth.[135]

When considering restoring neglected coppice—especially overmature stands that may not resprout—we have four primary options: do nothing and essentially abandon the stand; clear-cut the stand and replant as needed; "store" the coppice, singling stools and attempting to convert it to high forest; or a combination of felling, storing, and planting.[136]

Coppice workers generally take one of two approaches when restoring overstood coppice: either cut the oldest and lowest-quality area first, or begin with the area showing the least amount of decline. By starting with the lowest-quality stand, you'll help halt further deterioration, but the job often requires a greater investment in time and labor during harvest and restocking. In the meantime, the better-quality areas continue their decline. Starting with the better-quality stand requires

less initial investment, but the poorest-quality stands continue to deteriorate.[137]

Many of the wildlife benefits may be slow to reemerge after starting to bring a neglected stand back into rotation. It may even take several cutting cycles before flowering plants begin to reestablish themselves in the open understory. Invertebrates may have disappeared and need some time to discover and recolonize the habitat.[138]

Needless to say, much of this restoration work presents a challenging prospect for growers, physically and economically. Hopefully, as we see continued appreciation for the value of healthy, well-stocked, managed woodlands for people and wildlife, there will be an increase in incentives to encourage these efforts. Time will certainly tell, but one thing is for certain: They're not making any more land, and one of the key problems for 21st-century humans to solve is how to make more from less. May these traditional tools and practices inspire and inform the new and evolving state of the art.

And with that, we've arrived. Well…we've arrived at this point in our "rotation." Because for some of us, we've only just begun. And like most things biological, this iterative process just continues to grow, reset, and regrow once again. Coppice agroforestry is a tool that enables us to steer succession towards a particular set of goals. And hopefully by now, you have a firm understanding of the values and costs of these strategies and an appreciation of how and where they're appropriate.

The modern relevance of resprout silviculture invites us to rethink our relationship with wood: to consider new ways to engage with this most primal material that has nurtured our species since we've walked the planet. There is still much to learn, but hopefully this resource will help serve as a compass to guide us. But the path and destination are ultimately yours to choose.

Epilogue

So, we've reached the end of this phase of our journey. Thank you for joining me. I hope it has opened your mind to the potential woody plants can play in our landscapes, stimulated a new appreciation of their biology and their remarkable adaptations to damage and disturbance, given you some new ideas for uses of small-diameter roundwood, and helped catalyze your ability to design your own working landscape using some of these tools and practices.

And here we are—another successional cycle having come and gone. Now it's time for us to create some disturbance. And in so doing, keep in mind that, as an Oklahoma rancher friend of mine likes to say, "You gotta crack a few eggs to make an omelet." In other words, we're going to make some mistakes. And those mistakes are invaluable—as long as we learn from them.

Now more than ever, our ability to network, share, and learn from one another is unprecedented. Despite humans' dozens of generations of experience with resprout silviculture, there is still so much for us to learn. You should now feel well-equipped with many management tools at hand, but it's up to you to decide how

to put them into practice. Start small, build a foundation, pay careful attention, and accept the feedback from your investment. And most importantly, share those experiences with others. Because when we engage with woodlands, we're building a culture of participatory ecology as much as a supply of wood products to meet crucial needs.

In fact, when we consider the broader implications of resprout silviculture to human society as a whole, we might actually find that the greatest value lies in our connection to woody plants, their life cycles, the landscape, and our role in it, as well as our relationships with one another. We're growing community as we nurture and grow woody resprouts.

We are long removed from the wood-based economies of prehistory and the Middle Ages. Yet disruptions in global supply chains; rising material, labor, and energy costs; the erosion of ecological landscape function; lack of access to right livelihood; displacement of people from landscape, and many other pressing issues all point to a need to increase our resilience at the local and bioregional levels, and create ways to connect people of all backgrounds and means to an engaged working relationship with their ecosystems. And it's this very prospect that makes human-scale practices like resprout silviculture so promising for the future.

While our culture may have long forgotten how to form these alliances with woody plants, the trees certainly haven't. Trust them, learn from them, engage with informed, conscious intention, and you will be rewarded.

May the sprouts be with you!

Appendix 1: Additional Resources

Selected Billhook and Forestry Tool Suppliers (Courtesy of Bob Burgess)

Simon Grant Jones: simongrant-jones.com, hand-forged to order

Dave Budd: davebudd.com/otherwork.html, hand-forged to order

Spear and Jackson: spear-and-jackson.com/products/hand-contractor/range/agricultural-tools

Richard Carter: richardcarterltd.co.uk/

Morris (Dunsford): woodsmithstore.co.uk/shop, no direct sales to the public, no website

Leonelli (Italian): leonelliattrezzi.com/?lang=en

Panzeri (Italian): panzeri-tools.it/attrezzi_agricoli.htm

Rinaldi (Italian): flli-rinaldi.it/roncole-accessori.asp

Planting Tools, Materials, Irrigation

AM Leonard: amleo.com/c/leonard-hand-tools

BCC Sweden: forestry, plant propagation, and nursery tools and equipment; bccab.com/products-planting

Berry Hill Irrigation, VA: berryhilldrip.com

Brookdale Fruit Farm, Hollis, NH: brookdalefruitfarm.com

DripDepot, OR: dripdepot.com

Forestry Suppliers: tools and equipment for forestry, engineering and environmental science; forestry-suppliers.com/product_pages/categories.php

Nolt's Produce Supplies, PA: noltsproducesupplies.net

Pacforest Supply Company, OR: tools and equipment for forestry, tree planting, and natural resource management; pacforest.com

Stuewe and Sons, OR: seedling nursery containers; stuewe.com

Nurseries

Bailey Nursery: baileynurseries.com/programs/bareroot

Cold Stream Farm, MI: coldstreamfarm.net

Colorado State Forest Service: csfs.colostate.edu/seedling-tree-nursery/seedling-nursery-inventory

Grimo Nut Nursery, ON, Canada: grimonut.com

Kentucky Division of Forestry State Nurseries: eec.ky.gov/Natural-Resources/Forestry/state-nuseries-and-tree-seedlings/Pages/default.aspx

Mistletoe-Carter Seeds, CA: mcseeds.com/trees-shrubs (Purchased the former Lawyer's Nursery)

New Hampshire Department of Natural and Cultural Resources: nh.gov/nhnursery

Northwoods Nursery, OR: northwoodsnursery.com, wholesale only

Oklahoma Department of Agriculture Food and Forestry: forestry.ok.gov/order-seedlings

Virginia Department of Forestry: buyvatrees.com

Nevada Division of Forestry: forestry.nv.gov/ndf-state-forest-nurseries/washoe-state-tree-nursery

Fencing Products

Gallagher Electric Fence: gallagherelectricfencing.com

Premier 1: premier1supplies.com

Wellscroft Fence Company: wellscroft.com/

Appendix 2: Yield and Species-by-Use Tables

Table 2.1: Consolidated "yield" or mean annual increment data of various types of coppice systems from the US and Europe

This table consolidates and organizes a range of data from many sources. I've done my best to maintain consistency within the data, but note that there are many variables not easily included here, namely—tree spacing, density, and patterning; harvest rotation; stool age; soil types; aspect; site index; climate/precipitation, etc. Use this information with the understanding that many variables are at play and each site and context are unique. Also, realize that often these figures are projections based on data collected from sample plots and extrapolated to per hectare figures, and in some cases, they are based completely on models, not real world experimental data. Thus, there are inconsistencies that the data doesn't fully reflect. For more clarity, refer to the original source data.

References are listed along with their corresponding number in the Yield table bibliography. In most cases, the original source data was provided in *oven dry tonnes per hectare per year*—the measured or projected Mean Annual Increment (MAI) for a stand. I've chosen to present both the high and low figures from each study where available along with its converted imperial equivalent in *US tons per acre per year*. From this, I've used average density figures for each species (usually listed under the species name) to convert weight into volume (first cubic feet of wood per acre, and then cords per acre), since some readers will be interested in fuelwood yields. Remember—a cord is a vague number because air space between stacked wood can comprise 30+% of the total volume. Here, I've used 128 ft³/cord as the standard (a completely solid stack of wood with no voids—something that's not physically possible). In practice, a cord is more likely to contain between 70–90 ft³ of actual wood. This means the cord estimates should underestimate projected yields. Compiled by Mark Krawczyk.

Table 2.1: *Continued*

Species and Density	MAI Low Metric [Imperial]	MAI High Metric [Imperial]	Volume Low ft³ [Cords]	Volume High ft³ [Cords]	Location	Reference #
European Black Alder - 33 lbs/ft³	3.3 [1.47]	5.8 [2.59]	89.2 [0.70]	156.8 [1.22]	Kansas	4
Grey Alder (*Alnus incana*) - 30 lbs/ft³	3.18 [1.42]		94.6 [0.74]		Estonia	22
Red Alder - 28 lbs/ft³	6.3 [2.81]		200.7 [1.57]		Washington State	23
Red Alder and Black Cottonwood - 26 lbs/ft³	9.7 [4.33]		332.8 [2.60]		Washington State	23
Alnus, Populus, Salix (assuming 24 lbs/ft³ average)	2 [0.89]	13.5 [6.02]	74.3 [0.58]	501.8 [3.92]	Austria, Canada, Sweden, Denmark, Finland, USA, UK	13
Black Cottonwood - 24 lbs/ft³	5.2 [2.32]		193.3 [1.51]		England	23
Black Locust - 48 lbs/ft³	1.85 [0.83]	13.47 [6.01]	34.4 [0.27]	250.3 [1.96]	Indiana, Ohio, Illinois	14
Black Locust - 48 lbs/ft³	4.7 [2.10]	5.1 [2.27]	87.3 [0.68]	94.8 [0.74]	KY	23
Black Locust - 48 lbs/ft³	6 [2.68]		111.5 [0.87		Eastern Germany	17
Black Locust - 48 lbs/ft³	8.6 [3.84]	11 [4.91]	159.8 [1.25]	204.4 [1.60]	Kansas	4
Boxelder - 30 lbs/ft³	4.3[1.92]	5.2 [2.32]	127.9 [1.00]	154.6 [1.21]	Kansas	4
Catalpa - 29 lbs/ft³		7.2 [3.21]		221.5 [1.73]	Kansas	4
Cottonwood - 28 lbs/ft³	4.6 [2.05]	9.9 [4.42]	146.6 [1.14]	315.4 [2.46]	Kansas	4
Downy and Silver Birch - 39 lbs/ft³ (Projected)	2.5 [1.12]	2.9 [1.29]	57.2 [0.45]	66.3 [0.52]	Central Sweden	5
Mixed Broadleaf	2.24 [1]	2.95 [1.32]			Montherme, France	16
Oaks (Scarlet, Southern Red, White and Black) - 45 lbs/ft³	2.7 [1.20]		53.5 [0.4]		South Carolina	23
Poplar (species unspecified) - 24 lbs/ft³		10 [4.46]		371.7 [2.90]	Eastern Germany	17
Hybrid Aspen/Poplar, 5/10 years - 22 lbs/ft³	5.4 [2.41]	8.2 [3.66]	219.0 [1.71]	332.5 [2.60]	Bavaria, Germany	9
Hybrid Poplar (DN34 and NM6) - 22 lbs/ft³	3.81 [1.70]	8.30 [3.70]	154.5 [1.21]	336.4 [2.63]	Michigan	12
Poplar clone OP-367 - 22 lbs/ft³	2.1 [0.94]	3.4 [1.52]	85.2 [0.67]	137.9 [1.08]	Aiken, South Carolina	1
Poplar 'Fritii-Pauley' (low) and Poplar 'Boleare' (high) - 22 lbs/ft³	8.3 [3.70]	14.32 [6.39]	336.6 [2.63]	580.7 [4.54]	Les Hayers (low) and Conde sur Suippe (high) France	16
Populus NE-299 (*P. ·betulifolia* x *P. trichocarpa*) - 22 lbs/ft³	10.1 [4.50]	11.6 [5.17]	409.5 [3.20]	470.5 [3.67]	WI	20
Populus tremula (Simulation) - 28 lbs/ft⁵	5.4 [2.41]	8.8 [3.93]	172.1 [1.34]	280.7 [2.19]	Germany	7
Populus tremuloides (Modeled) - 26 lbs/ft³	4.6 [2.05]	15.6 [6.96]	157.8 [1.23]	535.3 [4.18]	Rhinelander, WI	2
P. trichocarpa x *P. deltoides* - 22 lbs/ft³	8 [3.57]	11.4 [5.09]	324.4 [2.53]	462.3 [3.61]	Boom, Antwerp, Belgium (waste disposal site)	8
Populus trichocarpa Torr. & A. Gray *Populus deltoides*	0.04 [0.02]	23.68 [10.56]	1.6 [0.01]	960.2 [7.50]	UK	15

Table 2.1: *Continued*

Species and Density	MAI Low Metric [Imperial]	MAI High Metric [Imperial]	Volume Low ft³ [Cords]	Volume High ft³ [Cords]	Location	Reference #
Hybrid Willow (*Salix sachalinensis* and *Salix miyabeana*) - 25 lbs/ft³		5.61 [2.50]		200.0 [1.56]	Michigan	12
Salix triandra - 25 lbs/ft³	5.2 [2.32]	9.87 [4.40]	185.5 [1.45]	352.2 [2.75]	Tomaszkowo, Poland	21
Salix viminalis - 25 lbs/ft³	8 [3.57]	11 [4.91]	285.5 [2.23]	392.5 [3.07]	Northamptonshire, UK	19
Salix viminalis x S. viminalis lanceolata - 25 lbs/ft³	18.22 [8.13]	29.56 [13.18]	650.1 [5.08]	1054.8 [8.24]	Tomaszkowo, Poland	21
Salix viminalis var. *gigantea* - 25 lbs/ft³	13.11 [5.85]	20.17 [9.00]	467.8 [3.65]	719.7 [5.62]	Tomaszkowo, Poland	21
Salix viminalis var. *regalis* - 25 lbs/ft³	13.55 [6.04]	25.24 [11.26]	483.5 [3.78]	900.6 [7.04]	Tomaszkowo, Poland	21
Willow - 25 lbs/ft³ (model)	8 [3.57]	17 [7.58]	285.5 [2.23]	606.6 [4.74]	Sweden	10
Willow (*Salix L.*) - 25 lbs/ft³	6 [2.68]	10 [4.46]	214.1 [1.67]	356.8 [2.79]	Eastern Germany	17
Silver Maple - 33 lbs/ft³	3.1 [1.38]	7.9 [3.52]	83.8 [0.65]	213.6 [1.67]	Kansas	4
Sweet Chestnut - 37 lbs/ft³	3.1 [1.38]	14.9 [6.65]	74.7 [0.58]	359.2 [2.81]	UK	18
Sweet Chestnut - 37 lbs/ft³				353.0 [2.76]	Arnea Forest Region, Greece	3
Sweet Chestnut - 37 lbs/ft³	3.4 [1.52]	7.5 [3.35]	82.0 [0.64	180.8 [1.41]	Parthenay (low) and Melle (high), France	16
Sweetgum - 34 lbs/ft³	6.34 [2.83]	19.6 [8.74]	166.3 [1.30]	514.2 [4.02]	Decatur County, GA	11
Sycamore - 34 lbs/ft³	3.4 [1.52]		89.2 [0.70]		Aiken, South Carolina	1
Sycamore - 34 lbs/ft³	4.3 [1.92]		112.8 [0.88]		Kentucky	23
Sycamore (2886 lb per acre (3233 kg/ha) per year in year one, 4 x 5 ft (1.2 x 1.5 m) spacing - 4541 lb (5088 kg/ha) - 34 lbs/ft³	3.24 [1.44]	5.09 [2.27]	84.9 [0.66]	133.6 [1.04]	Mississippi	6

Table 2.2: Durability and decay resistance of wood of select coppice species

Durability ratings sourced from Center for Wood Anatomy Research; Crawford, 2000b: **N** = Not durable; **D** = Durable; **VD** = Very Durable; **XD** = Extremely Durable; **W** = Durable under water.

Decay Resistance ratings from Forest Products Lab (Alden, 1995) and Burns and Honkala, 1990: *Su* = Susceptible to decay; *N* = Not resistant; *SR* = Slightly Resistant;

MR = Moderately Resistant; *R* = Resistant; *VR* = Very Resistant.

Thanks to Dave Jacke for organizing and editing this table, with help from Daniel Plane. Dave Jacke will separately publish a more robust version of this table.

Genus	Species	Common Name	Decay Resistance Durability
Acacia	melanoxylon	blackwood	D
Acer	macrophyllum	bigleaf maple	N-SR
Ailanthus	altissima	tree of heaven	R
Alnus	cordata	Italian alder	**VD-W**
Alnus	glutinosa	European alder	**VD-W**
Alnus	rubra	red alder	N-SR, D-W
Betula	lenta	black birch	N-SR
Betula	nigra	river birch	N-SR
Betula	papyrifera	paper birch	N-SR, **D**
Buxus	sempervirens	common box	D
Carya	cordiformis	bitternut hickory	N-SR
Carya	illinoinensis	pecan	N-SR
Carya	laciniosa	shellbark hickory	N-SR
Carya	ovata	shagbark hickory	N-SR
Carya	tomentosa	mockernut hickory	N-SR
Castanea	dentata	American chestnut	VR, **D**
Castanea	sativa	European chestnut	D
Catalpa	bignonoides	southern catalpa	VR, **VD**
Catalpa	speciosa	northern catalpa	VR, **XD**
Cornus	mas	Cornelian cherry	D
Crataegus	monogyna	English hawthorne	D
Cunninghamia	lanceolata	Chinese fir	D

Genus	Species	Common Name	Decay Resistance Durability
Diospyros	virginiana	common persimmon	VR/N
Eucalyptus	camaldulensis	river redgum	D
Eucalyptus	globulus	Tasmanian bluegum	MR/Su, **D**
Eucalyptus	leucoxylon	white ironbark	D
Fraxinus	americana	white ash	N-SR
Fraxinus	latifolia	Oregon ash	N-SR
Fraxinus	pennsylvanica	green ash	N-SR
Gleditsia	triacanthos	honey locust	R, **VD**
Juglans	cinerea	butternut	N-SR
Juglans	nigra	black walnut	VR, **VD**
Juglans	regia	English walnut	D
Liquidambar	styraciflua	sweetgum	N-SR
Maclura	pomifera	osage-orange	VR, **D**
Magnolia	fraseri	mountain magnolia	N-SR
Metasequoia	glyptostroboides	dawn redwood	**XD**
Morus	x hybrids	hybrid mulberry	**VD**
Morus	alba var. tatarica	Russian mulberry	**VD-W**
Morus	rubra	red mulberry	**VD**
Ostrya	virginiana	hophornbeam	D
Paulownia	tomentosa	princesstree	D
Pinus	rigida	pitch pine	SR-MR, **VD**
Populus	balsamifera	balsam poplar	N-SR

Table 2.2: *Continued*

Genus	Species	Common Name	Decay Resistance **Durability**
Populus	deltoides	eastern cottonwood	N-SR
Populus	tremuloides	quaking aspen	N-SR
Prunus	serotina	black cherry	VR
Quercus	alba	white oak	VR, **D**
Quercus	bicolor	swamp white oak	VR
Quercus	cerris	Turkey oak	**D**
Quercus	falcata	southern red oak	N-SR
Quercus	gambelii	Gambel oak	VR
Quercus	garryana	Oregon white oak	VR
Quercus	macrocarpa	bur oak	VR, **VD**
Quercus	michauxii	swamp chestnut oak	MR, **D**
Quercus	montana	mountain chestnut oak	VR, **D**
Quercus	petraea	sessile oak	**D**
Quercus	robur	English oak	**D**
Quercus	rubra	northern red oak	N-SR, **N**
Quercus	stellata	post oak	VR, **D**
Quercus	velutina	black oak	N-SR

Genus	Species	Common Name	Decay Resistance **Durability**
Quercus	virginiana	live oak	**VD**
Robinia	pseudoacacia	black locust	VR, **VD**
Sassafras	albidum	sassafras	VR, **D**
Sequoia	sempervirens	redwood	R-VR, **XD**
Sideroxylon	lycioides	buckthorn bully	MR
Taxodium	distichum	bald cypress	R-VR, **VD**
Taxus	baccata	English yew	**D**
Tilia	americana	American basswood	N-SR
Toona	sinensis	fragrant spring tree	MR, **D**
Ulmus	alata	winged elm	N-SR, **D-W**
Ulmus	americana	American elm	N-SR
Ulmus	glabra	Scotch elm	**D-W**
Ulmus	minor	smooth-leaved elm	**D-W**
Ulmus	procera	English elm	**D-W**
Ulmus	rubra	slippery elm	N-SR, **D**
Umbellularia	californica	California laurel	VR

Table 2.3: Selected woody basketry species and their uses

Parts used: BK, BKi = Bark, inner bark;
BR = Branches; RT = Root; SH = Shoots;
SK = Suckers; ST = Stems; TW = Twigs;
WD = Wood.

Thanks to Dave Jacke for compiling and making sense of this data, with the help of Daniel Plane. Dave Jacke will separately publish a more robust version of this table.

Genus	Species	Common Name	Parts Used	Notes
Acer	circinatum	vine maple	SH, WD	SH-quite pliable. Straight SH-open-work baskets. WD-baby basket frames.
Acer	glabrum var. douglasii	Douglas maple	BKi	BKi-soft basketry.
Acer	macrophyllum	bigleaf maple	BKi, ST	BKi-harvested in spring; withies, weft. Young ST-coarse twine for basket warp and weft. LV-basket liners.
Acer	rubrum	red maple	WD	WD-basket splints. Basketware.
Acer	saccharinum	silver maple	ST	
Adenostoma	fasciculatum	chamise	WD	
Alnus	rhombifolia	white alder	RT	
Alnus	rubra	red alder	RT, SH	RT-weft.
Amelanchier	alnifolia	Saskatoon serviceberry	ST	Young ST-rims, handles, stiffeners. TW, ST-basket hopper rim reinforcement (for acorn pounding). WD-stiffening baskets, handles. Withies. Basket rims. Seed beaters.
Amelanchier	utahensis	Utah serviceberry	WD	WD-basket rims.
Betula	glandulosa	resin birch	TW	TW-ribs of sweetgrass basket.
Betula	lenta	black birch	BK	BK-rice baskets, winnowing dishes.
Betula	papyrifera	paper birch	BK, BKi	BK-hoops and coverings for baskets. BKi.
Betula	pumila	bog birch	TW	TW-ribs.
Carya	laciniosa	shellbark hickory	BKi	BKi-finishing baskets.
Carya	ovata	shagbark hickory	BR	BR-split thinly.
Carya	tomentosa	mockernut hickory	BR, BKi	BR-split thinly. BKi-baskets and chair seats.
Ceanothus	cuneatus	buck brush, chaparral	ST	ST-burden basket rods. Warp of twined seed-beater baskets, cradle boards. Winnowing basket warp and rim sticks.
Ceanothus	integerrimus	deer brush	SH	Young flexible SH-circular withies. Rods-burden and fine coiled baskets. Warp and rim sticks.
Celtis	occidentalis	hackberry	ST	
Cercis	canadensis	eastern redbud	BK	BK of young shoots is used.
Cercis	occidentalis	California redbud	BKi, BR	BR-twined baskets, burden baskets, winnowing and twined sifter baskets, deep bottomless baskets, acorn meal baskets, watertight and coiled cooking baskets, cradle boards.

Table 2.3: *Continued*

Genus	Species	Common Name	Parts Used	Notes
Chilopsis	linearis	desert willow	TW, SH	Young pliable TW. SH-bark removed, used unsplit as rod foundations in coil basketry.
Cornus	sanguinea	common dogwood	SH, ST	SH-thin and whippy. Young ST.
Cornus	sericea	red osier dogwood	TW, ST, BR, BKi	TW-basket foundations. BR-pliable, basket rims, cylindrical basketry trap frames. ST-birch basket rims. BK-patterns.
Corylus	americana	American hazel	TW	TW-twig baskets, ribs for woven baskets.
Corylus	cornuta	beaked hazel	BR, TW	
Corylus	cornuta var. californica	California hazel	WD, ST, SH	WD-basket rims. ST-warp for sedge baskets, coarse seive baskets, vertical withies of saw grass baskets. Peeled ST-warp for twined and coiled baskets. Switches-large burden baskets, surf fish baskets, and other open work baskets. Coiled basket foundation. Dried SH, straight ST-Warp and weft.
Dirca	palustris	leatherwood	BR, BK, SH	BR, BK-very strong. SH-tough flexible. BK-sewing baskets, withies.
Forsythia	spp.	forsythia	ST	Weavers.
Fraxinus	americana	white ash	BR, ST, WD	BR-split small. ST-unsplit for handle bows.
Fraxinus	nigra	black ash	BKi, BR, ST, WD	Preferred spp. BR-split small. ST-unsplit for handle bows. WD-beaten to separate along growth rings, cut into strips.
Fraxinus	pennsylvanica	green ash	WD	WD-beaten to separate along growth rings, cut into strips.
Morus	alba	white mulberry	BKi, TW	BKi-soft basketry.
Morus	nigra	black mulberry	BKi	BKi-soft basketry.
Morus	rubra	red mulberry	BKi	BKi-soft basketry.
Philadelphus	lewisii	Lewis' mock orange	ST	ST rods-fine, coiled baskets. Pithy ST-light baby baskets, edging for birch baskets and cradle hoods.
Populus	angustifolia	narrowleaf cottonwood	SH	
Populus	balsamifera nigra x jackii	balsam poplar, Lombardy poplar, balm of Gilead	RT	
Populus	fremontii	Fremont cottonwood	BR, TW	Young TW-peeled and split.
Prunus	pensylvanica	pin cherry	BK	Outer BK-watertight, decay resistant, ornaments, baskets.
Prunus	virginiana var. demissa	western chokecherry	BK, SH	Withies-overlay twine weft bases. Shredded BK-basket
Quercus	alba	white oak	BR, WD	BR split thickly. Split WD for trugs.
Quercus	douglasii	blue oak	BR	BR-rims for twined work baskets.
Quercus	dumosa	coastal sage scrub oak	BR	BR-acorn storage baskets.
Quercus	falcata	southern red oak	BR, WD	BR-split thinly. WD split for trugs.
Quercus	kelloggii	black oak	SH	Withies-basket rims. Split SH-basket & scoop twining rods.
Quercus	macrocarpa	bur oak	BR, WD	BR-split thinly. WD split for trugs.

Table 2.3: *Continued*

Genus	Species	Common Name	Parts Used	Notes
Rhus	trilobata	sourberry	BR, ST, SH	BR-tough and slender; stripped of bark and split. Numerous uses and preparation methods including coarse baskets, water baskets, winnowing baskets, twined sifter baskets, seed beaters, cradle boards, foundation coils, repair work, fish baskets, burden baskets. Warp and weft.
Salix	alba alba ssp. caerulea	white willow cricket bat willow	BKi, BR, ST	Bki-soft basketry. ST-very flexible.
Salix	X americana babylonica eriocephala gracilistyla miyabeana petiolaris var. gracilis purpurea x rubra triandra viminalis	hybrid willow weeping willow Missouri River willow rosegold pussy willow Miyabe willow meadow willow purpleosier willow green-leaf willow almond leaf willow basket willow	BR, ST	BR-thin and whippy. Excellent. ST-very flexible.
Salix	amygdaloides	peachleaf willow	ST	ST-very flexible.
Salix	bebbiana	beak willow	BR, ST	BR-thin and whippy. Excellent. ST-pliable. ST-basket rims. BK-sewing birch BK onto frames.
Salix	caprea	goat willow	BR, ST, SH	ST-very flexible.
Salix	cinerea	grey willow	BR, SH	BR-thin and whippy. Excellent.
Salix	daphnoides	violet willow	BR, ST	BR-thin and whippy. Excellent. Purple bloom. ST-very flexible. Baskets, wattle and daub walls.
Salix	euxina	Black Sea crack willow	BR, ST	BR-somewhat brittle. ST-very flexible.
Salix	exigua	gray willow	TW, BK, BR	BR, SH. Cooking baskets, basket bodies. ST-main ribs (alternate for *Corylus*). BK-sewing birch onto basket frames. TW-warp in twined baskets. RT-inside of overlaid twined baskets. Switches-twined baskets and coiled baskets.
Salix	gooddingii	Goodding's black willow	ST, BR	ST, debarked-baskets tight enough to hold water. Small green BR-split in two, peeled, twisted, dried and used to sew coiled baskets. Outdoor storage baskets. Peeled, green BR split small-coiled baskets.
Salix	hookeriana	dune willow	BR, ST, BK	BR-thin and whippy. Excellent. ST-very flexible.
Salix	koriyanagi	kori-yanagi	BR, ST	BR-thin and whippy. Excellent. ST-red or russet-coloured, very flexible.
Salix	laevigata	red willow	BR, SH	BR-thin and whippy. Excellent.
Salix	lasiandra	Pacific willow	BR, ST, TW, BK	BR-thin and whippy. ST-very flexible. TW-withes for the three-rod foundation coils of baskets. BR-warp in twined baskets, foundation in coiled baskets. BK-thread.
Salix	lasiolepsis	arroyo willow	BR, ST, RT, SH	BR-thin and whippy. Excellent. ST-split for coiled baskets, weft in twined baskets, unsplit as the warp in twined baskets.

Table 2.3: *Continued*

Genus	Species	Common Name	Parts Used	Notes
Salix	matsudana	curly willow	ST	ST-wicker baskets.
Salix	nigra	black willow	BR, ST, TW, BKi	Young ST, Split TW-coiled basket foundation. BKi-sewing coiled baskets.
Salix	scouleriana	Scouler's willow	BKi, ST, RT	ST-very flexible. BK-sewing birch BK onto basket frames.
Salix	sitchensis	Sitka willow	BR, ST, RT	BR-thin and whippy. Excellent. ST-very flexible.
Salix	x fragilis	hybrid crack willow	BR, ST	ST-very flexible. A male form of the plant is used.
Tilia	americana	American basswood	BK, BKi	BKi-soft basketry. Reinforcement.
Tilia	cordata	littleleaf linden	BKi	BKi-15-30cm diameter for soft basketry.
Tilia	spp.	lindens	BKi, SK, ST	BKi-soft basketry.
Torreya	californica	California torreya	RT	RT-splints in basketry.
Ulmus	alata	winged elm	BKi	BKi-soft basketry.
Ulmus	americana rubra	American elm slippery elm	BK, BKi	BKi-soft basketry.
Vitex	agnus-castus	chaste tree	BR, ST	Young ST.
Vitex	negundo	Chinese chaste tree	ST	Young ST-basket making, wattle

Table 2.4: Firewood characteristics of select woody coppice species: density, heat content, and coaling qualities

When available, density figures are an average of several sources; most sources provided pounds/cubic foot, which we converted to kilograms/cubic meter. Heat content shown as MBTU/cf (= million BTUs per cubic foot) when available. "Coaling" qualities indicates how well a given wood produces long-lasting coals during the late stages of the burn cycle: E = Excellent, G= Good, F = Fair, P = Poor.

Thanks to Dave Jacke for compiling, organizing, and distilling this data, with the help of Daniel Plane and Aaron Guman. Dave Jacke will separately publish a more robust version of this table.

Genus	Species	Common Name	Lbs/cf Dry (Avg.)	Kg/m³ Dry (Avg.)	MBTU/cf Dry (Avg.)	Coaling Qualities
Acer	campestre	common maple	40.9	655.1		E
Acer	macrophyllum	bigleaf maple	33.8	541.4	0.27	E
Acer	negundo	boxelder	31.4	503.0		P, E
Acer	pensylvanicum	striped maple	32.0	512.6		E
Acer	rubrum	red maple	35.1	562.2	0.28	E
Acer	saccharinum	silver maple	32.2	515.8	0.24	E
Acer	saccharum	sugar maple	41.6	666.3	0.33	E
Aesculus	californica	California buckeye	25.0	400.5		
Aesculus	hippocastanum	horse chestnut	25.0	400.5		
Alnus	rubra	red alder	27.6	442.1	0.21	G
Amelanchier	alnifolia	Saskatoon serviceberry	52.0	832.9		
Amelanchier	alnifolia var. semiintegrifolia	Saskatoon serviceberry	52.0	832.9		
Amelanchier	arborea	downy serviceberry	52.0	832.9		
Amelanchier	canadensis	shadbush	52.0	832.9		
Amelanchier	canadensis var. obovalis	coastal serviceberry	52.0	832.9		
Amelanchier	laevis	Allegheny serviceberry	52.0	832.9		
Amelanchier	utahensis	Utah serviceberry	52.0	832.9		
Arbutus	menziesii	Pacific madrone	44.7	716.0		E
Avicennia	germinans	black mangrove	58.0	929.0		
Betula	lenta	black birch	45.0	720.8	0.34	G
Betula	papyrifera	paper birch	36.1	578.2	0.29	G
Betula	pubescens	downy birch	33.0	528.6	0.30	G
Carpinus	caroliniana	American hornbeam	42.1	674.4		E
Carya	aquatica	water hickory	43.0	688.8	0.35	E

Table 2.4: *Continued*

Genus	Species	Common Name	Lbs/cf Dry (Avg.)	Kg/m³ Dry (Avg.)	MBTU/cf Dry (Avg.)	Coaling Qualities
Carya	cordiformis	bitternut hickory	44.1	706.4	0.36	E
Carya	glabra	sweet pignut hickory	52.0	832.9	0.43	E
Carya	illinoinensis	pecan	46.0	736.8	0.36	G - E
Carya	laciniosa	shellbark hickory	48.0	768.9	0.37	E
Carya	myristiciformis	nutmeg hickory	42.0	672.8	0.34	E
Carya	ovata	shagbark hickory	47.5	760.9	0.38	E
Carya	tomentosa	mockernut hickory	46.7	748.0	0.38	E
Castanea	dentata	American chestnut	30.0	480.5	0.23	
Catalpa	bignonoides	southern catalpa	27.6	442.1		G
Catalpa	speciosa	northern catalpa	28.5	456.5		
Celtis	occidentalis	hackberry	36.0	576.6	0.27	G
Celtis	reticulata	netleaf hackberry	37.0	592.7		
Cornus	alternifolia	alternate-leaf dogwood	42.3	677.6		
Cornus	florida	flowering dogwood	51.0	816.9		
Cornus	nuttallii	Pacific dogwood	45.9	735.2		
Diospyros	virginiana	common persimmon	50.5	808.9		
Eucalyptus	camaldulensis	river redgum	35.0	560.6		
Eucalyptus	grandis	rose gum eucalyptus	50.7	812.1		
Fagus	grandifolia	American beech	41.4	663.1	0.33	E
Fagus	sylvatica	European beech	43.2	692.0		E
Frangula	purshiana	cascara buckthorn	36.0	576.6		
Fraxinus	americana	white ash	40.0	640.7	0.33	G
Fraxinus	excelsior	European ash				G
Fraxinus	latifolia	Oregon ash	38.0	608.7	0.29	G
Fraxinus	pennsylvanica	green ash	36.0	576.6	0.29	G
Fraxinus	profunda	pumpkin ash	36.0	576.6		G
Fraxinus	quadrangulata	blue ash	40.0	640.7	0.30	G
Gleditsia	triacanthos	honeylocust	44.4	711.2	0.33	E
Halesia	carolina	silverbell	32.0	512.6		
Hamamelis	virginiana	witch hazel	43.0	688.8		
Juglans	cinerea	butternut	25.1	402.1	0.20	G
Juglans	nigra	black walnut	36.6	586.3	0.29	G
Kalmia	latifolia	mountain-laurel	48.0	768.9		

Table 2.4: *Continued*

Genus	Species	Common Name	Lbs/cf Dry (Avg.)	Kg/m³ Dry (Avg.)	MBTU/cf Dry (Avg.)	Coaling Qualities
Liquidambar	styraciflua	sweetgum	34.0	544.6	0.27	
Liriodendron	tulipifera	tuliptree	28.0	448.5	0.22	F
Lithocarpus	densiflorus var. densiflorus	tanoak	40.8	653.5	0.33	
Maclura	pomifera	osage-orange	54.0	865.0	0.41	E
Magnolia	fraseri	mountain magnolia	31.0	496.6		
Magnolia	virginiana	sweetbay	29.0	464.5		
Malus	pumila	apple	42.8	685.6		G - E
Malus	sylvestris	European crab apple	47.0	752.8		E
Morus	alba	white mulberry	42.9	687.2		E
Morus	rubra	red mulberry	44.0	704.8		
Nyssa	sylvatica	blackgum	35.0	560.6	0.26	
Ostrya	virginiana	hophornbeam	47.9	767.3		
Oxydendrum	arboreum	sourwood	37.0	592.7		
Platanus	occidentalis	American sycamore	33.1	530.2	0.25	G
Populus	alba	white poplar	20.8	333.2	0.20	F
Populus	balsamifera	balsam poplar	23.5	376.4	0.18	
Populus	deltoides	eastern cottonwood	25.3	405.3	0.21	G
Populus	grandidentata	bigtooth aspen	26.3	421.3	0.21	
Populus	tremuloides	quaking aspen	25.1	402.1	0.20	G
Populus	trichocarpa	black cottonwood	24.0	384.4	0.19	G
Prunus	serotina	black cherry	34.1	546.2	0.26	E
Quercus	alba	white oak	45.6	730.4	0.37	E
Quercus	bicolor	swamp white oak	50.0	800.9	0.38	E
Quercus	coccinea	scarlet oak	47.0	752.8	0.35	E
Quercus	falcata	southern red oak	41.0	656.7	0.31	E
Quercus	garryana	Oregon white oak	44.7	716.0	0.40	E
Quercus	macrocarpa	bur oak	43.3	693.6	0.33	E
Quercus	michauxii	swamp chestnut oak	47.0	752.8	0.35	E
Quercus	montana	chestnut oak	46.0	736.8	0.34	E
Quercus	muehlenbergii	chinkapin oak	38.3	613.5		E
Quercus	nigra	water oak	44.0	704.8	0.33	E
Quercus	pagoda	cherrybark oak	47.0	752.8	0.35	E
Quercus	palustris	pin oak	44.0	704.8	0.33	E

Table 2.4: *Continued*

Genus	Species	Common Name	Lbs/cf Dry (Avg.)	Kg/m³ Dry (Avg.)	MBTU/cf Dry (Avg.)	Coaling Qualities
Quercus	rubra	northern red oak	40.7	651.9	0.35	E
Quercus	stellata	post oak	46.0	736.8	0.35	E
Quercus	velutina	black oak	43.0	688.8	0.32	E
Quercus	virginiana	live oak	57.9	927.4	0.46	E
Robinia	pseudoacacia	black locust	44.5	712.8	0.36	E
Salix	alba	white willow	26.4	422.9	0.20	P
Salix	nigra	black willow	26.0	416.5	0.21	
Sambucus	nigra ssp. cerulea	blue elder	36.0	576.6		
Sassafras	albidum	sassafras	31.0	496.6	0.24	
Taxodium	distichum	bald cypress			0.24	
Tilia	americana	American basswood	24.4	390.8	0.19	P
Ulmus	alata	winged elm	42.0	672.8		G
Ulmus	americana	American elm	32.5	520.6	0.27	G - E
Ulmus	crassifolia	cedar elm	41.0	656.7		G
Ulmus	laevis	Russian elm	34.0	544.6		
Ulmus	rubra	slippery elm	35.6	570.2	0.27	G - E
Umbellularia	californica	California laurel	39.4	631.1		

Table 2.5: Resprouting species known to produce quality charcoal

Please note that many other species can and have been used to make charcoal. This very select list of species have reported qualities that make them well-suited for charcoal production for various purposes.

Thanks to Dave Jacke for compiling and making sense of this data, with the help of Daniel Plane and Aaron Guman. Sources: Alden, 1995; Burns and Honkala, 1990; Crawford, 2000b; Forest Inventory and Analysis National Program, 2016; Plants for a Future; Tabor, 1994.

Genus	Species	Common Name	Notes
Acer	campestre	Common maple	Charcoal made from the wood is an excellent fuel.
Acer	saccharum	Sugar maple	This species is noted as producing very hot embers.
Alnus	rubra	Red alder	Makes high-grade charcoal.
Arbutus	menziesii	Pacific madrone	Pacific madrone makes a fine-grade charcoal. It was used for gunpowder, and preferred for it by the early Californians.
Betula	pubescens	Downy birch	A high-quality charcoal is obtained from the bark, and is used by artists and painters.
Carya	cordiformis	Bitternut hickory	Desirable for making charcoal.
Carya	laciniosa	Shellbark hickory	Used for charcoal.
Carya	ovata	Shagbark hickory	Often used to make excellent charcoal.
Fagus	grandifolia	American beech	Makes an excellent charcoal used in artwork.
Fagus	sylvatica	European beech	Yields a charcoal known as 'Carbo Ligni Pulveratus'.
Salix	alba	White willow	Used to make charcoal, which has medicinal uses.
Salix	nigra	Black willow	Makes a good charcoal, once used as a charcoal for gunpowder.

Table 2.6: Crude protein content of leaves of select woody plants, based on reviews of scientific literature published through 2012

Data may contain crude protein (CP) content of samples taken at various times of year, including dead or senesced leaves collected in autumn. Some data samples may have also included twigs or young green shoots, but generally, such data was excluded. Data for species with average leaf crude protein below 9.0% not shown. Crude protein is not by itself an indication of the edibility, palatability, lack of toxicity, or suitability of a potential fodder species for any given animal. Use this table as a springboard for further research. Use this information at your own risk.

Thanks to Dave Jacke for compiling and making sense of this data, with the help of Daniel Plane. Dave Jacke will separately publish a more robust version of this table, complete with citations for the numerous studies used to compile this table.

Genus	Species	Common Name	Avg % CP	% CP Low	% CP High	# of refs
Acer	rubrum	red maple	11.4	2.7	12.8	6
Acer	saccharinum	silver maple	14.0	11.7	16.3	3
Acer	saccharum	sugar maple	12.5	7.8	28.8	4
Ailanthus	altissima	tree-of-heaven	18.2	10.5	27.2	2
Albizia	julibrissin	mimosa	19.3	16.8	23.8	3
Alnus	glutinosa	European black alder	15.2	9.4	19.7	4
Alnus	incana	European gray alder	20.1	17.6	25.0	2
Alnus	rubra	red alder	14.2	12.7	15.6	2
Alnus	cordata	Italian alder	15.9	12.1	19.7	1
Amelanchier	utahensis	Utah serviceberry	12.0	11.0	13.0	1
Amorpha	fruticosa	false indigo	18.2	12.5	19.7	5
Betula	allegheniensis	yellow birch	23.5	12.5	34.4	1
Betula	alnoides	xi hua	19.4			1
Betula	lenta	black birch	13.5			1
Betula	pendula	European white birch	19.3	16.9	23.0	2
Betula	pubescens	downy birch	15.4	17.6	13.2	2
Caragana	jubata	shag-spine	20.3			1
Caragana	korshinskii	Korshinsk peashrub	17.0	13.3	19.1	3
Caragana	microphylla	litteleleaf peashrub	16.8			1
Castanea	sativa	European chestnut	14.5	12.4	17.0	3
Celtis	occidentalis	hackberry	11.5	8.7	13.5	1
Cornus	stolonifera	red-osier dogwood	14.4	11.3	17.5	2
Corylus	avellana	European filbert	10.5	8.5	12.1	1

Table 2.6: *Continued*

Genus	Species	Common Name	Avg % CP	% CP Low	% CP High	# of refs
Corylus	cornuta	beaked hazel	13.8	12.4	14.6	1
Elaeagnus	angustifolia	Russian olive	18.0	11.1	25.0	6
Elaeagnus	umbellata	autumn olive	17.9	13.8	21.9	1
Fagus	grandifolia	American beech	11.6	10.8	12.3	1
Fagus	sylvatica	European beech	16.1	11.9	17.5	3
Fraxinus	americana	white ash	14.4			1
Fraxinus	pensylvanica	green ash	9.9			1
Ginkgo	biloba	ginkgo	11.9			1
Gleditsia	triacanthos	honeylocust	13.7	10.9	17.7	6
Hippophae	rhamnoides	seabuckthorn	19.7	15.6	21.8	4
Liquidambar	styraciflua	sweetgum	10.9	9.0	16.9	3
Liriodendron	tulipifera	tulip-tree	12.4	7.9	16.8	5
Melia	azerdarach	Chinaberry	12.8			1
Morella	cerifera	southern bayberry	13.1	9.5	16.0	1
Morus	alba	white mulberry	20.5	10.7	35.9	24
Populus	alba	white poplar	14.8	13.0	16.5	2
Populus	deltoides	eastern cottonwood	12.0	5.6	18.5	5
Populus	nigra	black poplar	13.7	11.3	19.7	4
Populus	tremula	aspen	18.5	12.8	27.7	4
Populus	tremuloides	quaking aspen	13.8	5.4	26.9	5
Quercus	alba	white oak	9.2	3.4	12.9	3
Quercus	incana	bluejack oak	11.9			1
Quercus	nigra	water oak	13.0	10.3	19.1	1
Quercus	robur	English oak	14.7	11.5	18.2	2
Quercus	rubra	red oak	11.7	8.7	13.9	3
Quercus	stellata	post oak	12.3	12.1	12.4	1
Quercus	velutina	black oak	9.5	8.6	10.3	1
Quercus	virginiana	live oak	9.5	9.1	10.2	1
Robinia	pseudoacacia	black locust	19.8	11.9	27.3	23
Salix	babylonica	weeping willow	14.5	6.9	24.8	3
Salix	caprea	goat willow	18.8	16.5	22.3	2
Salix	euxina	Black Sea crack willow	15.5			1
Salix	humboldtiana	Humboldt's willow	12.7	6.9	18.4	1

Table 2.6: *Continued*

Genus	Species	Common Name	Avg % CP	% CP Low	% CP High	# of refs
Salix	nigra	black willow	10.5	8.3	13.3	2
Salix	pentandra	laurel willow	18.3			1
Salix	purpurea	purpleosier willow	23.7			1
Salix	udensis	fantail willow	9.9	8.2	11.4	1
Salix	viminalis	basket willow	18.3			1
Sassafras	albidum	sassafras	13.5	5.5	28.3	2
Sorbus	aucuparia	rowan	14.6	14.3	14.8	2
Ulmus	alata	winged elm	13.0	7.3	27.6	2
Ulmus	americana	American elm	12.6	4.8	16.3	3
Ulmus	crassifolia	cedar elm	10.9	8.7	12.1	1
Ulmus	glabra	Scotch elm	17.6			1
Ulmus	minor	smooth-leaved elm	15.0	12.1	19.7	2
Ulmus	rubra	red elm	9.9			1
Zanthoxylum	americana	prickly ash	17.2			1

Table 2.7: Estimated digestibility and crude protein content of leaves of select potential coppice fodder species for cattle, sheep, or goats

Data derives from studies on several continents under varied conditions. Methods of estimating digestibility vary between studies. Units and methods used are shown, but not defined here: see the references for details. Generally, the higher the digestibility percentage, the more energy and nutrients an animal can receive per bite. Leaf digestibility varies over the growing season, as can be seen. Digestibility data is not by itself an indication of the edibility, palatability, lack of toxicity, or suitability of a fodder species for any given animal. Use this table as a springboard for further research. Use this information at your own risk.

Thanks to Dave Jacke for compiling and making sense of this data, with the help of Daniel Plane. Dave Jacke will separately publish a more robust version of this table.

Abbreviations: IVDMD: in vitro dry matter digestibility; IVOMD: in vitro organic matter digestibility; ETDMD: estimated true dry matter digestibility; IsDMD 48: in situ dry matter digestibility at 48 hours; In vivo DMD: in vivo dry matter digestibility.

Genus	Species	Common Name	Animals Studied	Leaf Harvest Season	% Digestibility	Digestibility Units	Crude Protein as % of Dry Matter	Ref.
Acer	grandidentatum	bigtooth maple	Cattle	Mid-Sept.	58.5–60.6	IVOMD	9.5–10.2	2
Acer	platanoides	Norway maple	Sheep	July	67.2	IVOMD	18.5	4
Ailanthus	altissima	tree of heaven	Goats	Spring	78.6	IsDMD 48	27.2	1
				Winter	67.3	IsDMD 48	10.5	1
Albizia	julibrissin	mimosa	Cattle	July	70.31	IVOMD	17.2	3
Alnus	glutinosa	European alder	Sheep	July	35.9	IVOMD	19.7	4
Betula	pubescens	downy birch	Sheep	July	34.8	IVOMD	15.6	4
					52.0	IVDMD	17.6	7
Celtis	australis ssp. caucasica	Caucasian hackberry	Sheep	Fall (early)	46.3	IVDMD	10.1	13
Corylus	avellana	European filbert	Sheep	July	48.1	IVOMD	12.1	4
Elaeagnus	angustifolia	Russian olive	Goats	Spring	66.0	IsDMD 48	14.9	1
				Winter	71.2	IsDMD 48	13.9	1
				Winter	59.5	In vivo DMD		1
			Sheep	May, July, Sept	56.3–63.3	IVOMD	20.1–25.0	5
				April-October	60.0	IVOMD	20.6	6
Fraxinus	excelsior	European ash	Sheep	June	57.2	IVOMD	17.4	4
				September	67.8	IVOMD	17.3	4
					60.0	IVDMD	19.2	7

Table 2.7: *Continued*

Genus	Species	Common Name	Animals Studied	Leaf Harvest Season	% Digestibility	Digestibility Units	Crude Protein as % of Dry Matter	Ref.
Gleditsia	triacanthos	honey locust	Cattle	July	64.4	IVOMD	14.2	3
			Sheep	May, July, Sept	50.3–65.5	IVOMD	10.9–12.9	5
Liquidambar	styraciflua	sweetgum	Goats	1 sample/season	71.5–39.1	In vivo DMD		11
				Spring	82.1	ETDMD	16.9	10
				Summer	69.0	ETDMD	9.8	10
				Fall	76.1	ETDMD	9.9	10
				Winter	43.7	ETDMD	5.3	10
Melia	azedarach	Chinaberry	Sheep	Fall (early)	62.6	IVDMD	12.8	13
Morus	alba	white mulberry	Cattle	at 3 - 9 wks old	84.4–78.2	IVDMD	35.9–30.9	12
				January-April	72.1	IVDMD	15.1	9
					71.8	IVOMD	22.3	14
					66.4	IVDMD	22.3	14
			Goats	Spring	69.8	IsDMD 48	17.6	1
				Winter	77.7	IsDMD 48	13.7	1
				Winter	53.9	In vivo DMD		1
			Sheep	Spring	65.6–69.2	IVOMD	20.8–21.9	15
				Fall	56.3–61.4	IVOMD	18.9–21.9	15
				Fall (early)	52.3	IVDMD	14.8	13
Nyssa	sylvatica	blackgum	Goats	Spring	78.0	ETDMD	15.1	10
				Summer	73.0	ETDMD	8.8	10
				Fall	67.9	ETDMD	6.8	10
				Winter	56.8	ETDMD	6.4	10
				1 sample/season	80.9–54.7	In vivo DMD		11
Populus	deltoides	eastern cottonwood	Sheep		74.0	IVOMD	18.1	8
Populus	tremula	European aspen	Sheep	July	55.3	IVOMD	14.1	4
					61.0	IVDMD	14.2	7
Prunus	americana	American plum	Goats	1 sample/season	91.8–58.8	In vivo DMD		11
Prunus	americana & umbellata	American flatwood plums	Goats	Spring	88.4	ETDMD	26.2	10
				Summer	77.5	ETDMD	11.1	10
				Fall	83.6	ETDMD	13.1	10
				Winter	48.4	ETDMD	9.0	10
Quercus	incana	bluejack oak	Sheep	Fall (early)	42.4	IVDMD	11.9	13
Quercus	nigra	water oak	Goats	Spring	70.9	ETDMD	19.1	10

Table 2.7: *Continued*

Genus	Species	Common Name	Animals Studied	Leaf Harvest Season	% Digestibility	Digestibility Units	Crude Protein as % of Dry Matter	Ref.
Quercus	nigra	water oak	Goats	Summer	60.2	ETDMD	10.3	10
				Fall	61.2	ETDMD	12.2	10
				Winter	64.2	ETDMD	10.5	10
				1 sample/season	59.0–40.3	In vivo DMD		11
Robinia	pseudoacacia	black locust	Cattle	July	65.6	IVOMD	16.3	3
				July	68.2	IVOMD	21.9	3
			Goats	Spring	74.2	IsDMD 48	23.9	1
				Winter	65.7	IsDMD 48	14.5	1
				Winter	52.3	In vivo DMD		1
			Sheep	May, July, Sept	45.4–52.4	IVOMD	20.1–22.1	5
				Fall (early)	52.7	IVDMD	11.9	13
Salix	babylonica	weeping willow	Goats	Spring	73.8	IsDMD 48	12.5	1
				Winter	71.3	IsDMD 48	9.8	1
				Winter	61.7	In vivo DMD		1
Salix	caprea	goat willow	Sheep	June	72.0	IVOMD	22.3	4
					58.0	IVDMD	17.7	7
Salix	euxina	Black Sea crack willow	Sheep	September	53.7	IVOMD	15.5	4
Salix	pentandra	laurel willow	Sheep	June	48.4	IVOMD	18.3	4
Salix	purpurea	purpleosier willow	Sheep	September	39.5	IVOMD	23.7	4
Salix	viminalis	basket willow	Sheep	September	37.2	IVOMD	18.3	4
Sassafras	albidum	sassafras	Goats	Spring	75.1	ETDMD	28.3	10
				Summer	61.5	ETDMD	12.3	10
				Fall	69.1	ETDMD	9.6	10
				Winter	37.3	ETDMD	5.5	10
				1 sample/season	79.2–22.7	In vivo DMD		11
Sorbus	aucuparia	rowan	Sheep	July	74.3	IVOMD	14.8	4
					54.0	IVDMD	14.3	7
Ulmus	alata	winged elm	Goats	Spring	82.0	ETDMD	27.6	10
				Summer	53.5	ETDMD	11.2	10
				Fall	66.0	ETDMD	8.8	10
				Winter	36.3	ETDMD	7.3	10
				1 sample/season	89.7–26.6	In vivo DMD		11
Ulmus	glabra	Scotch elm	Sheep	July	74.0	IVOMD	17.6	4

Table 2.8: Edibility of select woody coppicing plants. This table contains information from a range of sources

Do not eat any plant parts shown below without proper identification and further research to confirm edibility. You use this information at your own risk. Thanks to Daniel Plane and Dave Jacke for compiling and making sense of this data. Dave Jacke will publish a more robust version of this table.

X = Toxicity issues exist: do further research. Parts Used: SD = Seed or nuts; FR = Fruit or pods; FL =

Flowers; LV = Leaves; SH = Shoots; TW = Twigs; S/T = Shoots and twigs; ST = Stems; Sap = Sap; BK, BKi, BKr = Bark, inner bark, root bark; RT = Roots; OP = Other plant parts. Uses: P = Protein, S = Starch; F = Fruit; V = Vegetable; T = Tea; B = Beverage; Cul = Culinary; O = Oil; E = Essential oil; H = Honey plant; OU = Other uses.

Genus	Species	Common Name	Toxicity	Notes
Acer	glabrum	Rocky Mountain maple		BKi. SH-cooked like asparagus. Seedlings-fresh, dried for later use. Dried, crushed LV-spice. SD 6 mm-dewinged and boiled. LV-wrapped as preservative around apples, rootcrops, etc.
Acer	negundo	boxelder	X	Sap-valued for drink, syrup. BKi-dried and ground to flour for breadstuff and soup thickener, boiled until sugar crystallizes out. Seedlings-fresh or dried for later use. SD 12 mm-dewinged and boiled. LV-wrapped as preservative around apples, rootcrops, etc.
Acer	rubrum	red maple		Young LV. Sap-drink, sugar, syrup. BKi-dried and ground to powder for thickening soups and breadstuff. Seedlings-fresh or dried for later use. SD 5 mm-dewinged and boiled. LV-wrapped as preservative around apples, rootcrops, etc.
Acer	saccharinum	silver maple		Young LV. Sap-drink, sugar, syrup. BKi-dried and ground to powder for thickening soups and breadstuff. Seedlings-fresh or dried for later use. SD 12 mm-dewinged and boiled. LV-wrapped as preservative around apples, rootcrops, etc.
Acer	saccharum	sugar maple		Sap-syrup, sugar. SD-de-winged and cooked in butter or milk. Young LV. Sap-drink, vinegar, tea 'water' (when brewing sassafras and spicebush). BKi-dried and ground to powder for thickening soups and breadstuff. Seedlings-fresh or dried for later use. SD 6 mm-dewinged and boiled. LV-wrapped as preservative around apples, rootcrops, etc.
Amelanchier	alnifolia	Saskatoon serviceberry		FR 10-15 mm, sweet-raw, cooked, jams, jellies, dried, in pancakes, muffins, pies, pemmican. LV-tea.
Betula	lenta	black birch	X	SH, dried LV and BKr from larger, reddish RT-tea. Young LV. Catkins. BKi-cooked, ground to flour for a thickener or breadstuff, birch beer. Sap-raw beverage, birch beer, wine, syrup, vinegar. Essential oil from TW, BK-wintergreen flavoring in sweets, baked goods, chewing gum, tea.
Caragana	arborescens	Siberian peashrub	X	Young, green SD pods-cooked and eaten as vegetable. Mature SD-bland, contain ≤12% fatty oil and ≤36% protein; cooked like dried beans or peas.
Carya	glabra	sweet pignut hickory		SD, small-raw or cooked, varies between bitter-astringent and sweet; stores for 6 months in cool, dry place. Sap-drink.

Table 2.8: *Continued*

Genus	Species	Common Name	Toxicity	Notes
Carya	illinoinensis	pecan		Oil from SD. LV-tea. SD up to 40 mm-raw, salted, in sweets, baked goods, made into milk that is used in soups, hominy, or fermented; stores for 6 months in cool, dry place.
Carya	laciniosa	shellbark hickory		SD up to 50 mm, sweet, hard, thick shell-raw, baked goods. Sweet sap-syrup, sugar; stores for 6 months in cool, dry place.
Carya	myristiciformis	nutmeg hickory		SD up to 30 mm, sweet, thick shelled-raw, cooked, resembles nutmeg; stores for 6 months in cool, dry place.
Carya	ovata	shagbark hickory		SD up to 40 mm, sweet-raw, roasted, in baked goods, made into milk and used like butter, ground into flour and used as a thickener; stores for 2 years in cool, dry place; improved cultivars exist. Sap, boiled WD chips and BK-syrup, sugar.
Castanea	dentata	American chestnut	X	SD, small and sweet, contains about 7% fat, 11% protein-raw, pureed, in baked goods, roasted to eat or for coffee. Oil from SD-crush nuts, boil, skim off oil, used as pudding topping. Kernel-chocolate substitute.
Castanea	mollissima	Chinese chestnut		SD-high in carbs and low in oils; raw, roasted, baked, pureed, boiled.
Castanea	pumila	chinquapin		SD-20 mm, 45% starch, 2.5% protein; raw, baked or roasted, small and very sweet.
Castanea	sativa	European chestnut	X	SD-raw, baked or roasted, boiled, pureed, dried, in sweets, ground to flour and used to thicken soups and in baked goods, used as coffee substitute, sugar extracted. Fine cheeses wrapped in LV. FL-source of dark amber honey traditionally used in Italy for baking.
Ceanothus	americanus	New Jersey tea		LV-gathered while in full bloom, shade-dried, used for tea.
Ceanothus	cuneatus	buck brush, chaparral		LV, FL-excellent tea.
Corylus	americana	American hazel		SD-ripe or unripe, raw or cooked, candied, ground to flour, soups, baked goods. Oil from SD. SD keeps 12 months unshelled in cool, dry place.
Corylus	avellana	European filbert	X	SD-16% protein, 60% fat; raw or roasted, candied, baked goods, sweets, ground to flour, nutmilk, flavoring in 'Frangelico' liquor. SD keeps 12 months unshelled in cool, dry place. LV-wrappers for 'sarmas' in Turkey. Oil from SD-raw, baked goods.
Corymbia	maculata var. citriodora	spotted gum		A sweet manna-like substance scraped off of LV. Honey plant. LV-spice.
Encelia	farinosa	brittle bush		Gum exudation from ends of mature ST-chewed, aromatic.
Eriodictyon	californicum	California yerba santa	X	Fresh or dried LV-aromatic tea. Extract of LV-flavoring sweets, soft drinks, and baked goods. Fresh LV-thirst quencher (raw).
Fraxinus	excelsior	European ash	X	Keys-pickled, used as condiment. Oil from SD-similar to sunflower oil. LV-tea, fermented beverage.
Fraxinus	ornus	manna ash	X	Gum exudation-sweetener, anti-caking agent.
Laurus	nobilis	bay laurel	X	FR-liqueur. Dried FR, fresh or dried LV-flavoring. LV-wrapping, tea; 1-3% oil, extracted for flavoring.
Populus	deltoides	eastern cottonwood	X	BKi-most likely dried and ground for soup thickener and breadstuff. SD-small and impractical. Buds. LV-protein-rich with a greater amino-acid content than wheat, corn, rice and barley; LV concentrate more nourishing, cheaper than meat, produced faster, possible major food source for humans.

Table 2.8: *Continued*

Genus	Species	Common Name	Toxicity	Notes
Populus	*deltoides* ssp. *monilifera*	plains cottonwood	X	BKi-most likely dried and ground for soup thickener and breadstuff. SH-cooked. Seedpod-chewed like gum. LV-protein rich; LV concentrate more nourishing, cheaper than meat, produced faster, possible major food source for humans.
Prunus	*avium*	sweet cherry	X	FR-raw, dried, candied, stewed, baked goods, sweets, jellies. LV-pickle flavoring. BK-tea. SD-raw, cooked. Gum exudation from trunk and branch wounds. FL-source of excellent, light colored honey.
Prunus	*dulcis*	almond	X	Oil from SD-cooking, flavoring. SD-raw, roasted, sprouted, baked goods, ground to powder, almond milk. FL-source of caramel-colored honey. Gum exudation from damaged ST.
Prunus	*persica*	peach	X	SD-raw, cooked. Oil from SD: 45% oil. Gum exudation from ST-chewed. FL-raw, cooked, tea, distilled. LV and essential oil from LV-flavoring. FR-fresh, dried, preserved, juiced, pickled, jams. Unripe FR-jams. SD kernel-flavoring, raw.
Quercus	*cerris*	Turkey oak		SD up to 25 mm-Leached and ground to flour (poor), roasted for coffee substitute. BR and ST gum exudation from insect damage-eaten as manna, boiled down to sweet syrup.
Quercus	*garryana*	Oregon white oak	X	SD up to 25 mm-leached, roasted, raw, ground to flour for soups and baked goods. Good host for the gourmet truffle, *Tuber melanosporum*.
Quercus	*petraea*	sessile oak	X	SD-leached, dried and ground to powder for use in stews and bread, famine food by one ref., roasted for coffee substitute. LV-winemaking. Edible gum from BK.
Quercus	*robur*	English oak	X	SD-washed, cooked for famine food, ground to flour, chopped and roasted for almond substitute, roasted for coffee substitute. LV-winemaking. BK-edible gum used as butter substitute in cooking.
Robinia	*neomexicana*	New Mexico locust	X	FL-raw, cooked, flavoring. Fresh SD pods-raw, cooked. SD-cooked. FL, Pods-cooked and dried for later use.
Robinia	*pseudoacacia*	black locust	X	FL-fried, pancakes, beverage; source of light colored acacia honey popular in eastern Europe. Young SD pods-boiled. SD 4 mm-cooked like peas or beans; 21% protein, 3% fat, 0% carbohydrates. Oil from seed? FR skin-strong narcotic and intoxicating drink. Piperonal can be extracted from the plant and used as vanilla substitute.
Salix	*alba*	white willow	X	LV, SH-raw or cooked. LV-tea. SH, LV, BKi-spring edibles; rich in vitamin C. BKi-raw or cooked, ground to flour for use as famine food.
Salix	*cinerea*	grey willow	X	SH, LV, BKi-spring edibles, rich in vitamin C.
Salix	*daphnoides*	violet willow	X	SH, young LV, BKi-spring edibles, rich in vitamin C. Rhizome tips-peeled, raw or cooked. SH, young catkins-raw or cooked. BKi-raw or cooked, ground to flour for use as famine food.
Salix	*exigua*	gray willow		LV-orange juice-like drink.
Salix	*nigra*	black willow	X	Young SH-raw or cooked. SH, LV, BKi-spring edibles, rich in vitamin C. BKi-ground to flour for use as famine food.
Salix	*pulchra*	tealeaf willow		Young LV, SH-raw or cooked, eaten with oil to make palatable, rich in vitamin C. Catkins. Dried LV-tea. BKi-raw or cooked, ground to flour for use as famine food.

Table 2.8: *Continued*

Genus	Species	Common Name	Toxicity	Notes
Salix	*purpurea*	purpleosier willow	X	Young SH-cooked. SH, LV, BKi-spring edibles, rich in vitamin C. BKi-ground to flour for use as famine food.
Salix	*sitchensis*	Sitka willow	X	Young SH-cooked. SH, LV, BKi-spring edibles, rich in vitamin C. BKi-ground to flour for use as famine food.
Sambucus	*nigra*	black elderberry	X	FR-raw, dried, jams, pies, condiments, dried, juiced, wine, flavoring; all FR products said to retain medicinal value of FR. FL-raw, fried, dried, flavoring, wine, tea. LV-green coloring for oils and fats.
Sambucus	*nigra* ssp. *canadensis*	elderberry	X	FR dried, raw, jellies, jams, pies, baked goods, sauces, soups, syrups, vinegar, wine. Unripe FR, unopened FL-pickled like capers. FL fritters, pancakes, muffins, wine, tea. Cvs available. Flowers and fruits on new wood.
Sambucus	*nigra* ssp. *cerulea*	blue elder	X	FR-raw, more often dried for jellies, pies, sauces, sherbets, soups, relishes, juices, wines; wine considered 'tonic.' FL-raw, fritters, pickled, pancakes, muffins, syrups, vinegar, tea.
Sassafras	*albidum*	sassafras	X	RT, FR-flavoring jelly, tea. BKr-boiled with sugar and water to make a condiment, tea. Winter buds, young LV tips, FL-raw or cooked. LV-raw, flavorful thickener (can be dried and ground to powder). SH-beer. FL, LV, RT-tea.
Tilia	*americana*	American basswood	X	Young buds, LV-raw or cooked. FL-raw, tea. Sap-drink, syrup. SD, paste of FR, FL-chocolate substitute. FL-source of light colored honey.
Tilia	*cordata*	littleleaf linden	X	Young LV-raw. FL, immature FR-ground to paste for chocolate substitute. FL-tea. Sap-drink, or syrup.
Tilia	*x europaea*	European linden	X	Manna. LV-raw. Sap-syrup, sweetener. FL-raw, wine, tea. FL, immature FL-source of distinct, light honey. FR-ground to paste for chocolate substitute.
Tilia	*x flavescens*	Glenleven Linden		
Tilia	*platyphyllos*	summer linden	X	Sap-drink, syrup. Young LV-raw. FL, immature FR-ground to paste for chocolate substitute. FL-tea.
Toona	*sinensis*	fragrant spring tree		SH, LV-garlic taste. LV-boiled, cooked in soy sauce, tea; ±6% protein. Young buds-blanched.
Ulmus	*americana*	American elm		LV-raw, cooked. BKi-coffee-like beverage.
Ulmus	*rubra*	slippery elm	X	LV, immature FR-raw or cooked. BKi-chewed as thirst quencher, powdered to flour for thickening soups and making bread, preservative (cooked with fats to keep them from going rancid), breading, tea.
Umbellularia	*californica*	California laurel	X	LV-condiment, seasoning; a commercial sweet bay substitute. SD-parched, roasted, ground to flour. FR-raw, cooked. LV-tea. BKr-coffee substitute.
Vitex	*agnus-castus*	chaste tree	X	FR, SD-pepper substitute. LV-spice

Table 2.9: Selected resprouting woody species useful as cut branches for ornamental flowers, fruits, foliage, and stems for the floral trade

Based on the work of Greer and Dole, 2009. Cv or cvs = cultivar(s); FFL = Forced flowers; FCat = Forced catkins; FFol = Forced foliage; FL = Flowers; Cat = Catkins; Fol = Foliage; Bud = Buds, usually flower buds; FR = Fruit; ST = Stems.

Thanks to Dave Jacke for compiling and making sense of this data, with the help of Daniel Plane. Dave Jacke will separately publish a more robust version of this table.

Genus	Species	Common Name	Forcing Flowers, Foliage, Catkins	Flowers, Catkins	Foliage	Fruit, Buds	Stems
Buxus	microphylla, sempervirens	boxwood, common box			Fol		
Cercis	canadensis, chinensis	eastern & Chinese redbuds	FFL, FFol	FL	Fol		
Chaenomeles	speciosa, x superba	flowering quinces	FFL	FL			
Cornus	alba, sericea	Tatarian & redosier dogwoods		FL		FR	ST
Cornus	mas	Cornelian cherry	FFL	FL		FR	
Cornus	sanguinea	common dogwood					ST
Corylus	americana, avellana	American hazel, European filbert	FCat	Cat	Fol		ST
Corylus	maxima var. purpurea	purple giant filbert		Cat	Fol		ST
Cotinus	spp.	smoketrees		FL	Fol		
Forsythia	x intermedia, ovata	hybrid & early forsythias	FFL	FL	Fol		
Hydrangea	arborescens, macrophylla, paniculata, quercifolia	smooth, bigleaf, peegee, and oakleaf hydrangea		FL			
Ilex	x altaclerensis, aquifolium, x attenuata, cornuta, x meserveae	altaclera holly, English holly, Foster holly (hybrids), Chinese holly, blue holly			Fol	FR	
Ilex	decidua, verticillata	possumhaw, winterberry				FR	
Kerria	japonica	kerria		FL			ST
Ligustrum	spp. & hybrids	privets			Fol	FR	
Magnolia	grandiflora	southern magnolia		FL	Fol	FR	
Malus	spp.		FFL	FL		FR	
Myrica	cerifera, pensylvanica	wax-myrtle, northern bayberry			Fol	FR	
Myrica					Fol	FR	
Nandina	domestica	heavenly bamboo			Fol	FR	
Philadelphus	spp.	mock oranges		FL			
Physocarpus	opulifolius	ninebark		FL	Fol	FR	

Table 2.9: *Continued*

Genus	Species	Common Name	Forcing Flowers, Foliage, Catkins	Flowers, Catkins	Foliage	Fruit, Buds	Stems
Pieris	formosa var. forrestii	Himalayan pieris		FL		Bud	
Pieris	japonica	Japanese pieris	FFL	FL		Bud	
Pieris	japonica x formosa var. forrestii			FL			
Pittosporum	tenuifolium, tobira	pittosporums			Fol		
Prunus	x blireana	blireana plum	FFL	FL			
Prunus	caroliniana	Carolina cherry laurel			Fol		
Prunus	cerasifera	purpleleaf plum	FFL	FL	Fol		
Prunus	dulcis	almond		FL		FR	
Prunus	glandulosa, incisa	flowering almond, Fuji cherry		FL			
Prunus	laurocerasus	common cherry laurel		FL	Fol		
Prunus	mume	Japanese flowering apricot	FFL	FL		FR	
Prunus	persica, sargentii	peach, Sargent cherry	FFL	FL			
Prunus	serrulata	Japanese flowering cherry	FFL	FL	Fol		
Prunus	subhirtella x yedoensis	'Hally Jollivette' flowering cherry		FL			
Prunus	triloba, x yedoensis	flowering almond, Yoshino cherry	FFL	FL			
Salix	alba	white willow	FFol				ST
Salix	chaenomeloides, gracilistyla	Japanese pussy willow	FCat	Cat			
Salix	irrorata, koriyanagi	bluestem willow, fodder willow		Cat			
Salix	matsudana, udensis	curly willow, fantail AKA fasciated willow					ST
Spiraea	betulifolia, japonica	birchleaf & Japanese spireas		FL			
Spiraea	prunifolia	bridal wreath spirea	FFL	FL			
Symphoricarpos	spp. & hybrids	snowberries				FR	
Vaccinium	ovatum	evergreen huckleberry			Fol		
Vaccinium	parvifolium	red huckleberry				Bud	ST
Viburnum	dentatum, nudum var. cassinoides	arrowwood, withe-rod				FR	
Viburnum	opulus	crampbark		FL			
Vitex	agnus-castus, unedo	chaste trees		FL		FR	
Weigela	florida	weigela			Fol		

Species-by-Use Tables Bibliography

Compiled by Dave Jacke, with help from Aaron Guman, and Daniel Plane. Except for the Yields table, compiled by Mark Krawczyk.

2.1 Yields

1 Coleman, Mark; Aubrey, Douglas. 2008. *Coppice culture for biomass production in southeastern United States.* In: Zalesny, Ronald S., Jr.; Mitchell, Rob; Richardson, Jim, eds. Biofuels, bioenergy, and bioproducts from sustainable agricultural and forest crops: proceedings of the short rotation crops international conference; 2008 August 19–20; Bloomington, MN. Gen. Tech. Rep. NRS-P-31. Newtown Square, PA: U.S. Department of Agriculture, Forest Service, Northern Research Station: 9.

2 Ek, Alan R.; Lenarz, John E.; Dudek, Albert. 1983. *Growth and yield of Populus coppice stands grown under intensive culture.* In: Hansen, Edward A., ed. Intensive plantation culture: 12 years research. Gen. Tech. Rep. NC-91. St. Paul, MN: U.S. Department of Agriculture, Forest Service, North Central Forest Experiment Station: 64–71.

3 European Forestry Commission Timber Committee. 1994. Seminar on harvesting and silviculture of degraded and coppice forests in the Mediterranean region Thessaloniki (Greece), 1–5 November 1994

4 Geyer, Wayne A. 2006. "Biomass production in the Central Great Plains USA under various coppice regimes." *Biomass & Bioenergy* 30: 778–783.

5 Johansson, Tord. (2008). Sprouting ability and biomass production of downy and silver birch stumps of different diameters. *Biomass & Bioenergy.* 32. 944–951.

6 Kennedy, Harvey E., Jr. 1980. *Coppice Sycamore Yields Through 9 Years. Res.* Note SO-254. New Orleans, LA: U.S. Department of Agriculture, Forest Service, Southern Forest Experiment Station. 4 p.

7 Lasch, Petra & Kollas, Chris & Rock, Joachim & Suckow, Felicitas. (2009). Potentials and impacts of shortrotation oppice plantation with aspen in Eastern Germany under conditions of climate change. *Regional Environmental Change.* 10. 83–94.

8 Laureysens, I & Bogaert, Jan & Blust, Ronny & Ceulemans, R. (2004). Biomass production of 17 poplar clones in a short-rotation coppice culture on a waste disposal site

and its relation to soil characteristics. *Forest Ecology and Management.* 187. 295–309.

9 M Liesebach, G von Wuehlisch, H.-J Muhs. 1999. Aspen for short-rotation coppice plantations on agricultural sites in Germany: Effects of spacing and rotation time on growth and biomass production of aspen progenies. *Forest Ecology and Management.* Volume 121, Issues 1–2. Pages 25–39.

10 Lindroth, A.; Bath, A. 1999. "Assessment of regional willow coppice yield in Sweden on basis of water availability." *Forest Ecology and Management.* 121: 57–65.

11 McCutchan, B.G. and E. L. Prewitt. 1982. Effect of harvest season and spacing on coppice sweetgum biomass yields. In Proceedings of the Fourth Central Hardwood Forest Conference, University of Kentucky, Lexington (R.N. Muller ed.) 113–124.

12 Miller, Raymond O.; Bender, Bradford A. 2008. *Growth and yield of poplar and willow hybrids in the central Upper Peninsula of Michigan.* In: Zalesny, Ronald S., Jr.; Mitchell, Rob; Richardson, Jim, eds. Biofuels, bioenergy, and bioproducts from sustainable agricultural and forest crops: proceedings of the short rotation crops international conference; 2008 August 19-20; Bloomington, MN. Gen. Tech. Rep. NRS-P-31. Newtown Square, PA: U.S. Department of Agriculture, Forest Service, Northern Research Station: 36.

13 C.P Mitchell, E.A Stevens, M.P Watters. 1999. Short-rotation forestry–operations, productivity and costs based on experience gained in the UK. *Forest Ecology and Management.* Volume 121, Issues 1–2. Pages 123–136.

14 Pope and Anderson. 1982. *Biomass Yields and Nutrient Removal in Short Rotation Black Locust Stands*

15 Rae, AM, K M Robinson, N R Street, and G Taylor. 2004. Morphological and physiological traits influencing biomass productivity in short-rotation coppice poplar. *Canadian Journal of Forest Research.* 34(7): 1488–1498.

16 Ranger and Nys. 1996. *Biomass and Nutrient Content of extensively and intensively managed coppice stands.* 02 Vol 69, Iss. 2. 91–110

17 Röhle, Heinz et al. "Establishment and expected yield of short-term rotation plantations in Eastern Germany." *Schweizerische Zeitschrift Fur Forstwesen* 159 (2008): 133–139.

18 Rollinson, TJD and J Evans. 1987. The Yield of Sweet Chestnut Coppice. *Forestry Commission, Bulletin* 64. England.

19 Souch, Claire & Martin, P.J. & Stephens, William & Spoor, G. (2004). Effects of soil compaction and mechanical damage at harvest on growth and biomass production of short rotation coppice willow. *Plant and Soil.* 263. 173–182.

20 Strong, Terry. 1989. *Rotation Length and Repeated Harvesting Influence Populus Coppice Production.* Research Note NC-350. St. Paul, MN: U.S. Dept. of Agriculture, Forest Service, North Central Forest Experiment Station

21 Szczukowski, S.; Stolarski, M.; Tworkowski, J.; Przyborowski, J.; Klasa, A. 2005.

Productivity of willow coppice plants grown in short rotations. *Plant Soil and Environment* 51(9): 423–430.

22 Uri, Veiko & Tullus, Hardi & Lõhmus, Krista. (2002). Biomass production and nutrient accumulation in shortrotation grey alder (Alnus incana (L.) Moench) plantation on abandoned agricultural land. *Forest Ecology and Management.* 161. 169–179.

23 Zimmerman, Richard William. 1984. *Growth and Development of Black Locust Stands.* University of Kentucky PhD Dissertation.

2.2 Durability

Alden, Harry A. 1995. *Hardwoods of North America.* General Technical Report FPL-GTR-83. Madison, WI: USDA Forest Service, Forest Products Laboratory.

Burns, Russell M. and Barbara H. Honkala. 1990. *Silvics of North America: Volume 2. Hardwoods.* Agriculture Handbook 654. Washington, DC: USDA Forest Service.

Center for Wood Anatomy Research. *Wood Properties Tech Sheets.* Madison, WI: USDA Forest Service, Forest Products Laboratory.

Crawford, Martin. 2000b. *Timber Trees for Temperate Climates,* 3rd ed. Devon, UK: Agroforestry Research Trust.

2.3 Basketry

Burns, Russell M. and Barbara H. Honkala. 1990. *Silvics of North America: Volume 2. Hardwoods.* Agriculture Handbook 654. Washington, DC: USDA Forest Service, accessed January 25, 2016.

Crawford, Martin. 2000a. *Plants for Basketry,* 2nd ed. Devon, UK: Agroforestry Research Trust.

Hopman, Ellen Evert. 1991. *Tree Medicine, Tree Magic.* Blaine, WA: Phoenix Publishing.

Plants for a Future. *PFAF Plant Database.* www.pfaf.org/user/ plantsearch.aspx.

Tabor, Raymond. 1994. *Traditional Woodland Crafts.* London: B.T. Batsford.

Vaughan, Susie. 2006. *Handmade Baskets from Nature's Colourful Materials.* Tunbridge Wells, UK: Search Press.

2.4 Firewood

Alden, Harry A. 1995. *Hardwoods of North America.* General Technical Report FPL-GTR-83. Madison, WI: USDA Forest Service, Forest Products Laboratory.

Burns, Russell M. and Barbara H. Honkala. 1990. *Silvics of North America: Volume 2. Hardwoods.* Agriculture Handbook 654. Washington, DC: USDA Forest Service.

Bushway, Stephen. 1992. *The New Woodburner's Handbook: A Guide to Safe, Healthy and Efficient Woodburning.* Pownal, VT: Storey Publishing.

Carey, Andrew B. and John D. Gill. 1980. *Firewood and Wildlife.* Forest Service Research Note-299. Broomall, PA: Northeastern Forest Experiment Station, USDA Forest Service, accessed January 18, 2016.

Crawford, Martin. 2000b. *Timber Trees for Temperate Climates,* 3rd ed. Devon, UK: Agroforestry Research Trust.

Folliott, Peter F. and Wayne T. Swank, Eds. 1986. *Potentials of Noncommercial Forest*

Biomass for Energy. Technical Bulletin No. 256. Tuscon, AZ: University of Arizona School of Renewable Natural Resources.

FirewoodResource.com. 2015. *Firewood BTU Ratings Charts for Common Tree Species,* accessed December 22, 2015.

Forest Inventory and Analysis National Program. 2016. *Forest Inventory and Analysis Database.* Washington, DC: USDA Forest Service, accessed January 18, 2016.

Hansen, Hugh J. 1980. *Fuelwood Facts.* Corvallis, OR: Extension Service, Oregon State University. ir.library.oregonstate.edu/xmlui/bitstream/handle/1957/24637/ECNO1023.pdf?sequence=1, accessed January 11, 2016.

Hill, Lewis. 1981. *Fast-Growing Firewood.* Bulletin A-69. Pownal, VT: Garden Way Publishing.

Ince, Peter J. 1979. *How to Estimate Recoverable Heat Energy in Wood or Bark Fuels.* Madison, WI: USDA Forest Products Laboratory, accessed December 12, 2016.

Kuhns, Mike and Tom Schmidt. 1988. *G88-881 Heating with Wood I. Species Characteristics and Volumes.* Lincoln, NB: University of Nebraska-Lincoln Extension. Paper 862, accessed January 11, 2016.

Nix, Steve. 2014. *Heating Properties of Firewood by Tree Species.* About.com Education, accessed December 22, 2015.

Plants for a Future. *PFAF Plant Database.* www.pfaf.org/user/ plantsearch.aspx.

Richberger, Wanda and Howard, Ronald A., Jr. 2015. *Firewood: From Woodlot to Woodpile.* Cooperative Extension, NY State College of Agriculture and Life Science, Cornell University.

Schnieder, Rollin D. 1990. G90-957 *Is Burning Wood Economical?* Paper 386. Lincoln, NB: University of Nebraska-Lincoln Extension.

Smedly, Ursula R. 2012. *Firewood Facts.* Guide G-102. Las Cruces, NM: NM State University Cooperative Extension.

Smith, Brad W. 1985. *Factors and Equations to Estimate Forest Biomass in the North Central Region.* Res. Pap. NC-268. St. Paul, MN: USDA Forest Service, North Central Forest Experiment Station, accessed January 11, 2016.

Smith, Sarah. Date unknown. *Firewood: Our Renewable Resource.* Forest Fact Sheet 8. Durham, NH: University of NH Cooperative Extension.

Tabor, Raymond. 1994. *Traditional Woodland Crafts.* London: B.T. Batsford.

The Chimney Sweep Online. 2011. *Firewood BTU Content Charts.* outreach.cnr.ncsu.edu/woodworkshops/documents/ChimneySweepsWoodBTUChart.pdf, accessed December 22, 2015.

University of Kentucky Forestry. 2015. *Comparison of Properties Affecting Firewood,* accessed December 22, 2015.

USDA Forest Service. Multiple dates. Fire Effects Information System, [Online]. USDA Forest Service, Rocky Mountain Research Station, Missoula Fire Sciences Laboratory (Producer), accessed January 25, 2016.

2.5 Charcoal

Alden, 1995; Burns and Honkala, 1990; Crawford, 2000b; Forest Inventory and Analysis National Program, 2016; *Plants for a Future*; Tabor 1994.

Alden, Harry A. 1995. *Hardwoods of North America*. General Technical Report FPL-GTR-83. Madison, WI: USDA Forest Service, Forest Products Laboratory.

Burns, Russell M. and Barbara H. Honkala. 1990. *Silvics of North America: Volume 2. Hardwoods*. Agriculture Handbook 654. Washington, DC: USDA Forest Service.

Crawford, Martin. 2000b. *Timber Trees for Temperate Climates*, 3rd ed. Devon, UK: Agroforestry Research Trust.

Forest Inventory and Analysis National Program. 2016. *Forest Inventory and Analysis Database*. Washington, DC: USDA Forest Service, accessed January 18, 2016.

Plants for a Future. *PFAF Plant Database*. www. pfaf.org/user/ plantsearch.aspx.

Tabor, Raymond. 1994. *Traditional Woodland Crafts*. London: B.T. Batsford.

2.7 Digestibility

1. Azim, A., A.G. Khan, J. Ahmad, M. Ayaz, and I.H. Mirza. 2002. "Nutritional Evaluation of Fodder Tree Leaves with Goats." Asian-*Australasian Journal of Animal Sciences* 15(1): 34–37.

2. Burritt, E.A., F.D. Provenza, and J.C. Malechek. 1987. "Changes in Concentrations of Tannins, Total Phenolics, Crude Protein, and in Vitro Digestibility of Browse Due to Mastication and Insalivation by Cattle." *Journal of Range Management* 40(5): 409–411.

3. Canbolat, Ö. 2012. "Determination of Potential Nutritive Value of Exotic Tree Leaves in Turkey." *Kafkas Üniversitesi Veteriner Fakültesi Dergisi* 18(3): 419–423.

4. Ciszuk, P. and M. Murphy. 1982. "Digestion of Crude Protein and Organic Matter of Leaves by Rumen Microbes in Vitro." *Swedish Journal of Agricultural Research* 12(1): 35–40.

5. Djumaeva, D., N. Djanibekov, P.L.G. Vlek, C. Martius, and J.P.A. Lamers. 2009. "Options for Optimizing Dairy Feed Rations with Foliage of Trees Grown in the Irrigated Drylands of Central Asia." *Research Journal of Agriculture and Biological Sciences* 5(5): 698–708.

6. Lamers, J.P.A. and A. Khamzina. 2010. "Seasonal Quality Profile and Production of Foliage from Trees Grown on Degraded Cropland in Arid Uzbekistan, Central Asia." *Journal of Animal Physiology and Animal Nutrition* 94(5): e77–e85.

7. Nedkvitne, J.J. and T.H. Garmo. 1986. "Conifer Woodland as Summer Grazing for Sheep." In *Grazing Research at Northern Latitudes*, 121–128. New York: Plenum.

8. Oppong, Samuel Kingsley. 1998. "Growth, Management and Nutritive Value of Willows (*Salix* Spp.) and Other Browse Species in Manawatu, New Zealand." PhD diss., Massey University, Palmerston North, New Zealand. muir.massey.ac.nz/handle/10179/2705.

9. Roothaert, R.L. 1999. "Feed Intake and Selection of Tree Fodder by Dairy Heifers."

Animal Feed Science and Technology 79 (1/2) (May 15): 1–13.

10. Short, Henry L., Robert M. Blair, and E.A. Epps, Jr. 1975. *Composition and Digestibility of Deer Browse in Southern Forests.* Research Paper. USDA Forest Service, Southern Forest Experiment Station.

11. Short, Henry L., Robert M. Blair, and Charles A. Segelquist. 1974. "Fiber Composition and Forage Digestibility by Small Ruminants." *Journal of Wildlife Management* 38(2): 197–209.

12. Saddul, D., R.A. Halim, J.B. Liang, and Z.A. Jelan. 2005. "Evaluation of Mulberry (Morus Alba) as Potential Feed Supplement for Ruminants: The Effect of Plant Maturity on in Situ Disappearance and in Vitro Intestinal Digestibility of Plant Fractions." *Asian-Australasian Journal of Animal Sciences* 18(11): 1569–1574.

13. Sultan, Javed Iqbal et al. 2008. "Nutritional Evaluation of Fodder Tree Leaves of Northern Grasslands of Pakistan." *Pakistan Journal of Botany* 40(6): 2503–2512.

14. Vu, C.C., M.W.A. Verstegen, W.H. Hendriks, and K.C. Pham. 2011. "The Nutritive Value of Mulberry Leaves (Morus Alba) and Partial Replacement of Cotton Seed in Rations on the Performance of Growing Vietnamese Cattle." *Asian-Australasian Journal of Animal Sciences* 24(9): 1233–1242.

15. Yao, J., B. Yan, X.Q. Wang, and J.X. Liu. 2000. "Nutritional Evaluation of Mulberry Leaves as Feeds for Ruminants." *Livestock Research for Rural Development* 12(2): 1–7.

2.8 Edibility

Burns, Russell M. and Barbara H. Honkala. 1990. *Silvics of North America: Volume 2. Hardwoods.* Agriculture Handbook 654. Washington, DC: USDA Forest Service.

Crawford, Martin. 2000c. *Useful Plants for Temperate Climates: Volume 1a: Trees and Volume 2a: Shrubs.* Totnes, Devon, UK: Agroforestry Research Trust.

Crawford, Martin. 2000d. *Useful Plants for Temperate Climates: Volume 2a: Shrubs.* Totnes, Devon, UK: Agroforestry Research Trust.

Facciola, Stephen. 1990. *Cornucopia: A Sourcebook of Edible Plants.* Vista, CA: Kampong Publications.

Hopman, Ellen Evert. 1991. *Tree Medicine, Tree Magic.* Blaine, WA: Phoenix Publishing.

Plants for a Future. *PFAF Plant Database.* www.pfaf.org/user/ plantsearch.aspx.

2.9 Woody Cuts

Greer, Lane and John M. Dole. 2009. *Woody Cut Stems for Growers and Florists: Production and Post-Harvest Handling of Branches for Flowers, Fruit, and Foliage.* Portland, OR: Timber Press.

Endnotes

Introduction: What Is Coppice?

1. Evelyn, *Sylva or a Discourse of Forest-trees*, 1664.
2. Butterfield. *Collins English Dictionary*, 2003.
3. www.worldwidewords.org/qa/qa-cop1.htm

Chapter 1: A Cultural History of Coppice Agroforestry

1. Miles, *Forestry in the English Landscape*, 1967, 43.
2. Hamilton, *The Journey: Walking the Road to Bethlehem*, 2011, 19.
3. Watkins, "Themes in the History of European Woods and Forests," 1998, 6.
4. Huntley, "European Post-glacial Forests," 1990, 507–518.
5. Ibid.
6. Rackham, *Trees & Woodland in the British Landscape*, 1990, 26–28.
7. Vera, *Grazing Ecology and Forest History*, 2000, 3.
8. Williams, *Deforesting the Earth*, 2003, 25.
9. Vera, 4.
10. Ibid., 9.
11. Ibid., 86.
12. Ibid., 156.
13. Ibid., 159.
14. McQuade and O'Donnell, "Late Mesolithic Fish Traps," 2007.
15. Wikipedia states agriculture arrived in Ireland around 4500 BCE; other sources say it was somewhat later, accessed January 28, 2013.
16. McQuade and O'Donnell, 574, 581.
17. Ibid., 581; Fischer, "Coastal Fishing in Stone Age Denmark," 2007.
18. Fischer, 62–3.
19. Ibid.
20. Edlin, *Trees, Woods, and Man*, 1966, 82, 85–6.
21. Rackham, 1990, 35.
22. J.M. Coles, "Man and Landscape in the Somerset Levels," 1978; Bryony Coles and J.M. Coles, *Sweet Track to Glastonbury*, 1986.
23. "The Day the Sweet Track Was Built," *New Scientist* 126, 30(1), 1990.
24. Bryony Coles, "Archeology Follows a Wet Track," 1987, 45.
25. Ibid.
26. Costen, *The Origins of Somerset*, 1992, 4–5.
27. Brunning and McDermott, "Chapter 21," 2013, 371.
28. J.M. Coles, "Archeology, Drainage, and Politics," 1983.
29. Bryony Coles, 1987, 45.
30. J.M. Coles, 1983, 205–6.

31. Brunning and McDermott, 371.

32. J.M. Coles, 1978, 89.

33. Miles, 35.

34. See Brunning and McDermott: p. 371 lists citations that support this claim. Regnell, "Charcoals from Uppåkra," 2003, quotes studies from regions other than Sweden. Also see studies cited in Schmidl et al., "Land Use in the Eastern Alps," 2005.

35. Regnell, 2003.

36. Schmidl et al., 462.

37. Hoogzaad, "Impact of Land Use and Climatic Change," 2009.

38. Andrič, "Holocene Vegetation Development," 2007.

39. Szabo, "Open Woodland in Europe," 2009, 2329.

40. This fact has become more well-documented and known recently. See Mann, *1491*, 2005; Lovgren, "'Guns, Germs and Steel,'" 2005.

41. Anderson, *Tending the Wild*, 2005, 125.

42. Ibid., 252.

43. Ibid., 2.

44. Ibid., 135.

45. Ibid., 153.

46. Ibid., 253.

47. Ibid., 8.

48. Ibid., 136.

49. Ibid., 217.

50. Ibid., 223.

51. Ibid., 239.

52. Ibid., 126.

53. Ibid., 148.

54. Ibid., 130.

55. Ibid., 139.

56. Ibid., 223.

57. Ibid., 228.

58. Ibid., 292.

59. Ibid., 235.

60. Ibid., 172.

61. Ibid., 232.

62. Ibid., 223.

63. Ibid., 188.

64. Ibid., 210.

65. Ibid., 215.

66. Ibid., 233.

67. Ibid., 9.

68. Gimpel, *The Medieval Machine*, 1976.

69. Meiggs, "Farm Forestry in the Ancient Mediterranean," 1989.

70. Farrell et al., "European Forest Ecosystems," 2000, 8.

71. Meiggs.

72. Ibid.

73. Rackham, 1990, 41.

74. Farrell et al., 9.

75. Salway, "Roman Britain," 1984, 1.

76. Rackham, 1990, 40–41.

77. Ibid., 41.

78. Gimpel argues that there was an industrial revolution in the Middle Ages that we have heretofore ignored. It makes sense: after the Roman Empire spread their innovations all over the West, of course people would innovate and develop what they learned. How else would Europe have developed so much, supporting the population growth it did during the period?

79. Szabo, 2009, 2329.

80. Rackham, 1990, 35, 39–41.

81. Ibid., 42–3.

82. Ibid., 48–55.

83. Ibid., 59–63.

84. Ibid., 70.

85. Ibid.

86. Ibid., 64.

87. Tabor, *Traditional Woodland Crafts*, 1994, 20.

88. Jenkins, *Traditional Country Craftsmen*, 1966, 10.

89. Law, *The Woodland Way*, 2001, 76.

90. Thomas, "Ecological Changes in Bernwood Forest," 1998, 233.

91. Jenkins, 13.

92. Rotherham, "The Implications of Perceptions," 2007.

93. Machar, "Coppice-with-standards," 2009, 306.

94. Rackham, 1990, 63.

95. Machar.

96. Vera, 134.

97. Ibid., 159.

98. Rotherham.

99. Jones, "The Rise, Decline and Extinction of Spring Wood Management in South-west Yorkshire," 1998.

100. J. Smith, *The History of Temperate Agroforestry*, 2010.

101. Dagley and Burman, "The Management of the Pollards of Epping Forest," 1996, 30–31.

102. Read, Frater, and Noble, "A Survey of the Condition of the Pollards," 1996, 50.

103. Bergmeier, Petermann, and Schröder, "Geobotanical Survey," 2010.

104. Rackham, "Savannas in Europe," 1998, 17.

105. Rackham, 1989.

106. Rackham, 1990, 185.

107. Maclean, *Hedges and Hedgelaying*, 2006, 10.

108. Rackham, 1990, 185.

109. Ibid., 10.

110. Ibid., 44.

111. Ibid., 114.

112. R.C. Thomas, 1998, 232.

113. Peter Szabo, "Ancient Woodland Boundaries in Europe," 2010a, 206.

114. Vera, 159.

115. Salway, 1.

116. Wikipedia, "Norman conquest of England."

117. Vera, 103–107, 128.

118. Ibid., 108.

119. Rackham, 1990, 166.

120. Vera, 124–128.

121. Ibid., 130–131, 166–167, 171–172.

122. Ibid., 128.

123. Rackham, 1990, 147.

124. Klingelhöfer, "Woodland on the Hampshire Downs," 1992, 172–173.

125. Rackham, 1990, 55.

126. Wikipedia, "Black Death."

127. Tabor, 13.

128. Lyle, *The Book of Masonry Stoves*, 1998, 44.

129. Vera, 178.

130. Lyle, 44.

131. Green, "Thoughts on Pollarding," 1996a, 1.

132. Linebaugh, *The Magna Carta Manifesto*, 2008, 94.

133. Rotherham, 105.

134. Sheafe Satterthwaite, land use historian, personal communication, January, 2014; this is based upon Sheafe's reading of a Royal Commission on Common Lands study from the 1940s.

135. Brooks and Agate, *Hedging*, 1998, 6; Rackham, 1990, 189.

136. Brooks and Agate, 6.

137. Linebaugh, 71–73.

138. Vera, 133–134.

139. Armstrong, *Woodcolliers and Charcoal Burning*, 1979, 9.

140. Ibid., 10.

141. Rackham, 1990, 86.

142. Ibid., 87.

143. Rotherham, 107.

144. Hammersley, "The Charcoal Iron Industry and Its Fuel," 1973.
145. Rackham, 1990, 77.
146. Vera, 162.
147. Gulliver, "What Were Woods Like in the 17th Century?" 1998, 135.
148. Rackham, 1990, 82.
149. John Brown, *Welsh Stick Chairs*, 1990, 35–36.
150. Tabor, 20.
151. Langsner, *Green Woodworking*, 1995, 19.
152. This whole sidebar derived from Szabo, "Driving Forces of Stability and Change in Woodland Structure, 2010b.
153. Evans, "Coppice Forestry," 1992, 20.
154. Rackham, 1990, 99.
155. Serrada, Allué, and San Miguel, "The Coppice System in Spain," 1992.
156. Burgi, "Habitat Alterations," 1998, 205.
157. Bergmeier, Petermann, and Schröder, 2010.
158. Smith, 2010.
159. Farrell et al., 10.
160. Burgi, 208.
161. Maclean, 8.
162. Stajić, Zlatanov, Dubravac, and Trajkov, "Past and Recent Coppice Forest Management," 2009, 11–12.
163. Ibid., 12–13.
164. Brooks and Agate, 9, 3.
165. Ernst, "An Ecological Revolution?" 1998.
166. Farrell et al., 5.
167. Ernst.
168. Farrell et al., 9.
169. Wittbecker, "Forest Practices Related to Forest Ecosystem Productivity," 1997, 45–46.
170. Rackham, 1990, 97.
171. Greene, *Vegetable Gardening*, 2012.
172. Whitney, *From Coastal Wilderness to Fruited Plain*, 1994, 211–212.
173. Cronon, *Changes in the Land*, 1983, 1.
174. Ibid., 20.
175. Wessels, *Reading the Forested Landscape*, 1997, 56.
176. Sloane, *A Museum of Early American Tools*, 2002, 42.
177. Cronon, 117.
178. Ibid., 120.
179. Beattie et al., *Working with Your Woodland*, 1983, 15.
180. Cronon, 120.
181. Ibid., 155.
182. Tabor, 17.
183. Cronon, 145
184. Beattie et al., 7.
185. Whitney, 225–6.
186. Cronon, 145.
187. Beattie et al., 10–11.
188. Ibid., 18–19.
189. Lexer et al., "CForSEE," 2009.
190. Abbott, *Living Wood*, 2002, 150.
191. Rotherham, 2007, 101.
192. Abbott, 151.
193. Ibid.
194. Oaks and Mills, *Coppicing & Coppice Crafts*, 2010, 10.
195. Ibid., 11.
196. Law, 2001, 77.
197. Green, 1996a, 1.
198. Smith, 2010.
199. Rackham, 1998, 33.
200. Serrada et al., 1992.
201. Stajić et al., 11.
202. Dekanic et al., 2009, 49–50.
203. Ibid., 50–51.
204. Borelli and Varela, *Mediterranean Oaks Network*, 2001, 29.
205. Lexer et al., 54.

206. Stajić et al., 9–19.
207. Ibid., 14.

Chapter 2: The Anatomy and Physiology of Woody Plants

1. Bell, *Plant Form*, 2008, 11.
2. Cech, "Balancing Conservation with Utilization," 1998, 64.
3. Shigo, *A New Tree Biology*, 1986, 4.
4. Ibid., 403.
5. Ibid., 194.
6. Ibid., 187.
7. Peter Thomas, *Trees*, 2000, 41.
8. Shigo, 145.
9. Ibid., 254–255; Peter Thomas, 41.
10. Hoadley, *Understanding Wood*, 2000, 16; Peter Thomas, 42–43.
11. Peter Thomas, 52.
12. Shigo, 256; Peter Thomas, 43.
13. Peter Thomas, 45.
14. Ibid.
15. Ibid., 43.
16. Ibid., 45.
17. Ibid., 46.
18. Hoadley, 22.
19. Shigo, 297. Kevin Smith, pers. comm., 7/30/21.
20. Peter Thomas, 54.
21. Shigo, 306.
22. Ibid., 296.
23. Pallardy, *Woody Plant Physiology*, 2008, 9.
24. Bell, 258.
25. Del Tredici, "Sprouting in Temperate Trees," 2001, 123.
26. Cline and Harrington, "Apical Dominance," 2006.
27. Harrington, Zasada, and Allen, "Biology of Red Alder," 1994, 11.
28. Kays and Canham, 1991.
29. Harrington, Zasada, and Allen, 11.
30. Jacke and Toensmeier, *Edible Forest Gardens*, 2005, 190–200; www.extension.iastate.edu/pages/tree/site/roots.html; www.vicstreeservice.com/tree-root-systems/
31. Shigo, 211.
32. Shigo.
33. Ibid., 73.
34. Peter Thomas, 175.
35. Ibid., 45, 175.
36. Shigo, 211–212.
37. Rackham, 1990, 11.
38. Shigo, 284.
39. Ibid.
40. Rackham, 1990, 11.
41. Ibid., 14.
42. Ibid., 15.
43. Oaks and Mills, 74.
44. Cronon, 37.
45. Shigo, 5.
46. Ibid., 34.
47. Ibid., 49.
48. Ibid., 42–46.
49. Ibid., 101.
50. Ibid., 293.
51. Ibid., 78.
52. Ibid., 425.
53. Ibid., 427–431, Kevin Smith, pers. comm., 7/31/2021.
54. Shigo, 91.
55. Ibid., 145.
56. Wessels, 2010, 14; Del Tredici, 2001, 122; Jeník, "Clonal Growth in Woody Plants," 1994, 291–293.
57. Keeley and Zedler, "Evolution of Life Histories in *Pinus*," 1998, 219.
58. Koop, "Vegetative Reproduction," 1987, 103–107.

59. Bond and Midgley, "Ecology of Sprouting," 2001, 49.
60. Bond and Midgley, "The Evolutionary Ecology of Sprouting," 2003, S103.
61. Jeník, 291.
62. Anderson, 236.
63. Ferrini, "Pollarding and Its Effects," 2007, 4.
64. Bond and Midgley, 2003, S108.
65. Del Tredici, 2001, 135.
66. Bond and Midgley, 2001, 47; Bellingham and Sparrow, "Resprouting as a Life History Strategy," 2000, 409; Guerrero-Campo et al., "Effect of Root System Morphology," 2006, 439; Del Tredici, 2001, 135.
67. Bellingham and Sparrow, 410, 412.
68. Ibid., 411–412.
69. Guerrero-Campo et al., 440; Bond and Midgley, 46.
70. James, "Lignotubers and Burls," 1984, 228–230.
71. Bond and Midgley, 48; Del Tredici, 2001, 122.
72. Bellingham and Sparrow, 410.
73. Bond and Midgely, 46.
74. Ibid., 45; Bellingham and Sparrow, 410.
75. Bond and Midgley, 45.
76. Ibid., 46.
77. Del Tredici, 2001, 122, 136.
78. Del Tredici, 2001, 131–132; Vogt and Cox, "Evidence for Hormonal Control," 1970, 165.
79. Del Tredici, 2001, 122.
80. Wikipedia, "North American beaver," accessed November 2, 2014.
81. Wessels, 1997, 100.
82. Ibid., 102.
83. Green, "Fungi, Trees and Pollards," 2006, 3.
84. Anderson, 214.
85. Fink, "The Occurrence of Adventitious and Preventitious Buds," 1983, 532.
86. Church and Godman, "The Formation and Development of Dormant Buds," 1966, 301.
87. Ibid., 306.
88. Ibid., 302; Wilson, "Red Maple Stump Sprouts," 1968, 9.
89. Fink, 1983, 532; Shigo, 245.
90. Meier et al., "Epicormic Buds in Trees," 2012.
91. Church and Godman, 305.
92. Del Tredici, 2001, 125.
93. Anderson, 214.
94. Fink, 1983, 541.
95. Oaks and Mills, 50.
96. Lonsdale, "Pollarding Success and Failure," 1996, 103.
97. Del Tredici, 2001, 121–122.
98. Ibid., 124–125.
99. Ibid., 125.
100. Ibid., 126–127; Bond and Midgley, 2003, S104.
101. Shigo, 490.
102. Del Tredici, 2001, 127–128.
103. Tiedemann et al., "Underground Systems," 1987, 1067; Whittock et al., "Genetic Control," 2003, 57.
104. Canadell and Zedler, "Underground Structures," 1995, 201.
105. James, 237, 241.
106. Ibid., 260.
107. Ibid., 248.
108. Canadell and Zedler, 204.
109. Del Tredici, 1995, 13.
110. Canadell and Zedler, 194.
111. Ibid., 200.
112. Whittock et al., 60.
113. Del Tredici, 2001, 128.
114. Koop, 104.
115. Del Tredici, 1995, 12.
116. Ibid., 13.
117. Del Tredici, 2001, 128–129.

118. Schier et al., "Vegetative Regeneration," 1985, 31.
119. Jones and Raynal, "Spatial Distribution," 1986, 1723.
120. Del Tredici, 1995, 12.
121. Del Tredici, 2001, 129.
122. Schier et al., 29; Brown and Kormanik, "Suppressed Buds," 1967, 338.
123. Brown and Kormanik, 344.
124. Ibid., 343.
125. Del Tredici, 2001, 123.
126. Anderson, 215.
127. Vogt and Cox, 165, 169.
128. Ibid., 165.
129. Wilson, 9.
130. Wenger, "The Sprouting of Sweetgum," 1953, 35.
131. Oaks and Mills, 51.
132. Del Tredici, 2001, 133.
133. Gracia and Retana, "Effect of Site Quality and Shading," 2004, 39–40, 44–46.
134. Shigo, 163.
135. Ibid., 42–46.
136. Ibid., 353.
137. Del Tredici, 2001, 131.
138. Sennerby-Forsse and Zsuffa, "Bud Structure and Resprouting," 1995, 224.
139. Evans, 18.
140. Oaks and Mills, 24, 51, 136; Agate, *Woodlands*, 1980, 95.
141. Del Tredici, 2001, 133.
142. Oaks and Mills, 51.
143. Johnson, "Growth and Structural Development," 1975, 413–415.
144. Roth and Hepting, "Origin and Development of Oak Stump Sprouts," 1943, 29.
145. Furze et al., "Whole-tree Nonstructural Carbohydrate Storage," 2018, 1466.
146. Kays and Canham, "Effects of Time and Frequency of Cutting," 1991, 524; Wenger, 35; Shigo, 490.
147. Furze et al., 1466–1468.
148. Ibid., 1470.
149. Kays and Canham, 528–529.
150. Ibid., 535–536.
151. Furze et al., 1472–1475.
152. Lonsdale, 100; James, 231.
153. Kruger and Reich, "Coppicing Alters Ecophysiology," 1993, 741, 747.
154. Ibid., 748.
155. Drake et al., "A Comparison of Growth," 2009, 672.
156. Wenger, 46; Gracia and Retana, 47.
157. Mattoon, "The Origin and Early Development of Chestnut Sprouts," 1909, 38.
158. Del Tredici, 2001, 125; Roth and Hepting, 27; Rogers and Johnson, "Thinning Sprout Clumps," 1989.
159. Shigo, 491, 496–497.

Chapter 3: Ecology of Coppice Systems

1. Shigo, 55.
2. Wittbecker, 52.
3. Hammond, "Standards for Ecologically Responsible Forest Use," 1997b, 206.
4. Eyre, *Forest Cover Types*, 1980, 1–2.
5. Ibid., v, 2, 4; David Brown, ed., *Biotic Communities*, 1994, 9, 13, 17.
6. Buckley, ed., *Ecology and Management of Coppice Woodlands*, 1992, preface.
7. Ibid., 4–5.
8. Peterken, "Coppices in the Lowland Landscape," 1992, 12.
9. Mitchell, "Growth Stages and Microclimate in Coppice and High Forest," 1992, 31.

10. Ibid., 33–34.
11. Barkham., "The Effects of Coppicing and Neglecy," 1992, 135.
12. Evans and Barkham, "Coppicing and Natural Disturbance," 1992, 81.
13. Ibid., 82.
14. Ibid., 82–83.
15. Goldsmith, "Coppicing: A Conservation Panacea?" 1992, 307.
16. Evans and Barkham, 88–89.
17. Phillips and Shure, "Patch-size Effects," 1990, 204.
18. Ibid.
19. Valbuena-Carabaña, González-Martíinez, and Gil, "Coppice Forests and Genetic Diversity," 2008, 226.
20. Ibid., 225–226, 229.
21. Barkham, 119.
22. Brown and Warr, "The Effects of Changing Management," 1992, 154.
23. Ibid., 147.
24. Ibid., 163; Rackham, 1990, 134.
25. Barkham, 141.
26. Ibid., 116.
27. Oaks and Mills, 76.
28. Peterken, 1992, 14.
29. Barkham, 139.
30. Ibid., 115.
31. Ibid., 138.
32. Evans and Barkham, 91.
33. Peterken, 1992, 4.
34. Buckley, 1992, preface.
35. Rayner, "The Tree as a Fungal Community," 1996, 6.
36. Kourik, Roots Demystified, 2008, 97.
37. Rayner, 7.
38. Kourik, 97, 99.
39. Rayner, 8.
40. Hrynkiewicz et al., "The Significance of Rotation Periods," 2010, 1943, 1945, 1948–1949.
41. Green, 2006, 1–4.
42. Mitchell, 46.
43. Ibid., 33.
44. Oaks and Mills, 72.
45. Mitchell, 34.
46. Ibid., 35.
47. Ibid., 36–38.
48. Ibid., 47–48.
49. Rackham, 1990, 16.
50. Adams, "The Bryophyte Flora of Pollards," 1996, 17.
51. Peterken, 1992, 8.
52. Ibid., 5–8.
53. Avon et al., "Does the Effect of Forest Roads Extend a Few Meters," 2010, 1546–1547.
54. Ibid., 1546–1547, 1550, 1553.
55. Ibid., 1550, 1553.
56. Fuller and Henderson, "Distribution of Breeding Songbirds," 1992, 73.
57. Fuller, and Moreton, "Breeding Bird Populations," 1987, 25.
58. Fuller, "Effects of Coppice Management," 1992, 170–172.
59. Sage and Robertson, "Wildlife and Game Potential," 1994, 43.
60. Fuller, 170.
61. Ibid., 170–172.
62. Ibid., 175.
63. Oaks and Mills, 82.
64. Fuller, 183.
65. Ibid., 181.
66. Ibid., 186.
67. Fuller and Henderson, 73–75, 77–78, 82.
68. Fuller and Moreton, 14, 18, 21.

69. Ibid., 24; Fuller, Stuttard, and Ray, "The Distribution of Breeding Songbirds," 1989, 265, 271–274.

70. Bealey and Robertson, "Coppice Management for Pheasants," 1992, 193–198; Sage and Robertson, 45.

71. Fuller, 188; Sage and Robertson, 44.

72. Fuller, 184–185.

73. Maclean, 52–53.

74. Evans and Barkham, 86; Gurnell, Hicks, and Whitbread, "The Effects of Coppice Management," 1992, 214.

75. Gurnell, Hicks, and Whitbread, 220, 222, 229.

76. Ibid., 217–218, 228.

77. Ibid., 214, 225.

78. Ibid., 225–227.

79. Holmes, "Ancient Trees," 1996, 19.

80. Ratcliffe, "The Interaction of Deer and Vegetation," 1992, 236.

81. Ibid., 241–242; Joys, Fuller, and Dolman, "Influences of Deer Browsing," 2004, 35.

82. Joys et al., 35.

83. Greatorex-Davies and Marrs, "The Quality of Coppice Woods," 1992, 271–272.

84. Ibid., 278, 280.

85. Ibid., 280.

86. Ibid., 281.

87. Barbalat and Getaz. "Influence de la remise en exploitation," 1999, 435–436.

88. Broome et al., "The Effect of Coppice Management," 2011, 729–730, 735–736, 742, 744.

89. Ibid., 729–730, 735–736, 742, 744.

90. Warren and Thomas, "Butterfly Responses to Coppicing," 1992, 264.

91. Oaks and Mills, 81.

92. Warren and Thomas, 249–250.

93. Ibid., 251, 254.

94. Ibid., 255–256.

95. Ibid., 259–260.

96. Ibid., 266.

97. Kirby, "Accumulation of Dead Wood," 1992, 99–100; Hammond, 1997a, 66.

98. Kirby, 100.

99. Ibid., 105.

100. Ibid., 107.

101. Ibid., 106.

102. Rubio and Escudero, "Clear-cut Effects on Chestnut Forest Soils," 2003, 196.

103. Evans, 24; Goldsmith, 307.

104. Rubio and Escudero, 2003, 196.

105. Adegbidi, "Nutrient Return Via Litterfall," 1994, 21–24; Hölscher, Schade, and Leuschner, "Effects of Coppicing," 2001, 159–160; Berthelot, Ranger, and Gelhaye, "Nutrient Uptake and Immobilization," 2000, 177.

106. Goldsmith, 307.

107. Hölscher et al., 157.

108. Rubio and Escudero, 201.

109. Yuksek and Yuksek, "Effects of Clear-cutting Alder Coppice," 2009, 2563.

110. Franklin, "The Productive Potential of Ancient Oak-Coppice Woodland in Britain," 1993.

111. Evans, 24.

112. Hölscher et al., 157, 159, 160.

113. Ibid., 155–162.

114. Shuai and Bruckman, "Soil Nitrogen and Carbon Pools," 2010, 1.

115. Berthelot et al., 177.

116. Hölscher et al., 157.

117. Berthelot et al., 178.

118. Blackmon, "Estimates of Nutrient Drain," 1979, 1, 3.

119. Ibid., 3.

120. Ibid., 3–4.
121. Adegbidi, xi.
122. Ranger and Nys, "Biomass and Nutrient Content," 1996, 91–92.
123. Ibid., 97, 102, 109.
124. Uri et al., "Long-term Effects on the Nitrogen Budget," 2011. 920–921.
125. Ibid., 920, 922, 924, 926.
126. Rubio and Escudero, 195–197.
127. Ibid., 200–201.
128. Ibid., 195, 202.
129. Lancho and González, "Sequestration of C in a Spanish Chestnut Coppice," 2004, 108.
130. Toensmeier, *The Carbon Farming Solution*, 2016, 29, 102–103.
131. Deckmyn et al., "Carbon Sequestration," 2004, 1482–1483, 1485.
132. Ibid., 1487.
133. Ibid.
134. Lancho and González, 111.
135. Kafaky, Khademi, and Mataji, "Carbon Storage and CO_2 Uptake," 2009, 85–87.
136. Cummings and Cook, "Soil-water Relations," 1992, 52.
137. Ibid., 53.
138. Ibid., 54, 61, 66, 69.
139. Ibid.
140. Ibid., 70–71.
141. Drake et. al., 663, 669.
142. Allen, Hall, and Rosier, "Transpiration by Two Poplar Varieties," 1999, 493, 500.
143. Ibid., 494.
144. Ibid., 498–500.
145. Goldsmith, 306.

Chapter 4: Woodland Management Systems

1. Brooks and Agate, 11
2. Ibid., 15.
3. Ibid., 8
4. Smith, 2010.
5. Wildlife Trust Wales, "Wood Pasture and Parkland," 2018.
6. Bergmeier et al., 2010.
7. Austad and Hague, "Pollarding in Western Norway," 2007, 1.
8. Ibid., 3.
9. Ibid., 2–3.
10. Ibid., 1.
11. Ibid., 3.
12. Ibid.
13. Ibid., 7.
14. Bergmeier et al., 2999.
15. Peterken, 1996, 402.
16. Roellig et al., "Reviving Wood-pastures," 2016, 185–195.
17. Chedzoy and Smallidge, *Silvopasturing in the Northeast*, 2011.
18. Garrett et al., "Hardwood Silvopasture Management," 2004, 27.
19. Ibid., 25.
20. Talvi, "Estonian Wooded Meadows and Wooded Pastures," 2010, 2.
21. Read, *A Study of Practical Pollarding*, 2003, 27.
22. Talvi, 2.
23. Kukk and Kull, "Wooded Meadows [Puisniidud]," 1997.
24. Talvi, 4.
25. Smith, 2010.
26. Rackham, 1998, 34.
27. Ibid., 33.
28. Talvi, 6.
29. Peterken, 1996, 403.
30. Ibid.
31. Zlatanov and Lexer, "Coppice Forestry," 2009, 6.
32. Evans, 19.

33. Dickman, "Silviculture and Biology," 2006, 696.
34. Ibid., 701–702.
35. Ibid.
36. Heller, Keoleian, and Volk, "Life Cycle Assessment," 2003, 152.
37. Dickman, 703.
38. Ibid., 699.
39. Ibid., 702.
40. Ibid., 703.
41. Stajić et al., 17.
42. Oaks and Mills, 57.
43. Agate, ed., *Woodlands*, 1980, 36.
44. Evans, 19.
45. Bane, "Coppice-with-Standards," 1998, 52–53.
46. Ibid.
47. Hawley and Hawes, *Manual of Forestry*, 1918, 64–66.
48. Society of American Foresters, 1994.
49. Zlatanov and Lexer, 6–7.
50. Stajić et al., 15.
51. Hawley and Hawes, 67.
52. Evans, 19.
53. Ibid., 25.
54. Fisher, *A Manual of Forestry*, 1896, 488.
55. Kirby and Watkins, eds., *Europe's Changing Woods and Forests*, 2015, 236–237.
56. Bergmeier et al., 2997.
57. Moreno et al., "Historical Ecology," 1998, 194.
58. Ibid., 195.
59. Ibid., 195–196.
60. Law, 2001, 13.
61. "Apples: Propagation by Stooling."
62. Cool Temperate, n.d., "The Own-Root Fruit Tree Project."
63. Toensmeier, 105–106.
64. Hallman, "Christmas Trees," 1999, 103–104.
65. Ibid.
66. Zlatanov and Lexer, 5.
67. Beattie et al., 132.
68. Wittbecker, 53, 55.
69. Camp, "Critical Elements of Forest Sustainability," 1997, 35.
70. Ibid., 35, 37.
71. *Uneven Aged Management*, 2012.
72. Beattie et al., 118.
73. Ibid., 123–126.
74. Ibid., 127–129.
75. Ibid., 129–131.
76. Ibid., 132–135.
77. Ibid., 137.
78. Ibid., 139–140.
79. Ibid., 141–145.
80. Ibid., 145–149.
81. Ibid., 149.
82. Agate, 42.
83. Beattie et al., 151.
84. Ibid., 152–153.
85. Ibid., 153–155.
86. Agate, 42.
87. Dey, Jensen, and Wallendorf, "Single-tree Harvesting," 2008, 32.

Chapter 5: Coppice Economics and Products

1. Ashton and Kelty, *The Practice of Silviculture: Applied Forest Ecology*, 2018, 13.
2. Alavalapati, Mercer, and Montambault, "Agroforestry Systems and Valuation Methodologies," 2005, 5.
3. Drescher, "The Battle for Sustainability," 1997, 58.
4. Alavalapati et al., 2–4.
5. Tabor, 7.
6. Teel and Buck, "From Wildcrafting to Intentional Cultivation," 2004, 10.

7. Collins, *Crafts in the English Countryside,* 2004, 110–111.
8. Ibid., 76.
9. Ibid., 82.
10. Teel and Buck, 12.
11. Ibid.
12. Law, 2001, 106.
13. Wolfe, "Research Challenges," 2000, 21.
14. Wood, "How to Price Craft Work," 2013.
15. Eaton, *Handicrafts of the Southern Highlands,* 1973, 43.
16. Seymour, *The Forgotten Crafts,* 1990, 10.
17. Chedzoy and Smallidge.
18. Nebraska Forest Service, "Lumber Market News," "Tables of Lumber Price Trends, 2006 and 1979–2006," and "Woody Biomass in Minnesota Impacts on Forestry," *Timber Talk* 45(1), 2007, 4–5; Mattson and Winsauer, "The Potential for Harvesting 'Puckerbrush' for Fuel," 1985, 1.
19. Germain and Lemieux, *Ramial Chipped Wood,* 2000, 1–2.
20. Ibid., 1, 4.
21. Ibid., 5–7.
22. Chedzoy and Smallidge.
23. Kirby and Watkins, 37.
24. Fisher, 534.
25. Burner et al., "Yield Components and Nutritive Value," 2008, 59, 61, 211; D.M. Burner et al., "Management Effects on Biomass," 2005, 207, 210–212.
26. Burner et al., 2008, 51–52, 55–56, 59.
27. Baptista and Launchbaugh, "Nutritive Value," 2001, 82–83.
28. National Academy of Sciences, *Firewood Crops,* 1980, v, 4, 70.
29. Ibid., 5.
30. Hoadley, 104.
31. Tabor, 70.
32. National Academy of Sciences, 24.
33. Tabor, 148.
34. National Academy of Sciences, 24.
35. Maclean, 162.
36. Tabor, 110.
37. Armstrong, Introduction.
38. Jenkins, 32; Armstrong, 66.
39. Jenkins, 32; Desai, Evans, and Levy, 1996, "Charcoal and Coppice," 19; Armstrong, 13.
40. Armstrong, 8, 64.
41. Desai et al., 29.
42. Ibid., 17, 24, 25, 27; Jenkins, 32.
43. Desai et al., 29.
44. Armstrong, 68.
45. Oaks and Mills, 131.
46. Hallman, 104.
47. Ibid.
48. Douglas and Hart, *Forest Farming,* 1978, xii.
49. World Bank, 2021.
50. Ontario Ministry of Agriculture and World Affairs, 2013.
51. Duke, *Handbook of Energy Crops,* 1983.
52. Cech, 64.
53. Law, 2001, 138.
54. Law, 2009, 130, 147.
55. Rutter, "The Potential of Hybrid Hazelnuts," 1999, 133–134.
56. Tabor, 89.
57. Wikipedia, "Barbed wire."
58. John Waller, pers. comm., April 2013.
59. Alistair Hayhurst, pers. comm., April 2013.
60. Collins.
61. Tabor, 122.
62. Collins, 142.
63. Oaks and Mills, 92.
64. Collins, 118.
65. Oaks and Mills, 110.

66. Collins, 6, 89, 91, 99.
67. Tabor, 123.
68. Ibid., 93.
69. Oaks and Mills, 132.
70. Tabor, 93.
71. Kuzovkina and Quigley, "Willows Beyond Wetlands," 2005, 194.
72. Teel and Buck, 16.
73. Labreque and Teodorescu, "Preliminary Evaluation," 2005, 95.
74. Ibid., 96–97.
75. Dorset Coppice Group, "Typical Products from Dorset Woodlands," 2005.
76. Collins.
77. Jenkins, 77.
78. Ibid., 81.
79. Tabor, 114.
80. Jenkins, 77; Dorset Coppice Group, 1.
81. Collins, 78.
82. Jenkins, 18.
83. Ibid., 19–21.
84. Ibid., 65.
85. Sundqvist, *Swedish Carving Techniques*, 1990, 132.
86. Collins, 78.
87. Agate, 110.
88. Dorset Coppice Group, 3.
89. Jenkins, 41.
90. Eaton, 168.
91. Langsner, 15.
92. Anderson, 216. (Margaret "Pegg" Mathewson, pers. comm., 1992)
93. Seymour, 38–39.
94. Langsner, 44.
95. Mack, *Making Rustic Furniture*, 1992, 101.
96. Ibid., 13, 106.
97. Ibid., 24, 36, 42, 54, 58.
98. Ibid., 141.
99. Ibid., 111.
100. Mack, Making Rustic Furniture, 120.
101. Ibid., 61.
102. Ibid.
103. Langsner, 13.
104. Nakamura, *Norwegian Farmhouses, Sod Roof*, 2009.
105. Oaks and Mills, 128, 130.
106. Langsner, 38, 44.
107. Langsner, 31.
108. Ibid., 37.
109. Ibid., 117.
110. Collins, 99.
111. Collins.
112. Jenkins, 35.
113. Ibid.; Agate, 110; Tabor, 98.
114. Tabor, 79, 96; Collins, 91, 117.
115. Jenkins, 35.
116. Alastair Hayhurst, pers. comm., April 2013.
117. Ibid.
118. Collins, 82, 120.
119. Ibid., 121.
120. Teel and Buck, 9–10.
121. Mack, 78.
122. Living Root Bridges, 2009.
123. Agate, 111.
124. Collins, 108.
125. Hawthorne, "Forest Cutting Plan Summary," 2005.
126. Machar, 306–308, 310.
127. Maxted et al., "Phytoextraction of Cadmium and Zinc," 2007, 162.
128. Kuzovkina and Quigley, 2005, 188–189; Paulson et al., 2003, 325.
129. Kuzovkina and Quigley, 2005, 190–192, 195.
130. Jossart, "Tertiary Waste Water Treatment," 2001, 1–2; Doran, *The Use of Willow Coppice as a Biofilter*, 2006, 1–2.

131. Dutton and Humphreys, "Assessing the Potential," 2005, 279, 291.
132. Leakey, "The Capacity for Vegetative Propagation," 1985, 114.
133. Bachmann, "Woody Ornamentals," 2002, 1.
134. Ibid., 2–3.
135. Freed, "Special Forest Products," 1997, 172.
136. Kuzovkina and Quigley, 2004, 419.
137. Bachmann, 4.
138. Kuzovkina and Quigley, 2004, 415, 419.
139. Ibid., 419.
140. Bachmann, 2–3.
141. Ibid., 2.
142. Hallman, Hatfield, and Macy. 1999. "Forest Farming in British Columbia," 1999, 47.
143. Teel and Buck, 9.
144. Cech, 65.
145. Ibid., 66–67.
146. Emery, "Historical Overview," 2002, 8, 18; Teel and Buck, 10.
147. Cech, 65; Teel and Buck, 9.
148. Teel and Buck, 14.
149. Collins, 125–126.
150. Ibid., 132.
151. Freed, 176; Teel and Buck, 22.
152. Secco, Pettenella, and Maso, "'Net-System' Models," 2009, 356, 358.
153. Ibid., 349, 351.

Chapter 6: Listening to the Land: The Art of Ecological Design

1. Beattie et al., 93.
2. Ibid., 102–3.
3. Ibid., 113.
4. Ibid., 79–80.
5. Jacke and Toensmeier, 2005.
6. Beattie et al., 22.
7. Ibid., 27.
8. Ibid., 23.
9. Geyer and Naughton, "Biomass Yield and Cost Analysis," 1980, 320.
10. Beattie et al.,. 23.
11. Ibid., 28.
12. Agate, 43.
13. Strobl, *A Silviculture Guide*, 2000, 107.
14. Beattie et al., 31.
15. Ibid., 107.
16. Tabor, 71.
17. Agate, 97.
18. Mummery, Tabor, and Homewood, *Guide to the Techniques of Coppice Management*, 1998, 14.
19. Oaks and Mills, 54.
20. Ibid.
21. Oaks and Mills, 58.
22. Mummery, Tabor, Homewood, 14; Tabor, 35.
23. Oaks and Mills, 58, 89.
24. Agate, 98.
25. Oaks and Mills, 119; Tabor, 16.
26. Agate, 98.
27. Law, 2001, 11.
28. Beattie et al., 194.
29. Oaks and Mills, 57, 80.
30. Beattie et al., 189–190.
31. Ibid., 216.
32. Ibid., 194–197.
33. Rackham, 1998, 37.
34. www.unl.edu/nac/agroforestrynotes/an22s04.pdf; www.unl.edu/nac/workingtrees/wts.pdf
35. Chedzoy and Smallidge.
36. Rackham, 1998, 38.
37. Ibid., 40.
38. Bargioni and Zanzi Sulli, "The Production of Fodder Trees," 1998, 43, 46.

39. Ibid., 43–44.
40. Ibid., 49.
41. Ibid., 45.
42. Ibid., 47.
43. Ibid., 48.
44. Ibid., 49–50.
45. Climate Technology Center and Network.
46. Kubkomawa et al., "Fodder Bank Establishment," 2019; West, "Assessment of Multi-species Fodder Bank," 2013.
47. Climate Technology Center and Network.
48. Ibid.
49. Maclean, 36, 65.
50. Ibid., 72; Brooks and Agate, 40.
51. Jim Jones, pers. comm., July 2021.
52. Brooks and Agate, 31, 39.
53. Ibid., 12.
54. Ibid.
55. Maclean, 50, 105.
56. Ibid., 85.
57. Ibid., 63; Brooks and Agate, 32.
58. Brooks and Agate, 3. Jim Jones, pers. comm., July 2021.
59. Brooks and Agate, 17.
60. Ibid., 40.
61. Stott, *Cultivation and Use of Basket Willows*, 2001, 6.
62. Ibid., 4.
63. Ibid.
64. Ibid., 3–5.
65. Ibid., 5.
66. Ibid., 8–9.
67. Ibid., 9.
68. Ibid., 10.
69. Beattie et al., 45.
70. Ibid., 48.
71. Ibid., 171.
72. Ibid., 170.
73. Key, "Invertebrate Conservation and Pollards," 1996, 21.
74. Ibid., 26; Oaks and Mills, 62, 79, 80; Beattie et al., 49, 170.

Chapter 7: Getting Started: Establishing Coppice on Your Land

1. Shigo, 353–357.
2. Agate, 95; Oaks and Mills, 59.
3. Stringer, "Development of Advanced Oak Regeneration," 1999.
4. Agate, 42.
5. Drengson and Taylor, eds., *Ecoforestry*, 1997, 28–29.
6. Sands and Abrams, "Effects of Stump Diameter," 2009.
7. Ibid.
8. McGee, "Size and Age of Tree," 1978, 1–2.
9. Shigo, 166.
10. Agate, 47.
11. Douglas and Hart, *Forest Farming*, 1978, 65.
12. Shigo, 163.
13. Ibid., 563; Law, 2001, 64.
14. Kourik, 124.
15. Agate, 105.
16. Ibid., 48; Brooks and Agate, 39.
17. Dirr and Heuser, *The Reference Manual of Woody Plant Propagation*, 2006, 13.
18. Agate, 49.
19. Kourik, 121, 123.
20. Dirr and Heuser, 9.
21. Silver, "Starting a Nursery Business," 2020.
22. Dirr and Heuser, 10.
23. Ibid., 11–12.
24. Hill, *Pruning Made Easy*, 1997, 38; Maguire and Harun, "Air Root Pruning," 2006.
25. Dirr and Heuser, 28.
26. Ibid., 20.

27. Ibid., 21–22.
28. Ibid., 24.
29. Ibid., 42–43.
30. Stott, 7–8.
31. Ibid., 8.
32. Dirr and Heuser, 2006, 25–26.
33. Ibid., 26.
34. Del Tredici, 1995, 18.
35. Agate, 41; Oaks and Mills, 71.
36. Agate, 41.
37. Douglas and Hart, 77.
38. Agate, 41.
39. Green, 1996b, "Bundle Planting," 91.
40. Douglas and Hart, 74.
41. Shigo, 15, 111.
42. Douglas and Hart, 1978, 74.
43. Agate, 47.
44. Ibid., 100.
45. Kourik, 130.
46. DEFRA, 2004, 7.
47. Ibid., 25.
48. Ibid., 6.
49. Ibid., 12.
50. Ibid., 27–28.
51. Maclean, 70.
52. Ibid., 66.
53. Ibid.
54. Ibid., 62.
55. Ibid., 63.
56. Agate, 58.
57. Maclean, 133.
58. Brooks and Agate, 53.

Chapter 8: Coppice Management and Harvest

1. Oaks and Mills, 54.
2. Bane, 1998.
3. Shigo, 183.
4. Hill, 27.
5. Mattoon, 40.
6. Wilson, 2.
7. Ducrey and Turrel, "Influence of Cutting Methods and Dates," 1992, 449, 453, 460.
8. Ibid., 463.
9. Oaks and Mills, 54.
10. Ibid., 52; Agate, 106.
11. Oaks and Mills, 52.
12. Pers. comm., 2010.
13. Mitchell et al., "Thinning a Coppice Regenerated Oak-Hickory Stand," 1985, 23.
14. Wendel, 1975, 4.
15. Oaks and Mills, 51.
16. Ibid., 121.
17. Ibid., 53.
18. Read, "A Brief Review of Pollards," 2007, 257.
19. Mitchell et al., 23.
20. Roth and Hepting, 35–36.
21. Ibid., 27.
22. Ibid., 33.
23. Agate, 105.
24. Pers. comm., 2004; Law, 2001, 75.
25. Agate, 105.
26. Solomon and Blum, "Stump Sprouting," 1967, 4–5.
27. Dey et al., 30.
28. Agate, 105.
29. Jenkins, 39.
30. Stott, 10–12.
31. Machatschek, *Laubgeschichten*, 2002, 155–159.
32. Agate, 85.
33. Austad and Hague, 3.
34. Atkinson, 86–88.
35. Hill, 28.
36. Shigo, 458–459, 467.
37. Rackham, 1998, 35; Agate, 85; Read, 2007, 257.

38. Green, 1996a, 3.
39. Shigo, 459; Lonsdale, 103.
40. Ferrini, 3.
41. Bargioni and Zanzi Sulli, 48.
42. Dagley and Burman, 36.
43. Lonsdale, 103.
44. Dagley and Burman, 34.
45. Read, 2007, 257.
46. Ibid.; Alfredo Bravo, pers. comm., 2011.
47. Rackham, 1998, 35.
48. Agate, 85.
49. Read, 2007, 257.
50. Ibid.
51. Chedzoy and Smallidge.
52. Bane, "From Grassroots to Tree Crowns," 2019; Jehne, "The Soil Carbon Sponge," 2019; Kravčík et al., *Water for the Recovery of the Climate*, 2007.
53. Austad et al., "Lauv som ressurs," 2003.
54. Austad and Haugue.
55. Slotte, "Lövtäkt i Sverige och på Åland," 2000.
56. Meuret and Provenza, "When Art and Science Meet," 2015.
57. Jim Jones, pers. comm., July 2021.
58. Maclean, 124.
59. Brooks and Agate, 56.
60. Maclean, 139.
61. Jim Jones, pers. comm., July 2021.
62. Brooks and Agate, 65.
63. Ibid., 62, 86; Maclean, 151.
64. Brooks and Agate, 62.
65. Ibid., 76; Maclean, 171.
66. Brooks and Agate, 67.
67. Tabor, 47.
68. Wikipedia.
69. Tabor, 47.
70. Wikipedia.
71. Tabor, 47.
72. Ibid.
73. Brooks and Agate, 28.
74. Mummery, Tabor, and Homewood, 12.
75. Fisher, 242.
76. Ducrey and Turrel, 449–451, 454.
77. Rackham, 1998, 6; Austad and Hague, 2007, 3.
78. Dunbar, 1989, 32.
79. Shigo, 438–441.
80. Agate, 80.
81. Ibid.
82. Ibid.
83. Mummery, Tabor, and Homewood, 5.
84. Ibid., 16.
85. Agate, 79; Oaks and Mills, 43.
86. Maine Employers Mutual Insurance Company, 1993, 11.
87. Agate, 81.
88. Oaks and Mills, 173.
89. Chevrier, pers. comm., 2010.
90. Agate, 83–84.
91. Ibid., 84.
92. Mummery, Tabor, and Homewood, 26.
93. Law, 2001, 78.
94. Langsner, 41.
95. Hoadley, 152.
96. Ibid.
97. Ibid., 155.
98. Langsner, 31.
99. Mummery, Tabor, and Homewood, 15.
100. Tabor, 70.
101. Ibid., 74.
102. Agate, 99.
103. Ibid., 100.
104. Tabor, 16.
105. Agate, 98.
106. Tabor, 72.

107. Agate, 103.

108. Brooks and Agate, 15; Blyth et al., *Farm Woodland Management*, 1983.

109. Jenkins, 39.

110. Evans, 24.

111. Lambert, *Tools and Devices*,1957, 6.

112. Rogers and Johnson, 6.12.

113. Wendel, 1975, 3, 5–12.

114. Johnson and Rogers, "Predicting Growth of Individual Stems," 1980, 420–433.

115. Long, Rebbeck, and Traylor, "Oak Stump Sprout Health," 2004, 209.

116. Mitchell et al., 1985, 19–24.

117. Rogers and Johnson, 6.12.

118. Brooks and Agate, 50–51.

119. Law, 2001, 70.

120. Booker and Tittensor, "Coppicing for Nature Conservation," 1992, 300.

121. Agate, 106.

122. Shigo, 157.

123. Oaks and Mills, 64.

124 Ibid, 65

125. Ibid., 67.

126. Ibid., 67.

127. Agate, 108.

128. Ibid.; Law, 2001, 80.

129. Agate, 108.

130. Ibid., 101.

131. Ibid., 109.

132. Tabor, 39.

133. Law, 2001, 11.

134. Agate, 101.

135. Ibid., 102.

136. Franklin.

137. Oaks and Mills, 55.

138. Agate, 102.

Glossary

accessory bud: an axillary bud subtending and typically smaller than its adjacent terminal or primary axillary bud; usually containing differentiated preformed structures. They do not grow unless or until there is a disturbance or change in conditions. Accessory buds sometimes develop initially in the axils of a terminal or axillary bud's bud scales. According to some authors, accessory buds are a primary source of latent or preventitious buds.

advanced regeneration: seedling or sapling trees that are either already present or emerging in a forest understory.

adventitious: growth that develops de novo or sometimes in an uncommon location. Often used to describe buds or new shoots that originate from undifferentiated lateral meristem, either on stems or roots, forming new shoots, or as a means of forming a new independent root system.

angiosperm: a seed-bearing plant with true flowers and seeds that form enclosed within an ovary.

anthropogenic: human-induced or human-generated.

apex: the growing tip of a shoot.

apical control: describing the biological process that limits lateral shoot growth in order to maintain the dominance of a primary lead shoot. Usually controlled by hormonal regulation. The process of apical control leads to the apical dominance of a shoot that forms at a stem's apex.

apical dominance: see **apical control**.

apical meristem: undifferentiated plant cells that can divide and grow to produce a shoot or root; any shoot or root may in turn have dormant or active buds that become apical meristems of their own when they grow.

arborsculpture: the horticultural practice of training and sculpting living plants to grow in a predetermined form, often for artistic purposes.

auxin: an essential plant hormone that regulates growth patterns and causes shoot elongation.

axil: the intersection point between a leaf petiole (stem) and the twig it attaches to. The leaf axils are located on the uphill side of this intersection, often a location where axillary or lateral buds form.

axillary bud: a germ of apical meristem located in the axil of the leaf (see axil); each axillary bud can develop into a new shoot.

axis: the longitudinal support on which organs or parts are arranged; the stem or root; the central line of any body.

BCE: Before the Common Era; denoting the time before the year 0 in the current global time system; the abbreviation used to be BC, "Before Christ."

bark: a sheath of protective tissues surrounding woody plant shoots, comprised of several layers including the periderm, cortex, and phloem or inner bark.

basal area: a common metric used by forestry professionals to measure forest stand density. Usually expressed as ft^2/acre (or m^2/ha), basal area is the sum total of the cross-sectional area of tree stems at breast height across one acre. This relative value helps to determine if stands are over-, under-, or well-stocked.

bast: fibrous material from the phloem (inner bark) of a plant, used as fiber in matting, cordage, basketry, etc.

between-habitat diversity: diversity at a landscape scale, indicating the total richness of species from different habitats in the same area.

bodger: a term used to describe a particular category of traditional craftsman, often unique to coppice woodlands. Usually involved in wood turning using a foot-operated spring pole lathe to make spindles and other turned goods in the woods.

bolling: the trunk or stem of a woody plant. Often used to specifically refer to the stem of a pollarded tree below any significant branches and pollard heads or knobs.

bote: from Middle or Old English: help, relief, benefit, advantage, profit, remedy; an allowance or right of use or benefit, usually of a specific resource for a specific application.

branch: an imprecise word, usually denoting an axis of woody growth more than one year old, but subordinate to the main axis or trunk of a tree or shrub.

branch bark ridge: a naturally occurring boundary formed by woody plants at the intersection between branch and trunk (or twig and branch).

brash: any brushwood or "slash" that remains following limbing coppice poles or rods.

brittle ecosystems: ecosystems adapted to irregularly distributed annual precipitation, such as annual wet and dry seasons.

bud: an undeveloped embryonic shoot that includes differentiated parts such as roots, shoots, leaves, or flowers; usually covered by bud scales to protect them from herbivory and the elements that are in fact modified leaves; may or may not be resting.

bud scale: specialized leaf tissues that form a protective cover around a bud.

butt rot: decay originating at the base of a woody sprout or stem.

callus: masses of undifferentiated and differentiated tissue actively dividing to form a protective cover on and around a wound on woody plants.

cambium: see vascular cambium.

canopy: the uppermost layer of vegetation in a forest stand. Also describing the individual woody plants that comprise this layer of vegetation.

canopy cover: a measure used to denote the density of a forest stand. Usually expressed as a percentage that indicates the estimated density of the forest canopy. A stand with 60% canopy cover only permits 40% of available sunlight to penetrate the canopy and reach the forest floor.

cant: a defined patch of woods (usually only used in relation to coppice) managed as one unit and usually harvested on a rotational basis; aka a stand.

cartbote: in Britain, a customary common right for wood to which a tenant is entitled for making and repairing carts and other instruments of husbandry.

clapper stick: an Indigenous musical instrument made from a long split stick (usually elderberry), hollowed out for ¾ of its length, which was struck against one's hand or trembled in the air to make clapping sounds during dances.

clear-fell: to cut an entire stand of trees to the ground; harvesting all the biomass of every stool in a coppice cant at once; a clear-cut. Can also be used at the scale of an individual stool; to clear-fell a stool is to cut all of the poles or rods that have formed on it.

codominant trees: in a forestry context, co-dominant trees' crowns form the canopy and receive light from above but little from their sides; their crowns are relatively crowded from the sides; compare dominant, intermediate, and suppressed trees.

common: *n.* land owned and shared in common use; the set of harmonious social relations produced by individuals jointly extracting the means of their subsistence from a collective pool; *v.* "to common," a customary activity of meeting subsistence needs from a collectively held resource through harmonious social relations.

compartmentalize: the process by which woody plants respond to a wound or other physical damage. The formation of natural boundaries by living cambium that resists the spread of decay upwards and downwards, inwards, and outwards. While this does lead to a loss of potential energy storage in affected sapwood, it allows the plant to form new tissues outboard of the wound, unaffected by the spread of decay.

coppice: the silvicultural practice of managing woody plants on regular cycles and harvesting the resultant resprouts. Often done on semi-regular cycles ranging from 1 to 30 years or more for specific material or economic purposes. From the French verb *couper*, "to cut."

coppice with standards: a variation on simple coppice management that integrates standard trees allowed to grow single-stemmed to full maturity, dispersed throughout a managed coppice woodland. Usually averaging densities of 5 to 20 per acre. Species often chosen for high-value lumber and mast (seed/nut/fruit) production for wildlife and livestock.

cork cambium: also known as the phellogen, a thin layer of lateral meristematic tissue that divides outwards, forming new cork cells, and in some species also dividing inwards to form the phelloderm. One of three layers comprising woody plant's outer bark or periderm.

cotyledon: a seed leaf. That is a leaf formed within the seed of a plant that first sprouts upon germination.

crown: woody plants' topmost organ, consisting of the uppermost portion of the trunk, branches, twigs, leaves, buds, flowers, and other structures. Also often referred to as the canopy. It serves as an armature supporting leaves and is the only part of the organism capable of manufacturing energy via photosynthesis.

current annual increment: a measure of the annual increase in wood volume for an individual tree or a stand of trees.

dbh: diameter at breast height (4.5 feet, or 1.37 m). A measurement standard for standing tree diameter that accommodates for the common flare at the base of growing trees. For consistency, we measure woody plants' diameter at breast height.

dehesa: a long managed, naturalized type of agroforestry landscape widespread throughout central and western Spain and much of Portugal (where it's referred to as *montado*). A managed silvopasture where livestock forage (cows, sheep, and especially pigs) in the understory below several oak species, including *Quercus suber* (cork oak) whose outer bark is the lone source of natural cork. Pastured pork finished on acorn-rich dehesa is among the finest and highest-value meats in the world (jamón ibérico de bellota).

deliquescent: branching such that the main axis or stem dissolves into many fine divisions, yielding a rounded or spreading crown.

dendritic: having a branching form; tree-like; may continue to have a main axis or trunk.

determinate growth: a growth habit where a shoot grows only as many leaves, nodes, and internodes in the current growing season as the apical bud has preformed in the previous growing season, after which shoot growth stops until the following season. Contrast with indeterminate growth and semi-determinate growth.

dicot: a major group of flowering plants (angiosperms) characterized by having two cotyledon leaves; contrast with monocots; includes all woody plants that are not gymnosperms.

diffuse porous: a classification of angiosperm wood type based on the relative evenness of cell size and structure across each growth ring. Contrast with ring-porous and semi-diffuse porous. Includes species like maple, beech, birch, and cherry. Sometimes referred to as even-grain.

disturbance: any relatively discrete event in time that disrupts ecosystem, community, or population structure and changes resources, substrate availability, or the physical environment.

disturbance anticipation buds: a term created in developing this book describing buds of preventitious origin; that is, buds formed when shoots first begin to grow and expand. They maintain a continuous trace back to the pith and often possess differentiated preformed structures within, remaining suppressed until needed.

dominant trees: in a forestry context, dominant tree's crowns reach above the level of the overall crown cover and receive full light from above and partly from the sides; they tend to exhibit larger size and more vigor than average trees in the stand, and have well-developed crowns, though they may be crowded on the sides.

dormant bud: in some cases, any bud that has yet to break dormancy and elongate. In our case, a common term used to describe buds of preventitious origin that formed at the time of shoot initiation, maintain a continuous trace back to the pith, and may remain suppressed until needed and buried slightly below the bark, keeping pace with cambial expansion. They have the potential to give rise to resprouts following damage, wounding, or harvest. Also sometimes known as latent or suppressed buds.

draw-felled: the selective harvest of individual rods from a coppice stool, while leaving the remainder to continue to grow.

dynamic mass: includes the actively growing parts of woody plants capable of storing energy. Young trees are essentially composed of 100% dynamic mass. This balance changes as they age.

earlywood: large, thin-walled vessel cells that form during the early stages of annual growth, particularly prominent in ring-porous hardwood species. This spongy layer is filled with large pores you can often see with the naked eye and enable the distribution of large volumes of stored root energy and water, driving vigorous early season growth.

enclosure movement: the formal privatization of land ownership that displaced community members from what was previously shared access to the commons or land held and managed collectively. While the process occurred gradually between the 13th and mid-19th century, it was formalized with the passing of a series of parliamentary acts in the 16th and 17th centuries. For one to claim ownership of land, it needed to be enclosed either by hedgerows, walls, or fences.

endemic: native to and found only in its given region.

epicormic bud: Viable buds formed at the time of shoot elongation that are lying latent on or within a non-succulent shoot. Epicormic literally means "upon the trunk." Their growth is usually triggered by some type of stress, damage, or disturbance.

estovers: in Britain, a customary common right for an allowance of wood made to a tenant; the freedom of a tenant to take necessary wood from the land occupied by that tenant.

even-aged (management): a forest stand where all trees are more or less in the same age class, usually the result of some type of disturbance event. As management, any silvicultural practice that aims to harvest a stand either all at once or sequentially to stimulate the regrowth of a stand that's all more or less the same age.

even-grained: wood with a cellular structure that's largely consistent in size and texture throughout each annual growth ring. Often synonymous with diffuse-porous wood. Contrast with uneven grain and the wood of ring-porous hardwoods.

excurrent: having a straight dominant main stem or central leader that runs from the ground to the apex with subordinate lateral branches, as in spruces, yielding a conical crown.

extractives: nonstructural components of wood usually concentrated within heartwood that are often toxic to fungi and reduce wood's permeability. Rot-resistant wood species generally possess high concentrations of extractives in their heartwood.

faggot: a bundle of small-diameter branch wood and brushy material usually used for firing masonry bread ovens, providing an inexpensive, hot-burning fuel source.

fealty: a feudal era oath of loyalty, often between vassal and lord. A profession of allegiance.

fibers: small-diameter cells comprising roughly 50% of wood's total volume, conferring strength to the wood; typically elongated with pointed ends and thick cell walls.

firebote: in Britain, a customary common right for an allowance of fuel, usually wood.

flush cut: a poorly executed pruning cut on a woody plant that passes through the branch collar, exposing the stem to pathogens and drying. Making a pruning cut flush against a woody plant's stem or branch.

form: the overall growth pattern and strategy of a plant, denoting its general physical size and architecture: tree, shrub, vine or herb, for example.

forest cover type: a system of forest ecosystem categorization created by the Society of American Foresters that describes different commonly occurring natural communities based on the dominant woody canopy species.

forest stand: a contiguous area that includes trees of similar species, age, size, and condition.

gap formation rate: the proportion of a given area of land annually affected by disturbance; usually express as a percentage of the total.

gross primary production: the sum to total biomass created by primary producers (plants and other organisms capable of manufacturing their own energy), mostly through photosynthesis over a particular period of time.

growth rings: visible layers of wood growth that correspond with what usually amounts to a single growth increment. Usually indicating one year's new growth. Can be used to correctly determine a tree's age and also learn about growing conditions in previous seasons.

gymnosperm: a naked-seeded plant, where the seed forms on a scale rather than within an ovary as in angiosperms. Conifers are gymnsosperms.

habit: patterns of plant growth within the general plant form, denoting the sequencing and configuration of budding, sprouting, elongation, and branching; applies to both above- and belowground plant architecture.

hardwood: often used generally to describe all broadleaf species of woody plants. Also used synonymously with "angiosperm." Also used to describe the density of wood. But not all hardwood species have hard wood.

heartwood: the core of some woody species' trunk, often marked by a stark change in color and often, though not always, an indicator of decay resistance. Heartwood cells are no longer living, therefore do not conduct or store water, minerals, or photosynthates. Often the most valuable wood in a tree.

hedgebote: in Britain, a customary common right for an allowance of wood to a tenant for repairing hedges or fences.

high forest: woodland consisting of even-aged or uneven-aged stands of large trees for timber production, with little or no coppice understory, though seedling trees for eventual replacement of the canopy may be present.

high-grading: in forestry, "taking the best and leaving the rest," that is, harvesting all the best trees and leaving the poorest; essentially mining the resource, and degrading it over time.

historical ecology: a field of study that examines the relationships between humans and their ecosystems, often over long time intervals.

houlette: a specialized tool shepherds used to harvest tree hay, similar to a sharpened spoon on a long handle.

housebote: in Britain, a customary common right for wood allowed to a tenant for repairing the house or for fuel.

improvement cut: an intermediate silvicultural treatment that focuses on reducing the total number of stems in an overcrowded or unmanaged forest stand, with a particular emphasis on the removal of undesirable species.

indeterminate growth: a growth habit where a shoot grows continuously during the growing season as long as conditions allow, producing leaves, nodes, and internodes from both preformed structures within a bud and neoformed structures produced by the apical meristem during the current growing season.

indicator plants: specific plant species whose presence suggests certain hydrological, soil, land use, or other characteristics.

intermediate treatments: a set of silvicultural practices intended to enhance the quality of a forest stand for future harvest. Often considered pre-commerical since materials hold little commercial value in modern fuelwood and lumber markets.

intermediate trees: in a forestry context, trees that are shorter than dominant or codominant trees, but extend into the canopy formed by codominants and dominants; they receive little light from above and none from the sides, and their crowns are small and usually crowded on the sides; compare dominant, codominant, and suppressed trees.

internode: a more or less elongated section of shoot or stem between two nodes.

Keyline: a system of holistic site design conceived by Australian P.A. Yeomans. Also a physical plane in the landscape: the contour line that passes through the keypoint. Also sometimes used to describe Yeomans' design for a specialized subsoil plow implement.

keystone species: any species whose population size, ecology, and behavior affects the population and ecology of all other species in the community.

knot: a formation that develops in wood as a result of the stem's annual cambial expansion enclosing a living or dead branch. Often considered a structural weakness and also commercially undesirable.

laid hedge: a hedgerow that's been rejuvenated by cutting ⅔ of the way through the stem of trees and shrubs and laying them diagonally to create a dense wall of vigorous resprouts. Often done on 10-to-15-year cycles.

landscape scale: contrast with patch scale; usually describing a broader level of spatial analysis, understanding, and planning often addressing numerous land-use objectives and ecological processes.

latent bud: often used synonymously with dormant or suppressed bud; a previously formed bud that remains suppressed; often triggered by some time of stress, damage, or disturbance; often of axillary or accessory bud origin.

lateral bud: also known as an axillary bud; buds that form along the length of a growing shoot.

lateral meristem: also called the cambium; actively dividing undifferentiated cells lying in a thin layer under the bark but outside the main woody stem of a tree or shrub that grows wood each year to the inside of the cambium and bark to the outside.

latewood: contrast with earlywood; small, dense, thick-walled wood cells that form later on in the growing season after the early spring flush of the large-pored earlywood. Ring-porous hardwood species exhibit a particularly stark contrast between earlywood and latewood cells.

leaf area index: a relative measure of half of the total leaf area (because it only measures one surface) coverage per unit ground surface area; used to quantify and compare total biomass and canopy density.

leaf hay: also known as tree hay; small-diameter twigs and leafy growth often harvested, dried, and stored as a winter or dormant season feed for livestock.

liberation cut: an intermediate silvicultural treatment that removes more mature trees, restricting the growth of sapling-aged and younger trees.

life history strategy: a framework that describes the timing and patterns of various key events in a species' life cycle, including life span, mature size, age of reproductive maturity, reproductive strategies, number of offspring produced.

lignotuber: a type of specialized organ that provides additional storage of starches and plant nutrients as well as a protected supply of dormant buds; often an adaptation to stressful growing conditions.

lynchet: an earthen bank formed on a hillside either intentionally or unintentionally and used as a boundary marker; these form at the upper or lower side of a ploughed field through erosional cutting or deposition that, over time, end up making the field more level; the word is a diminutive form of the word lynch which is an old English word for an "agricultural terrace."

mast: food sources for wildlife originating from woody plants; broken into two categories: hard and soft mast; hard mast includes nuts and seeds, while soft mast includes fruits and berries.

meristem: localized regions of active or potential cell division and enlargement within a plant, containing undifferentiated cells that may turn into various kinds of tissue. Meristems usually occur: at the apex of every shoot and every root (apical meristems); as future or potential apical meristems at nodes (axillary buds) or under the bark in various locations (dormant buds); and as a cylinder surrounding the stems and roots of dicots and gymnosperms under the bark (lateral meristem, or cambium), which includes cells that produce the stem wood (xylem) and bark (phellogen or phloem).

Mesolithic: Middle Stone Age, a brief period of human prehistory that formed the transition from Paleolithic hunter-gathering cultures to settled Neolithic farming communities; generally thought to have begun after the last Ice Age ended, but its dates vary from place to place. Some archeologists don't use the term at all, especially outside of Europe.

microclimate: a localized climate type that exists at varying scales; often the effect of aspect, elevation, large water bodies, etc.

monocot: roughly one quarter of all flowering plants (angiosperms) characterized by having a single cotyledon; mostly includes herbaceous species such as orchids and grasses; also includes palms; contrast with dicot.

monopodial: describing a type of plant growth where the terminal bud maintains apical dominance, forming a central leader arising from a single base, often leading to a pyramidal growth form.

montado: see dehesa; the Portuguese equivalent term for oak-dominated savannah covering much of the central and western Iberian peninsula.

mycrorrhizae/mycorrhizal: literally "fugal root"; species of fungi that grow in an often symbiotic relationship with nearby plants; the relationship between plant roots for fungus, often characterized by the exchange of photosynthesized sugars by plant roots for increased water and nutrient absorption by mycorrhizae.

natural community: a group of plants and animals that commonly co-occur on sites with similar ecological characteristics.

Neolithic: New Stone Age, a period of human prehistory that began with the invention of agriculture and settled human habitation and ended with the invention of metal tools in the Bronze, Iron, or Copper Age. Neolithic peoples tended to make polished stone tools (rather than knapped stone), animals were domesticated during this period, and pottery became particularly common.

net community/primary productivity: see **gross primary production**; the net carbon produced by ecosystems in excess of that necessary for respiration and life processes.

node: the place on a shoot where leaves or flowers attach and axillary buds form; see **internode**.

non-brittle ecosystems: ecosystems with evenly distributed precipitation, especially humid regions.

overstood: coppice stems or stands that have matured for longer than their intended or useful growth cycle; the sprouting ability of most trees declines with age and/or girth, so overstood stems will become less able to resprout once cut, and may also exhibit rot, crowding at the base, or other issues that reduce the usability of the poles.

Paleolithic: Early Stone Age, a period of human prehistory beginning with the use of the earliest stone tools about 2.6 million years ago, characterized by cultures consisting of small bands of hunter-gatherers at very low population density. The Paleolithic Age ended at different times in different places, varying by as much as several thousand years.

palynology/palynological: the study of, or evidence produced by, pollen, spores, and other microorganisms.

pannage: the traditional practice of releasing pigs into forest to feed on acorns and other types of mast.

PAR/photosynthetically active radiation: the spectrum of light plants use for photosynthesis; in the range of 400 to 700 nm.

parenchyma: thin-walled cellular tissues particularly concentrated in the soft portions of bark, pith, sapwood, and leaves, primarily involved with energy storage and distribution.

patch scale: a fundamental unit of landscape ecology with relatively uniform aspect, slope, soils, and vegetation; rotational coppice agroforestry management commonly occurs via patch-scale disturbances.

persistence: in regards to woody plant life history strategies, persistence describes a species' long-term ability to occupy a site; sprouting as a life history strategy enables plants to colonize the "persistence niche."

phellogen: one of several layers of bark, also known as the cork cambium; meristematic tissue located within woody plant's outer bark that divides outwards to create new cork cells (similar to vascular cambium) and in some species also divides inwards to form the phelloderm.

phloem: also known as the inner bark; the conduit through which photosynthesized sugars are transported downwards from the canopy to the stem and root system; formed as a result of outward division of the vascular cambium.

photosynthetically active radiation: see PAR.

phytoremediation: a type of biological remediation that uses plants to either break down or absorb pollutants in the environment.

pith: primary growth at the core of a new shoot or root, typically made from spongy parenchyma cells; sometimes crushed as a stem or root ages by the growth of secondary xylem. The trace that remains following apical meristem division.

pleacher: a tree stem managed in a laid hedge by severing ⅔ through it and laying diagonally to restore hedge density and vitality.

ploughbote, plowbote: in Britain, a customary common right for wood or timber allowed to a tenant for the repair of agricultural instruments.

polewood: a general term used to describe medium- to large-diameter coppice materials; usually describing roundwood used for a range of different products at least 2" (5 cm) in diameter, although from a silvicultural point of view, here we classify them as stems between 4-to-10-inch (10–25 cm) dbh; contrast with timber or lumber and rods that are smaller-diameter coppice-generated material.

pollard(ing): the practice of managing woody plants for sprouts located up off the ground often in a low canopy; a form of pruning or training often used to restrict the mature height of woody plants while producing fodder, fuelwood, shade, etc.; often done so resprouts lie beyond the reach of wildlife or grazing livestock.

pores: the exposed face of vessel cells in hardwood; especially visible in ring-porous hardwood species.

preformed: botanical structures created within a bud in the current growth flush that will lie dormant for a period before enlarging to their full size in the next growth flush.

preventitious buds: buds formed at shoot initiation that maintain a continuous trace to the pith; they remain viable but latent until growth is triggered by some kind of stress, damage, or disturbance; commonly synonymous with dormant or latent buds; contrast with adventitious.

primary growth: those portions of a shoot or root created by an apical meristem during the season of their creation; primary growth includes the pith, primary xylem, primary phloem, and endoderm, which takes the place of bark in the shoot's first year.

primordia: miniature partially differentiated leaves or other plant parts formed within or very near the meristem that later grow to full size and expression.

proleptic growth: growth of an axillary bud that occurs after a period of dormancy, in the growing season following that in which it formed.

provenance: describing the place of origin; in this case referring to the original source of plant genetic material.

ramial: small-diameter wood (less than 2.5 inches or 7 cm) in diameter; used to build soil organic matter and fertility and support diverse microbiological populations.

rays: flat bands of cellular tissue in wood that radiate from the pith outwards, connecting xylem and phloem and offering horizontal sap conduction and temporary carbohydrate storage; particularly prominent in some species. Also known as medullary rays.

recruitment: the addition of new organisms to a population.

regeneration treatments: silvicultural practices used to create a new age class of plants in a forest stand; includes clear-cuts, seed tree, shelterwood, selection, patch cuts, and coppice management.

release cutting: a type of intermediate silvicultural treatment that removes less desirable species and stems that are often overtopping young seedling regeneration.

resprout silviculture: a catch-all term used to describe the various types of woody plant management techniques used to produce resprouts; includes coppicing, pollarding, shredding, hedgelaying, stooling, etc.

rhizome: a modified underground plant stem capable of forming roots and shoots along its nodes.

ring-porous: a category of broadleaf woody plants (angiosperms) whose wood is characterized by two distinctly different layers in a single growth ring (see **earlywood** and **latewood**).

rive: a controlled process of splitting, used as a way to process roundwood by splitting it longitudinally along its fibers.

riven: wood that's been carefully split along the grain by riving.

rod: a small-diameter stem growing on a short-rotation coppice stool; often used for various types of woven crafts.

root collar: the place on a woody plant where aboveground stem wood meets and transitions to root wood, typically at or just below the soil surface; large numbers of dormant buds ring the stem at this location in many trees and shrubs.

sail: upright stakes through which cleft and/ or uncleft rods are woven to make a wattle hurdle; sometimes written as zale.

sap: water within the plant, containing dissolved nutrients and plant chemicals and circulating throughout the vascular system.

sapling: the youngest stage of woody plant growth; the size classification varies between sources but here we consider them anything under 0.5 inch (1.25 cm) in diameter.

saprophytic: in this case used to describe a category of fungi that feed by decomposing organic matter.

sapwood: xylem tissues that actively convey water and minerals upwards through the plant stem; sometimes limited to just the few outermost growth rings, although in young trees, the entire stem is sapwood.

sawlog (small, medium, large): commonly used to describe stem wood harvested for conversion into lumber; also used as a size classification for trees and often broken into sub-categories of small, medium, and large based on dbh (10 to 15 inches, 15 to 20 inches, and over 20 inches) (25–38, 38–60, and 60+ cm).

sclerophyllous: plants that have hard, thick, stiff, leathery, and usually small, evergreen leaves, as an adaptation to stress—typically sunny, dry conditions.

secondary growth/thickening: those portions of a shoot or root formed from lateral meristem in the seasons after their initial creation.

seedling: a young woody plant in the forest understory; here we consider plants between 0.5-to-4-inches (1.25–10 cm) dbh seedlings; also often used to describe a plant propagated by seed as opposed to clonally or by coppicing.

seedling sprouts: young trees that persist in a forest understory by repeated sprouting as a result of some type of stress or damage; many young seedlings persist for years, biding their time for a canopy gap to form.

semi-determinate growth: a growth habit where shoot growth occurs in two or more rhythmic flushes in a given growing season, with a period of bud formation and dormancy between them; each growth flush produces leaves, nodes, and internodes only from preformed structures in each successive bud. Contrast with determinate growth and indeterminate growth.

semi-ring/diffuse porous: a category of woody species whose wood contains large earlywood pores and smaller latewood pores but lack a distinct break between the two.

senescence: the loss of productive vigor and an overall decrease in functionality; the final stage in most woody plants' life cycles.

serotinous: describing the delayed or gradual opening or release; common in fire-adapted species; serotinous cones from some conifers will only open to release seed in the presence of fire.

shelterwood: a silvicultural treatment used to regenerate an even-aged stand incrementally, by removing 20% to 40% of the canopy in 2 or 3 interventions, often over the course of 5 to 20 years.

shoot: the main constructional unit of a plant, consisting of the stem, leaves, buds, flowers, and other botanical structures that arise from a single apical meristem. Any new bud that begins to grow forms its own shoot.

shredding: the practice of removing side branches and tops from woody plants on relatively short cycles to generate fodder for livestock, craft materials, and small-diameter fuelwood.

shrub: a woody plant with one stem maturing under 10 feet (3 m) high, or a multistemmed woody plant maturing up to 33 feet (10 m) tall, whose stem or stems increase in diameter each year through secondary thickening of the lateral meristem.

silviculture: the growing and cultivation of trees; forestry/forest management.

silvopasture: an agroforestry practice integrating grazing management with various forms of forestry or woody agriculture.

singling: thinning a coppice stool to a single stem; often done to select a new standard tree in a coppice-with-standards system.

site index: a common measure of site productivity that describes trees' projected height in feet at a given age (usually 50 years); often closely related to soil type as well as several other characteristics.

snedding: the process of removing side branches and brushy tops from harvested rods and poles; often referred to as limbing in the United States.

softwood: generally used to describe coniferous species (angiosperms).

spar: a U-shaped, staple-like wood product, often made from coppiced hazel used to attach roofing thatch to the roof framing.

spermatophyte: a category of seed-producing plants that includes all angiosperms and gymnosperms.

standard: a normal single-stemmed tree or shrub often grown among coppice stools to produce a long-term timber yield.

stele: the core of a root or stem capable of transporting water, minerals, and photosynthates; a specialized tissue capable of producing buds and maintaining a direct connection between preventitious buds and the pith.

stochastic: random or unpredictable processes.

stolon: an aboveground stem capable of lateral growth radiating away from the parent plant and generating new shoots and roots; contrast with rhizomes which form belowground.

stooling: a vegetative propagation technique used to produce clones of a parent plant; the result of coppicing a stool, mounding soil around the base of the developing sprouts and allowing them to form adventitious roots before cutting free and replanting. Sometimes also referred to as stool layering.

stump culture: a horticultural practice used on conifers that, while not true coppicing, allows for the harvest of treetops while stimulating the emergence of a new leader capable of forming another tree for harvest; sometimes referred to as coppiced Christmas trees.

suberin: the waxy compound located in plant cell walls and especially in bark; the substance that makes bark corky.

succession: in ecology, the sequence of events following disturbance; the progression of ecosystem transformation as populations of flora and fauna adapt and change over time.

sucker: a sprout originating from woody plants' roots, often in response to some type of wound or disturbance.

suppressed trees: in a forestry context, trees whose crowns lie under the canopy and do not receive direct light from above or the sides.

sylleptic growth: growth of an axillary bud immediately after it forms, in the same growing season or growth flush.

symbiosis: literally "same life"; a mutualistic relationship between organisms.

sympodial: a plant growth habit leading to a zig-zag pattern and a deliquescent form; it occurs once a terminal bud has ceased growing and new shoot elongation continues from a subordinate lateral bud.

terminal bud: a germ of apical meristem at the terminus or growing tip of a shoot that goes dormant for a cold or dry season and resumes growth the following growing season on the same growth axis.

thatching spar: see **spar**.

timber: in our case, wood from large-diameter standard trees allowed to grow to maturity and harvested for lumber; contrast with wood.

traditional ecological knowledge: the assemblage of information and experience amassed and curated by Indigenous peoples and other human traditions; often relating directly to sustainable resource management, vernacular architecture, ecological agriculture, etc.

tree: a woody plant maturing at more than 10 feet (3 m) tall with one or a few elongated main stems that support a well-defined crown consisting of branches, twigs, leaves, perennating buds, and flowers, with the trunk(s) increasing in diameter each year through secondary thickening of the lateral meristem.

tree hay: see **leaf hay**.

trunk: the main stem of a woody plant; usually the most commercially valuable part of the plant.

tule: any of many species of bulrush (*Juncus* spp.), especially in California or the southwestern US.

twig: a woody or soon-to-be-woody stem one year old or less.

tyloses: bubble-like structures that form in the cell cavities of some woody species that impede the transmission of liquids and offer some degree of rot resistance.

underwood: woody plants growing under an overstory or canopy.

undifferentiated plant cells: germinal cells that are able to divide and then change their form and structure to serve one of a variety of functions depending on the form they take.

uneven-aged (management): a forest or woodland management strategy allowing trees within a stand to be in different age classes through one or another form of selective cutting, allows continuous or periodic yields while maintaining tree cover at all times.

uneven grain: see ring-porous; describing wood formed with two dramatically different layers of cells per growth ring: porous, fast-grown earlywood and dense, summer-produced latewood.

usufruct: the right to use and enjoy the resources and benefits of something owned by another, as long as the resource is not damaged or reduced in value.

vascular cambium (or cambium): a thin cylindrical layer of living meristematic tissue that surrounds the entirety of a live woody plant; located between the wood and inner bark, it divides outwards and inwards and is the means by which woody plants increase in girth.

vegetative reproduction: clonal forms of plant reproduction that do not rely on seed; either producing new stems on an existing root system as in coppicing or using stem or root cuttings to generate a new genetically identical individual; also known as asexual reproduction; contrast with sexual/seed reproduction.

vessels: a vertically oriented end-to-end array of wood cells with open end walls that conduct water/sap and provide structural integrity; comprising roughly 30% of a woody plant's mass; we see their exposed ends as "pores" on a stem cross-section.

wattle: a woven framework of sticks comprised of a series of vertical uprights, between which are woven small-diameter horizontal stems. Historically most commonly crafted using hazel wood.

wattle and daub: a traditional form of wall construction that uses a framework of woven small-diameter rods (the wattle) as a substrate upon which a layer of daub (usually earthen plaster) is applied. The precursor to lath and plaster wall systems.

wildwood: the British term used to describe what North Americans call old-growth forest.

withes, withies: straight, flexible woody plant stems used for woven crafts, especially baskets.

within-habitat diversity: the diversity of organisms and species within a given habitat type, such as an estuary (high diversity), an oak-hickory forest (moderate diversity), or a suburban lawn (low diversity).

wood: the hard fibrous material forming the main substance of the trunk or branches of shrubs and trees; the secondary xylem lying beneath the bark, composed mainly of lignin and cellulose; also a general term used historically to describe polewood resources, **contrast with timber.**

woodland ride: a managed access way, often located within coppice stands. Commonly recognized as hosting more diverse herbaceous vegetation due to increased available sunlight.

wood pasture: an agroforestry system featuring a canopy of trees, often managed by coppicing or pollarding, overtopping grazed pasture below; often synonymous with various forms of modern silvopasture.

wooded meadows: a traditional agroforestry system with a sparse canopy of woody plants often managed by pollarding or coppicing, and an herbaceous understory managed for hay production.

woody cuts: a horticultural practice used to manage woody plants for decorative stems for the floral industry. Production often involves various forms of resprout silvicultural practices to maintain a living plant for perennial production.

woody plant: perennial plants capable of secondary thickening; that is, they produce wood as a structural tissue and add a new layer of growth to the entire organism each growing season, increasing in thickness; includes trees, shrubs, and woody vines.

woodbank: an earthwork used to protect valuable timber or agricultural resources; usually consisting of a ditch surrounding the perimeter of the protected area with spoils mounded up to form a mound or wall.

woodward: the manager of a given wood.

xylem: one of the two main types of transport tissues in plants, responsible for the upward flow of water and minerals from roots. In woody plants, the living sapwood.

Bibliography

Abbott, Mike. 2002. *Living Wood: From Buying a Woodland to Making a Chair,* 4th ed. White River Junction, VT: Chelsea Green.

Adams, Paul. 1996. "The Bryophyte Flora of Pollards and Pollard Woodland with Particular Reference to Eastern England." In *Pollard and Veteran Tree Management II,* Helen Read, ed., 12–16. London: Corporation of London.

Adegbidi, Hector Guy. 1994. "Nutrient Return Via Litterfall and Removal During Harvest in a One-year Rotation Bioenergy Plantation." PhD diss., College of Environmental Science and Forestry, State University of New York.

Agate, Elizabeth, ed. 1980. *Woodlands: A Practical Handbook.* Doncaster, UK: British Trust for Conservation Volunteers.

Ahrens, Glenn R. and Michael Newton. 2008. "Root Dynamics in Sprouting Tanoak Forest of Southwestern Oregon." *Can. J. For. Res.* 38: 1855–1866.

Alaska Department of Fish & Game. 2013. *Revegetation Techniques: Bundles (Fascines).* Juneau: Alaska Dept of Fish & Game. www.adfg.alaska.gov/static/lands/habitatrestoration/streambankprotection/pdfs/csbs_livebundle.pdf, accessed April 11, 2013.

Alavalapati, Janaki R.R., D. Evan Mercer, and Jensen R. Montambault. 2005. "Agroforestry Systems and Valuation Methodologies: An Overview." In *Valuing Agroforestry Systems: Methods and Applications,* Janaki R.R. Alavalapati, and D. Evan Mercer, eds., 1–8. *Advances in Agroforestry,* vol. 2.

Allen, Simon J., Robin L. Hall and Paul T. W. Rosier. 1999. "Transpiration by Two Poplar Varieties Grown as Coppice for Biomass Production." *Tree Physiology* 19(8): 493–501.

Anderson, Bob. 2007. *Woody Cuts.* Crop Profile Series. Lexington: Center for Crop Diversification, Cooperative Extension Service, University of Kentucky. www.uky.edu/Ag/CCD/introsheets/woodycuts.pdf.

Anderson, M. Kat. 2005. *Tending the Wild: Native American Knowledge and the Management of California's Natural Resources.* Oakland: University of California Press.

Andrič, M. 2007. "Holocene Vegetation Development in Bela Krajina (Slovenia) and the Impact of First Farmers on the Landscape." *Holocene* 17(6): 763–776.

"Apples: Propagation by Stooling." www.suttonelms.org.uk/apple72.html.

Archaux, Frédéric, Richard Chevalier, and Alain Berthelot. 2010. "Towards Practices Favourable to Plant Diversity in Hybrid Poplar Plantations." *Forest Ecology and Management* 259(12): 2410–2417.

Armstrong, Lyn. 1979. *Woodcolliers and Charcoal Burning.* Weald and Downland Open Air Museum. Horsham, Sussex: Coach Publishing House.

Mark S. Ashton and Matthew J. Kelty. 2018. *The Practice of Silviculture: Applied Forest Ecology.* New York: John Wiley & Sons.

Atkinson, Martin. 1996. "Creating New Pollards at Hatfield Forest, Essex." In Pollard and Veteran Tree Management II, Helen Read, ed., 86–88. London: Corporation of London.

Austad, Ingvild, Anders Braanaas, and Marvin Haltvik. 2003. "Lauv som ressurs; Ny bruk av gammel kunnskap." HSF rapport nr. 4/30.

Austad, Ingvild, L.N. Hamre, K. Rydgren, and A. Norderhaug. 2003. "Production in Wooded Hay Meadows." *Transactions on Ecology and the Environment* 64: 1091–1101.

Austad, Ingvild and Leif Hauge. 2007. "Pollarding in Western Norway." In *1er colloque Européen sur les trognes, Vendôme,* 26, 27 et 28 Octobre 2006. Boursay, France: La Maison Botanique. Available at www.maison-botanique.fr/colloque-europeen.php.

Austad, Ingvild, Mary Holmedal Losvik, Knut Rydgren, Liv Norunn Hamrek, and Ann Norderhaug. n.d. *The Wooded Hay Meadow: A Sustainable Production System?* Microsoft PowerPoint presentation. Sogndal, Norway: Sogn og Fjordane University College. pan.cultland.org/files/WS 5_files/Ingvild Austad – Wooded hay meadows.pdf, accessed June 6, 2012.

Avon, Catherine, Laurent Berges, Yann Dumas, and Jean-Luc Dupouey. 2010. "Does the Effect of Forest Roads Extend a Few Meters or More into the Adjacent Forest? A Study on Understory Plant Diversity in Managed Oak Stands." *Forest Ecology and Management* 259(8): 1546–1555.

Bachmann, Janet. 2002. "Woody Ornamentals for Cut Flower Growers." Fayetteville, AR: Appropriate Technology Transfer for Rural Areas, National Center for Appropriate Technology. https://attra.ncat.org/attra-pub/summaries/summary.php?pub=43, accessed August 21, 2014.

Bally, Battina. 1999. "Energy Wood Production in Coppice with Standards and Coppice Forests in Switzerland." *Schweiz. Z. Forstwes* 150(4): 142–147. In German.

Bane, Peter. 1998. "Coppice-with-Standards: New Forestry with Ancient Roots." *Permaculture Activist* 40: 52–53. Reprinted in *Overstory* 47, Holualoa, HI, accessed December 17, 2010.

Bane, Peter. 2019. "From Grassroots to Tree Crowns: Organizing to Cool the Climate." Conference presentation, Soil and Nutrition Conference from Bionutrient Association (Southbridge, MA, November 16, 2019). YouTube audiotape available, accessed February 28, 2021.

Baptista, R. and K. L. Launchbaugh. 2001. "Nutritive Value and Aversion of Honey Mesquite Leaves to Sheep." *Journal of Range Management* 54(1): 82–88.

Barbalat, Sylvie and Daniel Getaz. 1999. "Influence de la remise en exploitation de taillis-sous-futaie sur la faune entomologique." *Schweiz. Z. Forstwes.* 150(11): 429–436.

Bargioni, Elena and Alessandra Zanzi Sulli. 1998. "The Production of Fodder Trees in Valdagno, Vicenza, Italy." In The Ecological History of European Forests, Keith Kirby and Charles Watkins, eds., 43–52. New York: CAB International.

Barker, S. 1998. "The History of the Coniston Woodlands, Cumbria, UK." In *The Ecological History of European Forests,* Keith Kirby and Charles Watkins, eds., 167–183. New York: CAB International.

Barkham, J.P. 1992. "The Effects of Coppicing and Neglect on the Performance of the Perennial Ground Flora." In *Ecology and Management of Coppice Woodlands,* G.P. Buckley, ed., 115–146. New York: Chapman & Hall.

Basket Makers Association and Institute of Arable Crop Research, Long Ashton Research Station. 2001. *Cultivation and Use of Basket Willows 2001: A Guide to Growing Basket Willows.* Surrey, UK: Basketmakers Association.

Battisti, Corrado and Giuliano Fanelli. 2011. "Does Human-induced Heterogeneity Differently Affect Diversity in Vascular Plants and Breeding Birds? Evidences from Three Mediterranean Forest Patches." *Rendiconti Lincei. Scienze Fisiche e Naturali* 22(1):25–30

Bealey, C.E. and P.A. Robertson. 1992. "Coppice Management for Pheasants." In *Ecology and Management of Coppice Woodlands,* G.P. Buckley, ed., 193–210. New York: Chapman & Hall.

Beattie, Mollie, Charles Thompson, and Lynn Levine. 1983. *Working with Your Woodland: A Landowner's Guide.* Lebanon, NH: University Press of New England.

Bell, Adrian D. 2008. *Plant Form: An Illustrated Guide to Flowering Plant Morphology.* Illustrated by Alan Bryan. Portland, OR: Timber Press.

Bellingham, Peter J. and Ashley D. Sparrow. 2000. "Resprouting as a Life History Strategy in Woody Plant Communities. *Oikos* 89(2): 409–416.

Benes, Jiri, Oldrich Cizek, Jozef Dovala, and Martin Konvicka. 2006. "Intensive Game Keeping, Coppicing and Butterflies: The Story of Milovicky Wood, Czech Republic." *Forest Ecology and Management* 237(1–3): 353–365.

Bengtsson, Jan, Sven G. Nilsson, Alain Franc, and Paolo Menozzi. 2000. "Biodiversity, Disturbances, Ecosystem Function and Management of European Forests." *Forest Ecology and Management* 132(1): 39–50.

Berglund, Björn E. 2008. "Satoyama, Traditional Farming Landscape in Japan, Compared to Scandinavia." *Japan Review* 20: 53–68.

Bergmeier, Erwin, Jörg Petermann, and Eckhard Schröder. 2010. "Geobotanical Survey of Wood-pasture Habitats in Europe: Diversity, Threats and Conservation." *Biodivers Conserv* 19: 2995–3014.

Berthelot, Alain, Jacques Ranger, and Dominique Gelhaye. 2000. "Nutrient

Uptake and Immobilization in a Short-rotation Coppice Stand of Hybrid Poplars in North-west France." Forest Ecology and Management 128(3): 167–179.

Bertolotto, S. and R. Cevasco. 2000. "The 'Alnoculture' System in the Ligurian Eastern Apennines: Archive Evidence." In *Methods and Approaches in Forest History*, 189–202. Papers selected from a conference held in Florence, Italy, 1998.

Blackmon, B.G. 1979. "Estimates of Nutrient Drain by Dormant-season Harvests of Coppice American Sycamore." New Orleans: USDA Forest Service, Southern Forest Experiment Station. Res. Note SO-245, accessed December 15, 2010.

Blyth, John, Julian Evans, William E.S. Mutch, and Caroline Sidwell. 1983. *Farm Woodland Management*. London, UK: Farming Press.

Bond, William J. and Jeremy J. Midgley. 2001. "Ecology of Sprouting in Woody Plants: The Persistence Niche." *Trends in Ecology and Evolution* 16(1): 45–51.

Bond, William J. and Jeremy J. Midgley. 2003. "The Evolutionary Ecology of Sprouting in Woody Plants." *International Journal of Plant Sciences* 164(3 Suppl.): S103–S114.

Booker, John and Ruth Tittensor. 1992. "Coppicing for Nature Conservation: The Practical Reality." In *Ecology and Management of Coppice Woodlands*. Buckley, G.P. (ed.), 299–305. London: Chapman & Hall.

Borelli, S. and M.C. Varela. 2001. *Mediterranean Oaks Network, Report of the first meeting, 12–14 October 2000, Antalya, Turkey*. Rome: Inter national Plant Genetic Resources Institute.

Bosela, Michael J. and Frank W. Ewers. 1997. "The Mode of Origin of Root Buds and Root Sprouts in the Clonal Tree *Sassafras albidium* (*Lauraceae*)." *American Journal of Botany* 84(11): 1466–1481.

Bratkovich, Stephen M. 1991. "Shiitake Mushroom Production on Small Diameter Oak Logs in Ohio." In McCormick, Larry H., and Kurt W. Gottschalk, eds. *Proceedings of the Eighth Central Hardwood Forest Conference* (Pennsylvania State University, University Park, 1991), 543–549. USDA Forest Service, Northeast Forest Experiment Station, and Bureau of Forestry, Pennsylvania Dept. of Env. Resources, accessed December 7, 2010.

Brooks, Alan and Elizabeth Agate. 1998. *Hedging: A Practical Handbook*. Wallingford, UK: British Trust for Conservation Volunteers.

Broome, Alice, Susan Clarke, Andrew Peace, and Mark Parsons. 2011. "The Effect of Coppice Management on Moth Assemblages in an English Woodland." *Biodiversity and Conservation* 20(4): 729–749.

Brown, A.H.F. and Susan Warr. 1992. "The Effects of Changing Management on Seed Banks in ancient Coppices." In *Ecology and Management of Coppice Woodlands*, G.P. Buckley (ed.), 147–166. New York: Chapman & Hall.

Brown, Claud L. and Paul P. Kormanik. 1967. "Suppressed Buds on Lateral Roots of *Liquidambar styraciflua*." *Botanical Gazette* 129(3/4): 208–211.

Brown, David, ed. 1994. *Biotic Communities: Southwestern United States and Northwestern*

Mexico. Salt Lake City: University of Utah Press.

Brown, John. 1990. *Welsh Stick Chairs: A Workshop Guide to the Windsor Chair.* Wales, UK: Abercastle Publications.

Brunning, Richard and Conor McDermott. 2013. "Chapter 21: Trackways and Roads Across the Wetlands." In *The Oxford Handbook of Wetland Archeology*, Francesco Menotti and Aidan O'Sullivan, eds., 359–384. New York: Oxford University Press.

Buckley, G.P., ed. 1992. *Ecology and Management of Coppice Woodlands.* New York: Chapman & Hall.

Burgi, M. 1998. "Habitat Alterations Caused by Long-term Changes in Forest Use in Northeastern Switzerland." In *The Ecological History of European Forests,* Keith Kirby and Charles Watkins, eds., 203–212. New York: CAB International.

Burner, David M., Danielle J. Carrier, David P. Belesky, Daniel H. Pote, Adrian Ares, and E. C. Clausen. 2008. "Yield Components and Nutritive Value of *Robinia pseudoacacia* and *Albizia julibrissin* in Arkansas, USA." *Agroforestry Systems* 72(1): 51–62.

Burner, D.M., D.H. Pote, and A. Ares. 2005. "Management Effects on Biomass and Foliar Nutritive Value of *Robinia pseudoacacia* and *Gleditsia triacanthos f. inermis* in Arkansas, USA." *Agroforestry Systems* 65(3): 207–214.

Butler, Jill. 2010. "Ancient, Working Pollards and Europe's Silvo-pastoral Systems: Back to the Future." In *The End of Tradition? Part 2: Commons: Current Management and Problems,*

Ian D. Rotherham, Mauro Agnoletti, and Christine Handley, eds. *Landscape Archaeology and Ecology* 8: 54–57. Sheffield, UK: Wildtrack Publishing.

Butterfield, Jeremy. 2003. *Collins English Dictionary, Complete and Unabridged,* 6th ed. Glasgow, UK: HarperCollins.

Camp, Orville. 1997. "Critical Elements of Forest Sustainability." In *Ecoforestry: The Art and Science of Sustainable Forest Use,* Alan Drengson and Duncan Taylor, eds., 35–41. Stony Creek, CT: New Society.

Canadell, Josep and Paul H. Zedler. 1995. "Underground Structures of Woody Plants in Mediterranean Ecosystems of Australia, California, and Chile." In *Ecology and Biogeography of Mediterranean Ecosystems in Chile, California, and Australia,* Mary T. Kalin Arroyo, Paul H. Zedler, and Marilyn D. Fox, eds., 177–210. *Ecological Studies* 108. New York: Springer-Verlag.

Casals, P., T. Baiges, G. Bota, C. Chocarro, F. de Bello, R. Fanlo, M.T. Sebastià, and M. Taull. 2009. "Silvopastoral Systems in the Northeastern Iberian Peninsula: A Multifunctional Perspective." In *Agroforestry in Europe: Current Status and Future Prospects,* A. Rigueiro-Rodríguez, María Rosa Mosquera-Losada, and Jim McAdam, eds., 161–182. *Advances in Agroforestry*, vol. 6. Springer Science.

Castro, M. 2009. "Silvopastoral Systems in Portugal: Current Status and Future Prospects." In *Agroforestry in Europe: Current Status and Future Prospects,* A. Rigueiro-Rodríguez, María Rosa Mosquera-Losada, and Jim McAdam, eds., 111–126. *Advances in Agroforestry*, vol. 6. Springer Science.

Cech, Richard A. 1998. "Balancing Conservation with Utilization: Restoring Populations of Commercially Valuable Medicinal Herbs in Forests and Agroforests." In *Proceedings of the North American Conference on Enterprise Development Through Agroforestry: Farming the Agroforest for Specialty Products* (Minneapolis, October 4–7, 1998), Scott J. Josiah, ed., 64–68. Saint Paul: University of Minnesota.

Chedzoy, Brett J., and Peter J. Smallidge. 2011. *Silvopasturing in the Northeast: An Introduction to Opportunities and Strategies for Integrating Livestock in Private Woodlands.* Ithaca, NY: Cornell University Cooperative Extension.

Church, T.W. and R.M. Godman. 1966. "The Formation and Development of Dormant Buds in Sugar Maple." *Forest Science* 12(3): 301–306.

Ciabocco, G., L. Boccia, and M. N. Ripa. 2009. "Energy Dissipation of Rockfalls by Coppice Structures." *Natural Hazards and Earth Systems Science* 9(3): 993–1001.

Clason, T.R. and S.H. Sharrow. 2000. "Silvopastoral Practices." In *North American Agroforestry: An Integrated Science and Practice*, H.E. Garrett, W.J. Rietveld, and R.F. Fisher, eds., 119–147. Madison, WI: American Society of Agronomy.

Clément, Vincent. 2008. "Spanish Wood Pasture: Origin and Durability of an Historical Wooded Landscape in Mediterranean Europe." *Environment and History* 14(1): 67–87.

Climate Technology Center and Network. "Fodder Banks." www.ctc-n.org/technologies/fodder-banks.

Cline, Morris G. and Constance A. Harrington. 2006. "Apical Dominance and Apical Control in Multiple Flushing of Temperate Woody Species." *Canadian Journal of Forest Research* 37(1): 74–83.

Coleman, Mark D., J.G. Isebrands, David N. Tolsted, and Virginia R. Tolbert. 2004. "Comparing Soil Carbon of Short Rotation Poplar Plantations with Agricultural Crops and Woodlots in North Central United States" *Environmental Management* 33, Supplement 1: S299–S308.

Coles, Bryony. 1987. "Archeology Follows a Wet Track." *New Scientist* 15, 42–46.

Coles, Bryony and J.M. Coles. 1986. *Sweet Track to Glastonbury: The Somerset Levels in Prehistory.* London: Thames and Hudson.

Coles, J.M. 1983. "Archeology, Drainage, and Politics in the Somerset Levels." *Journal of the Royal Society of Arts* 131(5320): 199–213.

Coles, J.M. 1978. "Man and Landscape in the Somerset Levels." *In Effect of Man on the Landscape: The Lowland Zone*, Susan Limbreye, ed., 86–89. York: Council for British Archaeology.

Collins, E.J.T., ed. 2004. *Crafts in the English Countryside: Towards a Future.* West Yorkshire, UK: Countryside Agency Publications.

Comis, Don. 2005. "A Sylvan Scene in Appalachia: Mixing Trees with Pastures Has Benefits." *Agricultural Research*, August 2005.

Cooke, A.S. and K.H. Lakhani. 1996. "Damage to Coppice Regrowth by Muntjac Deer

Muntiacus reevesi and Protection with Electric Fencing." *Biological Conservation* 75(3): 231–238.

Cool Temperate: Plants and Services for a Sustainable World. n.d. "The Own-Root Fruit Tree Project."

Cooper-Ellis, S., D. R. Foster, G. Carlton, and A. Lezberg. 1999. "Forest Response to Catastrophic Wind: Results from an Experimental Hurricane." *Ecology* 80(8): 2683–2696.

Costen, M.D. 1992. *The Origins of Somerset.* Manchester: Manchester University Press.

Cronon, William. 1983. *Changes in the Land: Indians, Colonists, and the Ecology of New England.* New York: Hill and Wang.

Cronquist, A. 1981. *An Integrated System of Classification of Flowering Plants.* New York: Columbia University Press.

Cummings, Ian and Hadrian Cook. 1992. "Soil-water Relations in an Ancient Coppice Woodland." *In Ecology and Management of Coppice Woodlands*, G.P. Buckley, ed., 52–76. New York: Chapman & Hall.

Cutini, Andrea. 2001. "New Management Options in Chestnut Coppices: An Evaluation on Ecological Bases." *Forest Ecology and Management* 141(3): 165–174.

Dagley, Jeremy. 2007. "Pollarding in Epping Forest." In *1er colloque Européen sur les trognes, Vendôme, 26, 27 et 28 Octobre 2006.* Boursay, France: La Maison Botanique. www.maison-botanique.fr/colloque-europeen.php.

Dagley, J. and P. Burman. 1996. "The Management of the Pollards of Epping Forest: Its History and Revival." In *Pollard and Veteran Tree Management II.* Helen Read, ed., 29–41. London: Corporation of London.

DeBell, Dean S. and Peter A. Giordano. 1994. "Growth Patterns of Red Alder." In *The Biology and Management of Red Alder,* David E. Hibbs, Dean S. DeBell, and Robert F. Tarrant, eds., 116–130. Corvallis, OR: Oregon State University Press.

DeBruyne, S.A., C.M. Feldhake, J.A. Burger, and J.H. Filke. 2011. "Tree Effects on Forage Growth and Soil Water in an Appalachian Silvopasture." *Agroforestry Systems* 83(2):189–200.

Deckmyn, G., I. Laureysens, J. Garcia, B. Muys, R. Ceulemans. 2004. "Poplar Growth and Yield in Short Rotation Coppice: Model Simulations Using the Process Model SECRETS." *Biomass and Bioenergy* 26(3): 221–227.

Deckmyn, G., B. Muys, J. Garcia Quijano, and R. Ceulemans. 2004. "Carbon Sequestration Following Afforestation of Agricultural Soils: Comparing Oak/Beech Forest to Short-Rotation Poplar Coppice Combining a Process and a Carbon Accounting Model." *Global Change Biology* 10(9): 1482–1491.

DEFRA. 2004. *Growing Short Rotation Coppice: Best Practice Guidelines for Applicants to DEFRA's Energy Crops Scheme.* Crewe, UK: Department of Environment, Food and Rural Affairs, Organic and Energy Crops National Implementation Team. http://dendrom.de/daten/downloads/defra_src-guide.pdf.

Dekanic, Stjepan, Tomislav Dubravac, Manfred J. Lexer, Branko Stajic, Tzvetan Zlatanov, and Pande Trajkov. 2009. "European Forest

Types for Coppice Forests in Croatia." *Silva Balcanica* 10(1): 47–62.

Del Tredici, Peter. 1995. "Shoots from Roots: A Horticultural Review." *Arnoldia* 55(3): 11–19.

Del Tredici, Peter. 2001. "Sprouting in Temperate Trees: A Morphological and Ecological Review." *Botanical Review* 67(2): 121–140.

Desai, Pooran, Martin Evans, and Simon Levy. 1996. "Charcoal and Coppice: A Background to the Industries and Their Potential in the Weald." Report of the Coppice and Charcoal Initiative. Carshalton, UK: Bioregional Development Group.

Dey, Daniel C. and Randy G. Jensen. 2002. "Stump Sprouting Potential of Oaks in Missouri Ozark Forests Managed by Even- and Uneven-aged Silviculture." In *Proceedings of the Second Missouri Ozark Forest Ecosystem Project Symposium: Post-treatment Results of the Landscape Experiment,* S.R. Shifley and J.M. Kabrick, eds., 102–113. St. Paul: USDA Forest Service, North Central Forest Experiment Station.

Dey, Daniel C., Randy G. Jensen, and Michael J. Wallendorf. 2008. "Single-tree Harvesting Reduces Survival and Growth of Oak Stump Sprouts in the Missouri Ozark Highlands." In *Proceedings, 16th Central Hardwood Forest Conference* (April 8–9, 2008, West Lafayette, IN), Douglass F. Jacobs and Charles H. Michler, eds., 26–37. Newton Square, PA: USDA Forest Service, Northern Research Station.

Dickman, Donald I. 2006. "Silviculture and Biology of Short-rotation Woody Crops in Temperate Regions: Then and Now." *Biomass and Bioenergy* 30(8–9): 696–705.

Dickmann, Donald I. and Yulia Kuzovkina. 2014. "Poplars and Willows of the World, with an Emphasis on Silviculturally Important Species." In P*oplars and Willows: Trees for Society and the Environment,* J.G. Isebrands and J. Richardson, eds., 8–91. Boston: CAB International.

Dirr, Michael A. 1990. *Manual of Woody Landscape Plants: Their Identification, Ornamental Characteristics, Culture, Propagation and Uses,* 4th ed. Champaign, IL: Stipes Publishing.

Dirr, Michael A. and Charles W. Heuser, Jr. 2006. *The Reference Manual of Woody Plant Propagation: From Seed to Tissue Culture,* 2nd ed. Timber Press.

Doran, M. 2006. *The Use of Willow Coppice as a Biofilter.* PowerPoint presented at the XXIII FIG (International Federation of Surveyors) Congress (Munich, Germany, October 8–13, 2006). www.fig.net/pub/fig2006/ppt/ts06/ts06_02_doran_ppt_0878.pdf.

Dorset Coppice Group. 2005. "Typical Products from Dorset Woodlands." Durweston, UK: Dorset Coppice Group. http://dorsetcoppicegroup.co.uk/index.php?option=com_content&task=view&id=23&Itemid=44.

Douglas, James Sholto and Robert Hart. 1978. *Forest Farming: Towards a Solution to Problems of World Hunger and Conservation.* Emmaus, PA: Rodale Press.

Drake, Paul L., Daniel S. Mendham, Don A. White, and Gary N. Ogden. 2009. "A

Comparison of Growth, Photosynthetic Capacity and Water Stress in *Eucalyptus Globulus* Coppice Regrowth and Seedlings During Early Development." *Tree Physiology* 29(5): 663–674.

Drengson, Alan and Duncan Taylor eds. 1997. *Ecoforestry: The Art and Science of Sustainable Forest Use.* Stony Creek, CT: New Society.

Drescher, Jim. 1997. "The Battle for Sustainability." In *Ecoforestry: The Art and Science of Sustainable Forest Use,* Alan Drengson and Duncan Taylor, eds., 57–62. Stony Creek, CT: New Society.

Ducrey, M. and M. Turrel. 1992. "Influence of Cutting Methods and Dates on Stump Sprouting in Holm Oak (*Quercus ilex L.*) Coppice." *Annals of Forest Science* 49(5): 449–464.

Duke, James A. 1983. *Handbook of Energy Crops.* unpublished. https://hort.purdue.edunewcropduke_energyGleditsia_triac-anthos.html#Yields%20and%20Economics, accessed January 12, 2016.

Dunbar, Michael. 1989. *Restoring, Tuning Using Classic Woodworking Tools.* New York: Sterling.

Dutton, M. and P. Humphreys. 2005. "Assessing the Potential of Short Rotation Coppice (SRC) for Cleanup of Radionuclide-Contaminated Sites." *International Journal of Phytoremediation* 7(4): 279–93.

Dwyer, John P., Daniel C. Dey, and William B. Kurtz. 1993. "Profitability of Precommercially Thinning Oak Stump Sprouts." In *Proceedings of the Ninth Central Hardwood Forest Conference* (Purdue University, West Lafayette, IN, 1993), Andrew R. Gillespie, George R. Parker, and Phillip E. Pope, eds., 373–380. USDA Forest Service, North Central Research Station.

Eaton, Allen H. 1973. *Handicrafts of the Southern Highlands.* New York: Dover.

Ecoagriculture Partners. n.d. *The Dehesa and the Montado: Ecoagriculture Land Management Systems in Spain and Portugal.* Ecoagriculture Snapshots, No. 9. www.ecoagriculture.org/documents/files/doc_68.pdf.

Edlin, H.L. 1966. *Trees, Woods, and Man,* 2nd ed. London: Collins.

El Omari, Bouchra, Xavier Aranda, Dolors Verdaguer, Gemma Pascual, and Isabel Fleck. 2003. "Resource Remobilization in *Quercus ilex* L. Resprouts." *Plant and Soil* 252(2): 349–357.

Emery, Marla R. 2002. "Historical Overview of Nontimber Forest Product Uses in the Northeastern United States." In *Nontimber Forest Products in the United States,* Eric T. Jones, Rebecca J. McLain, and James Weigand, eds., 3–25. Lawrence, KS: University Press of Kansas.

Ernst, Christopher. 1998. "An Ecological Revolution? The 'Schlagwaldwirtschaft' in Western Germany in the Eighteenth and Nineteenth Centuries." In *European Woods and Forests: Studies in Cultural History,* Charles Watkins, ed., 83–92. New York: CAB International.

European Environment Agency. *EUNIS: The European Nature Information System.* http://eunis.eea.europa.eu/index.jsp. Copenhagen, Denmark: European Environment Agency.

Evans, Julian. 1992. "Coppice Forestry: An Overview." In *Ecology and Management of Coppice Woodlands,* G.P. Buckley, ed., 18–27. New York: Chapman & Hall.

Evans, Martin N. and John P. Barkham. 1992. "Coppicing and Natural Disturbance in Temperate Woodlands: A Review." In *Ecology and Management of Coppice Woodlands,* G.P. Buckley, ed., 79–98. New York: Chapman & Hall.

Evelyn, John. 1664. *Sylva or a Discourse of Forest-trees and the Propagation of Timber in His Majesty's Dominions.*

Eyre, F.H. 1980. *Forest Cover Types of the United States and Canada.* Washington, DC: Society of American Foresters.

Farrell, Edward P., Erwin Fuhrer, Dermot Ryan, Folke Andersson, Reinhard Huttl, and Pietro Piussu. 2000. "European Forest Ecosystems: Building the Future on the Legacy of the Past." *Forest Ecology and Management* 132(1): 5–20.

Feldhake, Charles M. 2006. "Appalachian Silvopasture Research." *Temperate Agroforester* No. 4. Columbia, MO: Association for Temperate Agroforestry.

Feldhake, Charles M., David P. Belesky, and James P.S. Neel. 2006. "Photosynthetically Active Radiation Relationship to Forage Yield in Central Appalachian Silvopastures." In *Proceedings of the 60th Southern Pasture and Forage Crop Improvement Conference* (Auburn University, AL, April 11–13, 2006), 20–25. http://spfcic.tamu.edu/proceedings/2006/SPFCIC%202006%20Proceedings.pdf.

Feldhake, Charles M., D.P. Belesky, and E.L. Mathias. 2008. "Forage Production Under and Adjacent to *Robinia pseudoacacia* in Central Appalachia, West Virginia." In *Toward Agroforestry Design: An Ecological Approach,* S. Jose and A.M. Gordon, eds., 55–66. *Advances in Agroforestry,* vol. 4. Springer.

Ferrini, Francesco. 2007. "Pollarding and Its Effects on Tree Physiology: A Look to Mature and Senescent Tree Management in Italy," In *1er colloque Européen sur les trognes, Vendôme, 26, 27 et 28 Octobre 2006,* 1–8. Boursay, France: La Maison Botanique. www.maisonbotanique.fr/colloque-europeen.php

Fink, Siegfried. 1983. "The Occurrence of Adventitious and Preventitious Buds Within the Bark of Some Temperate and Tropical Trees." *American Journal of Botany* 70(4): 532–542.

Fink, Siegfried. 1984. "Some Cases of Delayed or Induced Development of Axillary Buds from Persisting Detached Meristems in Conifers." *American Journal of Botany* 71(1): 44-51.

Fischer, Anders. 2007. "Coastal Fishing in Stone Age Denmark: Evidence from Below and Above the Present Sea Level and from Human Bones." In *Shell Middens in Atlantic Europe,* N. Milner, G. Bailey and O. Craig, eds., 54–69. Oxford: Oxbow.

Fisher, W.R. 1896. *A Manual of Forestry: Forest Utilization.* An English translation of *Die Forstbenutzung* by Dr. Karl Gayer. London, UK: Bradbury, Agnew and Co.

Flora of North America Editorial Committee, eds. 1993+. *Flora of North America North of Mexico.* 16+ vols. New York and Oxford. http://floranorthamerica.org.

Florida Forest Stewardship Program. 2006. "Uneven-aged Management: A 'Natural' Approach to Timber." Gainesville, FL: Institute of Food and Agricultural Sciences Extension, University of Florida.

Fonti, Patrick and Fulvio Giudici. 2002. "Production of Parquet Flooring Using Chestnut Timber from Coppices in Southern Switzerland." *Schweiz. Z. Forstwes* 153(1): 10–16. In Italian.

Fralish, James S. and Scott B. Franklin. 2002. *Taxonomy and Ecology of Woody Plants in North American Forests (Excluding Mexico and Subtropical Florida)*. New York: John Wiley & Sons.

Franklin, Daniel J. 1993. "The Productive Potential of Ancient Oak-Coppice Woodland in Britain." *Rural Development Forestry Network Paper* 15d. Also available in French and Spanish.

Franklin, Jennifer and David Mercker. 2009. *Tree Growth Characteristics*. Knoxville, TN: University of Tennessee Institute for Agriculture.

Freed, James. 1997. "Special Forest Products: Past, Present and Future." In *Ecoforestry: The Art and Science of Sustainable Woodland Use,* Alan Drengson and Duncan Taylor, eds. Stony Creek, CT: New Society.

Fross, David and Dieter Wilken. 2006. *Ceanothus*. Portland, OR: Timber Press.

Fuller, R.J. 1992. "Effects of Coppice Management on Woodland Breeding Birds." In *Ecology and Management of Coppice Woodlands,* G.P. Buckley, ed., 169–192. New York: Chapman & Hall.

Fuller, R.J. and A.C.B. Henderson. 1992. "Distribution of Breeding Songbirds in Bradfield Woods, Suffolk, in Relation to Vegetation and Coppice Management." *Bird Study* 39(2): 73–88.

Fuller, R.J. and B.D. Moreton. 1987. "Breeding Bird Populations of Kentish Sweet Chestnut (*Castanea sativa*) Coppice in Relation to Age and Structure of the Coppice." *Journal of Applied Ecology* 24(1): 13–27.

Fuller, R.J., P. Stuttard, and C.M. Ray. 1989. "The Distribution of Breeding Songbirds Within Mixed Coppice Woodland in Kent, England, in Relation to Vegetation Age and Structure." *Annales Zoologici Fennici* 26(3): 265–275.

Furze, Morgan E., Brett A. Huggett, Donald M. Aubrecht, Claire D. Stolz, Mariah S. Carbone, and Andrew D. Richardson. 2018. "Whole-tree Nonstructural Carbohydrate Storage and Seasonal Dynamics in Five Temperate Species." *New Phytologist* 221(3): 1466–1477.

Gabriel, Steve. 2018. *Silvopasture*. White River Junction, VT: Chelsea Green.

Garrett, H.E. "Gene," ed. 2009. *North American Agroforestry: An Integrated Science and Practice,* 2nd ed. Madison, WI: American Society of Agronomy.

Garrett, H.E., M.S. Kerley, K.P. Ladyman, W.D. Walter, L.D. Godsey, J.W. Van Sambeek, and D.K Brauer. 2004. "Hardwood Silvopasture Management in North America." *Agroforestry Systems* 61(1):21–33.

Germain, Diane and Gilles Lemieux. 2000. *Ramial Chipped Wood: The Clue to a Sustainable Fertile Soil*. Quebec: Laval

University, Coordination Group on Ramial Chipped Wood.

Geyer, Wayne A. and Gary G. Naughton. 1980. "Biomass Yield and Cost Analysis (4th Year) of Various Tree Species Grown Under a Short Rotation Management Scheme in Eastern Kansas." In *Proceedings of the Third Central Hardwood Forest Conference* (University of Missouri, Columbia, September 16–17, 1980), Harold E. Garrett, Gene S. Cox, and Stephen G. Boyce, eds., 309–314. USDA Forest Service, North Central Research Station. www.ncrs. fs.fed.us/pubs/ch, accessed December 7, 2010.

Gilman, Edward, F. 2002. *An Illustrated Guide to Pruning*. Boston, MA: Cengage Learning.

Gimpel, Jean. 1976. *The Medieval Machine: The Industrial Revolution of the Middle Ages*. New York: Penguin.

Goldsmith, F.B. 1992. "Coppicing: A Conservation Panacea?" In *Ecology and Management of Coppice Woodlands*, G.P. Buckley, ed., 306–312. New York: Chapman & Hall.

Gracia, Marc, and Javier Retana. 2004. "Effect of Site Quality and Shading on Sprouting Patterns of Holm Oak Coppices." *Forest Ecology and Management* 188(1–3): 39–49.

Greatorex-Davies, J.N., and R.H. Marrs. 1992. "The Quality of Coppice Woods as Habitats for Invertebrates." In *Ecology and Management of Coppice Woodlands*, G.P. Buckley, ed., 271–296. New York: Chapman & Hall.

Green, E.E. 1996a. "Thoughts on Pollarding." In *Pollard and Veteran Tree Management II*, Helen Read, ed., 1–5. London: Corporation of London.

Green, E.E. 1996b. "Bundle Planting," In *Pollard and Veteran Tree Management II*, Helen Read, ed., 91. London: Corporation of London.

Green, E.E. 2006. "Fungi, Trees and Pollards," In *1er colloque Européen sur les trognes, Vendôme, 26, 27 et 28 Octobre 2006*. Boursay, France: La Maison Botanique. www.maison-botanique.fr/colloque-europeen.php

Greene, Wesley. 2012. *Vegetable Gardening the Colonial Williamsburg Way: 18th Century Methods for Today's Organic Gardener*. Emmaus, PA: Rodale Books.

Guerrero-Campo, Joaguin, Sara Palacio, Carmen Perez-Rontome, and Gabriel Montserrat-Marti. 2006. "Effect of Root System Morphology on Root-sprouting and Shoot-rooting Abilities in 123 Plant Species from Eroded Lands in North-east Spain." *Annals of Botany* 98(2): 439–447.

Gulliver, Richard. 1998. "What Were Woods Like in the 17th Century? Examples from the Helmsley Estate, Northeast Yorkshire." In *The Ecological History of European Forests*, Keith Kirby and Charles Watkins, eds., 135–154. New York: CAB International.

Gurnell, John, Martin Hicks, and Steve Whitbread. 1992. "The Effects of Coppice Management on Small Mammal Populations." In *Ecology and Management of Coppice Woodlands*, G.P. Buckley, ed., 213–232. New York: Chapman & Hall.

Hæggström, Carl-Adam. 1998. "Pollard Meadows: Multiple Use of Human-made Nature." In *The Ecological History of European Forests*, Keith Kirby and Charles Watkins, eds., 33–42. New York: CAB International.

Hæggström, Carl-Adam. 2007. "Pollards in Art." In *1er colloque Européen sur les trognes, Vendôme, 26, 27 et 28 Octobre 2006*. Boursay, France: La Maison Botanique. www.maison-botanique.fr/colloque-europeen.php

Hallman, Richard D. 1999. "Christmas Trees: Plantations to Agroforestry Systems." In *Proceedings of the North American Conference on Enterprise Development Through Agroforestry: Farming the Agroforest for Specialty Products* (Minneapolis, October 4–7, 1998), Scott J. Josiah, ed., 103–105. Saint Paul: Center for Integrated Natural Resources and Agricultural Management, University of Minnesota. www.nfs.unl.edu/SpecialtyForest/sfpresources.asp, accessed December 9, 2010.

Hallman, Richard D., Jill Hatfield, and Harold E. Macy. 1999. "Forest Farming in British Columbia." In *Proceedings of the North American Conference on Enterprise Development Through Agroforestry: Farming the Agroforest for Specialty Products* (Minneapolis, October 4–7, 1998), Scott J. Josiah, ed., 47–50. Saint Paul: Center for Integrated Natural Resources and Agricultural Management, University of Minnesota. www.nfs.unl.edu/SpecialtyForest/sfpresources.asp, accessed December 9, 2010.

Hamilton, Adam. 2011. *The Journey: Walking the Road to Bethlehem*. Nashville: Abingdon Press.

Hammersley, G. 1973. "The Charcoal Iron Industry and Its Fuel." *Economic History Review*. Second Series 26: 593–613.

Hammond, Herb. 1997a. "Why Is Ecologically Responsible Forest Use Necessary?" In *Ecoforestry: The Art and Science of Sustainable Woodland Use*, Alan Drengson and Duncan Taylor, eds., 63–67. Stony Creek, CT: New Society.

Hammond, Herb. 1997b. "Standards for Ecologically Responsible Forest Use." In *Ecoforestry: The Art and Science of Sustainable Woodland Use*, Alan Drengson and Duncan Taylor, eds., 204–210. Stony Creek, CT: New Society.

Hammond, Herb and Hammond, Susan. 1997. "What Is Certification?" In *Ecoforestry: The Art and Science of Sustainable Woodland Use*, Alan Drengson and Duncan Taylor, eds., 196–199. Stony Creek, CT: New Society.

Hampshire County Council. 2013. *Grants for Coppice Management*. www.coppice-products.co.uk/Grants.htm, accessed April 11, 2013.

Hampshire County Council. 1995. *Hazel Coppice: Past, Present and Future*.

Harper, Duane E. and Brent H. McCown. 1991. "Microcoppice: A New Strategy for Red Oak Clonal Propagation." In *Proceedings of the Eighth Central Hardwood Forest Conference* (Pennsylvania State University, University Park), Larry H. McCormick and Kurt W. Gottschalk, eds., 586. NEFES General Technical Report NE-148. Upper Darby, PA: USDA Forest Service, Northeast Forest Experiment Station. www.ncrs.fs.fed.us/pubs/ch.

Harrington, CA, J.C. Zasada, and E.A. Allen. 1994. "Biology of Red Alder (*Alnus rubra* Bong.)." In *The Biology and Management of*

Red Alder, David E. Hibbs, Dean S. DeBell, and Robert F. Tarrant, eds., 3–22. Corvallis, OR: Oregon State University Press.

Hartig, R. 1894. *Textbook of the Diseases of Trees*. London: Country Life.

Hawley, Ralph Chipman and Austin Foster Hawes. 1918. *Manual of Forestry for the Northeastern United States, Being Volume I of "Forestry in New England, Revised,"* 2nd ed. New York: John Wiley & Sons.

Hawthorne, Brian. 2005. "Forest Cutting Plan Summary, Fox Den WMA, Middlefield, MA." Pittsfield, MA: Massachusetts Division of Fisheries and Wildlife.

Heller, Martin C., Gregory A. Keoleian, and Timothy A. Volk. 2003. "Life Cycle Assessment of a Willow Bioenergy Cropping System." *Biomass and Bioenergy* 25(2): 147–165.

Hibbs, David E., Dean S. DeBell, and Robert F. Tarrant, eds. 1994. *The Biology and Management of Red Alder*. Corvallis: Oregon State University Press.

Hill, Lewis. 1997. *Pruning Made Easy*. North Adams, MA: Storey.

Hightshoe, Gary L. 1988. *Native Trees, Shrubs, and Vines for Urban and Rural America: A Planting Design Manual for Environmental Designers*. New York: Van Nostrand Reinhold.

Hoadley, R. Bruce. 2000. *Understanding Wood: A Craftsman's Guide to Wood Technology*. Newtown, CT: Taunton Press.

Holmes, Mike. 1996. "Ancient Trees: Their Importance to Bats." In *Pollard and Veteran Tree Management II*, Helen Read, ed., 19–20. London: Corporation of London.

Hölscher, Dirk, Elke Schade, and Christoph Leuschner. 2001. "Effects of Coppicing in Temperate Deciduous Forests on Ecosystem Nutrient Pools and Soil Fertility." *Basic and Applied Ecology* 2(2): 155–164.

Hoogzaad, Yvonne P.G. 2009. "Impact of Land Use and Climatic Change on the Late Holocene Landscape of the Gasserplatz Area: Pollen and Macrofossil Analysis of Profile Gasserplatz (Feldkirch, Vorarlberg, Austria)." Master's thesis, Earth Sciences, UvA Amsterdam, December 2009.

Horn, Henry S. 1971. *The Adaptive Geometry of Trees*. Princeton, NJ: Princeton University Press.

Hrynkiewicz, Katarzyna, Christel Baum, Peter Leinweber, Martin Weih, and Ioannis Dimitriou. 2010. "The Significance of Rotation Periods for Mycorrhiza Formation in Short Rotation Coppice." *Forest Ecology and Management* 260(11): 1943–1949

Huntley, Brian. 1990. "European Post-glacial Forests: Compositional Changes in Response to Climatic Change." *Journal of Vegetation Science* 1(4): 507–518.

Integrated Taxonomic Information System (ITIS). www.itis.gov.

Isebrands, J.G. and J. Richardson, eds. 2014. *Poplars and Willows: Trees for Society and the Environment*. Boston: CAB International.

Jacke, Dave and Eric Toensmeier. 2005. *Edible Forest Gardens*, 2 vols. White River Junction, VT: Chelsea Green.

James, Susanne. 1984. "Lignotubers and Burls: Their Structure, Function and Ecological

Significance in Mediterranean Ecosystems." *Botanical Review* 50(3): 226–255.

Jehne, Walter. 2019. "The Soil Carbon Sponge." Conference presentation, *Soil Carbon Sponge Gathering from Didi Pershouse and Walter Jehne* (Fairlee, VT, August 15–18, 2019).

Jeník, Jan. 1994. "Clonal Growth in Woody Plants: A Review." *Folia Geobotanica Phytotax* 29(2): 291–306.

Jenkins, J. Geraint. 1966. *Traditional Country Craftsmen.* London: Routledge & Kegan Paul.

Joffre, R., S. Rambal, and J.P. Ratte. 1999. "The Dehesa System of Southern Spain and Portugal as a Natural Ecosystem Mimic." *Agroforest Syst* 45: 57–79.

Johnson, Paul S. 1975. "Growth and Structural Development of Red Oak Sprout Clumps." *Forest Science* 21(4): 413–418.

Johnson, Paul S. and Robert Rogers. 1980. "Predicting Growth of Individual Stems Within Red Oak Sprout Clumps." In *Proceedings of the Third Central Hardwood Forest Conference* (University of Missouri, Columbia, September 16–17, 1980), Harold E. Garrett, Gene S. Cox, and Stephen G. Boyce, eds., 420–439. USDA Forest Service, North Central Research Station. www.ncrs.fs.fed.us/pubs/ch.

Jones, Melvyn. 1998. "The Rise, Decline and Extinction of Spring Wood Management in South-west Yorkshire." In *European Woods and Forests: Studies in Cultural History,* Charles Watkins, ed., 55–71. New York: CAB International.

Jones, Robert H. and Dudley J. Raynal. 1986. "Spatial Distribution and Development of

Root Sprouts in *Fagus grandifolia* (*Fagaceae*)." *Journal of American Botany* 73(12): 1723–1731.

Jose, S., A.R. Gillespie, and S.G. Pallardy. 2004. "Interspecific Interactions in Temperate Agroforestry." *Agroforestry Systems* 61: 237–255.

Jossart J-M. 2001. "Tertiary Waste Water Treatment Using Short Rotation Willow Coppice in Belgium." In *Proceedings of the Fifth Biomass Conference of the Americas* (Orlando, FL, September 2001).

Joys, A.C., R.J. Fuller, and P.M. Dolman. 2004. "Influences of Deer Browsing, Coppice History, and Standard Trees on the Growth and Development of Vegetation Structure in Coppiced Woods in Lowland England." *Forest Ecology and Management* 202(1–3): 23–37.

Kafaky, Sasan Babaei, Amin Khademi, and Asadolahe Mataji. 2009. "Carbon Storage and CO_2 Uptake in Oak Coppice Stand (Case Study: North-West of Iran)." In *Proceedings of the 7th WSEAS International Conference on Environment, Ecosystems and Development (EED '09)* (Puerto De La Cruz, Tenerife, Canary Islands, December 14–16, 2009), Cornelia A. Bulucea, Valeri Mladenov, Emil Pop, Monica Leba, Nikos Mastorakis, eds., 85–91. Athens, Greece: World Scientific and Engineering Academy and Society Press.

Kays, Jonathan S. and Charles D. Canham. 1991. "Effects of Time and Frequency of Cutting on Hardwood Root Reserves and Sprout Growth." *Forest Science* 37(2): 524–539.

Keeley, Jon E. and Paul H. Zedler. 1998. "Evolution of Life Histories in *Pinus*." In

Ecology and Biogeography of Pinus, D.M. Richardson, ed., 219–251. Cambridge: Cambridge University Press.

Key, Roger S. 1996. "Invertebrate Conservation and Pollards." In *Pollard and Veteran Tree Management II,* Helen Read, ed., 21–28. London: Corporation of London.

Keyser, Richard. 2009. "The Transformation of Traditional Woodland Management: Commercial Sylviculture in Medieval Champagne." *French Historical Studies* 32(3): 353–384.

Kirby, K.J. 1992. "Accumulation of Dead Wood: A Missing Ingredient in Coppicing?" In *Ecology and Management of Coppice Woodlands,* G.P. Buckley, ed., 99–112. New York: Chapman & Hall.

Kirby, Keith and Charles Watkins, eds. 1998. The *Ecological History of European Forests.* New York: CAB International.

Kirby, Keith and Charles Watkins, eds. 2015. *Europe's Changing Woods and Forests: From Wildwood to Managed Landscapes.* CABI.

Klingelhöfer, Eric. 1992. "Woodland on the Hampshire Downs: The Morphology of a Medieval Landscape." *Forest & Conservation History* 36(4): 172–178.

Koop, H. 1987. "Vegetative Reproduction of Trees in Some European Natural Forests." *Vegetatio* 72(2): 103–110.

Kormanik, Paul P. and Claud L. Brown. 1967. "Root Buds and the Development of Root Suckers in Sweetgum." *Forest Science* 13(4): 338–345.

Kourik, Robert. 2008. *Roots Demystified.* Occidental, CA: Metamorphic Press.

Kozlowski, T.T. 1971. *Growth and Development of Trees.* New York: Academic Press.

Kozlowski, Theodore T. and Stephen G. Pallardy. 1997. *Growth Control in Woody Plants.* New York: Academic Press.

Kravčík, Michal, Jan Pokorný, Juraj Kohutiar, Martin Kováč, and Eugen Tóth. 2007. *Water for the Recovery of the Climate. A New Water Paradigm.* Žilina, Slovakia: Krupa Print.

Krome, Margaret, Teresa Maurer, and Katie Wied. 2009. *Building Sustainable Farms, Ranches and Communities: Federal Programs for Sustainable Agriculture, Forestry, Entrepreneurship, Conservation, and Community Development.* Fayetteville, AR: ATTRA National Sustainable Agriculture Information Service.

Kruger, E. and Peter B. Reich. December 1993. "Coppicing Alters Ecophysiology of *Quercus rubra* Saplings in Wisconsin Forest Openings." *Physiologia Plantarum* 89(4): 741–750.

Kubkomawa, H.I., A.M. Kenneth-Chukwu, J.L. Krumah, I.N. Yerima, Z. Audu, and W.D. Nafarnda. 2019. "Fodder Bank Establishment and Management for Dry Season Maintenance of Small Scale Livestock Industry: A Review." *Nigerian Journal of Animal Production* 45(4): 211–221.

Kukk, Toomas and Kalevi Kull. 1997. "Wooded Meadows [Puisniidud]." *Estonia Maritima* 2: 1–249.

Kuzovkina, Y.A. 2008. "Establishment and Maintenance of Living Structures Made of Willow (*Salix*) Stems." *Arboriculture & Urban Forestry* 34(5): 290–295.

Kuzovkina, Y.A., M. Knee, and M.F. Quigley. 2004. "Cadmium and Copper Uptake and Translocation of Five *Salix* L. Species." *International Journal of Phytoremediation* 6(3): 269–287.

Kuzovkina Y.A. and M.F. Quigley. 2004. "Selection of Willows for Floral and Stem Quality and Continuous Production Sequence in Temperate North America." *HortTechnology* 14(3): 415–419.

Kuzovkina, Yulia A. and Martin F. Quigley. 2005. "Willows Beyond Wetlands: Uses of *Salix* L. Species for Environmental Projects." *Water, Air, & Soil Pollution* 62(1–4): 183–204.

Kyriazopoulos, A.P., E.M. Abraham, Z.M. Parissi, G. Korakis, and Z. Abas. 2010. "Floristic Diversity of an Open Coppice Oak Forest as Affected by Grazing." In *The Contributions of Grasslands to the Conservation of Mediterranean Biodiversity*, C. Porqueddu and S. Ríos, eds., 247–250. Zaragoza: CIHEAM (International Centre for Advanced Mediterranean Agronomic Studies), CIBIO, FAO, SEEP. Options Méditerranéennes: Série A. Séminaires Méditeranéens, n. 92.

Labreque, Michel and Traian Ion Teodorescu. 2005. "Preliminary Evaluation of a Living Willow (*Salix* spp.) Sound Barrier Along a Highway in Quebec, Canada." *Journal of Arboriculture* 31(2): 95–98.

Lambert, F. 1957. *Tools and Devices for Coppice Crafts*. Chatham, UK: W. &J. Mackay & Co.

Lamson, Neil I. 1988. "Precommercial Thinning and Pruning of Appalachian Stump Sprouts: 10 Year Results." *Southern Journal of Applied Forestry* 12(1): 23–27.

Lancho, J. F. Gallardo and M. I. González. 2004. "Sequestration of C in a Spanish Chestnut Coppice." *Investigacion Agraria: Sistemas y Recursos Forestales* 13(+1) October 2004: 108–113. In Spanish.

Langer, Vibeke. 2001. "The Potential of Leys and Short Rotation Coppice Hedges as Reservoirs for Parasitoids of Cereal Aphids in Organic Agriculture." *Agriculture, Ecosystems and Environment* 87(1): 81–92.

Langsner, Drew. 1995. *Green Woodworking: A Hands-On Approach*. Asheville, NC: Lark Books.

Laubin, Reginald and Gladys Laubin. 1977. *The Indian Tipi: Its History, Construction, and Use*. Norman, OK: University of Oklahoma Press.

Law, Ben. 2010. *The Woodland House*. East Meon, UK: Permanent Publications.

Law, Ben. 2001. *The Woodland Way: A Permaculture Approach to Sustainable Woodland Management*. East Meon, UK: Permanent Publications.

Law, Ben. 2009. *The Woodland Year*. East Meon, UK: Permanent Publications.

Leakey, R.R.B. 1985. "The Capacity for Vegetative Propagation in Trees." In *Attributes of Trees as Crop Plants*, M.G.R. Cannell and J.E. Jackson, eds., 110–133. Abbotts Ripton, UK: Institute of Terrestrial Ecology.

Lehmkuhler, Jeff. 2006. "Livestock Performance and General Considerations for Cattle Management in Temperate Silvopastoral Systems." In *Proceedings of the 60th Southern Pasture and Forage Crop Improvement*

Conference (Auburn University, AL, April 11–13, 2006): 26–34.

Lehmkuhler, J.W., E.E.D. Felton, D.A. Schmidt, K.J. Bader, H.E. Garrett, and M.S. Kerley. 2003. "Tree Protection Methods During the Silvopastoral-System Establishment in Midwestern USA: Cattle Performance and Tree Damage." *Agroforestry Systems* 59(1): 35–42.

Lewis, David. 2007. "Energy Generation: New Markets for Old Woods?" in ROOTS Rural Research Conference 2007 (RICS Headquarters, London, UK, April 17, 2007).

Lexer, M.J. et al. 2009. "CForSEE: Multifunctional Management of Coppice Forests." In *Scientific Results of the SEE-ERA. NET Pilot Joint Call,* Jana Machacova and Katarina Rohsman, eds., 49–56. Vienna: Centre for Social Innovation.

Linebaugh, Peter. 2008. *The Magna Carta Manifesto: Liberties and Commons for All.* Berkeley: University of California Press.

Living Root Bridges. 2009. http://rootbridges. blogspot.com, accessed June 14, 2014.

Londo, Marc, Jos Dekker, and Wim ter Keurs. "Willow Short Rotation Coppice for Energy and Birdlife: An Exploration of Potentials in Relation to Management." *Biomass and Bioenergy* 28(3): 281–293.

Lonsdale, D. 1996. "Pollarding Success and Failure: Some Principles to Consider." In *Pollard and Veteran Tree Management II,* Helen Read, ed., 100–104. London: Corporation of London.

Long, Robert P., Joanne Rebbeck, and Zachary P. Traylor. 2004. "Oak Stump Sprout Health and Survival Following Thinning and Prescribed Burning in Southern Ohio." In *Proceedings of the 14th Central Hardwood Forest Conference* (Wooster, OH, March 16–19), Daniel A. Yaussy, David M. Hix, Robert P. Long, and P. Charles Goebel, eds., 209. Gen. Tech. Rep. NE-316. Newtown Square, PA: USDA Forest Service, Northeastern Research Station.

Lovgren, Stefan. 2005. "'Guns, Germs and Steel': Jared Diamond on Geography as Power." *National Geographic* News, July 6, 2005.

Lowell, K. E., Garrett, H.E., and Mitchell, R.J. 1989. "Early Thinning Can Improve Your Stand of Coppice-regenerated Oak." In *Proceedings of the Seventh Central Hardwood Forest Conference* (Southern Illinois University at Carbondale, March 5–8, 1989), George Rink and Carl A. Budelsky, eds., 53–58. General Technical Report NC-132. St. Paul, MN: USDA Forest Service, North Central Forest Experiment Station.

Lushaj, B.M. and V. Tabaku. 2010. "Conversion of Old, Abandoned Chestnut Forest into Simple Coppice and from Simple Coppice Forest into Orchards in Tropoja, Albania." In *Proceedings of the 1st European Congress on Chestnut, Castanea* (Cuneo-Torino, Italy, 2009), G. Bounous and G.L. Beccaro, eds., 251–258. Acta Hort 866.

Lyle, David. 1998. *The Book of Masonry Stoves: Rediscovering an Old Way of Warming.* White River Junction, VT: Chelsea Green.

Lynch, Ann M. and Bassett, John R. 1987. "Oak Stump Sprouting on Dry Sites in Northern

Lower Michigan." *Northern Journal of Applied Forestry* 4(3): 142–145.

Mabey, Richard. 1996. *Flora Britannica.* London: Chatto & Windus.

Machar, I. 2009. "Coppice-with-standards in Floodplain Forests: A New Subject for Nature Protection." *Journal of Forest Science* 55(7): 306–311.

Machatschek, Michael. 2002. *Laubgeschichten: Gebrauchswissen Einer Alten Baumwirtschaft, Speise- Und Futterlaubkultur.* Germany: Bohlau Verlag GmbH & Co. Kg.

Mack, Daniel. 1992. *Making Rustic Furniture: The Tradition, Spirit, and Techniques with Dozens of Project Ideas.* New York: Sterling.

Maclean, Murray. 2006. *Hedges and Hedgelaying: A Guide to Planting, Management, and Conservation.* Marlborough, UK: Crowood Press.

Maguire, Teresa and Harun, Raja. 2007. "Air Root Pruning to Accelerate the Growth of *Elaeagnus x ebbingei* from Vegetative Cuttings." *Combined Proceedings International Plant Propagators' Society* 57: 456–61.

Mann, Charles. 2005. *1491: New Revelations of the Americas Before Columbus.* New York: Knopf.

Marañón, T. and J.F. Ojeda. 1998. "Ecology and History of a Wooded Landscape in Southern Spain." In *The Ecological History of European Forests,* Keith Kirby and Charles Watkins, eds., 107–116. New York: CAB International.

Martin, C. Wayne and Louise M. Tritton. 1991. "Role of Sprouts in Regeneration of a Whole-tree Clearcut in Central Hardwoods of Connecticut." In *Proceedings of the Eighth Central Hardwood Forest Conference* (Pennsylvania State University, University Park, 1991), Larry H. McCormick and Kurt W. Gottschalk, eds., 305–320. USDA Forest Service, Northeast Forest Experiment Station, Bureau of Forestry, Pennsylvania Dept. of Env. Resources.

Matsubayashi, Takeshi. 2000. "Arrangement of Woody Species of Coppice Forest in Relation to Landform in the Takadate Hills, Northeastern Japan." *Science Reports of Tohoku University,* 7th Series (Geography) 50(2): 149–160.

Matthews, John D. 1989. *Silvicultural Systems.* New York: Oxford University Press.

Mattoon, W.R. 1909. "The Origin and Early Development of Chestnut Sprouts." *Forestry Quarterly Volume* 7 1909: 34–47.

Mattson, James A. and Sharon A. Winsauer. 1985. "The Potential for Harvesting 'Puckerbrush' for Fuel." Research Paper NC-262. St. Paul, MN: USDA Forest Service, North Central Forest Experiment Station.

Matz, George. 1997. "The Worth of a Birch." In *Ecoforestry: The Art and Science of Sustainable Woodland Use,* Alan Drengson and Duncan Taylor, eds., 123–128. Gabriola Island, BC: New Society.

Maxted, A.P., C.R. Black, H.M. West, N.M.J. Crout, S.P. McGrath, and S.D. Young. 2007. "Phytoextraction of Cadmium and Zinc by *Salix* from Soil Historically Amended with Sewage Sludge." *Plant and Soil* 290(1–2): 157–172.

McGee, C.E. 1978. "Size and Age of Tree Affect White Oak Stump Sprouting." Southern

Forest Experiment Station Research Note, SO-239, 1–2.

McKenney, Daniel W., Denys Yemshanov, Saul Fraleigh, Darren Allen, and Fernando Preto. 2011. "An Economic Assessment of the Use of Short-rotation Coppice Woody Biomass to Heat Greenhouses in Southern Canada." *Biomass and Bioenergy* 35(1): 374–384.

McQuade, M. and L. O'Donnell. 2007. "Late Mesolithic Fish Traps from the Liffey Estuary, Dublin, Ireland." *Antiquity* 81(313): 569–584.

McShea, William J. and William M Healy, eds. 2002. *Oak Forest Ecosystems: Ecology and Management for Wildlife*. Baltimore: Johns Hopkins University Press.

Meier, Andrew R., Michael R. Saunders, and Charles H. Michler. 2012. "Epicormic Buds in Trees: A Review of Bud Establishment, Development and Dormancy Release." *Tree Physiology* 32(5): 565–584.

Meiggs, Russell. 1989. "Farm Forestry in the Ancient Mediterranean." *Rural Development Forestry Network Paper 8b* Summer 1989.

Meuret, Michel and Fred Provenza. 2015. "When Art and Science Meet: Integrating Knowledge of French Herders with Science of Foraging Behavior." *Rangeland Ecology & Management* 68(1):1–17.

Miles, Roger. 1967. *Forestry in the English Landscape: A Study of the Cultivation of Trees and Their Relationship to Natural Amenity and Plantation Design*. London: Faber & Faber.

Mitchell, P.L. 1992. "Growth Stages and Microclimate in Coppice and High Forest." In *Ecology and Management of Coppice Woodlands*, G.P. Buckley, ed., 31–51. New York: Chapman & Hall.

Mitchell, R.J., Musbach, R.A., Lowell, K., Garrett, H.E., and Cox, G.S. 1985. "Thinning a Coppice Regenerated Oak-Hickory Stand: Thirty Years of Growth and Development." In *Fifth Central Hardwood Forest Conference* (University of Illinois at Urbana-Champaign, April 15–17, 1985), Jeffrey O. Dawson and Kimberly A. Majerus, eds., 19–24. Urbana-Champaign, IL: Department of Forestry, University of Illinois.

Møller, Fanny G., Rasmus Bartholdy Jensen, and Anders Busse Nielsen. 2007. "Potentials for Coppice-based Systems in Urban Forestry: Landscape Laboratory in Herning, Denmark." Presentation at 10th European Forum on Urban Forestry (EFUF) (Gelsenkirchen, Germany, May 16–19, 2007). Abstract and PowerPoint at www.wald-und-holz.nrw.de/walderleben/lernen-und-erleben/industriewald-ruhrgebiet/downloads.html.

Moreno, D., R. Cevasco, S. Bertolotto, and G. Poggi. 1998. "Historical Ecology and Post-medieval Management Practices in Alderwoods (*Alnus incana* (L.) Moench) in the Northern Apennines, Italy." In *The Ecological History of European Forests*, Keith Kirby and Charles Watkins, eds., 185–202. New York: CAB International.

Mosquera-Losada, M.R., M. Pinto-Tobalina, and A. Rigueiro-Rodriguez. 2005. "The Herbaceous Component in Temperate Silvopastoral Systems." In *Silvopastoralism and Sustainable Land Management*, M.R.

Mosquera-Losada, A. Rigueiro-Rodriguez, and J. McAdam, eds., 93–100. Cambridge, MA: CABI.

Mroz, Glenn David. 1983. "An Evaluation of Whole Tree Harvest Effects on Northern Hardwoods Soils: Site Relationships and Coppice Regrowth (Michigan)." PhD diss., North Carolina State University.

Mummery, Cyril, Raymond Tabor, and N. Homewood. 1998. *Guide to the Techniques of Coppice Management.* Watch Over Essex series. Felixstowe Ferry, UK: Woodland Craft Supplies.

Nabakov, Peter and Robert Easton. 1989. *Native American Architecture.* New York: Oxford University Press.

Nakamura, Kotaro. 2009. *Norwegian Farmhouses, Sod Roof,* www.kotarox.com/Norway/Sodroof.htm.

National Academy of Sciences. 1980. *Firewood Crops: Shrub and Tree Species for Energy Production.* Washington, DC: National Academy of Sciences.

Naughton, G.G., W.A. Geyer, and E. Chambers IV. 2006. "Making Syrup from Black Walnut Sap." *Transactions of the Kansas Academy of Science* 109(3/4): 214–220.

Nebraska Forest Service. 2007. "Lumber Market News," "Tables of Lumber Price Trends, 2006 and 1979–2006," and "Woody Biomass in Minnesota Impacts on Forestry." *Timber Talk* 45(1).

New Scientist. 1990. "The Day the Sweet Track Was Built." *New Scientist* 126, 30(1).

Newsholme, Christopher. 1992. *Willows: The Genus Salix.* Portland, OR: Timber Press.

Oaks, Rebecca and Edward Mills. 2010. *Coppicing & Coppice Crafts: A Comprehensive Guide.* Wiltshire, UK: Crowood Press.

Ontario Ministry of Agriculture and World Affairs. 2013. www.omafra.gov.on.ca/english/stats/hort/glance/table11.htm, accessed December 4, 2016.

Paillet, Frederick L. 1984. "Growth-form and Ecology of American Chestnut Sprout Clones in Northeastern Massachusetts." *Bull. Torrey Bot.* Club 111(3): 316–328.

Pallardy, Stephen G. 2008. *Woody Plant Physiology,* 3rd ed. Burlington, MA: Elsevier.

Paulson, Mark, Paul Bardos, Joop Harmsen, Julian Wilczek, Malcolm Barton, and David Edwards. 2003. "The Practical Use of Short Rotation Coppice in Land Restoration." *Land Contamination & Reclamation* 11(3): 323–338.

Pausas, Juli G. 1997. "Resprouting of *Quercus suber* in NE Spain After Fire." *Journal of Vegetation Science* 8(5): 703–706.

Pavlov, Georgi. 2016. *Farm Roads: Assessment and Design Criteria.*

Pellegrino, Elisa, Claudia Di Bene, Cristiano Tozzinia, and Enrico Bonaria. 2011. "Impact on Soil Quality of a 10-year-old Short-rotation Coppice Poplar Stand Compared with Intensive Agricultural and Uncultivated Systems in a Mediterranean Area." *Agriculture, Ecosystems and Environment* 140: 245–254.

PESI: Pan-European Species directories Infrastructure. 2014. *EU-NOMEN Taxonomic Database.* www.eu-nomen.eu/portal.

Peterken, G.F. 1992. "Coppices in the Lowland Landscape." In *Ecology and Management of*

Coppice Woodlands, G.P. Buckley, ed., 3–17. New York: Chapman & Hall.

Peterken, George, F. 1996. *Natural Woodland: Ecology and Conservation in Northern Temperate Regions.* Cambridge, UK: Cambridge University Press.

Phillips, Donald and Donald Shure. 1990. "Patch-size Effects on Early Succession in Southern Appalachian Forests." *Ecology* 71(1): 204–212.

The Plant List. 2013. Version 1.0 and Version 1.1. www.theplantlist.org/.

Plants for a Future. *PFAF Plant Database.* www. pfaf.org/user/ plantsearch.aspx.

Quelch, Peter. 2010. "Upland Wood Pastures." In *The End of Tradition? Part 2: Commons: Current Management and Problems,* Ian D. Rotherham, Mauro Agnoletti, and Christine Handley, eds. *Landscape Archaeology and Ecology* 8: 172–177. Sheffield, UK: Wildtrack Publishing.

Rackham, Oliver. 1989. "Hedges and Hedgerow Trees in Britain: A Thousand Years of Agroforestry." *Rural Development Forestry Network Paper 8c.*

Rackham, Oliver. 1980. "The Medieval Landscape of Essex." In *Archaeology in Essex to A.D. 1500: In Memory of Ken Newton,* D.G. Buckley, ed., 103–107. London: Council for British Archeology.

Rackham, Oliver. 1998. "Savannas in Europe." In *The Ecological History of European Forests,* Keith Kirby and Charles Watkins, eds., 1–24. New York: CAB International.

Rackham, Oliver. 1990. *Trees & Woodland in the British Landscape: The Complete History of Britain's Trees, Woods & Hedgerows.* London: Phoenix Giants.

Ranger, J. and C. Nys. 1996. "Biomass and Nutrient Content of Extensively and Intensively Managed Coppice Stands." *Forestry* 69(2): 91–110.

Ratcliffe, Philip R. 1992. "The Interaction of Deer and Vegetation in Coppice Woods." In *Ecology and Management of Coppice Woodlands,* G.P. Buckley, ed., 233–245. New York: Chapman & Hall.

Rayner, Alan. 1996. "The Tree as a Fungal Community." In *Pollard and Veteran Tree Management II,* Helen Read, ed., 6–9. London: Corporation of London.

Read, Helen. 2007. "A Brief Review of Pollards and Pollarding in Europe." In *1er colloque Européen sur les trognes, Vendôme, 26, 27 et 28 Octobre 2006.* Boursay, France: La Maison Botanique. Available at www.maisonbotanique.fr/ colloque-europeen.php.

Read, Helen. 2003. *A Study of Practical Pollarding Techniques in Northern Europe.*

Read, Helen, Mark Frater, and Daniel Noble. 1996. "A Survey of the Condition of the Pollards at Burnham Beeches and Results of Some Experiments in Cutting Them." In *Pollard and Veteran Tree Management II,* Helen Read, ed., 50–54. London: Corporation of London.

Reed, Paul. 1996. "The Ecological Value of Tree Management and Its Significance to Species of Epiphytic Moss." In *Pollard and Veteran Tree Management II,* Helen Read, ed., 17–18. London: Corporation of London.

Regnell, Mats. 2003. "Charcoals from Uppåkra as Indicators of Leaf Fodder." In *Centrality Regionality: The Social Structure of Southern Sweden During the Iron Age,* Lars Larsson and Birgitta Hårdh, eds., 105–115. Acta Archeologica Lundensia, series 8, vol. 40.

Roellig, Marlene, Laura M. E. Sutcliffe, Marek Sammul, Henrik von Wehrden, Jens Newig, and Joern Fischer. 2016. "Reviving Wood-pastures for Biodiversity and People: A Case Study from Western Estonia." In *Ambio* 45(2): 185–195.

Rogers, Robert and Paul S. Johnson. 1989. "Thinning Sprout Clumps." In *Central Hardwood Notes,* Jay G. Hutchinson, ed., 6.12. St. Paul, MN: USDA Forest Service, North Central Forest Experiment Station.

Rollinson, T.J.D. and J. Evans. 1987. *The Yield of Sweet Chestnut Coppice.* Bulletin 64. London: HMSO, Forestry Commission.

Roth, Elmer R. and George H. Hepting. 1943. "Origin and Development of Oak Stump Sprouts as Affecting Their Likelihood to Decay." *Journal of Forestry* 41(1): 27–36.

Roth, Elmer R. and Bailey Sleeth. 1939. *Butt Rot in Unburned Sprout Oak Stands.* Technical Bulletin No. 684. Washington, DC: USDA.

Rotherham, Ian D. 2007. "The Implications of Perceptions and Cultural Knowledge Loss for the Management of Wooded Landscapes: A UK Case Study." *Forest Ecology and Management* 249(1–2): 100–115.

Royal Botanic Garden, Edinburgh. *Flora Europaea Data Set.* http://rbg-web2.rbge.org.uk/FE/fe.html.

Rubio, Augustin and Adrian Escudero. 2003. "Clear-cut Effects on Chestnut Forest Soils Under Stressful Conditions: Lengthening of Time-rotation." *Forest Ecology and Management* 183: 195–204.

Rupérez, A. 1957. *La encina y sus tratamientos (Quercus ilex and Its Silviculture).* Ediciones Selvícolas, Madrid.

Rutter, Philip A. 1999. "The Potential of Hybrid Hazelnuts in Agroforestry and Woody Agriculture Systems." In *Proceedings of the North American Conference on Enterprise Development Through Agroforestry: Farming the Agroforest for Specialty Products* (Minneapolis, October 4–7, 1998), Scott J. Josiah, ed., 133–134. Saint Paul: Center for Integrated Natural Resources and Agricultural Management, University of Minnesota. www.badgersett.com/basic%20 haz.html.

Sage, R.B. and P.A. Robertson. 1994. "Wildlife and Game Potential of Short Rotation Coppice in the U.K." *Biomass and Bioenergy* 6(112): 41–48.

Salway, Peter. 1984. "Roman Britain (c 55 BC – c AD 440)." In *The Oxford Illustrated History of Britain,* Kenneth O. Morgan, ed. New York: Oxford University Press.

San Miguel-Ayanz, A., 2004. "Mediterranean European Silvopastoral Systems." International Congress on Silvopastoralism and Sustainable Management, Lugo, Spain, April, 2004.

Sander, Ivan L. 1979. "Regenerating Oaks with the Shelterwood System." In *Proceedings Regenerating Oaks in Upland Hardwood Forests* (J.S. Wright Forestry Conference, Purdue University, Lafeyette, IN, February

22–23, 1979), Harvey A. Holt and Burnell C. Fischer, eds.

Sands, Benjamin A. and Marc D. Abrams. 2009. "Effects of Stump Diameter on Sprout Number and Size for Three Oak Species in a Pennsylvania Clearcut." *Northern Journal of Applied Forestry* 26(3): 122–125.

Sargent, Charles Sprague. 1965. *Manual of Trees of North America (Exclusive of Mexico)*, 2nd rev. ed., two volumes. New York: Dover Publications.

Schier, George A., John R. Jones, and Robert P. Winokur. 1985. "Vegetative Regeneration." In *Aspen: Ecology and Management in the Western United States,* Norbert V. DeByle and Robert P. Winokur, eds., 29–33. USDA Forest Service General Technical Report RM-119. Fort Collins, CO: Rocky Mountain Forest and Range Experiment Station.

Schmidl, A., W. Kofler, N. Oeggl-Wahlmuller, and K. Oeggl. 2005. "Land Use in the Eastern Alps During the Bronze Age: An Archaeobotanical Case Study of a Hilltop Settlement in the Montafon (Western Austria)." *Archaeometry* 47(2): 455–470.

Schmidt, Hans-Peter. 2012. "55 Uses of Biochar." *Ithaka Journal* 1: 286–289.

Schoenian, Susan. 2008. "General Guidelines for Feeding Sheep and Goats." *Small Ruminant Info Sheet.* Maryland Small Ruminant Page, Sheepandgoat.com. Clear Spring, MD.

Secco, Laura, Davide Pettenella, and Daria Maso. 2009. "'Net-System' Models Versus Traditional Models in NWFP Marketing: The Case of Mushrooms." *Small-Scale Forestry* 8(3): 349–365.

Sennerby-Forsse, L. and L. Zsuffa. 1995. "Bud Structure and Resprouting in Coppiced Stools of *Salix viminalis* L., *S. eriocephala* Michx., and *S. amygdaloides* Anders." *Trees* 9(4): 224–234.

Serrada, R., M. Allué, and A. San Miguel. 1992. "The Coppice System in Spain: Current Situation, State of Art and Major Areas to be Investigated." *Annali dell'Istituto Sperimentale del la Selvicoltura* 23: 266–275.

Seymour, John. 1990. *The Forgotten Crafts: A Practical Guide.* New York: Portland House.

Sharrow, S.H., D. Brauer, and T.R. Clason. 2009. "Silvopastoral Practices." In *North American Agroforestry: An Integrated Science and Practice,* 2nd ed., H. "Gene" Garrett, ed., 105–131. Madison, WI: American Society of Agronomy.

Shepard, Mark L. 1999. "Producing Forest-based Food Products in Permaculture Systems." In *Proceedings of the North American Conference on Enterprise Development Through Agroforestry: Farming the Agroforest for Specialty Products* (Minneapolis, October 4–7, 1998), Scott J. Josiah, ed., 115–122. Saint Paul: Center for Integrated Natural Resources and Agricultural Management, University of Minnesota.

Shigo, Alex L. 1986. *A New Tree Biology: Facts, Photos, and Philosophies on Trees and Their Problems and Proper Care.* Durham, NH: Shigo and Trees, Associates.

Shuai, Yan and Viktor J. Bruckman. 2010. "Soil Nitrogen and Carbon Pools under High and Coppice Forests in the Vienna Woods."

Poster presented at the 23rd World Congress of the International Union of Forest Research Organizations (Seoul, South Korea, August 23–28, 2010). Vienna: Commission for Interdisciplinary Ecological Studies, Austrian Academy of Sciences. www.oeaw.ac.at/kioes/iufro2010/IUFRO_Shuai.pdf.

Silver, Akiva. 2020. "Starting a Nursery Business." www.twisted-tree.net/starting-a-nursery-business.

Sloane, Eric. 2002. *A Museum of Early American Tools*. Mineola, NY: Dover Publications.

Sloane, Eric. 1965. *A Reverence for Wood*. New York: Wildred Funk.

Slotte, Håkan. 2001. "Harvesting of Leaf-hay Shaped the Swedish Landscape." *Landscape Ecology* 16(8): 691–702.

Slotte, Håkan. 2000. "Lövtäkt i Sverige och på Åland; Metoder och påverkan på landskapet." PhD diss., Swedish University of Agricultural Sciences, Uppsala.

Smith, J. 2010. *The History of Temperate Agroforestry*. Hamstead Marshall, UK: Progressive Farming Trust Limited.

Society of American Foresters. 1994. *Silviculture Terminology*. Silviculture Working Group document., September, 1994. Published online by Center for Invasive Species and Ecosystem Health, University of Georgia, Tifton, GA. www.bugwood.org/silviculture/terminology.html.

Solomon, Dale S. and Barton M. Blum. 1967. "Stump Sprouting of Four Northern Hardwoods." Research Paper NE-59. Upper Darby, PA: USDA Forest Service, Northeastern Forest Experiment Station.

Sotir, R.B. and J.C. Fischenich. 2001. *Live and Inert Fascine Streambank Erosion Control*. Ecosystem Management and Restoration Research Program, Technical Notes Collection (ERDC TN-EMRRP-SR-31). Vicksburg, MS: U.S. Army Engineer Research and Development Center.

Sourdril, Anne, Gaetan du Bus de Warnaffe, Marc Deconchat, Gerard Balent, and Eric de Garine. 2006. "From Farm Forestry to Farm and Forestry in South-western France as a Result of Changes in a 'House-centered' Social Structure." *Small-scale Forest Economics, Management and Policy* 5(1): 127–144.

Stajić, Branko, Tzvetan Zlatanov, Tomislav Dubravac, and Pande Trajkov. 2009. "Past and Recent Coppice Forest Management in Some Regions of South Eastern Europe." *Silva Balcanica* 10(1): 9–19.

Stanturf, John A. and C. Jeffrey Portwood. 1999. "Economics of Afforestation with Eastern Cottonwood (*Populus deltoides*) on Agricultural Land in the Lower Mississippi Alluvial Valley." In *Proceedings of the Tenth Biennial Southern Silvicultural Research Conference* (Shreveport, LA, February 18, 1999), James D. Haywood, ed., 66–72. General Technical Report SRS-30. Asheville, NC: USDA Forest Service, Southern Research Station.

Stevens, P.F. 2015. *Angiosperm Phylogeny Website*. Version 13, May 12, 2015. www.mobot.org/MOBOT/research/APweb.

Stone, Jr., E.L. and M.H. Stone. 1943. "Dormant Buds in Certain Species of *Pinus*." *American Journal of Botany* 30(5): 346–351.

Stone, Jr., E.L. and M.H. Stone. 1954. "Root Collar Sprouts in Pine." *Journal of Forestry* 52(7): 487–491.

Stott, K.G. 2001. *Cultivation and Use of Basket Willows.* Bristol, UK: Basketmakers Association and IACR Long Ashton Research Station.

Stringer, Jeffrey W. 1999. "Development of Advanced Oak Regeneration from Two-age Reserve Trees." In *Proceedings of the 12th Central Hardwood Forest Conference* (Lexington, KY, February 28, March1–2, 1999), Jeffrey W. Stringer and David L. Loftis, eds., 291–293. Gen. Tech. Rep. SRS-24. Asheville, NC: USDA Forest Service, Southern Research Station.

Strobl, Silvia. 2000. *A Silviculture Guide to Managing Southern Ontario Forests.* Ontario Ministry of Natural Resources. https://dr6j45jk9xcmk.cloudfront.net/documents/2819/silv-guide-southern-on.pdf.

Sundqvist, Willie. 1990. *Swedish Carving Techniques.* Newtown, CT: Taunton Press.

Suttonelms.org.uk. n.d. "Apples: Propagation by Stooling."

Szabo, Peter. 2010a. "Ancient Woodland Boundaries in Europe." *Journal of Historical Geography* 36(2): 205–214.

Szabo, Peter. 2010b. "Driving Forces of Stability and Change in Woodland Structure: A Case Study from the Czech Lowlands." *Forest Ecology and Management* 259: 651.

Szabo, Peter. 2009. "Open Woodland in Europe in the Mesolithic and in the Middle Ages: Can There Be a Connection?" *Forest Ecology and Management,* 257(12): 2327–2330.

Tabor, Raymond. 1994. *Traditional Woodland Crafts.* London: B.T. Batsford.

Takeuchi, K. 2003. "Satoyama Landscapes as Managed Nature." In *Satoyama: The Traditional Rural Landscape of Japan,* K. Takeuchi, R.D. Brown, I. Washitani, A. Tsunekawa, and M. Yokohari, eds., 9–16. New York: Springer-Verlag.

Talvi, Tiina. 2010. "Estonian Wooded Meadows and Wooded Pastures." www.keskkonnaamet.ee/sites/default/files/maahooldus/wooded_meadow_eng.pdf.

Teel, Wayne S. and Louise E. Buck. 2004. "From Wildcrafting to Intentional Cultivation: The Potential for Producing Specialty Forest Products in Agroforestry Systems in Temperate North America." In *Proceedings of the North American Conference on Enterprise Development Through Agroforestry: Farming the Agroforest for Specialty Products* (Minneapolis, October 4–7, 1998), Scott J. Josiah, ed., 7–24. Saint Paul: Center for Integrated Natural Resources and Agricultural Management, University of Minnesota.

Tewari, D.D. 2000. "Valuation of Non-timber Forest Products (NTFPs): Models, Problems, and Issues." *Journal of Sustainable Forestry* 11(4): 47–68.

Thomas, Peter. 2000. *Trees: Their Natural History.* New York: Cambridge University Press.

Thomas, R.C. 1998. "Ecological Changes in Bernwood Forest: Woodland Management During the Present Millennium." In *The Ecological History of European Forests,* Keith

Kirby and Charles Watkins, eds., 225–240. New York: CAB International.

Thompson, Liz, Eric Sorenson, and Robert H. Zaino. 2019. *Wetland, Woodland, Wildland*. Vermont Fish and Wildlife, the Nature Conservancy, and the Vermont Land Trust.

Tiedemann, A.R., W. P. Clary, and R. J. Barbour. 1987. "Underground Systems of Gambel Oak (*Quercus gambelii*) in Central Utah." *American Journal of Botany* 74(7): 1065–1071.

Toensmeier, Eric. 2016. *The Carbon Farming Solution: A Global Toolkit of Perennial Crops and Regenerative Agriculture Practices for Climate Change Mitigation and Food Security*. White River Junction, VT: Chelsea Green.

Tolunay, Ahmet, Mehmet Korkmaz, and Hasan Alkan. 2007. "Definition and Classification of Traditional Agroforestry Practices in the West Mediterranean Region of Turkey." *International Journal of Agricultural Research* 2(1): 22–32.

Uneven Aged Management. 2012. Gainesville, FL: Institute of Food and Agricultural Sciences Extension, University of Florida. www.sfrc.ufl.edu/extension/florida.

USDA, ARS, National Genetic Resources Program. *Germplasm Resources Information Network (GRIN)* [Online Database]. National Germplasm Resources Laboratory, Beltsville, MD. www.ars-grin.gov/cgi-bin/npgs/html/genform.pl.

USDA, NRCS. 2014. *The PLANTS Database*. http://plants.usda.gov. National Plant Data Team, Greensboro, NC.

Uri, Veiko, Krista Lõhmus, Ülo Mander, Ivika Ostonen, Jürgen Aosaar, Martin Maddison, Heljä-Sisko Helmisaari, and Jürgen Augustin. 2011. "Long-term Effects on the Nitrogen Budget of a Short-rotation Grey Alder (*Alnus incana* (L.) Moench) Forest on Abandoned Agricultural Land." *Ecological Engineering* 37: 920–930.

Valbuena-Carabaña, M., S.C. González-Martíinez, and L. Gil. 2008. "Coppice Forests and Genetic Diversity: A Case Study in *Quercus pyrenaica* Willd. from Central Spain." *Forest Ecology and Management* 254: 225–232.

Vandenhove H., F. Goor, S. Timofeyev, A. Grebenkov, and Y. Thiry. 2004. "Short Rotation Coppice as Alternative Land Use for Chernobyl-contaminated Areas of Belarus." *International Journal of Phytoremediation* 6(2): 139–156.

Vera, F.W.M. 2000. *Grazing Ecology and Forest History*. New York: CABI.

Vogt, Albert R. and Gene S. Cox. 1970. "Evidence for the Hormonal Control of Stump Sprouting by Oak." *Forest Science* 16: 165–171.

Waller, Martyn. 2010. "Ashtead Common, the Evolution of a Cultural Landscape: A Spatially Precise Vegetation Record for the Last 2000 Years from Southeast England." *The Holocene* 20(5): 733–746.

Warren, M.S. and J.A. Thomas. 1992. "Butterfly Responses to Coppicing." In *Ecology and Management of Coppice Woodlands*, G.P. Buckley, ed., 249–270. New York: Chapman & Hall.

Watkins, Charles. 1998. "Themes in the History of European Woods and Forests." In *European*

Woods and Forests: Studies in Cultural History, Charles Watkins, ed., 1–10. New York: CAB International.

Weigel, Dale R., Daniel C. Dey, and Chao-Ying Joanne Peng. 2011. "Stump Sprout Dominance Probabilities of Five Oak Species in Southern Indiana 20 Years After Clearcut Harvesting." In *Proceedings of the 17th Central Hardwood Forest Conference* (Lexington, KY, April 5–7, 2010), Songlin Fei, John M. Lhotka, Jeffrey W. Stringer, Kurt W. Gottschalk, and Gary W. Miller, eds., 10–22. Gen. Tech. Rep. GTR-NRS-P-78. Newtown Square, PA: USDA Forest Service, Northern Research Station.

Wendel, G.W. 1975. "Stump Sprout Growth and Quality of Several Appalachian Hardwood Species After Clearcutting." Res. Pap. NE-329. Upper Darby, PA: USDA Forest Service, Northeastern Forest Experiment Station.

Wenger, Karl F. 1953. "The Sprouting of Sweetgum in Relation to Season of Cutting and Carbohydrate Content." *Plant Physiology* 28(1): 35–49.

Wessels, Tom. 2010. *Forest Forensics: A Field Guide to Reading the Forested Landscape.* Woodstock, VT: Countryman Press.

Wessels, Tom. 1997. *Reading the Forested Landscape: A Natural History of New England.* Woodstock, VT: Countryman Press.

West, Justin. 2013. "Assessment of Multi-species Fodder Bank Cropping Systems to Improve on Farm Protein Production, Food Security, and Ecological Resilience." UC Berkeley Center for Latin American Studies.

Whittock, S.P., L.A. Apiolaza, C.M. Kelly, and B.M. Potts. 2003. "Genetic Control of Coppice and Lignotuber Development in *Eucalyptus globulus.*" *Australian Journal of Botany* 51: 57–67.

Whitney, Gordon G. 1994. *From Coastal Wilderness to Fruited Plain.* Cambridge, UK: Cambridge University Press.

Wikipedia. "Barbed wire."

Wikipedia. "Billhook,"

Wikipedia. "Black Death."

Wikipedia. "Norman conquest of England."

Wildlife Trust Wales. 2018. "Wood Pasture and Parkland." www.wtwales.org/wildlife/habitats/wood-pasture-and-parkland.

Williams, Michael. 2003. *Deforesting the Earth: From Prehistory to Global Crisis.* Chicago: University of Chicago Press.

Wilson, B.F. 1968. "Red Maple Stump Sprout Development the First Year." *Harvard Forest Paper* 18. Petersham, MA: Harvard University.

Wittbecker, Alan. 1997. "Forest Practices Related to Forest Ecosystem Productivity." In *Ecoforestry: The Art and Science of Sustainable Forest Use,* Alan Drengson and Duncan Taylor, eds., 42–56. Stony Creek, CT: New Society.

Witters, Nele, Stijn Van Slycken, Ann Ruttens, Kristin Adriaensen, Erik Meers, Linda Meiresonne, Filip M.G. Tack, Theo Thewys, Erik Laes, and Jaco Vangronsveld. 2009. "Short-rotation Coppice of Willow for Phytoremediation of a Metal-contaminated Agricultural Area: A Sustainability Assessment." *BioEnergy Research* 2(3): 144–152.

Wolfe, Ron. 2000. "Research Challenges for Structural Use of Small-Diameter Round Timbers." *Forest Products Journal* 50(2): 21–29.

Wolfe, Ronald W. 1999. "Round Timbers and Ties." In *Wood Handbook: Wood as an Engineering Material,* 18.1–18.9. General technical report FPL-GTR-113. Madison, WI: USDA Forest Service, Forest Products Laboratory.

Wood, Robin. 2013. "How to Price Craft Work: Business Advice for Craftspeople." www.robin-wood.co.uk/wood-craft-blog.

World Bank. 2021. https://data.worldbank.org/indicator/AG.YLD.CREL.KG, accessed August 18, 2017.

Yoshida, Toshiya and Tomohiko Kamitani. 1999. "Growth of a Shade-intolerant Tree Species, Phellodendron amurense, as a Component of a Mixed-species Coppice Forest of Central Japan." *Forest Ecology and Management* 113: 57–65.

Yoshida, Toshiya and Tomohiko Kamitani. 1997. "The Stand Dynamics of a Mixed Coppice Forest of Shade-tolerant and Intermediate Species." *Forest Ecology and Management* 95: 35–43.

Yuksek, Turan and Filiz Yuksek. 2009. "Effects of Clear-cutting Alder Coppice on Surface Soil Properties and Aboveground Herbaceous Plant Biomass." *Communications in Soil Science and Plant Analysis* 40(15–16): 2562–2578.

Zeller, H., A.M Häring., and N. Utke. 2009. "Investing in Short Rotation Coppice: Alternative Energy Crop or an Albatross Around the Neck?" In *Proceedings of the 17th International Farm Management Congress: Agriculture: Food, Fiber and Energy for the Future* (Illinois State University, Bloomington/Normal, July 19–24, 2009), Harold H. Guither, Jean L. Merry, and Carroll E. Merry, eds., 328–338.

Zlatanov, Tzvetan and Manfred J. Lexer. 2009. "Coppice Forestry in Southeastern Europe: Problems and Future Prospects." *Silva Balcanica,* 10(1): 5–8.

Index

About the Author

Mark Krawczyk is an applied ecologist, educator, and grower practicing permaculture design, agroforestry, natural building, traditional woodworking, and small-scale forestry. He owns and operates Keyline Vermont LLC, providing farmers, homeowners, and homesteaders with education, design, and consulting services. He and his wife also manage Valley Clayplain Forest Farm, a 52-acre agroforestry farm in New Haven, Vermont.

ABOUT NEW SOCIETY PUBLISHERS

New Society Publishers is an activist, solutions-oriented publisher focused on publishing books to build a more just and sustainable future. Our books offer tips, tools, and insights from leading experts in a wide range of areas.

We're proud to hold to the highest environmental and social standards of any publisher in North America. When you buy New Society books, you are part of the solution!

At New Society Publishers, we care deeply about *what* we publish—but also about *how* we do business.

- All our books are printed on 100% post-consumer recycled paper, processed chlorine-free, with low-VOC vegetable-based inks (since 2002). We print all our books in North America (never overseas)
- Our corporate structure is an innovative employee shareholder agreement, so we're one-third employee-owned (since 2015)
- We've created a Statement of Ethics (2021). The intent of this Statement is to act as a framework to guide our actions and facilitate feedback for continuous improvement of our work
- We're carbon-neutral (since 2006)
- We're certified as a B Corporation (since 2016)
- We're Signatories to the UN's Sustainable Development Goals (SDG) Publishers Compact (2020–2030, the Decade of Action)

To download our full catalog, sign up for our quarterly newsletter, and to learn more about New Society Publishers, please visit newsociety.com

---

ENVIRONMENTAL BENEFITS STATEMENT

New Society Publishers saved the following resources by printing the pages of this book on chlorine free paper made with 100% post-consumer waste.

TREES	WATER	ENERGY	SOLID WASTE	GREENHOUSE GASES
149	12,000	62	510	64,800
FULLY GROWN	GALLONS	MILLION BTUs	POUNDS	POUNDS

Environmental impact estimates were made using the Environmental Paper Network Paper Calculator 4.0. For more information visit www.papercalculator.org